Poverty Mosaics: Realities and Prospects in Small-Scale Fisheries

Svein Jentoft • Arne Eide
Editors

Poverty Mosaics: Realities and Prospects in Small-Scale Fisheries

In collaboration with:
Maria-Victoria Gunnarsdottir, Maarten Bavinck,
Ratana Chuenpagdee and Jesper Raakjær

Springer

Editors
Svein Jentoft
Faculty of Biosciences,
Fisheries and Economics
Norwegian College of Fishery Science
University of Tromsø
Tromsø, Norway
svein.jentoft@uit.no

Arne Eide
Faculty of Biosciences,
Fisheries and Economics
Norwegian College of Fishery Science
University of Tromsø
Tromsø, Norway
arne.eide@uit.no

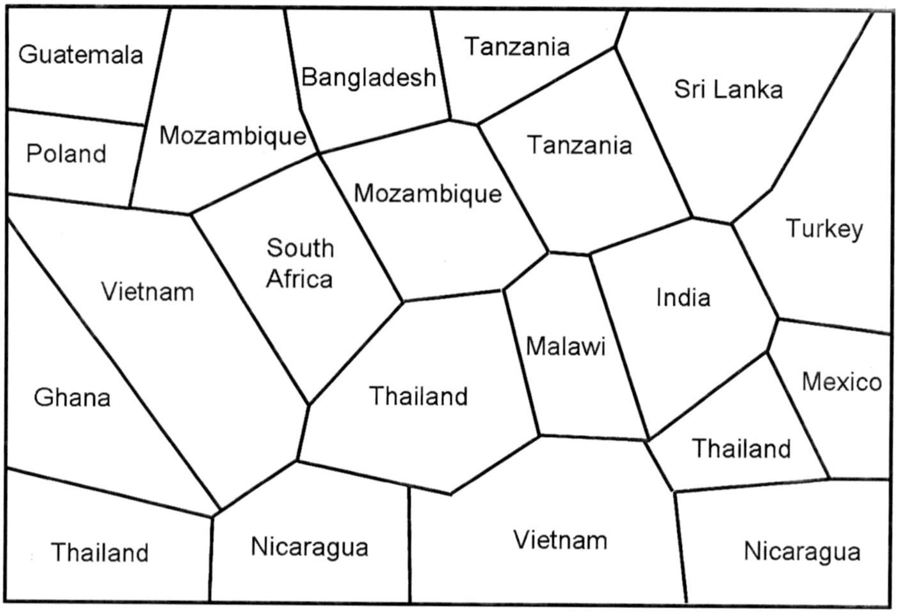

Key to Cover Photo – All Photos Courtesy of Chapter Authors

ISBN 978-94-007-1581-3 e-ISBN 978-94-007-1582-0
DOI 10.1007/978-94-007-1582-0
Springer Dordrecht Heidelberg London New York

Library of Congress Control Number: 2011930867

Printed on acid-free paper

Springer is part of Springer Science+Business Media (www.springer.com)

Foreword

My Lord, since you have banished poverty
From this fair land, I feel it is my duty
To lay an information that the outlaw
Has taken refuge in my humble home

Old Sanskrit verse written many centuries ago
by an anonymous poet

The poverty of fishing communities is proverbial. My unexpected tryst with it took place one hot summer afternoon in 1973 when I walked into a fishing village in Kerala State in India to meet a friend. She took me on the ritual walk to the sea shore. There, I witnessed the enthralling sight of several tens of brown-sailed wooden rafts called *kattumarams* beaching on the golden sands.

The sea shore was buzzing with activity. Scantily dressed, able-bodied fishers dragged soft palm leaf baskets brimming with fish onto the beach. Fishers, merchants, auctioneers, wives and children jostled each other, negotiating different kinds of deals with the fish being brought ashore.

Seeing my friend, one of the fishers broke out from the crowd and came toward us – his body still wet and glistening in the sun. He was Nelson, president of the newly formed fisher cooperative of Marianad – the village of Mary.

When I was introduced to him as a person who had studied a lot about sales and marketing, Nelson interjected boldly saying "Why don't you come and stay with us here and help us sell our fish?" The tenor of his proposal seemed just an honest request from a simple fisher. But for me, it was a challenge to make a sea change in life.

Coming from an urban middle-class background and having grown up far away from the sea, I found the circumstances in this fishing village of Marianad at once romantic and disturbing. The quality of life – congested housing conditions, poor sanitation, high morbidity, low level of literacy, alcoholism and the significant impact all these features had on women and girls in the community – left much to be desired and hugely incongruous against the lush swaying palms, white sands, blue sea and amazing sunsets. I quickly realized that only those who do not earn a livelihood from the sea can wax eloquently and romantically about it! Despite the potential resources "freely available" to the fishers at sea, the surrounding environmental, social and economic conditions on land denied them the chances to lead the good life.

In the 4 years that I subsequently spent in this village and surrounding areas organizing fish marketing cooperatives, I had to unlearn and relearn much! My understanding of the necessary conditions for eliminating poverty in small-scale fishing communities grew by leaps and bounds. The most important insights I gained were the following:

The inter-connection between wealth and poverty
The need for conscious awareness among the fishers about this link and their resolve to break it
The primacy of credit as a foundational requirement for investment in fishing equipment to raise productivity and incomes
The need to disassociate the source of credit from the control of the market for fish
The vital requirement for fishers to control the first sale transaction
The importance of obtaining preferential rights to coastal lands for providing houses to fishers and post-harvest activity
The role of democratically and transparently functioning people's organizations which handle the sale of fish and fishing requisites
The need to establish social control over the export market for fish
The fishers' huge fund of knowledge of the sea and its resources
The pivotal but invisible function of women in shoring up – and in effect subsidizing the fish economy – through their fishery and household-related tasks
The key role of women in the households and communities which finally do make the transition out of poverty
The wisdom of investment in education – particularly for the girl child
The crucial nature of bottom-up collective action as the bedrock of change

Readers will note that the issues of improving technology and access to and control over the resources in the sea were not included in the above list of insights pertaining to eliminating poverty. The dominant ethos at that juncture was that the humble *kattumaram*, combined with the simple but ecologically sophisticated and diverse types of fishing gear, need never be replaced; and that the sharks on land were more dangerous than those at sea! However, both these propositions quickly waned. And quite dramatically at that!

By the early 1980s, the ingress of locally-owned trawlers into the coastal waters – following heightened international market demand for shrimp and cuttlefish – shifted the discourse about the cause of poverty from factors associated with "getting price for fish" to simply "getting fish." The focus, in a sense, shifted from the land to the sea. It emphasized the need for fighting to keep the traditional rights of access; and the importance of improved, but "appropriate" technology, which would help to get fishers to the fishing grounds, and back to the shore faster.

This paradigm shift also provided the point of departure for the militant small-scale fisher unions, which had coalesced by then, to undertake more politically-organized adversarial collective action against the state; and the new actors claiming access to the sea in pursuit of quick profits. In typical response of a capitalist state, the offer for greater welfare measures to "banish poverty" and a bias toward small-scale fishing in the state budgets were announced and largely implemented. However, the more contentious management issues of creating trawler-free zones, closed seasons for trawlers, a ban on absentee ownership of trawlers, to name a few, were relegated to committees and commissions to research and study before decisions were to be taken.

The political party and governance flip-flops ensured that few of these basic fishery management measures were ever honestly implemented. The ruling political dispensations of the day – they changed from left to center-right coalitions – balanced on knife-edge the need for votes from the numerous members of the fishing communities with the need for party funds from the small groups of trawler owners.

Much has been written, by no less a person than Nobel Laureate Amartya Sen, about Kerala State's unique public action model of development that yielded a higher quality of life despite low levels of per capita income. But fishers in Kerala remained "outliers" through this development process. The range of their absolute incomes has certainly improved. Their adoption of technological innovations expanded their physical realms of operation in the sea. Their enhanced reach to a reasonably funded state welfare system has resulted in coverage of medical expenses, better housing and the possibility of a pension. Their access to education through a system of financial grants has contributed to the improved capabilities of their children, and relative betterment of their households and community context.

But the grim absolute fact is that small-scale fishers in Kerala today still continue to be relatively poor – in terms of income and capabilities.

I have narrated my own experience as an introduction here, largely to add one more narrative to this timely and unique publication *Poverty Mosaics*. It is also to forewarn that the "outlaw" can still lurk around, even in what may be considered the best of circumstances from which it could be totally banished.

The PovFish project was a research project financed by the Research Council of Norway, and organized by a multidisciplinary research group at the Centre of Marine Resource Management at the Norwegian College of Fishery Science, University of Tromsø (MaReMa). The project has been a novel and very illuminating exercise undertaken by a diverse group of researchers from various disciplinary backgrounds and nationalities. Unlike my brief account of circumstances surrounding the fishers in Kerala, researchers of the PovFish team have elaborated their respective case studies in profound and meticulous detail.

The case studies from 15 countries across the globe illustrate the wide range of contexts, causes, and consequences of poverty among fishing communities; and describe policy suggestions which have emanated as proposals for solutions.

Poverty Mosaics provides us with very nuanced and poignant analytical descriptions of the very diverse life and struggles of small-scale fishers, the world over. Every one of the case studies brings a refreshing insight, highlights a salient feature, uses a different research methodology, adopts a specific "theoretical" framework, analyses a particular state of affairs. These context-specific pieces have been skilfully patterned into a mosaic which helps the specialist and the general reader to glean the common threads.

A refreshingly original feature of this collection is that the "voice of the fishers" gets through to the reader due to the conscious participatory research process adopted by this distinguished team of researchers. They also share the distinction of being not only top-rate researchers, but also concerned scholars with empathy for the small-scale fishers whom they are writing about. Being from the countries which they studied, knowing the language and idiom of the community, and yet being able to stand apart and view the phenomenon of poverty and vulnerability has made for good quality social science investigation.

The volume is also very creatively organized by placing the chapters into sections interestingly titled: *Positioning, Understanding, Coping, Changing and Imagining.*

Fisheries are the most commercially-oriented, primary sector of the global economy. In numerous developing countries, fish exports are the fastest growing or account for the largest *net* foreign exchange earnings. If even half this gain could reach those who labored to catch and process this fish, their income poverty would be largely addressed, and result in less scope for poverty research among them!

But alas, the structure of fish trade and the current nature of control over it do not permit this to happen. ***Positioning*** treats us to a theoretical and historical perspective with adequate empirical details, and an analytical description of the course of international fisheries development and the discourse on poverty. This provides an appropriate backdrop for situating in context the case studies which follow.

Poverty as understood and articulated by the poor, in their own terms, is vital to comprehending what needs to be done to alter their context. Fishers are big risk takers and optimists. They are not folk who speak or act in terms of averages. Their fishing outcomes fluctuate between the zenith of the bumper catch, and the nadir of drawing a blank after a hard day or night at sea. Their habits and lives mirror this. ***Understanding*** reinforces these features for us. It makes us acutely aware that poverty must be comprehended in terms of the concrete reality, and not merely in generalized empirical terms such as defining a poverty line, and taking a head-count of those below it.

The perpetual harvest and common pool nature of capture fisheries, in which most of the developing world's fishers are engaged, provides the possibility of "putting food on the fisher-household's table" in a more regular manner than in subsistence agriculture. Capture fisheries is also a realm of autonomy and freedom of expression. This gives scope to innovate; to remain optimistic and content with life; and where necessary to use the "weapons of the weak" to defend their rights. ***Coping*** gives us hope that fishers do not resign to fate in an occupation fraught with uncertainty and high risk. While they are challenged by its hardships, they will rarely give up easily.

Public action, by the state and by the fishers themselves, is one option to eliminate poverty. But change does not come out of thin air or strong waves. It must be consciously fostered. There must be vision for this change, and a vanguard to champion the cause. Human capacity and the ability to reason must be built. It needs leadership ideally from within, and support from without. New institutional arrangements and organizational forms have to be envisioned. ***Changing*** takes up these issues and points to the need for altering the circumstances through reform or revolution, if the current situation is to alter for the benefit of the fishers.

Fishery can rhyme with both poverty and wealth. Also, transitions out of poverty can be achieved by a variety of pathways. These include: techno-economical options, allocation of rights, organizational innovations, and collective socio-political actions.

Such measures can be effective solutions to poverty prevention and reduction. However, finally it is changing the structure of distribution of all resources and power which alone ensures greater equity, fairness and removal of injustice. This alone is the key to poverty eradication. ***Imagining*** provides us the collective sense of the authors on these issues. It is about what small-scale fishers can possibly be. What is most urgent and what needs to be done first for them to take control of their livelihoods and transit out of poverty? How can one transcend from the fight for freedom *from* dependence, exploitation by others and the ruin of resources, and rally for freedom *toward* more self-reliance, justice, sustainable use of resources and participation in decision making?

Poverty Mosaics thus contributes abundantly to our knowledge of the dynamics of poverty and its concrete manifestations among small-scale fishers around the world. From perceptions of poverty to prescriptions for policy, this book sails us through a lot of uncharted waters and gives us new meanings and methods of analysis. The book is a holistic text for those who wish to expand their knowledge on issues of poverty, and how it manifests itself in human society.

For the more academically-oriented reader, the book draws on, and contributes to, "sustainable livelihoods theory," "governance theory," "rural development theory," "integrated coastal zone management" and "ecosystem-based management." These theories emphasize socio-cultural and institutional issues concerning rural development, such as coping strategies, social and ecological resilience, empowerment, social capital and trust. For social activists, fishery managers, and others who wish to unravel the reality which they see in small-scale fishing communities, this is a book they cannot afford to set aside. Given the growing global interest of the international community to take a new look at empowering small-scale fisheries and fishing communities, the content of this book is ideal for use by policy makers, fishery development workers, fishery department officers and fishery politicians as a source and reference document for training and capacity-building programs.

In brief, this is a book which will not only contribute to understand and interpret the multi-faceted dimensions of small-scale fisheries; it will also facilitate in surprising ways to change the lives of the small-scale fishers for the better.

Read on!

Founder, International Collective in Support of Fishworkers John Kurien
Former Professor, Centre for Development Studies,
Trivandrum, India

Acknowledgments

The PovFish project (Full Title: Unravelling the Vicious Circle – Poverty Alleviation and Sustainable Livelihoods in Small-Scale Fisheries) was funded by the Norwegian Research Council through its Poverty and Peace (POVPEACE) research program (http://povfish.maremacentre.com/). Without this funding, the project's main product – this book – would not have been possible.

The project was initiated while the two scientific editors were affiliated with MaReMa – the Centre of Marine Resource Management – an interdisciplinary research unit of the Norwegian College of Fishery Science, at the University of Tromsø, Norway. The kick-off meeting was held in March 2008, and the project lasted through 2010.

We are extremely grateful for the cooperation and contributions from the 27 people around the world who became part of the PovFish team. It has been a great pleasure working with all of them. We were a collegial network before the project started, and we became a closely integrated group during the collective work on this book. The enthusiasm and positive approach of the PovFish team members soon became contagious. They not only concentrated on their own PovFish research, but they also showed great interest in the work of other members by reading and commenting on other authors' chapters.

We extend our appreciation to their home institutions who offered their support for our project. Gratitude goes to our hosts at the Department of Aquatic Sciences and Fisheries at the University of Dar-es-Salaam; and the Tanzania Fisheries Research Institute, Mwanza Centre, where we had our mid-term meeting in October 2009. Similarly, we are thankful to everyone from the Coastal Development Centre, Faculty of Fisheries, Kasetsart University, in Bangkok, Thailand, who helped organize the World Small-Scale Fisheries Conference in October 2010. At this conference, some of our PovFish team members were key-note speakers, and two special sessions were held where the PovFish research team presented their findings.

We are thankful to all the members of our project steering group who were dedicated in their task of seeing this project through to completion. From the beginning, we enjoyed the support of Ms. Chandrika Sharma, the director of the International Collective in Support of Fish Workers (ICSF), and Ms. Kirsten Bjøru, then at the

Norwegian Agency of Development Cooperation (Norad). Some people also took on extra responsibilities much beyond their steering group membership roles: Drs. Maarten Bavinck, at the MARE Centre for Maritime Research, Amsterdam University, The Netherlands; Ratana Chuenpagdee, Department of Geography, Memorial University of Newfoundland, Canada; and Jesper Raakjær, Innovative Fisheries Management, Ålborg University, Denmark. All provided constructive advice and wholehearted support throughout the project, including the structuring of this book. These three colleagues are also contributors to chapters within this volume.

We are sincerely thankful for the contribution from our technical editor, Ms. Maria-Victoria Gunnarsdottir who joined the team during the last phase of this project, editing each chapter, and working with all contributing authors to finalize the volume for submission to the publisher. Maria came in at the stage when the two scientific editors had reached a state of exhaustion. We are also grateful to Theresa Heath for supplying additional English editing services.

Sincere gratitude is extended to Mr. Kristoffer Kokvold, whose skills in managing international projects, including administrative and budgetary management, have been invaluable. Involving many people and organizations from several different countries, all with varied administrative and institutional functions, the PovFish project was organizationally complex and demanding. If we should ever again initiate a project of this magnitude, we have learned that we would need Kristoffer's help again.

Special thanks are extended to our graphic designer, Ms. Frøydis Strand, for design of our maps; and Bjørn Hatteng who created the photo mosaic for the cover of this book.

For decades, Dr. John Kurien has provided moral and intellectual guidance for social science researchers on small-scale fisheries. We are proud and honoured that he has written the foreword for this book. His involvement in small-scale fisheries globally is an invaluable inspiration.

The dedication and spirit of the PovFish team allowed this book to become a reality. We extend our deepest gratitude and congratulations to the entire team for a job superbly done, and for a contribution we can all be proud of.

Tromsø, Norway
15 February, 2011

Svein Jentoft and Arne Eide

Contents

List of Acronyms

Chapter 2

CPUE	Catch Per Unit of Effort
EEZ	Exclusive Economic Zones

Chapter 3

GRT	Gross Register Tonnage

Chapter 4

FAO	Food and Agriculture Organization of the United Nations

Chapter 5: Bangladesh

ESBN	Estuarine Set Bag Net
MSBN	Marine Set Bag Net
MMD	Mercantile Marine Department

Chapter 6: Tanzania

FGD	Focus Group Discussions
BMU	Beach Management Unit

Chapter 7: Poland

GOPS	District Social Services Center (GOPS)
NGOs	Non-Governmental Organizations (NGOs)

Chapter 8: Ghana

CPUE	Catch Per Unit Effort
DfID	Department for International Development
FAO	Food and Agriculture Organization of the United Nations
MFRD	Marine Fisheries Research Division
NCU	National Coordination Unit of the SFLP
SFLP	Sustainable Fisheries Livelihood Programme

Chapter 9: India

MT	Metric Tonnes
CITES	Convention on International Trade in Endangered Species

Chapter 10: Mexico

DB	Dzilam de Bravo
PAN	Partido Accion Nacional
PRI	Partido Revolucionario Institucional
PRD	Partido de la Revolucion Democratica
SAGARPA	La Secretaría de Agricultura, Ganadería, Desarrollo Rural, Pesca y Alimentación
SF	San Felipe

Chapter 11: Turkey

AKP	Adalet ve Kalkınma Partisi (Justice and Progress Party Bağ-Kur Esnaf Ve SanatkâRlar Ve Diger Bagimsiz ÇAlisanlar Sosyal Sigortalar Kurumu) is a publicly administered social security foundation for the self-employed, including farmers and fishers
GNP	Gross National Product
GDP	Gross Domestic Product
HDI	Human Development Index
SSK	Sosyal Sigortalar Kurumu – the public social insurance foundation for all salaried employees outside of state civil servants
TL	Turkish Lira
TURKSTAT	Turkish Statistical Institute
UNDP	United Nations Development Programme
USD	United States Dollar
WB	World Bank

Chapter 12: Malawi

DoF	Department of Fisheries
HDI	Human Development Index
HPI	Human Poverty Index
MGDS	Malawi Growth and Development Strategy
MDG	Millennium Development Goals
MSY	Maximum Sustainable Yield
PPP	Purchasing Power Parity
SAP	Structural Adjustment Programme
UNDP	United Nations Development Programme
WB	World Bank

Chapter 13: Nicaragua

CAMPLab	Coastal Area Monitoring Project and Laboratory
CCARC	Caribbean Central American Research Council
CIDCA	The Center for Research and Documentation of the Atlantic Coast
DIPAL	*Proyecto para el Desarrollo Integral de la Pesca Artesanal de la Región Autónoma Atlántico Sur.* A bilateral funding initiative between The Netherlands and Nicaragua
IDRC	International Development Research Centre, Canada

INPESCA Nicaraguan Institute for Fisheries (Instituto Nicaragüense de la Pesca)
RAAN Región Autónoma del Atlántico Norte (North Atlantic Autonomous Region)
RAAS Región Autónoma del Atlántico Sur (South Atlantic Autonomous Region)
UCA Central American University
UNDP United Nations Development Program

Chapter 14: Thailand
CHARM Coastal Habitat and Resource Management
GRT Gross Registered Tonnage
GDP Gross Domestic Product

Chapter 15: Vietnam
MARD Ministry of Agriculture and Rural Development
VND Vietnam Dong

Chapter 16: South Africa
ANC African National Congress
COSATU Confederation of South Africa Trade Union
DAFF Department of Agriculture, Forestry and Fisheries
DEAT Department of Environmental Affairs and Tourism
HACCP Hazard Analysis and Critical Control Points
IDC Industrial Development Corporation of South Africa
ITQ Individual Transferable Quota
MCM Marine and Coastal Management
MLRA Marine Living Resources Act
MPA Marine Protected Area
RDP Reconstruction and Development Plan
TAE Total Allowable Effort
TURF Territorial User Rights Fisheries
VMS Vehicle Monitoring System

Chapter 17: Sri Lanka
ISDR International Strategy for Disaster Reduction
MEY Maximum Economic Yield
MSY Maximum Sustainable Yield
MFAR Ministry of Fisheries and Aquatic Resources
NIFNE National Institute for Fisheries and Nautical Engineering
NTRB Non-motorized Traditional Boat
OAE Open Access Equilibrium
OCDC Overseas Cooperative Development Council
OFRP Fibreglass Boat with Outboard Engine
SCACO Social Capital Approach to Coping with Vulnerability
TC Total Cost
TR Total Revenue

Chapter 18: Mozambique

CAP	Comissão Administração Pesquera (Committee of Fisheries Management)
FCC	Fishing Community Council
FMP	Fisheries Master Plan
GNP	Gross National Product
IDPPE	Instituto De Desenvolvimento De Pesca De Pequena Escala (Institute for the Development of Small Scale Fisheries)
IFAD	International Fund for Agricultural Development
MPA	Marine Protected Area
PARPA	Plano de Acção para a Redução da Pobreza Absoluta
PRSP	Poverty Reduction Strategy Paper
PESPA	Plano Estratégico para o Subsector da Pesca Artesanal or Strategic Plan for the Artisanal Fisheries Sub-sector
SAP	Structural Adjustments Policy

Chapter 19: Guatemala

BRD	By-catch Reduction Devices
ICSED	Inter-American Centre for Sustainable Ecosystems Development
MCS	Monitoring, Control and Surveillance
MEY	Maximize Economic Yield
MSY	Maximum Sustainable Yield

List of Figures

List of Tables

Global Map of Case Studies

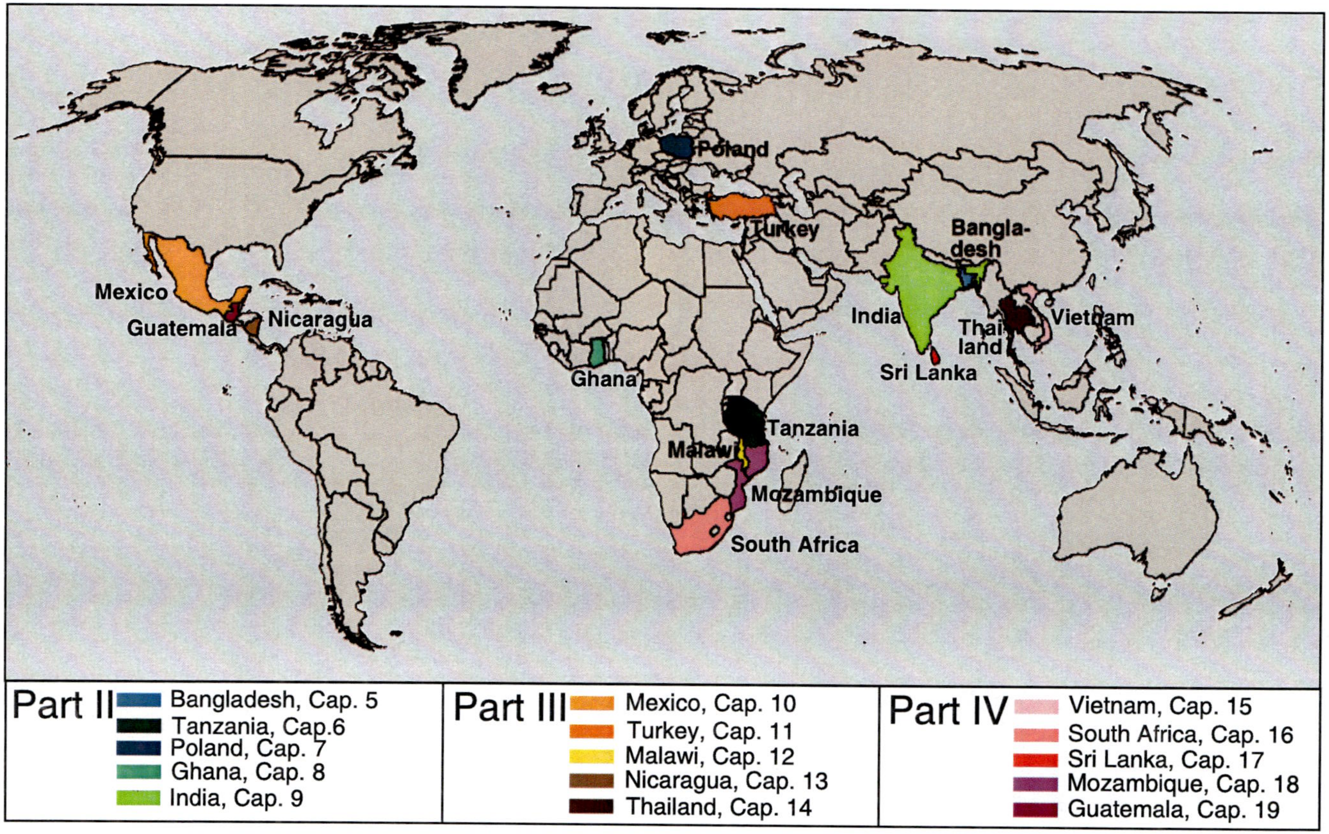

Chapter 1
Setting the Stage

Svein Jentoft and Arne Eide

> *People and the sea the world over are similar in many respects; this is what makes social science possible. People and the sea are also different in many respects; this is what makes social science necessary. There are still many areas of ignorance both with regard to the similarities and the dissimilarities. Therefore, we should not be too self-satisfied about our current stocks of knowledge. I believe that fashioning more nuanced maps to identify and address these areas of ignorance is the best approach to begin this new phase of research.*
>
> Kurien 2002, p. 24

Abstract This book is about small-scale fisheries and the many poor and vulnerable people who draw their livelihoods from this sector. The focus is on what fishing means to them, their adaptations to shifting environments, and how fisheries contribute to food security and well-being. It is also about institutions and governance of small-scale fisheries, and how they influence the coping capabilities of the people in addressing poverty and vulnerability. Drawing on case studies from 15 countries around the world – from Latin America, Africa, Asia, and Europe – the book presents a remarkable mosaic of small-scale fishers' stories, situations, and coping strategies. Small-scale fisheries are variable, and therefore hard to define. What is small in one country is not necessarily small in another. Countries often have their own way of categorizing small-scale fisheries. Their diversity, fluctuations, and change complicate statistical comparison. This implies that policies and development initiatives aiming to alleviate poverty and create sustainable growth need to be tailored to the particular problems, circumstances, and opportunities that small-scale fisheries are facing, wherever they exist.

S. Jentoft (✉) • A. Eide
Norwegian College of Fishery Science, University of Tromsø, 9037 Tromsø, Norway
e-mail: svein.jentoft@uit.no; arne.eide@uit.no

S. Jentoft and A. Eide (eds.), *Poverty Mosaics: Realities and Prospects in Small-Scale Fisheries*, DOI 10.1007/978-94-007-1582-0_1,

1.1 Introduction

Fisheries have an important role to play in meeting the first UN Millennium Development Goal to eradicate extreme poverty and hunger (UN 2010). Today, fish proteins represent more than 15% of the world's total animal protein intake. For many countries, fisheries are sources of economic growth, but fisheries also support a significant number of poor people. An estimate by the WorldFish Centre and FAO (2005, p. 1) suggests that there might be: "some 23 million fishery-dependent people living on less than US$1 per day." Thus, as the fishing industry helps to feed the world, it also has an inherent poverty problem to deal with.

This book is about small-scale fisheries and the many poor and vulnerable people who draw their livelihoods from this sector. The focus is on what fishing means to them, their adaptations to shifting environments, and how fisheries contribute to food security and well-being. It is also about institutions and governance of small-scale fisheries, and how they influence the coping capabilities of the people in addressing poverty and vulnerability. Drawing on case studies from 15 countries around the world – from Latin America, Africa, Asia, and Europe – the book presents a remarkable mosaic of small-scale fishers' stories, situations, and coping strategies. Small-scale fisheries are variable, and therefore hard to define. What is small in one country is not necessarily small in another.

Fisheries vary according to technological criteria such as boat size and gear type. They also vary with regard to capital use, economic performance, and market linkages. The nature of activities such as subsistence and/or commercial, number of crew, and travel time differs from fishery to fishery and from place to place. Small-scale fisheries obviously also play key roles within the society. In addition to providing primary support, they often represent an important safety valve when livelihoods in non-fishing (e.g., agricultural) communities are under threat. Small-scale fisheries are also extremely diverse from a cultural point of view. Normative orientations and social values vary to the extent that small-scale fishers may have different ideas about their profession, what they aspire to, and how they relate to each other as a community. The images of how they see themselves and how they fit into the larger picture of nature and society may also be very different.

For these reasons, countries often have their own way of categorizing small-scale fisheries. Their diversity, fluctuations, and change complicate statistical comparison. This implies that policies and development initiatives aiming to alleviate poverty and create sustainable growth need to be tailored to the particular problems, circumstances, and opportunities that small-scale fisheries are facing, wherever they exist. Context is always important. Still, small-scale fisheries display many similarities. Therefore, it makes sense to talk about small-scale fisheries as a separate sector, distinctly different from large-scale industrial fisheries, both often competing for the same resources and political attention.

Within modern management thinking, it seems easier to deal with industrialized, large-scale fisheries than with traditional, small-scale fisheries (Berkes et al. 2001). The sheer number of small-scale fishers and their enormous variety and combinations of fishing technologies and practices globally and locally represent a governability challenge (Kooiman et al. 2005); one that requires considerable sensitivity and dexterity, which often exceeds what the central government is capable of. Therefore, decentralization and involvement of small-scale fishers in co-management arrangements is now perceived as a way out, as noted in the FAO's Code of Conduct for Responsible Fisheries.[1] However, implementing the Code represents a governability problem in itself, as governments may have different resources, views, and ambitions for this sector. It is therefore not surprising that an international comparison regarding implementation shows very mixed results (Pitcher et al. 2008).

Despite their potential contribution to poverty alleviation, small-scale fisheries all over the world are subject to major challenges and threats, be they resource degradation, conflicts over resources and coastal space, globalization of markets, and climate change. Those who are already poor are likely to feel these pressures more than others, as they are usually more vulnerable and less resilient. If, for instance, those who are seeking opportunities in the fishery as a consequence of poverty are excluded from exploiting the resource, they will be relegated to an even worse situation. This is the moral dilemma that fisheries resource management aiming to reduce fishing pressure must confront. How can policies and management conserve the fish and protect the fisher at the same time? Or is social injustice the price we must pay if we want to sustain the resource base, as Garrett Hardin (1969) believed?

Small-scale fisheries have long been a focus of social researchers. Commencing with Raymond Firth's (1946) seminal work on Malay fishers, there is now considerable academic literature on this sector. The focus has been on small-scale fisheries as a livelihood, a source of income, and as a culture (Acheson 1981). Resource management and governance is a more recent interest, with McGoodwin's (1990) book as an influential contribution. Not only are small-scale fisheries important as a contributor to food security and economic development for the poor, they are also intriguing because of their technological, social and cultural diversity, complexity, and dynamics. As Béné (2003) observes, fishing communities represent an "inexhaustible mine" for social research, including that of small-scale fisheries and poverty, particularly with regard to issues such as collective action, power relationships, and decision-making.

The WorldFish Centre/FAO report (2005) concludes that there is a need for more in-depth case studies of how and why small-scale fisheries are becoming increasingly marginalized and impoverished. This is exactly what this book aims to do. As such, it illustrates that small-scale fisheries are variable and diverse. Indeed, globally, they form a *mosaic* of adaptations, characteristics, and problems. Small-scale fishers are also unequally poor and vulnerable; and, they are poor and vulnerable for many and often different reasons. People's perceptions of what poverty means are also not uniform.

Therefore, we need to *understand* the many dimensions of poverty and vulnerability – the political, economic, social, cultural, and ecological *contexts* that small-scale fisheries operate within – and we must learn how people within this sector are *coping*, adapting, and transforming their situation, either for the better or possibly for the worse. These are all basic knowledge requirements before we can start

[1] http://www.fao.org/fishery/ccrf/en

imagining a future where small-scale fishing people are free from poverty and vulnerability; and before we can begin *changing* the current state of affairs in order to arrive at that future. This book is structured to follow these steps.

1.2 The PovFish Project

The approach of the PovFish project[2] captured in this volume is to combine an overall, global assessment of small-scale fisheries and economic development, with in-depth case studies of small-scale fisheries from 15 countries around the world: Guatemala, Nicaragua, and Mexico, in Latin America; Poland and Turkey in Europe; Ghana, Malawi, Mozambique, South Africa, and Tanzania in Africa; and Bangladesh, India, Sri Lanka, Thailand, and Vietnam in Asia.

Notably, the countries are primarily the *locus* and not the research focus (cf. Arensberg 1961). This means that the case studies in this volume do not attempt to provide a comprehensive overview of small-scale fisheries in those countries. Instead, they investigate specific issues pertaining to small-scale fisheries, poverty, vulnerability, and sustainable development in the social and cultural contexts of particular communities and regions. This is also why the book is not organized with a regional presentation, but rather follows a thematic order.

The aim of the PovFish project and this volume is to highlight the dimensions constituting the biological, economic, social, and cultural diversity of small-scale fisheries; their dynamics and stressors; and the complexity of poverty as it is experienced by small-scale fisher folk (which includes the dependants of those who actually fish) around the world. The project is concerned with how poverty and marginalization is evolving in this sector. It focuses on how small-scale commercial and subsistence fishing people individually and collectively cope with poverty and vulnerability. One issue is how people are attracted to fisheries as a way of making a living and, in some instances, as a way out of poverty. Another is about how poverty may be a consequence of overfishing and resource depletion.

In the PovFish project, we are interested in how poverty and use of natural resources are interconnected. How does overfishing, environmental degradation, or natural disasters affect people's livelihoods? How are small-scale fishers victims of large-scale fishing activities? How do poor people cope and adapt, individually and collectively, to resource crises and poverty? How are culture and ethnicity expressed in the ways people relate to natural environments and cope with economic deprivation? What institutions are created to mitigate negative impacts on resources and livelihoods, and do they distribute wealth? Do institutions help to alleviate poverty

[2]Unravelling the Vicious Circle - Poverty Alleviation and Sustainable Livelihoods in Small-Scale Fisheries (PovFish), was a project executed through the Centre of Marine Resource Management – MaReMa – Norwegian College of Fishery Science, University of Tromsø, Norway. http://povfish.maremacentre.com/

and vulnerability, or not? What are the conditions and opportunities for converting the vicious circle of poverty into a virtuous one?

These are questions that the PovFish project was initiated to address, and which the authors of the chapters in this volume are discussing, each in their own way. They have also exchanged their knowledge at joint sessions throughout the project process, including the World Small-Scale Fisheries Congress in Bangkok, 17–22 October 2010. The PovFish team members, all of whom contributed to this volume, have a high command of the topics in the settings they are studying, either as local researchers and/or from long-term engagement and research experience.

1.3 This Volume

In this book, theoretical perspectives and methodological approaches differ from chapter to chapter, depending on the issue investigated and on the disciplinary background of the authors, which includes biology, economics, sociology, and social anthropology. Some of the chapters have multiple authors from different disciplines, while in some instances individual authors have an interdisciplinary background. The multi- and cross-disciplinary quality of this book is, in our view, a clear strength. Indeed, it could be argued that this is how poverty and vulnerability research should be approached (Hulme and Toye 2006).

Case study sites and methodological approaches were selected by the researchers. Some apply qualitative methods (fieldwork observations, interviews); others use quantitative methods (surveys); while some employ both. The case studies also allow comparative analyses, syntheses, and generalizations, which are put forward by Chuenpagdee and Jentoft, Chap. 3; Jentoft and Midré, Chap. 4; and the concluding Chap. 20 by Jentoft et al. Theoretically, their approach is more generative than deductive (Glaser and Strauss 1967), as they develop their theoretical arguments and propositions from their own empirical observations and findings, as much as from academic literature.

Part I: Positioning of this volume is done in three chapters that provide a global overview and background for the subsequent parts and chapters. First, Chap. 2 by Arne Eide, Maarten Bavinck, and Jesper Raakjær presents an overview of the status and trends in global fisheries, and discusses how these affect the population of small-scale fishers. Although the sector harbors millions of people who are poor, marginalized and vulnerable, the authors of this chapter argue that their situation must also be understood within the context of the role that fisheries resources have played in wealth creation. Rather than supporting the "trickledown" theory, visualized by the aphorism "a rising tide lifts all boats," they are in favor of economic policies which prioritize distribution of wealth to small-scale fisheries.

In Chap. 3, Ratana Chuenpagdee and Svein Jentoft present a chain analysis framework for poverty research in fisheries, arguing that poverty needs to be understood and addressed by linking the aquatic environment with the harvest and the post-harvest systems. The chain analysis is then illustrated with examples from the case studies. They also compare the 15 countries represented in this volume as they figure in various global indexes measuring well-being and poverty for the countries as a whole, and for their small-scale fisheries.

Chapter 4, by Svein Jentoft and Georges Midré, introduces the case studies and synthesizes the findings and arguments. The chapter also relates the contributions of the volume's authors to the literature on poverty and sustainability in general, and to the research literature on poverty and vulnerability in small-scale fisheries. What are poverty and vulnerability, how should they be conceptualized, and what governability challenges do they represent?

Part II: Understanding aims to give the reader a sense of the contextual diversity of small-scale fisheries, and a deeper understanding of the nature of poverty and vulnerability that people experience in different parts of the world. First, in Chap. 5, the reader is introduced to fisher folk in Bangladesh, where Mohammad Mahmudul Islam portrays the lives and livelihoods of fishing people on the Bay of Bengal. His main theme is about how they are recurrent victims of natural disaster and social pressure, and how their resilience is continually tested while they live on the margin of existence. He argues that alleviating poverty in this situation requires, first and foremost, dealing with people's vulnerability.

In Chap. 6, Paul Onyango reports from a small fishing community on Lake Victoria, Tanzania. In comparison with Islam's chapter, Onyango discusses people's reasons for joining the fisheries, and how they live their fishing life on a daily basis. Onyango argues that despite the poverty situation, small-scale fisheries are offering a rich way of life, and that policymakers and fisheries managers need to recognize the full meaning and satisfaction that small-scale fishers attach to their occupation. Otherwise, policies designed to regulate fishing activities may easily fail.

From Africa, the journey heads north to Poland and the Vistula Lagoon, where Boguslaw Marciniak, in Chap. 7, presents the new political reality. The fall of communism and later the integration of Poland into the European Union have brought not only environmental pressure but also social marginalization, to the extent that many people find it hard to sustain their traditional fishing livelihoods. As a consequence, young people leave their communities to seek opportunities elsewhere; while those remaining are unemployed and dependent on government assistance in order to get by. Fisher folk in the district that Marciniak studied are perhaps not poor in an absolute sense, when compared with counterparts in Bangladesh and Tanzania, but they are certainly poor and marginalized when compared to other social groups in their surroundings.

Returning to the African continent, Marloes Kraan, in Chap. 8, portrays the way small-scale fishers from Ghana have expanded their livelihood space by migrating over large distances. Her chapter describes the ethnic Anlo-Ewe beach seine fishers, and argues that for them small-scale fishing is definitely not an "occupation of last resort" but one to which they attach their cultural identity. This fishery has traditionally been a lucrative one, but is now threatened by overfishing and poverty. She believes that strong governance efforts are needed, that indigenous institutions have a role to play, and rather than moving people into alternative livelihoods, a more realistic approach is to develop income-generating activities supplementary to fishing.

Maarten Bavinck's Chap. 9 describing the coastal fishery in Tamil Nadu, India, is another illustration of the global diversity of small-scale fisheries and poverty, which is characterized by an enormous increase in the fishing population and its subsequent governance implications. As in the case of Ghana, this area is rich with traditional institutions and legal systems that are installed to regulate fisheries, and to make sure that people get their fair share. The technological modernization of the fisheries instigated by the government as part of their Blue Revolution agenda has not only had impacts on the fishery, but also on the social structure and location of communities and fishing activities. Bavinck holds that it would be a distortion of fact to say that Indian fisheries are synonymous with poverty, but that the hardship that still exists in small-scale fisheries is closely connected with the wealth that has been created within the fishery sector. Attention must be directed toward social mobility and the exclusion of small-scale fishers that has followed in its wake.

Part III: Coping shifts the focus from depicting the nature of small-scale fishing and the deprivation and marginalization of its people, to the coping strategies that people employ to sustain their livelihoods and their resilience. This section first carries the reader to Mexico's Yucatan Peninsula and to two fishing communities where people have to deal with frequent hurricanes that damage the coast (Chap. 10). When interviewed by Silvia Salas and her colleagues (Maiken Bjørkan, Felipe Bobadilla, and Miguel A. Cabrera), the fishers acknowledged their vulnerability but do not necessarily see themselves as poor; at least, when they compare themselves with others, and as long as they have enough to eat. Still they are experiencing declining catch levels and therefore lower incomes. Notably, those who have joined a local fishing cooperative feel less vulnerable because it means preparedness and greater security, which suggests that in order to strengthen the adaptive capacity of fisher folk, it is important to focus on their organization.

In Chap. 11, Ståle Knudsen and Hakan Koçak present the story of the sea snail fishery on the Black Sea coast of Turkey – an examination of people's adaptability to a "boom and bust" fishery, which has given many local fishers an opportunity to move out of poverty. Still, they argue that marginalization and poverty of small-scale fishers, particularly among certain ethnic groups, is a fact that is more related to structural issues such as agricultural and educational policies, cultural stigma, and the politics of relations to state representatives and business patrons, than to the vicious circle of poverty and overfishing. They further assert that fisheries management in this region must go hand in hand with social policies, in order to reduce overall poverty and inequality in the fisheries.

Also in the case of Malawi and Lake Malombe, described by Mafaniso Hara in Chap. 12, fishers' incomes and livelihoods used to depend on one species, the *chambo*. However, this fishery has witnessed a decline in recent years, which has forced people to adapt by moving to other species, and has required government to take action. He tells about the impressive flexibility and inventiveness of fishers in their response to this decline. As the government of Malawi is poorly resourced to carry out management tasks, Hara argues that there is a need to bring fishers into a co-management scheme, such that they are involved in research, data collection and enforcement. He also believes that fishers need to change some cultural attitudes toward the fishery being naturally open-access.

Chapter 13 by Miguel Gonzalez brings the reader across the Atlantic again, this time to Nicaragua and a community situated at the banks of the Pearl Lagoon basin on the Caribbean Coast. He locates small-scale livelihood fishing into the wider historic, economic, and political contexts of people of diverse ethnic origins. Fishing is not the only thing they do, and people, therefore, also need secure access to land and forest resources. Rights are, however, not guaranteed; overfishing and cattle ranching have led to environmental degradation and to social conflict. Still, these fishers have acted to secure their communal land and aquatic rights, organize themselves, and, by implementing informal community-based actions inspired by sustainable principles, manage the resources of the Lagoon.

Chapter 14 by Ratana Chuenpagdee and Kungwan Juntarashote carries the reader to Thailand, where small-scale fishing communities cannot easily be classified as poor. While small-scale fishing communities are faced with changes brought about by industrialized fishing, and coastal development in the form of a booming tourist industry, some fishing communities seem to possess a high capability to adapt by diversifying fishing activities, as well as engaging in other livelihood activities such as farming and tourism. Organizing local market cooperatives is another coping strategy. Conflicts with large-scale fishers and with fishers using illegal gears are, however, causing worries among local fishers, which calls for better governance.

Part IV: Changing is, first and foremost, about the factors that may improve or worsen poverty and vulnerability, and what governance and policy reforms are needed to turn the situation around to benefit small-scale fisheries. In Chap. 15, Kim Anh Nguyen and Ola Flaaten report from Thanh Phong, a small-scale fishing community in the Mekong Delta of Vietnam, where poverty alleviation policies have proven largely ineffective. Illegal fishing gears are being used, and have negative impacts on the marine system. Still, there is awareness among local fishers that resources are limited and that illegal fishing is exhausting them even further. However, their dependence on fisheries' resources and the absence of alternative incomes has forced them to continue illegal fishing practices. This is obviously a threat to the security of their livelihoods causing more poverty, which is higher than the national average to begin with.

Chapter 16, by Moenieba Isaacs, is situated in South African fisheries and discusses the extent to which the poor and dispossessed are benefitting from the governance process targeting small-scale fisheries in that country. Her findings, based on fieldwork in two coastal communities, suggest that the new policies have not succeeded in reducing the vulnerability and poverty of small-scale fishers. This is mainly because local elites have been able to direct the policy process to their advantage, as the poor are too unorganized to represent a powerful voice. In frustration of lost fishing rights, people have resorted to poaching, even within protected areas. Isaacs therefore calls for a paradigm shift that addresses the disparity between policies and practices, so that these are better attuned to the relationship between poverty reduction, environmental sustainability, and an allocation of fishing rights that ensures greater social justice.

The story that Oscar Amarasinghe and Maarten Bavinck tell in Chap. 17 is from the Hambantota District on the southern coast of Sri Lanka. Their main focus is on the role of cooperatives in connecting small-scale fishing communities to external resources. For fishers disadvantaged by weak credit arrangements, product and insurance markets, increasing costs of fishing equipment, and deficient educational and training services, cooperatives have made a very positive difference to their livelihood security. But not all fishing cooperatives in Sri Lanka are functioning well. Some have focused on promoting welfare at the expense of resource conservation. Thus for cooperatives to work properly, there is still room for improvement.

In Chap. 18, we are back on the African continent where Ana Menezes, Arne Eide, and Jesper Raakjær describe how Mozambique has undergone major regime shifts and policy changes over the last few decades. After independence, governments changed approaches toward the development of coastal fisheries. Infrastructure is improved and basic common goods are more available than they used to be; but still more than 70% of the coastal population lives below the poverty line. Co-management organizations have been established at the community level, but have benefitted the better-off fishers to a larger extent than the poor. Fisheries resources continue to serve as an important economic buffer for the poor, but the development of the small-scale fishery sector is hampered by poor infrastructure, and by emerging conflicts regarding the use of the coastal zone.

Chapter 19 by Hector Andrade and Georges Midré brings the reader back to Latin America to the small-scale fishers in the Amatique Bay on the Caribbean coast of Guatemala. Similar to Nicaragua, Guatemala figures among the countries in Latin America with the highest rates of poverty and inequality. Poverty is especially profound among indigenous and displaced people in rural areas. The open-access fishery does not discriminate against any particular group and, as in the case of Mozambique, works as a poverty-mitigating buffer for people who have no other source of livelihood. But, fish resources are under pressure from high fishing intensity, unsustainable fishing practices, and ineffective management. Still, fishers have negotiated an agreement on how to distribute access among themselves.

Part V: Imagining, the concluding Chap. 20 by Svein Jentoft, Arne Eide, Maarten Bavinck, Ratana Chuenpagdee, and Jesper Raakjær, is formed as a joint statement from members of the PovFish research team, summarizing the main findings and arguments, and the lessons that should be learned from this volume. This chapter addresses questions such as: What does the future hold for small-scale fisheries; and what policies, governance principles, and interventions can make a difference?

References

Acheson JM (1981) Anthropology of fishing. Annu Rev Anthropol 10:275–316

Arensberg CM (1961) The community as an object and as sample. Am Anthropol 63(2): 241–264

Béné C (2003) When fishery rhymes with poverty: a first step beyond the old paradigm on poverty in small-scale fisheries. World Dev 31(6):949–975

Berkes F, Mahon R, McConney P, Pollnac R, Pomery R (2001) Managing small-scale fisheries. International Development Research Centre, Ottawa, http://web.idrc.ca/es//ev-9328-201-1-DO_TOPIC.html

Firth R (1946) Malay fishermen: their peasant economy. Kegan Paul, Trench, Trubner, London

Glaser BG, Strauss AL (1967) The discovery of grounded theory: strategies for qualitative research. Aldine Publishing, New York

Hardin G (1969) The tragedy of the commons. Science 162:1243–1248

Hulme D, Toye J (2006) The case for cross-disciplinary social science research on poverty, inequality and well-being. J Dev Stud 42(7):1085–1107

Kooiman J, Bavinck M, Jentoft S, Pullin R (eds.) (2005) Fish for life: interactive governance for fisheries. Amsterdam University Press, Amsterdam

Kurien J (2002) People and the sea: a 'tropical-majority' world perspective. Mar Stud 1(1):9–26

McGoodwin JR (1990) Crisis in the world's fisheries: people, problems and policies. Stanford University Press, Stanford

Pitcher TJ, Kalikoski D, Short K, Varkey D, Pramod G (2008) An evaluation of progress in implementing ecosystem-based management of fisheries in 33 countries. Mar Policy 33:223–232

UN (2010) http://www.un.org/millenniumgoals/

WorldFish Center/FAO (2005) Towards an interdisciplinary approach to the assessment of small-scale fisheries and its role in food security and poverty alleviation and sustainable resource use. Concept note. WorldFish Center and the FAO FishCode-STF project, Penang and Rome

Part I
Positioning

The first chapter of this section presents a global overview of fisheries, particularly small-scale, and their contribution to economic development, food supply and security, and to the livelihoods of millions of people around the world. For these people, fishing, fish processing and trading are sometimes the best, or in some instances, indeed the only alternative source of income and employment. Small-scale fisheries, more so than large-scale, industrial fisheries, tend to be more important for domestic needs than for international export trade. For many countries, fisheries resources are essential for wealth creation, export earnings and for the viability of coastal and rural communities. It is argued that in one economic reality, fisheries serve as a safety valve; but in favorable economic, social and/or institutional conditions, fisheries can generate considerable economic growth not only to small-scale fishing communities but to the nation as a whole. In Chaps. 2 and 3, the 15 countries represented in this volume are compared with regard to their contribution to the overall national economies and human welfare.

The small-scale sector plays a central role in feeding and employing the poor and vulnerable; but it is also a sector often ignored by political authorities leading to social and economic marginalization and exclusion of those depending on it. Small-scale fisheries are variedly defined. What is small in one country or fishery may be big in another. Small-scale fisheries also have some ecological, technical, economic, social and cultural attributes that distinguish them from large-scale fisheries.

It is argued in Chap. 3 that in order to alleviate poverty, we need to understand the systemic nature of small-scale fisheries and poverty. Poverty and vulnerability must be analysed from a "chain perspective" linking pre-harvest, harvest and post-harvest functions. This chain of interaction and exchange is also embedded in and conditioned by a social and ecologic system from which it draws support and to which it contributes.

Chapter 4 provides an overview and synthesis of the literature on both general poverty and vulnerability, and on poverty related to small-scale fisheries. The latter may have been under-researched in the past, but has in recent years seen a considerable surge within the social sciences. This chapter situates the case studies that follow by presenting their theoretical perspectives, arguments and key findings. Conclusions show that poverty is both absolute and relative; it is relational and therefore also gendered. It must be studied as a process that may well keep people trapped over time, and often over generations, with few prospects for economic improvement, freedom and development.

Chapter 2
Avoiding Poverty: Distributing Wealth in Fisheries

Arne Eide, Maarten Bavinck, and Jesper Raakjær

Abstract Aquatic resources contribute to economic growth, food security, and the livelihoods of millions of fishers around the world. This is evidenced by the industrialization of capture fisheries in the twentieth century, which has generated enormous wealth. Rather than supporting a policy aimed at maximizing economic efficiency though, this chapter argues for the distribution of wealth among small-scale fishers. After all, the small-scale fisheries function as a safety valve for a host of rural poor, for whom alternative livelihoods are not available.

2.1 Introduction

Aquatic resources are a valuable global asset. Whether they are large-scale or small-scale, fishers carry out their activities either to produce food for their own consumption, fish products for a market, or to engage in recreation. Often these objectives

A. Eide (✉)
Norwegian College of Fishery Science, University of Tromsø, 9037 Tromsø, Norway
e-mail: arne.eide@uit.no

M. Bavinck
Department of Human Geography, Planning, and International Development Studies, University of Amsterdam, Nieuwe Prinsengracht 130, 1018VZ Amsterdam, The Netherlands
e-mail: j.m.bavinck@uva.nl

J. Raakjær
Innovative Fisheries Management (IFM), Aalborg University Research Centre, Department of Development and Planning, Aalborg University, Aalborg, Denmark
e-mail: jr@ifm.aau.dk

S. Jentoft and A. Eide (eds.), *Poverty Mosaics: Realities and Prospects in Small-Scale Fisheries*, DOI 10.1007/978-94-007-1582-0_2,

coexist; and frequently, they also constitute different stages in a development process. Almost 250 years ago, Adam Smith described the different facets of fisheries in this way:

> Hunting and fishing, the most important employments of mankind in the rude state of society, become in its advanced state their most agreeable amusements, and they pursue for pleasure what they once followed from necessity. In the advanced state of society, therefore, they are all very poor people who follow as a trade, what other people pursue as a pastime. Fishermen have been so since the time of Theocritus (Smith 1776, section I.I.4).

Smith's concept of fishers being poor has continued into the present age. In recent years, however, the emphasis is on the connection between poverty and resource depletion (Béné 2003). This discourse has identified open-access to aquatic resources and so-called Malthusian overfishing (Pauly 1994) as the core of the problem. To the contrary, we argue that rather than forming an obstacle to successful management, access to aquatic resources ensures food security and self-sufficiency for large numbers of poor people living in coastal areas. Common pool natural resources are the employer of last resort, suggesting relief, not disaster. Free access to fish resources gives comfort and satisfaction when the alternative is penury and hunger. On the basis of such observations, MacKenzie (1979) concludes that the poor fisher is fishing because he is poor, not the other way around.

We are not suggesting, however, that fishing today is inevitably equated with poverty. Poverty certainly occurs, sometimes quite prominently so, but at the same time fisheries have become a source of enormous economic wealth. This apparent paradox reflects a dualist reality in which many people exploit aquatic resources primarily to avoid hardship, while others – a smaller category – acquire further riches. Both groups are taking advantage of the economic value of the natural resource, though this value takes on different forms in different contexts.

We limit our discussion to capture fisheries in the context of marine and inland waters. We investigate how poverty and wealth relate to fisheries, and how both figure in the context of economic development. The first section sketches the development of fisheries in North and South geographies, and emphasizes their various trajectories. We subsequently explain the relevance of economic growth in capture fisheries, but also the importance of taking distribution into account.

2.2 Global Changes in Fisheries: Southern Growth and Northern Decline

The worldwide changes that have taken place in fisheries from the beginning of the twentieth century have been characterized as the "great fish race" (Butcher 2004); an era of industrialization of the ocean (Smith 2000) or the "blue revolution" (Bailey 1988; Bavinck 2011). We note that this process of industrialization – to borrow Smith's terminology, which involved the development of technology and markets – has created enormous economic wealth and, in Butcher's (2004) perception, a

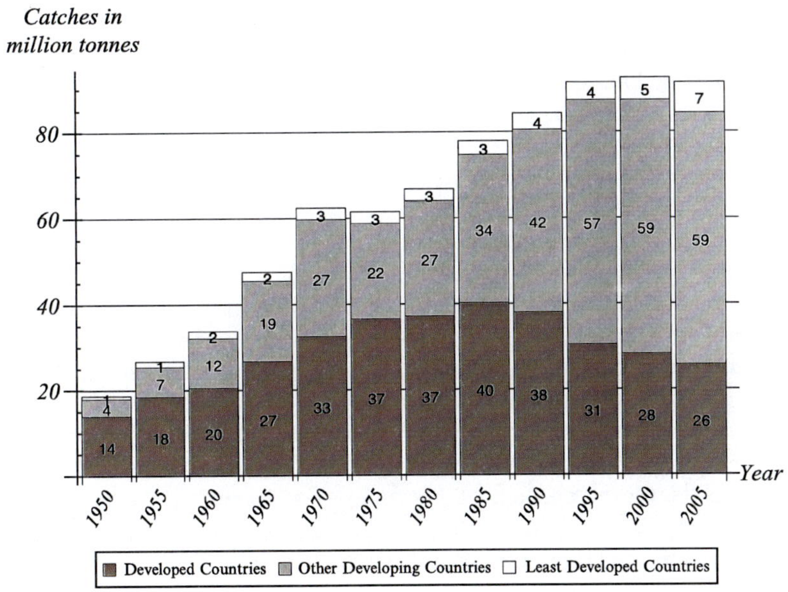

Fig. 2.1 World marine and inland catches distributed between developed and developing countries. The latter is split into two categories: least developed and other developing countries (Source: FAOs online capture statistics. Accessed 20 January 2011 at http://www.fao.org/fishery/statistics/global-capture-production/query/en)

"golden age."[1] A recent report (World Bank 2008) estimates that capture fisheries now represent a monetary annual value of US$ 50 billion, with many "sunken billions" still remaining to be reaped – if management regimes are improved.

Global catches today are five times the catches of 50 years ago (Fig. 2.1), and have been stable at around 90 million tons over the last decades. The most recent FAO report (2009) on the state of world fisheries and aquaculture explains that fish is now the main source of animal protein for 20% of the world's population, and that more than 37% of the global catches are traded internationally. An increasing share of world fish production (77% in 2006) is used for direct human consumption. In addition, a world average of about 10 kg per capita fish consumption in 1960 had increased to almost 17 kg per capita in 2005 (FAO 2009).

These are impressive results. Such aggregate production and consumption figures, however, mask three important trends that have been taking place. The first

[1] The industrialization of capture fisheries (Platteau 1989; Thorpe and Bennett 2001; Bavinck 2011) that took place during the twentieth century had two phases. Phase 1 commenced late in the nineteenth century and was centered in Europe, North America, and Japan. Post-colonial governments initiated phase 2 in the period after WWII. The technologies in both phases were identical, and consisted of engine-driven harvesting technology, new gear types, refrigeration, and large-scale infrastructure (such as harbors). Markets developed alongside increases in production, with most landing sites now being connected to national and international markets.

is a shift from industrialized countries to the developing world. The introduction of Exclusive Economic Zones (EEZ), stock collapses, economic growth of developing countries, and an increased demand for fish products in the world are factors explaining the current dominance of developing countries in the global fish trade. Today, about 80% of the world's fisheries production takes place in developing countries; their share of the world fish trade value has passed 50%, and is steadily increasing. This pattern is even more striking when it comes to inland fisheries: 95% of global inland capture production now comes from developing countries (FAO 2009).

The second trend that has occurred – most notably in the developing world – is the partition between a large-scale, industrialized fishery, and a small-scale fishery. This division is mainly the consequence of the development pattern pursued by governments in the post-independence era, namely the creation of a modern, technically efficient, harbor-based fishery, in proximity to – and in conflict with – many already existing small-scale fisheries (Platteau 1989; Bavinck 2005, 2011). Although, as the various chapters of this volume testify, the latter have been far from stagnant, there is still a remarkable gap between the two subsectors in terms of investment levels, catch per unit of effort (CPUE), and employment, with small-scale fisheries employing a far greater portion of the fishing population than the large-scale subsector (also see Johnson 2006).

This brings us to the third trend, which relates to demography. While employment in fishing is decreasing, and the number of recreational fishers now exceeds professional ones in most industrialized countries, fisheries employment is growing steadily in the rest of the world. FAO statistics (FAO 2009) show 35 million part-time and full-time ocean fishers, in addition to 4 million occasional fishers and fish farmers on a global level in 2006. Less than 0.9 million of these are located in industrialized countries. The total production per fisher is around 20 tons per year in Europe and North America, while the global production per fisher is 3.3 tons per year, indicating a much more capital intensive fishery in the industrialized, as compared to the developing, countries.

Unlike the prices of many other export products from developing countries, real fish prices are also increasing. The World Bank (2008, p. 8, Fig. 8) has thus calculated that the real export unit value (US$/ton) of fishery products in the world more than doubled between 1976 and 2004. In conjunction with increasing production, these price rises saw the total export value minus the total import value of fish products from developing countries grow from USD 1.8 billion to USD 24.6 billion in the period 1976–2006 (FAO 2009). Fish products now represent a significant monetary worth equivalent to the combined export values of coffee, rubber, cocoa, meat, and sugar. Analyzing these trends, Delgado and others (2003) note that rising consumer demand has resulted in a long and virtually uninterrupted increase in prices, which probably benefits fishers of all types and geographical origins. Fishing costs and income vary greatly by type of fishery and locality, however, and reasoning based on global aggregated figures has its limitations.

Figure 2.2 presents capture fisheries in the period 1950–2008 for the countries that are discussed in this volume. These data suggest that developing countries have

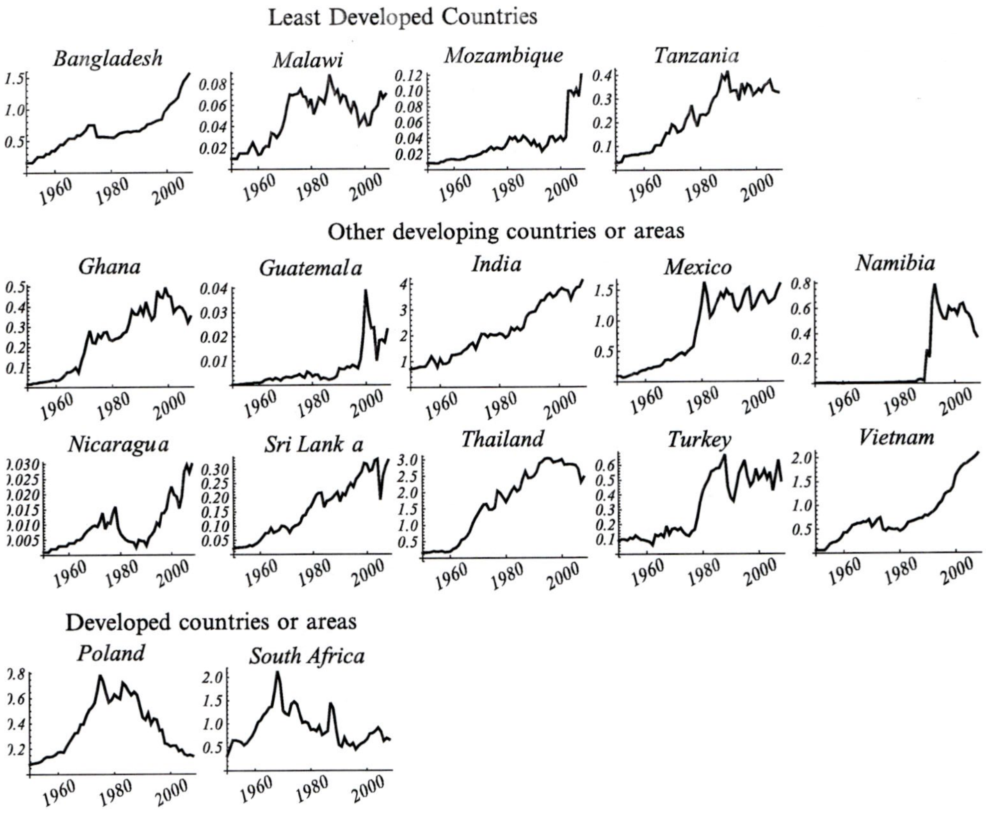

Fig. 2.2 Total catches of the countries discussed in this book in the period 1950–2008. The development level according to the FAO categorization (as in Fig. 2.1) places two of the countries (Poland and South Africa) in the developed group, and Bangladesh, Malawi, Mozambique and Tanzania in the least developed. The graphs include both marine and inland catches; graphs of these countries' marine catches only are shown in Fig. 3.1, Chap. 3 (Source: FAOs online capture statistics. Accessed 20 January 2011 at http://www.fao.org/fishery/statistics/global-capture-production/query/en)

witnessed spectacular growth of fish production in the past decades, with leveling off (or some decline) in recent years. However, developed countries, represented by Poland and South Africa in this volume, have simultaneously witnessed severe reductions in fish production.

About 90% of those employed in world fisheries are classified as small-scale fishers (FAO 2007, p. 2). An additional 100 million people are estimated to be employed in other occupations associated with fisheries, primarily in processing and trading. Total global employment, directly or indirectly associated with small-scale fisheries and aquaculture, was estimated to be about 135 million in 2002 and the number is increasing (Table 2.1).

Table 2.1 presents an overview of capture fisheries employment in the period 1970–2005.[2] Note that the growth of world fish catches and the number of fishers outdistanced the growth of the world's population in a period when the latter almost doubled. Table 2.1 shows that while the world's population increased by 75%, the number of capture fishers in the world increased by 178% between 1970 and 2005, more than twice the growth of the world's population. The distribution of fishers over geographical regions, however, is severely skewed, with Asia (where the real increase in numbers of fishers is found – including 84% of the world's capture fishers), and Africa making up the bulk of the remainder.[3] Europe and America show a declining share of the world's fishers, from 11% in 1970 to below 6% in 2005 (Table 2.1).[4]

Where did additional growth in the number of people employed in capture fisheries (mainly in Asia, but also in Africa) come from? Assuming that population growth among fishers is roughly similar to general population growth, and that most fishers and their descendants have stayed in capture fishing, the surplus must have come from outside the sector. Our hypothesis is that most of these people have been attracted by the possibility of obtaining a share of the wealth available in fisheries. This situation would seem to pertain to countries with a stable administration, but also to countries where people have been driven to the coast by civil war (as, for example, in the case of Mozambique, Chap. 18).

[2] There are no figures for the period before 1970, which constitutes the nucleus of the industrialization era, and is therefore likely to have created the largest impact on employment too. Similar to global catch data, which are regularly disputed, employment figures are also imprecise (FAO 1999). The figures in Table 2.1, however, appear at least to indicate a general trend.

[3] The stabilization that has apparently taken place in the fishing population from 2000 to 2005 (see Table 2.1) suggests that capital is also substituting for labor at a higher rate in developing countries. However, this requires further investigation.

[4] Low and declining employment figures in the fisheries of the North do not indicate a marginal role in global fisheries. FAO (2009, Table 7) points out, for example, that production per person in Europe and North America is eight to ten times the production rate in Asia and Africa.

Table 2.1 Number in millions of capture fishers and total populations 1970–2005, and the relation between the two[a]

	Regions	1970	1980	1990	2000	2005	Percentage increase 1970–2005
Fishers in	Africa	1.360	1.570	1.770	3.524	3.478	156
	Americas	0.906	1.080	1.421	1.382	1.367	51
	Asia	9.301	13.558	20.028	27.412	28.572	207
	Europe	0.436	0.372	0.634	0.768	0.663	52
	Oceania	0.046	0.050	0.054	0.044	0.050	8
	World	12.261	16.259	23.905	33.199	34.131	178
Population of	Africa	366.795	482.232	638.726	819.462	921.070	151
	Americas	517.755	616.751	725.000	839.884	891.689	72
	Asia	2079.820	2566.922	3112.180	3698.295	3936.540	89
	Europe	693.082	739.232	777.460	726.569	729.420	5
	Oceania	19.636	22.944	26.926	31.163	33.560	71
	World	3677.089	4428.081	5280.292	6115.373	6512.279	77
Percentage of fishers in total population	Africa	0.37	0.33	0.28	0.43	0.38	2
	Americas	0.17	0.18	0.20	0.16	0.15	−12
	Asia	0.44	0.53	0.64	0.74	0.73	62
	Europe	0.06	0.05	0.08	0.11	0.09	44
	Oceania	0.23	0.22	0.20	0.14	0.15	−36
	World	0.33	0.37	0.45	0.54	0.52	57

[a]Fisher figures calculated from FAO (1999 and 2009, p. 23, Table 5). General population figures from UN (2009)

2.3 Fisheries Development and Economic Growth

Capture fishing is one of the oldest sources of livelihood in human existence and displays an amazing technical variety (Von Brandt 1984). It is also a profession with distinctive social characteristics (Acheson 1981; Van Ginkel 2001). A collection of human *niches* (Tuomi-Nikula 1985) formed around the capture of various species of fish as they are distributed over marine space. Generally speaking, there are two conditions for developing a fishery: (1) The availability of aquatic resources; and (2) Techniques and knowledge on how to harvest and utilize them.

At a subsistence level, aquatic resources constitute primary or secondary sources of nutrition for a large number of households. Seasons and years of lower productivity or reduced availability prompt migration and a pursuit of alternative livelihoods. Despite these fluctuations, aquatic resources are a staple asset for people who depend on them for subsistence (rather than monetary income).

By principle, common pool resources as utilized in the capture fishery are open to anyone with the knowledge and equipment to harvest them. Finite resources do, however, limit the number of people who can derive a fisheries livelihood. Further restrictions are also imposed by the fishing technology available, or lack thereof. At the other end of the value chain, new technology and the discovery of previously untapped aquatic resources increase the population that is potentially supplied with fish products.

The development of subsistence fisheries involves commercialization and access to markets. For this to take place, issues relating to preservation, product quality and logistics need to be addressed. Dry and salted fish products have historically been transported over large distances and stored for long periods, even years. Fresh fish products, however, necessitate other preservation methods. Improved infrastructure and refrigeration technologies have radically changed conditions for the trade of seafood in the twentieth and twenty-first centuries. These changes have had major repercussions for the fisheries.

Access to markets induces subsistence fishers to harvest beyond their sustenance needs; excess catches are converted into capital, and economic wealth is thereby generated. Economic theory argues that as wealth accumulates, capital becomes less expensive and labor more expensive; hence, capital often substitutes for labor. Increased buying power results in possibilities for employing labor in other economic sectors, and in higher prices for other products (including fisheries products). This type of economic development has taken place in many industrialized countries (Béné et al. 2010), where many former fishers have entered new professions, and employment in fisheries has thus declined.

As the demand for fish products augments, the resulting increases in fishing effort in unregulated fisheries may at some point lead to a decline in fish production, a situation referred to as biological overfishing. Economic overfishing occurs at an even earlier point in time, as resource rent starts to decline at levels of fishing effort below those leading to a decline in surplus biomass growth (Gordon 1954). Biological overfishing today represents the main reason for contemporary fisheries management. The aim of resource-motivated management is to reduce fishing effort in order to increase sustainable catches, which also will eventually reduce the cost per unit of harvest.

Given that the national objective in many developing countries is to promote and encourage economic growth, governments' first – and valid – challenge is to create conditions for the commercialization of fisheries to take place. In the case of subsistence fisheries, access to markets represents a critical factor. First of all, it is necessary to make markets available through infrastructure, trade agreements, and preservation technology; none of which are easily established by regular market mechanisms. Extension services promote market production at the producer end of the value chain, and the introduction of new technology is one of the objectives of this mechanism. Economic growth, however, is not obtained from improved technology alone – a demand for fish products must also be present. All market failures hindering economic growth therefore need to be addressed.

As is corroborated by various chapters in this book, small-scale fisheries have many faces. They are also embedded in societies, economic structures, and historical trajectories of varying character, which do not necessarily follow the trends of commercialization and capital replacing labor described above. Small-scale fisheries have defied many predictions just by staying around, and it is likely that they will be present for many years to come. The current state of world fisheries has induced various views about how to move forward. While Cunningham et al. (2009) still seem to advocate resource rent optimization as the best way to alleviate poverty, Béné et al. (2010) counter that this method is neither realistic nor relevant to the majority of small-scale fisheries in developing countries.

If the distribution of wealth is not taken into consideration, successful economic growth may worsen, rather than improve, the situation for the poor. We emphasize that even though fish stock resources may sustain subsistence capture fisheries, ensuring food security for those exploiting the resource, development of markets is necessary to secure economic benefits. The issue of resource distribution, moreover, becomes crucial when economic wealth is incipient. It represents another management challenge, as fair and equitable distribution is not likely to occur in an unregulated market (Atkinson 1997). The final challenge, of course, is not to develop higher economic resilience at the expense of the natural resource base.

2.4 Distribution of Common Goods

Our starting point is that coastal societies are fortunate to possess aquatic resources that can provide large economic benefits. However, this raises such questions as: What should be the objective of resource exploitation? And who should benefit from its use?

One contemporary line of argument is that fish resources are to be exploited in order to maximize economic wealth for society as a whole, and that maximization is attained by favoring efficiency, taking externalities and opportunity costs into consideration. Since income in the fishery sector is linked to the rest of the economy, increasing wages outside the fishery affects the supply of labor inside (Eide 2009). Béné et al. (2010) provide the example of Norway, where employment in fisheries today is one-tenth of employment after World War II. This is not a consequence of regulation but of regular economic growth. Capital has thus replaced labor in the production of fishing effort.

The authors of the report "The sunken billions – the economic justification for fisheries reform" (World Bank 2008), point out the economic irrationality presently evident in global fisheries. Governments strive to introduce reforms that will result in the capture of the "sunken billions" that wait in aquatic ecosystems. A first item of attention is the massive public subsidies pumped into the global fishing fleet (primarily directed toward industrial fishing), which exacerbate overcapacity and contribute greater pressure on fish resources (Sumaila et al. 2008). Subsidies have a negative impact on small-scale fisheries by diminishing their fishing opportunities. Furthermore, the small-scale fleet has also lost economic competitiveness to the large-scale fleet, because the small-scale sector is not subsidized at a level even remotely comparable to the former.

A second bone of contention is the introduction of property rights to fish resources. Individual property right systems may improve profitability, but will most likely be devastating to the small-scale sector and concentrate wealth in a few hands (World Bank 2005). We tend therefore to side with Hersoug (2007), who enquires: How do we secure rights-based fisheries to the right people? This position builds upon the understanding that fish resources represent an important safety valve when livelihoods in non-fishing (e.g., agricultural) sectors are under threat (Jul-Larsen et al. 2003). Small-scale fishers are often the most dependent on the common pool nature of fisheries, and thus will suffer the most from changes in property regimes. Such changes will also prevent, once and for all, aquatic resources from being a safety valve in times of economic hardship.

In these conditions, is it not better to prioritize the livelihoods of the millions depending on small-scale fishing rather than maximizing wealth creation for a relatively small and already privileged group of better-off fishers? Even if the total value obtained from the fisheries can be increased by reducing the number of fishers, most of those who are removed will be worse off, since the wealth created will not reach them and alternative livelihoods are scarcely available.[5] Closing the commons is therefore a risky business, especially for those who are left outside.

2.5 Conclusions

In the vast literature dealing with the quantification of poverty, the dominant approach is to specify a poverty line which defines the level of income necessary for subsistence. Recent studies (Pinkovskiy and Sala-i-Martin 2009) point out that world poverty is declining based on this viewpoint. The establishment of industrial

[5] It is important in this regard to reiterate that solutions to prevent or reduce poverty in fisheries in developing countries also need to be found outside the sector, where alternative employment is to be created for those who have sought in fisheries an employer of last resort (Cunningham 1999).

fisheries in developing countries has probably been one of several factors contributing to this trend; others include the availability of health services, global trade, and general economic development. On a global scale, poverty (referring to a poverty line of one US$ per day) has declined by 80% over a period of 36 years, in spite of a significant increase in total population (Table 2.1), reducing the number of poor from 403 million in 1970 to 152 million in 2006.

Loury (1981) notes, however, that when markets are imperfect – as they often are, particularly in developing countries – redistributive policies may improve economic growth. Sen (1976) points out conversely that *poverty* and *inequality* are related terms. Inequality occurs at different levels. In fisheries, we see it not only between industrial and small-scale fishers, but also within the small-scale fishing sector itself. This is not necessarily condemned by such fishers, who view a measure of inequality to be a logical result of differences in skill and in fortune (Acheson 1981; Van Ginkel 2001). Many scholars studying small-scale fisheries, however, have noted the nearly ubiquitous presence of community mechanisms to support those who are unable to secure a decent livelihood – widows, the unfortunate, the destitute (see, for example, Kurien and Paul 2001). Such support is predicated on notions of fundamental human rights. Small-scale fishers also challenge, however, those who take an inordinate share of what the sea has to offer. This has resulted in a large and continuing body of conflicts between small-scale and industrial fishers all over the world, especially in terms of inshore resources.

Developing countries face two main challenges today with regard to their fisheries: (1) to continue their efforts to generate economic development, which will hopefully result in better living conditions for their populations; and (2) to nurture the present small-scale fisheries, which provide, in the interim, large measures of employment, nutrition, and hope. The latter approach means having an eye for allocating the resource according to the needs of those employed in the sector.

Recent trends in fisheries demonstrate that the economic value embedded in aquatic resources allows society to employ and feed an increasing number of fisher families, thus spending natural and economic wealth for disadvantaged people to survive. We argue that the distribution of these riches among a larger population of small-scale fishers is the better policy direction to employ in the context of economic development and poverty reduction.

References

Acheson JM (1981) Anthropology of fishing. Annu Rev Anthropol 10:275–316

Atkinson AB (1997) Bringing income distribution in from the cold. Econ J 107(March):297–321

Bailey C (1988) The political economy of marine fisheries development in Indonesia. Indonesia 46:25–38

Bavinck M (2005) Understanding fisheries conflicts in the South: a legal pluralist perspective. Soc Nat Resour 18(9):805–820

Bavinck M (2011) The mega-engineering of ocean fisheries: a century of expansion and rapidly closing frontiers. In: Brunn SD (ed.) Engineering earth: the impacts of mega-engineering projects. Kluwer, Dordrecht, pp 257–273

Béné C (2003) When fishery rhymes with poverty: a first step beyond the old paradigm on poverty in small-scale fisheries. World Dev 31(6):949–975

Béné C, Hersoug B, Allison EH (2010) Not by rent alone: analysing the pro-poor functions of small-scale fisheries in developing countries. Dev Policy Rev 28(3):325–358

Butcher JG (2004) The closing of the frontier: a history of the marine fisheries of Southeast Asia c. 1850–2000. KITLV Press, Leiden

Cunningham S (1999) Towards the increased policy relevance of fisheries research. FAO FIPP/C964

Cunningham S, Neiland AE, Arbuckle M, Bostock T (2009) Wealth-based fisheries management: using fisheries wealth to orchestrate sound fisheries policy in practice. Mar Resour Econ 24:271–287

Delgado CL, Wada N, Rosegrant MW, Meijer S, Ahmed M (2003) Fish to 2020: supply and demand in changing global markets. International Food Policy Research Institute (in association with WorldFish Centre), Washington DC

Eide A (2009) Economic principles: an economic perspective on fishing. In: Cochrane KL, Garcia SM (eds.) A fishery manager's guidebook. FAO/Wiley-Blackwell, Rome/Oxford, pp 75–104, chap 4

FAO (1999) Number of fishers 1970–1996. Fishery Information Data and Statistics Unit. Food and Agriculture Organization, Rome

FAO (2007) Increasing the contribution of small-scale fisheries to poverty alleviation and food security. FAO fisheries technical paper 481. FAO, Rome. ISBN 978-92-5-105664-6

FAO (2009) The state of world fisheries and aquaculture 2008. Food and Agriculture Organization, Rome

Gordon HS (1954) The economic theory of a common property resource: the fishery. J Polit Econ 63:124–142

Hersoug B (2007) Opening the tragedy? In: Sizing up: property rights and fisheries management: a collection of articles from SAMUDRA report. ICSF, Chennai, pp 87–91

Johnson D (2006) Category, narrative, and value in the governance of small-scale fisheries. Mar Policy 30:747–756

Jul-Larsen, E., Kolding, J., Overå, R., Raakjær Nielsen, J., and van Zwieten, P. A. M. (2003). Management, co-management or no management? Major dilemmas in southern African freshwater fisheries. Synthesis report. FAO Fisheries Technical Paper 426/1.

Kurien J, Paul A (2001) Social security nets for marine fisheries. The growth and changing composition of social security programmes in the fisheries sector of Kerala State, India. Working paper 318. Centre for Development Studies, Thiruvananthapuram

Loury GC (1981) Intergenerational transfers and the distribution of earnings. Econometrica 49(4):843–867

MacKenzie W (1979) Rational fishery management in a depressed region: the Atlantic groundfishery. J Fish Res Board Can 36:811–826

Pauly D (1994) From growth to Malthusian overfishing: stages of fisheries resources misuse. Tradit Mar Resour Manage Knowl Info Bull 3:7–14

Pinkovskiy M, Sala-i-Martin X (2009) Parametric estimations of the world distribution of income. NBER working paper series 15433. National Bureau of Economic Research, Cambridge

Platteau JP (1989) The dynamics of fisheries development in developing countries: a general overview. Dev Change 20(4):565–599

Sen A (1976) Poverty: an ordinal approach to measurement. Econometrica 44(2):219–231

Smith A (1776) An Inquiry into the Nature and Causes of the Wealth of Nations. In: Cannan E (ed.) Library of Economics and Liberty. Methuen and Co., Ltd. 1904, London. http://www.econlib.org/library/Smith/smWN4.html. Accessed 24 Mar 2008

Smith HD (2000) The industrialization of the world ocean. Ocean Coast Manage 43:11–28

Sumaila UR, Teh L, Watson R, Tyedmers P, Pauly D (2008) Fuel price increase, subsidies, overcapacity, and resource sustainability. ICES J Mar Sci 65:832–840. doi:10.1093/icesjms/fsn070

Thorpe A, Bennett E (2001) Globalisation and the sustainability of world fisheries: a view from Latin America. Mar Resour Econ 16:143–164

Tuomi-Nikula O (1985) The cultural-ecological aspect of culture change. Studia Fennica 30:147–163

UN (2009) World population prospects: the 2008 revision. Population Division of the Department of Economic and Social Affairs of the United Nations Secretariat. http://esa.un.org/unpp. Accessed 5 July 2009

van Ginkel RJ (2001) Inshore fishermen: cultural dimensions of a maritime occupation. In: Symes D, Phillipson J (eds.) Inshore fisheries management. Kluwer Academic, Dordrecht, pp 177–193

von Brandt A (1984) Fish catching methods of the world. Fishing News Books, Farnham, Surrey

World Bank (2005) Turning the tide. Saving fish and fishers: building sustainable and equitable fisheries and governance. http://siteresources.worldbank.org/INTARD/Resources/seaweb_FINAL_pt.1.pdf and seaweb_FINAL_pt.2.pdf

World Bank (2008) The sunken billions – the economic justification for fisheries reform. World Bank with FAO Agriculture and Rural Development Department, Washington DC

Chapter 3
Situating Poverty: A Chain Analysis of Small-Scale Fisheries

Ratana Chuenpagdee and Svein Jentoft

Abstract In this chapter, we argue that poverty in small-scale fisheries needs to be examined within the context of the fisheries chain which links the aquatic environment with the natural and social systems at the harvest and post-harvest processes. It is within this context that the factors and conditions underlying poverty may be found and resolved. Such an examination widens the considerations about external and internal sources and drivers of poverty. Poverty in small-scale fisheries extends beyond local communities to regional and national levels. For this reason, we begin by presenting small-scale fisheries globally, with an emphasis on the 15 countries included in this volume. Next, we investigate causes of the vulnerability of small-scale fishing communities to poverty, based on several indices. With some references to the case studies that follow in later chapters, we suggest how the drivers and consequences of poverty in small-scale fisheries may be examined from the chain perspective.

3.1 Introduction

At the meeting of the Commission on Sustainable Development held in May 2009, the United Nations Secretary General, Mr. Ban Ki-moon, warned of the crisis that the world is facing in terms of food security. He suggested that without immediate

R. Chuenpagdee (✉)
International Coastal Network, Department of Geography, Memorial University of Newfoundland, St. John's, NL A1B 3X9, Canada
e-mail: ratanac@mun.ca

S. Jentoft
Norwegian College of Fishery Science, University of Tromsø, 9037 Tromsø, Norway
e-mail: svein.jentoft@uit.no

S. Jentoft and A. Eide (eds.), *Poverty Mosaics: Realities and Prospects in Small-Scale Fisheries*, DOI 10.1007/978-94-007-1582-0_3,

and concerted actions, we risk slipping into "poverty, degradation and despair." As the world is trying to reach the target set in the Millennium Development Goals, which is to eradicate 50% of the extreme poverty by 2015 (UN 2009), a new perspective is needed to understand and address the problem.

One such perspective is offered by Narayan and colleagues in the *Voices of the Poor* book series (Narayan et al. 2000, Chap. 3). They assert that people who attempt to find solutions for poverty are often not poor, while the true poverty "experts," i.e., the poor people themselves, are never asked about what poverty means and how best to eradicate it. Indeed, the meaning of poverty is often considered intuitive (Adams et al. 2004), which leads researchers and policymakers to focus on the complexity of poverty measurement rather than reflecting fundamentally on what poverty means for those who are poor. This often results in polices and development agendas that are disconnected from the problem, as it is experienced on the ground. When poverty alleviation programs do not produce desirable outcomes, it may therefore be due to the lack of direct involvement in problem identification and solution creation by the most important stakeholder group, i.e., the poor people.

We submit that in addition to aligning an understanding of poverty with the values and knowledge of the poor, a broader perspective is needed. This implies looking at poverty as part of the linked social and ecological systems where associated characteristics of poverty and the underlying conditions may be found. We recognize that these systems and their linkages are dynamic and locally specific, and that poverty alleviation needs to be situated in the local socio-political context, but we argue that it must also consider macro-level issues that lie outside the fisheries realm. This is why we suggest a fisheries chain approach to examine poverty, which links "pre-harvest," "harvest," and "post-harvest" processes and inter-dependencies (Kooiman et al. 2005). Factors external and indirectly related to the fisheries ecosystem are accounted for in the pre-harvest system, while the effects of globalized markets are examined in the post-harvest part of the chain. Systematic examination across the production chain extends the discussion about sources and conditions of poverty, and also acknowledges the horizontal linkages in terms of the aquatic and terrestrial environments. As will be shown in the case studies, fisheries are not the only source of livelihoods for fishing people. Rather, coastal people take advantage of opportunities wherever they are found.

In the following, we first situate poverty in the small-scale fisheries of the 15 countries discussed in this book, to highlight areas of the world where fishing communities are likely to be most susceptible to poverty. Next, we examine factors causing poverty and opportunities for responses and interventions at all levels using the chain approach. The chapter concludes with some preliminary discussion about governance implications and global challenges in coming together to act responsibly in avoiding the vicious cycle that Mr. Ban Ki-moon talked about, as earlier mentioned, and which will be revisited in the later chapters of this volume.

3.2 Small-Scale Fisheries in the Global Context

The characteristics of small-scale fisheries differ from one place to the next. Nevertheless, when looking broadly, Chuenpagdee et al. (2006) show that several countries use similar measures in describing this sector (Table 3.1). For instance, they found that about 90 out of 140 coastal countries use boat size to differentiate between small- and large-scale fisheries. Many small-scale fisheries involve a boat of between 5 and 7 m in size, while some countries consider larger boats, but generally no more than 12–15 m, as small-scale. Other defining features are gross registered tonnage (hereafter GRT, which is mostly less than 10 for small-scale fishing boats), engine size (if used; between 40 and 75 HP), and types of boats and fishing activities.

A variety of boats characterize small-scale fisheries such as canoe, dinghy, non-motorized boat, wooden boat, and traditional boat with no deck. While several fishing gears are employed in both the small- and large-scale fishing sectors, fishing activities such as shellfish gathering, beach seining, diving, and fishing using set bag net, hand-line, and small trap are mostly considered small-scale. Small-scale fishing is also described by different names, such as artisanal, traditional, coastal, inshore, peasant fishing, emphasizing their social and cultural aspects rather than their technological characteristics. These alternative labels for small-scale fisheries are used in several of the chapters that follow.

Fisheries information specific to small-scale fisheries is not readily available. For example, global landing data reported by coastal nations to FAO do not differentiate whether catches come from the large-scale or small-scale fishing sector. In other words, there is no information about how much of the 93 million tons of the global fisheries production (capture and culture, inland and marine) reported by FAO in 2005 (FAO 2005a) comes from the small-scale fishing sector, except for a suggestion

Table 3.1 General characterization of small-scale fisheries – based on how 90 nations define this sector. Examples are boat size, engine power, gear type, distance from shore and crew number

Key characteristics	Common definition (range)
Boat size	Between 5 and 7 m; less than 10, 12, or 15 m (2–24 m)
Boat GRT	Less than 10 GRT (3–50 GRT)
Size of engine	Less than 60 HP; between 40 and 75 HP (15–400 HP)
Boat type	Canoe, dinghy, non-motorized boat, wooden boat, boat with no deck, traditional boat
Gear type	Coastal gathering, fishing on foot, beach seine, small ring net, handline, dive, traps
Distance from shore	Between 5 and 9 km; within 13 km; up to 22 km
Water depth	Less than 10, 50, or 100 m depth
Nature of activity	Subsistence, ethnic group, traditional, local, artisanal
Number of crew	2–3; 5–6
Travel time	2–3 h from landing sites

Source: Chuenpagdee et al. 2006

that about 80% is from developing countries. However, FAO provides an estimate that about 90% of the total 43 million fishers and fish farmers around the world are small-scale.

Contributing to knowledge about the small-scale fisheries sector, Chuenpagdee et al. (2006) compiled available information and existing data sources pertaining to marine capture fisheries in 140 coastal nations. Using rule-based algorithms, they estimated that about 12 million people are engaged in full-time small-scale fishing, contributing to the total marine capture fisheries catches of about 25 million tons. These estimates are comparable with FAO data, when assuming that about half of the total world catches come from marine capture fisheries, as suggested by FAO statistics. Since the estimates provided by Chuenpagdee et al. (2006) are country-specific, they are used as a basis for the discussion about the 15 countries included in this volume.

Chuenpagdee et al. (2006) submit that the contribution of small-scale fisheries could range anywhere between 15% and 30% of the total world fisheries production, depending on how much of these catches is assumed to have been included in the FAO statistics. If all of the estimated 25 million tons from small-scale fisheries had already been reported in the FAO data, their contribution to the world total is then about 30%. If, on the other hand, none of them was included, the total marine capture fisheries production would be increased by 25 million tons, and the contribution from small-scale fisheries would be about 26%. The final scenario is that the FAO marine capture fisheries data include about half of the estimated catches from the small-scale fisheries sector. In this case, small-scale fisheries contribute about 15% of the world total.

No matter which assumption is applied, the contribution of small-scale fisheries to local food security in the global context is highly significant, especially considering that they produce food for direct human consumption (i.e., only a minimal amount goes to fish meal), employ more people, are more energy efficient, and are generally more sustainable than industrial, large-scale fisheries (Zeller and Pauly 2005; Pauly 2006; Jacquet and Pauly 2008). The lack of a global information system on small-scale fisheries, and of thorough understanding about the fisheries at the local scale, add to the difficulty in addressing issues and concerns specific to small-scale fishing people, such as poverty. They are easily ignored by decision-makers and thus marginalized in global fisheries policies. As a consequence, their actual and potential role in poverty alleviation is not sufficiently recognized (Béné et al. 2007).

As an overview of the fisheries in the 15 countries included in this volume, we present the time series catch data from 1950 to 2005 in the following (Fig. 3.1a, b). The figures show India and Vietnam as the two major fishing nations among the high production countries. Whereas India's catches dropped significantly since 2000 (due to sharp decline in catches from trawls, drift nets, beach seines, and hand-lines), Vietnam continued to show increasing growth. In other cases, catches fluctuated throughout the 50-year duration but in a somewhat stabilized manner. Of the six "low production" countries, dramatic changes are observed in Sri Lanka, but others show similar fluctuation patterns as in the high production countries presented in Fig. 3.1a.

Further analysis shows that Poland is the only country discussed in this book in which fisheries are dominated by trawls (Table 3.2). The majority of catches in

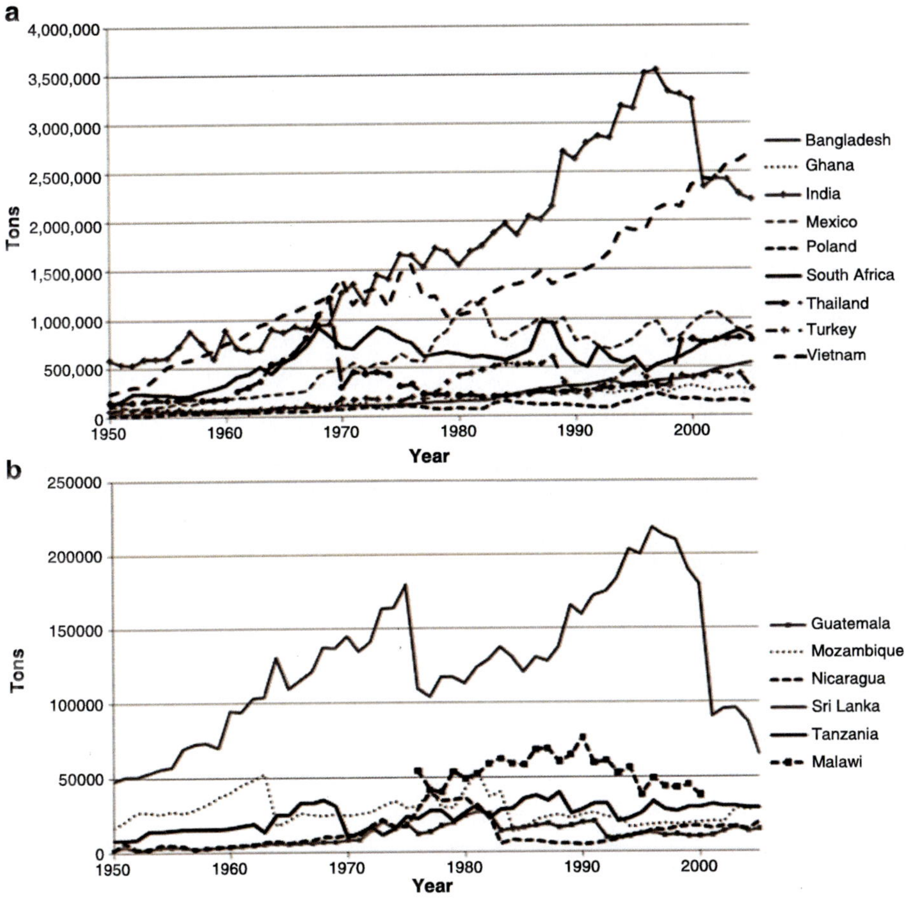

Fig. 3.1 Marine fishery catches within the Exclusive Economic Zone of the 15 countries included in this volume, differentiated into two groups of countries. (**a**) Countries with catches higher than 100,000 tons. (**b**) Countries with catches lower than 100,000 tons. In the case of Malawi, the data represents the national inland fisheries based on Weyl (2005) from 1976 to 2000 (Source: Sea Around Us Project – www.seaaroundus.org – for all countries except Malawi)

Mexico and South Africa are from purse seines, while those of Bangladesh and Vietnam are from gill nets. In contrast, data for Turkey, Ghana, Guatemala, and Tanzania show more than 40% of the catches coming from other gears, which are presumably small-scale, for the most part. These other gears include seine nets, beach seines, hand-lines, and pots and traps. Countries like Mozambique and Nicaragua represent a fishery dominated by both commercial trawling and other small-scale gears. If gill nets are considered small-scale, then fisheries in Thailand, India, and Sri Lanka look similar to the previous two. On the whole, under these assumptions, all countries are dominated by small-scale fisheries except Poland, Mexico, South Africa, and to a lesser extent, Mozambique.

Table 3.2 Percentages of catches by fishing gear

Country	Trawls	Purse seines	Gill nets	Others
Poland	67.2	5.0	18.5	9.2
Mexico	10.9	65.6	5.7	17.9
South Africa	17.5	53.9	5.5	23.1
Bangladesh	3.7	0.2	89.9	6.1
Vietnam	18.0	2.8	56.4	22.8
Turkey	27.4	28.3	4.0	40.3
Ghana	19.4	28.6	11.2	40.8
Guatemala	16.6	32.5	2.7	48.1
Tanzania	13.7	5.3	31.6	49.5
Mozambique	51.9	0.5	10.9	36.7
Nicaragua	40.8	0.1	1.2	57.9
Thailand	35.1	9.1	38.6	17.3
India	32.6	4.7	35.4	27.2
Sri Lanka	27.7	4.9	39.3	28.0

Source: Sea Around Us Project (www.seaaroundus.org)

3.3 Poverty Indicators for Small-Scale Fisheries

Globally, nations can be categorized and compared using the Human Development Index (HDI), which measures the well-being of countries based on life expectancy, literacy, education, standard of living, and Gross Domestic Product (GDP) per capita. Of the 15 case studies included in this volume, Malawi and Mozambique are in the low-HDI category (column A, Table 3.3). Tanzania and Bangladesh were in the low-HDI category in 2000, but are considered medium-HDI in the 2007 report. Similarly, Mexico and Turkey, formerly classified as medium-HDI, are now in the high-HDI category, along with Poland. The other countries are in the medium-HDI category.

HDI offers a partial picture about the poverty of nations and what being a citizen of those countries may entail. Others have tried to categorize nations into poverty levels using different indices. For instance, with the help of satellite imagery and on-the-ground survey information, Elvidge et al. (2009) derive indices by dividing population counts with emitted nocturnal lights (referred to as "population/light index" in column B, Table 3.3). The use of "light as a proxy for wealth" assumes that areas with high population counts in developing countries but which are poorly lit are likely to be poorer than countries with more nocturnal lights. Applying their indices to the 15 countries included in the study, Poland, Turkey, and Mexico (all of which are in the high-HDI category) are the least poor (lowest population/light index). In contrast, Mozambique and Malawi (of the low-HDI category) have more than 80% of the people living in darkness; indicating a high level of poverty (high population/light index, column B in Table 3.3). With the exception of Tanzania, poverty levels according to the light index in all medium-HDI countries are considered to be in the medium range, from 28% to 56%. For Tanzania, the 2007 United Nations data moved it from low-HDI to the medium category, despite the fact that the poverty level, according to the population/light index, is the highest among the 15 countries at 84%.

Table 3.3 Categorization of the 15 case study countries by different indicators

Continent	Country	HDI category[a] (A)	Pop./light index (%)[b] (B)	Gini index[c] (C)	Catch/fisher[d] (D)
Africa	Ghana	M	56	40.8	1.59
	Malawi	L	81	39.0	n/a
	Mozambique	L	80	47.3	2.00
	South Africa	M	34	57.8	1.72
	Tanzania[e]	M	84	34.6	2.50
Asia	Bangladesh[e]	M	53	33.4	0.44
	India	M	41	36.8	1.46
	Sri Lanka	M	28	40.2	0.45
	Thailand	M	39	42.0	5.45
	Vietnam	M	37	34.4	0.66
Latin America	Guatemala	M	35	55.1	2.80
	Mexico[f]	H	14	46.1	5.99
	Nicaragua	M	46	43.1	0.32
Europe	Poland	H	11	34.5	2.96
	Turkey[f]	H	24	43.6	1.71

Sources:
[a]UNDP (2000, 2009)
[b]Elvidge et al. (2009)
[c]UNDP (2008)
[d]Chuenpagdee et al. (2006)
[e]HDI category for these countries have changed from low to medium in accord with the 2007 data (UNDP 2009)
[f]HDI category for these countries have changed from medium to high in accord with the 2007 data (UNDP 2009)

The other indicator that is often used in the discussion about poverty is the Gini Index, which is a measure of income inequality. The Gini coefficient can range from 0 to 1 or can be presented as a percentage (as in Table 3.3, column C). A low Gini coefficient indicates a more equal distribution (with 0 corresponding to complete equality), while higher coefficients indicate more unequal distribution (Gini 1921). The index can be used to compare income distribution across countries, complementing the GDP, thus controlling for the fact that poverty may not be improving for the majority of the population if the Gini coefficient is rising as well as GDP.

Of the countries covered in this volume, Bangladesh, Vietnam, Poland, and Tanzania have the lowest score on the Gini Index (thus suggesting the highest income equality), while Guatemala and South Africa are on the high end of the Index (column C, Table 3.3). If we were to assume that the narrow gap between the rich and the poor is an indication of greater ability to mobilize resources to address poverty, we may see that countries at the lower end of the Gini Index are likely to be less vulnerable to poverty, compared to those on the high end. However, their GDP may be so low that there are no resources to mobilize. On the other hand, countries with a high Gini Index and high GDP, like Poland in our case examples, may have more untapped potential to address poverty through re-distribution.

When specifically considering the small-scale fishing sector, estimates by Chuenpagdee et al. (2006) can be used to further differentiate countries according to the catch per small-scale fisher ratio (column D, Table 3.3). The catch per fisher is calculated using the available national catch and fisher data from various sources, such as FAO country profiles, reports, and research papers. Estimates for countries without data are extrapolated using the calculated inshore fishing areas where small-scale fishing are assumed to take place, but within the same HDI category (see details in Chuenpagdee et al. 2006). The assumption here is that fishers in countries with a high estimated catch per small-scale fisher ratio (like Thailand and Mexico) are likely to be less poor than those with a low catch/fisher ratio.

Countries can be classified into three levels of "poverty likelihood" of small-scale fishers, based on the four indices presented in Table 3.3. For instance, low-HDI countries are considered to have a high level of poverty likelihood, while high-HDI countries are classified as low poverty likelihood. According to the population/light ratio (column B, Table 3.3), three countries on the low end of this index – Mexico, Poland, and Turkey – are considered to have a "low" level of poverty likelihood. The high poverty likelihood countries are Tanzania, Malawi, and Mozambique, which have a population/light index higher than or equal to 80%. Next, the average Gini Index of the 15 countries is 41.9% with a standard deviation of 7.34. Thus, countries with a Gini Index higher than 49.2 are considered to have high poverty likelihood, while those with a Gini Index lower than 34.6 are grouped under low poverty likelihood. Finally, the criterion for classifying countries according to the catch/fisher ratio is arbitrarily set using two extreme values as opposed to an average. For instance, a country with a catch/fisher ratio less than one is considered to have a high level of poverty likelihood. Countries at the other end of the scale, where poverty likelihood is low, have a catch/fisher ratio greater than 5. Under these criteria, Thailand and Mexico have low poverty likelihood, while Bangladesh, Sri Lanka, Vietnam, and Nicaragua are in the high poverty likelihood category (Table 3.4). Malawi is not considered using the catch/fisher ratio because of the lack of data but, as will be later shown, the overall poverty likelihood for Malawi is calculated based on three indices instead of four.

The overall poverty likelihood is calculated using a simple, non-weighted average of the four indices, where values of 1, 2, and 3 are given to low, medium, and high poverty likelihood, respectively. As shown in the last column in Table 3.4, the two countries where small-scale fishers are likely to be poorer than others are Malawi and Mozambique. This is also confirmed by the case studies from these two countries (Hara, Chap. 12; Menezes et al., Chap. 18). The large trawl fishing sector in Mozambique (Table 3.2) may present additional stress on poor fishers, if these two sectors are competing for the same resources.

Small-scale fishers in Mexico, Poland, and Turkey, on the other hand, are likely to be less poor. Although the overall level of poverty likelihood is equivalent to poverty likelihood according to the HDI category, a weighing scheme can of course be used, which may then change the overall poverty likelihood picture. For instance, more weight can be given to the catch/fisher ratio since this is the most specific index for small-scale fisheries.

Table 3.4 Countries categorized by HDI and "poverty likelihood" based on different indices

		Relative level of "poverty likelihood" based on:				
Continent	Country	HDI	Pop/light	Gini index	Catch/fisher	Overall
Africa	Ghana	M	M	M	M	M
	Malawi	H	H	M	n/a	H
	Mozambique	H	H	M	M	H
	South Africa	M	M	H	M	M
	Tanzania	M	H	L	M	M
Asia	Bangladesh	M	M	L	H	M
	India	M	M	M	M	M
	Sri Lanka	M	M	M	H	M
	Thailand	M	M	M	L	M
	Vietnam	M	M	L	H	M
Latin America	Guatemala	M	M	H	M	M
	Mexico	L	L	M	L	L
	Nicaragua	M	M	M	H	M
Europe	Poland	L	L	L	M	L
	Turkey	L	L	M	M	L

We have argued in this chapter that poverty in fisheries cannot be resolved without direct attention to the small-scale fisheries sector. This position is well supported by the FAO (2005b) *Technical Guidelines for Responsible Fisheries* that focuses exclusively on the contributions of small-scale fisheries to poverty alleviation and food security. According to the data presented above, countries vary in the likelihood that their small-scale fishing people are poor, therefore implying that different policy responses are required. For instance, poverty *alleviation* efforts are most urgent in countries with potentially high poverty likelihood, like Malawi and Mozambique; while poverty *prevention* programs may be suitable in countries with medium and high poverty likelihood. This also means that, at the initial stage, the debate about the "why" of poverty and the "how" of poverty alleviation is not as helpful in areas with severe poverty where interventions are urgently needed. Financial and human resources must be allocated for immediate remediation while long-term, conservation- or development-oriented solutions are being explored. As submitted by Sen and Anand (2000), there is not much point talking about sustaining life opportunities in the future (i.e., the conservation aim) if the present ones are miserable and indigent.

It may be argued that the generalization about the low level of poverty likelihood among small-scale fishers in high-HDI countries is an oversimplification. As the case studies show, poverty can also be an issue for fishing communities in relatively higher developed countries. This is also the experience of developed countries not represented in this volume, such as Canada. The collapse of the Northern cod fishery in the early 1990s, and the consequent moratorium, was symbolically considered as the end of "the way of life" known by the people of Newfoundland for hundreds of years (Hamilton and Butler 2001). In other words, it was not only the economic activities that were disrupted, but also the social fabric inherent to the fishing culture and tradition. The Canadian government soon put an income adjustment package in place, and unemployment insurance for fishers, which is still ongoing, had been available to some since the late 1950s (Wright 2001). This level of support from government distinguishes high-HDI countries from medium- and low-HDI countries in alleviating poverty. Not all countries in the world are in the position to provide financial support to the fishing communities, no matter how important fisheries are to the nation, and regardless of their ability to distribute wealth and minimize income disparity. That said, the Gini Index (Table 3.3) suggests that redistribution of wealth may also be possible in some countries in the south to alleviate domestic poverty.

We hold in this chapter that macro analysis of poverty in small-scale fisheries can bring us only one step closer to effectively alleviating poverty. There are certainly policy issues that need attention at the international level, where efforts to provide overall frameworks for small-scale fisheries are ongoing, such as those of FAO, as well as others pertaining to international trade, markets, environmental monitoring, and control. However, there is also a need for an in-depth analysis of specific situations in which small-scale fishers find themselves. Poverty is a lived experience by people who struggle to cope with the challenges that they face in their daily life. In order to understand what poverty means and what measures can make a difference to people, we need to dig deeper than gross statistical analysis allows us to do. For that, we require empirical data at the level of households, communities and the fishing sector, and an analytical framework that provides a systematic treatment of such experiences, as described in the next section.

3.4 The Chain Analysis of Poverty in Fisheries

Following Kooiman et al. (2005) in their conceptualization of the "fish chain" as an analytical framework to examine fisheries governance, we employ the chain analysis to examine the characteristics of poverty issues in small-scale fisheries (Fig. 3.2). In the following, we make use of this framework by referencing the case studies in the analysis according to these various issues. The framework also provides a road map to help compare the 15 case studies in this book.

Theoretically, Kooiman et al. (2005) posit that challenges and concerns directly related to fisheries, as well as opportunities for improving fisheries governance, can be found throughout the entire production chain, from the bio-physical environments to the harvest and the post-harvest systems. The characteristics of the natural and social fisheries systems and their interconnectivity in the fish chain, along with how these systems interact with the governing system, are the substance of governance that needs to be properly examined and understood. These features also differentiate small- from large-scale fisheries.

For example, the majority of the world's small-scale fisheries are in tropical waters where the natural and social systems are highly diverse, complex and dynamic. Despite the fact that small-scale fisheries cover a smaller area than large-scale

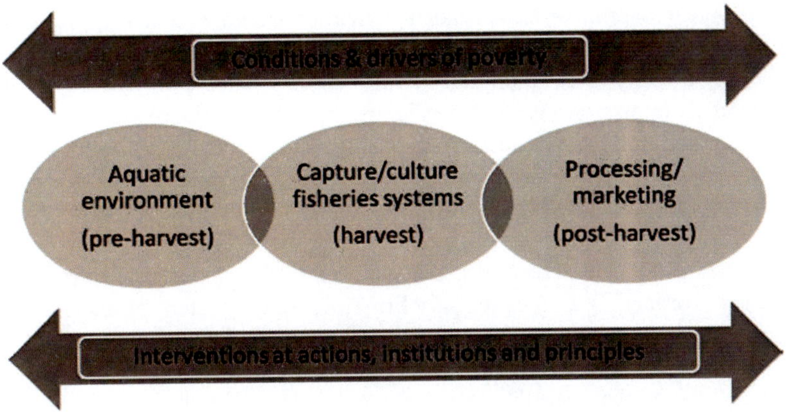

Fig. 3.2 The chain analysis of poverty in fisheries

fisheries, their systemic characteristics amount to high governance challenges or a low level of "governability" from a top-down perspective. These characteristics are what make poverty a vicious circle in some cases, but they are also where opportunities for innovative poverty alleviation initiatives may lie.

Different governing activities are required to address poverty. According to the interactive governance theory (Kooiman et al. 2005), the "first-order" deals largely with the routine, day-to-day problem-solving normally performed as part of the government mandate, while the "second-order" concerns the appropriate design and formulation of effective institutions, mechanisms, and measures (including rules and regulations) that are responsible for addressing the problems. Addressing poverty as a first-order problem can only take us so far. When resources become over-exploited and heavily damaged, threatening the livelihoods of fishing people and pushing them into poverty, common policy interventions are gear and technology improvement to increase catches, controlling prices, or providing fuel subsidies.

Not only are these solutions short-term, they often lead to bigger and more serious problems later on, like loss of biodiversity and dependency on external support. Instead, second-order governance focuses on creating regulations that ban destructive fishing methods and practices in order to enhance conservation. It also involves helping fishers organize local markets, and providing training on post-harvest and value-added products. As earlier mentioned, building local capacity to participate in management is another example of second-order governance. Treating poverty using second-order governance shifts the focus from reactive problem-oriented to proactive planning-oriented solutions.

This governance lens implies that efforts must be balanced to realize small-scale fisheries' often untapped and underestimated potential (Thorpe et al. 2004) to contribute to economic growth and food security, with that of resource conservation and management (cf. Béné et al. 2009). This is certainly no small challenge. Poverty and environmental degradation are "twin evils" (Heady 2000, p. 243), where poverty may induce people to overharvest, thus causing environmental degradation, such as loss of biodiversity, but can also affect efforts to eradicate poverty (cf. Adams et al. 2004). This is why poverty alleviation can be seen as a "wicked problem," and why the Brundtland Commission (WCED 1987) argued that sustainable development and poverty alleviation have a delicate relationship.

Ultimately, poverty issues require "meta-order" governance. As suggested by Kooiman and Jentoft (2009), this highest order of governance is concerned with values, norms, and principles, which determine governance goals and underlie how institutions are designed (second-order) and how rules and regulations are implemented (first-order). The Precautionary Principle for ecosystem-based management and the Code of Conduct for Responsible Fisheries belong to this order (see Jentoft et al., Chap. 20). Jentoft et al. (2010) add "images" as another element to be considered at the meta-level. Similar to other meta-order elements, although often not explicit, images drive many of the decisions that take place in fisheries.

Huxley's idea of the world's fisheries as an inexhaustible resource, and Hardin's description of the tragedy of the commons are both images that have had profound consequences for how fisheries have been governed. Likewise, images about poverty and small-scale fisheries can be powerfully used to steer governance in certain directions. The notion that small-scale fisheries "rhyme with poverty" (Béné 2003) creates an image of destitution among fishers, while the "sufficiency economy philosophy" that fishing communities in Thailand seem to embrace (Chuenpagdee and Juntarashote, Chap. 14) offers hope and signifies the importance of life satisfaction. At the meta-level, poverty problems can thus be addressed through an understanding of the underlying images, values, norms, and principles held by fishing actors, including governments, and also a determination of how closely they align with actual behavior, choices, and decisions (Onyango and Jentoft 2010).

Returning to the chain analysis, Fig. 3.2 shows a cross examination of the natural and the social systems at the pre-harvest (aquatic environment), harvest, and post-harvest levels, as well as the governance order. The chain analysis recognizes the interconnectivity in the fisheries production systems, as well as the external conditions and environment, which may give rise to poverty. It offers a lens to examine poverty in small-scale fisheries beyond the commonly known causes, such as overfishing and user conflicts with other sectors. For example, the linkages suggest that what affects the aquatic environment will influence the natural productivity and eventually the fish catches, which has consequences for the income of fishers and their well-being. Similarly, high debt to middlepersons may drive fishers to use destructive fishing gears in order to increase their catches, which in turn damage fisheries habitats and the productivity of the aquatic ecosystem.

Although our emphasis is on small-scale fisheries, poverty issues cannot be understood and addressed only within this sector. Rather, activities taking place in the ocean, including large-scale, industrialized fisheries, oil and gas exploration, shipping and transportation, agriculture, and coastal tourism can impact the aquatic ecosystem that small-scale fishers depend on. Further, we recognize that no single analytical framework can provide an all-inclusive understanding of complex problems such as poverty. The chain analysis is employed to help explore the systemic and dynamic mechanisms that give rise to poverty and what factors may inhibit alleviation efforts.

Examining poverty using the chain perspective means acknowledging that the causes, drivers, and factors affecting poverty are found throughout the chain. Likewise, different governing interventions and responses can take place at all orders. As suggested in Table 3.5, responses at the first order are mostly remediating actions, while the second order involves the design and establishment of institutions relevant to each part of the chain. At the third, i.e., meta-order, the guiding norms and principles are likely applicable through the entire chain, as will be discussed in the last chapter. Table 3.5 puts the issues along the fisheries production chain with the case studies from different countries, and presents examples of the required policy interventions suggested in these chapters. The last order of governance is covered by most of the case studies and also revisited in the concluding chapter.

3.5 Conclusions

The chapter underscores the importance of understanding poverty as it relates to both the global and the local contexts, and how governance policies to address poverty need to take place not only in the fisheries system but also in the surrounding environment. It is through this broad perspective that innovative solutions on poverty prevention and impact mitigation may be found.

The examples from the case studies show that, despite their complexity and distinctiveness, small-scale fishing communities around the world face very similar challenges in terms of activities and phenomena taking place within the fishing sector and outside that impact their aquatic environments. More often than not, they impede the ability of fishers and their families to attain a satisfactory level of food, income, and well-being. These impacts are not easy to assess because it can be difficult to determine the actual causes, like in the case of non-point source pollution. Further, the build-up of effects over time, and the possible time lag before they can be observed add to the complications of assessment.

In nearly all water bodies, several rules and regulations are set to manage fish harvesting. The general suitability of their application to small- and large-scale fishing sectors, as well as in tropical and temperate waters, has already been questioned (Salas et al. 2007). However, the relationship between these rules and regulations and poverty in small-scale fisheries has yet to be examined. For instance, critical assessment of the impact of quota allocation and rights-based fisheries management on the well-being of small-scale fisheries is required. This is particularly relevant in areas known for legal pluralism in fisheries, such as India (Jentoft et al. 2009). Decisions related to resource allocation and user rights need to be founded on principles such as social justice and equity, and those that express the social values that are active in the particular context to which they are meant to apply.

Numerous opportunities to alleviate poverty can be found in the post-harvest chain, many of which are the responsibility of governments, such as building

Table 3.5 Examples of factors affecting poverty in the fish chain and possible governance responses

System and governance	Pre-harvest	Harvest	Post-harvest
Natural and social systems	Environmental conditions, natural disasters and perturbations affecting fisheries productivity (*Bangladesh, Ghana, Guatemala, Mexico, Nicaragua, Poland*)	Impact of fishing gear, user conflict, safety at sea, cost of fishing (*Bangladesh, India, Ghana, Guatemala, Malawi, Mozambique, Nicaragua, Sri Lanka, Tanzania*)	Market access, processing technology (*Bangladesh, Mozambique, Thailand*)
First-order governance (actions)	Stock enhancement, aquaculture, reforestation (*Malawi, Turkey, Tanzania, Thailand, Vietnam*)	Monitoring, surveillance and control, fishing subsidies, licensing, gear/species restrictions (*South Africa, Thailand*)	Low interest loans, infrastructure development (*Bangladesh, Mozambique, Sri Lanka, Vietnam*)
Second-order governance (institutions)	Environmental standards, awareness programs, education (*Bangladesh, Thailand*)	Co-management system, stock assessment and monitoring system, structural policy (*India, Malawi, Mozambique, Poland, Sri Lanka, Tanzania, Turkey, Vietnam*)	Price setting scheme, auction system, relationship with middlepersons (*Bangladesh, Mexico, Sri Lanka, Tanzania, Thailand*)
Third-order governance (principles)	Conservation, precaution, responsibility, equity, fairness, transparency (*most countries*)		

infrastructure and transportation systems to improve access to markets. The examples from the case studies show that training in proper handling and processing of fish products are effective means to enhance fishers' income. Further, building local capacity in social organization, business and financing help reduce the dependency of small-scale fishers on buyers and middlepersons.

The case studies included in this volume illuminate how susceptible small-scale fishers may be to poverty, and what measures are needed to increase their well-being. But small-scale fishers also live under conditions that make them vulnerable to natural disaster and social pressure. In many instances, which will be detailed in the chapters that follow, we also see examples of how some of them have been able to reduce their vulnerability through habitat protection and conservation initiatives, collaboration and partnership with governments, and self-organization. Sharing these experiences between fishing communities and with governments and non-governmental organizations can contribute to building resilience and increasing adaptive capacity. Mechanisms and ways to encourage such exchange should be developed to reduce both poverty and vulnerability. For instance, these experiences can be integrated in a global database and information system for small-scale fisheries called for by Chuenpagdee et al. (2006).

At the end of the day, the prospect for poverty alleviation rests upon our willingness to fully comprehend what poverty means to the poor, to acknowledge that the causes and effects of poverty are connected through the fish chain, and that solutions to poverty must be found both inside and outside of the fisheries sector, and at global, national, regional, and local levels. Small-scale fisheries need political, financial, and institutional support, and some countries have more urgent needs than others. But, as many of the case studies in this volume show, there can be no doubt that the poverty and vulnerability confronting small-scale fisheries requires local initiatives and involvement. Thinking outside the box by employing an alternative lens, like the chain analysis and the interactive governance framework, and working with the true poverty "experts," can help us avoid slipping into a cycle of poverty, degradation, and despair that Mr. Ban Ki-moon warned about.

References

Adams WM, Aveling R, Brockington D, Dickson B, Elliott J, Hutton J, Roe D, Vira B, Wolmer W (2004) Biodiversity conservation and the eradication of poverty. Science 306(5699):1146–1149

Béné C (2003) When fishery rhymes with poverty: a first step beyond the old paradigm on poverty in small-scale fisheries. World Dev 31:949–975

Béné C, Macfadyen G, Allison, EH (2007) Increasing the contribution of small-scale fisheries to poverty alleviation and food security. FAO Fisheries Technical Paper, 481. Food and Agriculture Organization of the United Nations, Rome

Béné C, Steel E, Kambala Luadia B, Gordon A (2009) Fish as "bank in the water": evidence from chronic-poor communities in Congo. Food Policy 34:104–118

Chuenpagdee R, Liguori L, Palomares MLD, Pauly D (2006) Bottom-up, global estimates of small-scale fisheries catches. Fisheries Centre Research Report, vol 14(8), 110 p. http://www.fisheries.ubc.ca/publications/

Elvidge CD, Sutton PC, Ghosh T, Tuttle BT, Baugh KE, Bhaduri B, Bright E (2009) A global poverty map derived from satellite data. Comput Geosci 35:1652–1660
FAO (2005a) The state of world fisheries and aquaculture 2000. FAO Publication, Rome
FAO (2005b) Technical guidelines for responsible fisheries. FAO Publication, Rome
Gini C (1921) Measurement of inequality of incomes. Econ J 31(121):124–126
Hamilton LC, Butler MJ (2001) Outport adaptations: social indicators through Newfoundland's cod crisis. Hum Ecol Rev 8(2):1–11
Heady C (2000) Natural resource sustainability and poverty reduction. Environ Dev Econ 5:241–258
Jacquet JL, Pauly D (2008) Funding priorities: big barriers to small-scale fisheries. Conserv Policy 4:832–835
Jentoft S, Bavinck M, Johnson D, Kaleekal T (2009) Co-management and legal pluralism: how an analytical problem becomes an institutional one. Hum Organ 68(1):27–38
Jentoft S, Chuenpagdee R, Bundy A, Mahon R (2010) Pyramids and roses: alternative images for the governance of fisheries systems. Mar Policy 34:1315–1321
Kooiman J, Jentoft S (2009) Meta-governance: values, norms and principles, and the making of hard choices. Public Adm 87:818–836
Kooiman J, Bavinck M, Jentoft S, Pullin RSV (eds) (2005) Fish for life: interactive governance for fisheries. Amsterdam University Press, Amsterdam
Narayan D, Chambers R, Shah MK, Petesch P (2000) Voices of the poor: crying out for change. Oxford University Press/The World Bank, New York
Onyango P, Jentoft S (2010) Assessing poverty in small-scale fisheries. Fish Fish 11:250–263
Pauly D (2006) Major trends in small-scale marine fisheries, with emphasis on developing countries, and some implications for the social sciences. Marit Stud (MAST) 4(2):7–22
Salas S, Chuenpagdee R, Seijo JC, Charles A (2007) Challenges in the assessment and management of small-scale fisheries in Latin America and the Caribbean. Fish Res 87(1):5–16
Sen A, Anand S (2000) The income component of the human development index. J Hum Dev 1(1):83–106
Thorpe A, Reid C, van Anrooy R, Brugere C (2004) African poverty reduction strategy programmes and the fisheries sector: current situation and opportunities. Afr Dev Rev 16(2):328–362
UN (2009) The millennium development goals report 2009. United Nations Department of Economic and Social Affairs, New York
UNDP (2000) Human development report 2000. Human rights and human development. United Nations Development Programme Publication, New York
UNDP (2008) Human development report 2007/2008. Fighting climate change: human solidarity in a divided world. United Nations Development Programme Publication, New York
UNDP (2009) Human development report 2009. Overcoming barriers: human mobility and development. United Nations Development Programme Publication, New York
WCED (1987) Our common future. World Commission on Environment and Development. Oxford University Press, Oxford
Weyl OLF (2005) Capture fisheries in Malawi and their contribution to national fish supply. Aquaculture Development in Malawi (ADiM) report. Enviro-Fish Africa (Pty) Ltd., Grahamstown
Wright MC (2001) A fishery for modern times: the state and the industrialization of the Newfoundland fishery, 1934–1968. Oxford University Press, Oxford
Zeller D, Pauly D (2005) Good news, bad news: global fisheries discards are declining but so are total catches. Fish Fish 6:156–159

Chapter 4
The Meaning of Poverty: Conceptual Issues in Small-Scale Fisheries Research

Svein Jentoft and Georges Midré

Even the present is not guaranteed.
Not to speak of the future.
He lives of today. That is the poor.

Indigenous leader, Guatemala

Abstract This chapter synthesizes the conceptual issues and arguments advanced in the 15 case studies presented in this volume. It also relates the arguments and findings to major issues in poverty research and debates, and to the literature on small-scale fisheries and poverty. Moreover, the chapter discusses the many dimensions of poverty, and how poverty and vulnerability undermine the role of small-scale fisheries as providers of sustainable livelihoods, food security, and economic development. At the same time, it recognizes that poverty, vulnerability, and development are essentially "wicked" problems, difficult to define and solve, partly because they have no simple technical solution. This has important policy and governance implications; therefore listening to the poor and involving them in decision-making is essential.

S. Jentoft (✉)
Norwegian College of Fishery Science, University of Tromsø, 9037 Tromsø, Norway
e-mail: svein.jentoft@uit.no

G. Midré
Department of Sociology, Political Science and Community Planning,
University of Tromsø, 9037 Tromsø, Norway
e-mail: georges.midre@uit.no

S. Jentoft and A. Eide (eds.), *Poverty Mosaics: Realities and Prospects in Small-Scale Fisheries*, DOI 10.1007/978-94-007-1582-0_4,

4.1 Introduction

During the last 20 years, a considerable number of studies have explicitly focused on how different social groups and classes define poverty and characterize the poor. Some investigate elite discourses (Veit-Wilson 2000; Midré and Flores 2002; Reis and Moore 2005). By definition, social elites have a political and economic power that non-elites do not have (Verba et al. 1987). They also have symbolic power in the sense that their values, opinions, and worldviews have an impact on the less powerful (Bourdieu 1986). The way in which political and economic elites conceptualize poverty also has consequences for what kinds of policies are established for the people dependent on assistance. This is also why interactive governance theory (Kooiman et al. 2005; Jentoft et al. 2010a) emphasizes the role that metaphors and "images" of people in powerful positions, such as fisheries administration, play in policymaking, and why images of governance should be subject to closer inspection.

Pinker (1999) writes that this has considerable political significance if poverty is mainly perceived as connected to structural processes in society, the functioning of its economic system, to institutions, and the exploitation of one class by the other, or to personal characteristics of the poor (see also O'Connor 2001). We could add here the dominant paradigm in fisheries, which sees poverty as an outcome of overfishing, resource degradation, and the "tragedy of the commons" (Hardin 1969), to the extent that small-scale fisheries "rhymes with poverty" (Béné 2003).

Elite perceptions and understandings of poverty are also reflected in the documents produced by governments and international organizations fighting against poverty. This can be explained by their need for information about the overall levels of poverty and its social distribution, but often leads to what has been called "bureaucratic definitions of poverty." Chambers (1997, p. 46) writes that poverty comes "[...] to be defined, not by the changing and varied wants and needs of the poor, but by the static and standardized wants and needs of professionals. Conceptually, professionals are then caught in their own reductionist poverty trap. Poverty becomes what has been measured."

A similar criticism has been raised by Narayan and colleagues (2000), in the *Voices of the Poor* book series, who argue that the true poverty "experts," i.e., the poor themselves, are rarely asked what poverty means to them and how best to eliminate it. Several of the authors in this volume also note this about the poor small-scale fishing people whom they interviewed (see Chuenpagdee and Juntarashote, Chap. 14; Nguyen and Flaaten, Chap. 15; Salas et al., Chap. 10).

Drawing from the academic discourse on poverty and poverty alleviation, this chapter introduces some of the conceptual issues in poverty research in general, and focuses particularly on small-scale fisheries. The chapter also introduces the case studies that follow in this volume by synthesizing their analytical perspectives and arguments. We specifically discuss the multi-dimensional aspects of poverty, and how poverty and vulnerability not only erode the well-being of small-scale fishers and their dependants, but also undermine the role of small-scale fisheries as a major provider of food security and economic development (see Eide et al., Chap. 2).

At the same time, we argue that it is essential to recognize that poverty and poverty alleviation (which should be interpreted as including both poverty *prevention* and *reduction* (cf. Béné 2004a)) are essentially what Rittel and Webber (1973) called "wicked" problems. These are problems where definitions are contested and where solutions are not always obvious or easy to implement. This is partly because they have no simple technical solution (such as technology transfer (Macfadyen and Corcoran 2002)), and partly because they raise issues that are basically of a moral nature (Hardin 1969). They are therefore often ideologically and politically controversial (Dean and Melrose 1999). This has important policy and governance implications, but is also part of the explanation why poverty remains a pervasive issue waiting to be solved.

4.2 Listening to the Poor

Poverty research has been criticized as distant and abstract, and lacking meaningful descriptions of how poverty actually translates for the poor. An example of how to accommodate this critique is the above mentioned *Voices of the Poor* project funded by the World Bank, and which included more than 60,000 informants in 60 countries (Narayan et al. 2000). There are also a number of smaller projects with a similar aim (cf. Dean and Melrose 1999; Midré 2005). Although researchers have become more receptive to the "voices of the poor" in identifying the causes of poverty, the approach has also met with some skepticism. For instance, some have questioned if the project really has been able to capture and represent the voices of the poor (Moser and McIlwayne 2001; Pender 2002; Cornwall and Fujita 2007). The involvement of the World Bank has generated the expectation that it will reduce what is seen as a bias toward market liberalism and the down-scaling of public institutions and support mechanisms (Matthews 2002; Okin 2003).

4.2.1 Emic vs. Etic Perspective

The approach of "sounding out" the poor is also chosen by the chapter authors of this volume who have been interviewing fishing people in their homes and communities, and by going fishing with them. How does poverty and vulnerability look from the perspective of poor fishers and those who rely on fishing? How do they define it and explain it? The data reveal that fisher folk have their own views of what constitutes poverty and vulnerability, and of what it would take to reduce poverty in their communities (see Knudsen and Koçak, Chap. 11). It is argued by Onyango (Chap. 6) that the "emic" perspective of the poor is a good starting point for policy development. Ideas do not necessarily have to come from the outside. Ideas of why people are poor and vulnerable, and how to address them, are flourishing in the community that Onyango is studying. People who are poor have their own perspectives on their quandaries, and policymakers should start by listening to what they have to say. As one Thai fisher interviewed by Chuenpagdee and Juntarashote (Chap. 14) said: "Poor people have brain but lack capital."

However, trying to understand how poor people perceive poverty and vulnerability does not mean that research based on concepts of the kind Chambers and others are criticizing is without merit. The "etic" perspective is also needed. Poverty must be analyzed at multiple levels (Thorpe 2004; Thorpe et al. 2007). Macro analysis and cross-sectoral comparisons are always useful for policy development, particularly at national and global levels. For the UN organizations, international NGOs and national governments, even crude estimations of poverty levels would be helpful. They would, for instance, assist poor countries to acquire resources for significant anti-poverty programs.

However, poverty must also be understood in its particular local contexts, which is a central thesis of this volume. In small-scale fisheries, these contexts are diverse and complex and they change, sometimes rapidly, due to forces both inside and outside the community and the fishing industry (Chuenpagdee and Jentoft, Chap. 3).

4.2.2 Case Studies

The 15 case studies in this volume are important supplements to the national and international quantitative descriptions of poverty in terms of income or consumption on a large scale that are given in the two previous chapters (Eide et al., Chap. 2; Chuenpagdee and Jentoft, Chap. 3). They also illustrate arguments by, for instance, Thorpe et al. (2007) and Béné (2004b) that poverty in small-scale fisheries is multi-dimensional and has many faces, of which low income is only one. Indeed, low income should be seen as a symptom rather than a cause of poverty. Poverty is a composite concept, which includes a number of elements, such as poor health, illiteracy, political marginalization, social exclusion, and high exposure to risk. Small-scale fishers often lose out as a consequence of policies aimed at modernization, rationalization, and conservation. The "trickle-down" effect does not always work as assumed and promised (cf. Isaacs et al. 2007; Béné et al. 2010).

Small-scale fishers are often also victims of abuse in terms of corruption (for instance development aid that gets lost before it reaches them), exploitation by middlepersons, piracy, and violent crime (Islam, Chap. 5). They are often pushed aside by industrial vessels who do not respect traditional tenure or restrictions set by governments (Amarasinghe and Bavinck, Chap. 17; Bavinck, Chap. 9; Jentoft et al. 2010b; Menezes et al., Chap. 18; Tvedten and Hersoug 1992). Access to fisheries resources is also crucial for livelihood security. Without it, and the rights to support it (including landing rights, rights to settle – see Bavinck, Chap. 9), fisheries-dependent people are in a similar situation as those in India hit by famine, described by Sen (2000). In Sen's case, the critical issue is not lack of food per se, but the lack of control and ownership of the food supply, a situation he calls "entitlement failure" (2000, p. 162ff.). Thus, Béné et al. (2004b, p. 77) argue that fishers who are really poor "are those with limited or no access to (fishing) resources."

It is essential to recognize that the poor, including those who fish, do not conform to one homogenous category of people. Both Hara and Marciniak (Chaps. 12 and 7) argue that as far as poverty in small-scale fisheries is concerned, it is important to distinguish between owners and crew, men and women. One also needs to be sensitive to the fact that one group may draw benefits from the other in ways that indicate subordination and exploitation (cf. Isaacs, Chap. 16). This is also why a relational perspective on poverty is important.

4.2.3 Methods

Poverty research suffers from a constant risk of bias. Interviews and observations typically take place in a situation created and controlled by the researcher. The researcher may then easily miss a deeper reflection of the problems confronting the poor. This may also happen in the researcher's translation of the concepts of informants (Moser and McIlwayne 2001). Poverty is a sensitive issue that people are not always willing to talk about, due to the stigma often associated with it. As several of the PovFish researchers experienced during their fieldwork, in the eyes of the poor, they were sometimes seen as government representatives and therefore among those who they blame for their problems. There is no easy way out of this predicament, but it does speak to the degree of thoroughness of the investigation, the particular methods employed, and the ethical issues involved in poverty research. Spending time with the people to talk about their lives, rather than their poverty per se, is extremely important. The familiarity of the researchers with the situations of the poor people, along with their knowledge of external determinants such as policy instruments, market pressures, and ecosystem conditions, is a good starting point.

This volume is strengthened by the fact that the case study chapters are all written by people who are native to the countries of study, and/or who have years of experience researching fisheries and the fishing communities in their areas. This does not make them immune to the criticism that researchers cannot really represent the "voices of the poor," but it does counteract the danger of misinterpreting data because they do not know the context, the language, the social code, and local idioms.

An issue that has drawn comment in the aftermath of the *Voices of the Poor* program is the salient position of the concept of "social capital." It is not a novel concept, but one that has inspired substantial social science literature highlighting trust and cooperation (Bourdieu 1986; Putnam 1995; Halpern 2005). It is also a central analytical concept in Chap. 17 by Amarasinghe and Bavinck; and Islam, Chap. 5. The concept adds dimensions to the social relationships of family and community. It also emphasizes the need to assess more of the linkages between people and communities across larger temporal and spatial scales. This is illustrated in the case of Bangladesh (Islam, Chap. 5), where fishers who have connections with people and institutions outside their community seem to be able to acquire a larger repertoire of livelihood resources as compared to those who do not have such a network.

As Granovetter (1973) points out, weaker ties have strength because they expand action space and the gamut of entitlements. However, as noted by Harriss (2002) with regard to the World Bank publications, the analytical potential of the concept of social capital largely disappeared and thereby isolated poverty from its structural and political causes.

4.3 Poverty Mosaics

4.3.1 Poverty as a Trap

To be poor is to have a limited set of options, to be left with little choice of how to live and work. Poverty therefore is often perceived as a "trap"; once in it, escape is difficult. For instance, for their own good as well as for the future well-being of the family, poor people may well recognize the importance of sending their children to school. Still, they may not be able to afford it, sometimes because the child's labor is needed. This is a situation described by Islam in the case of Bangladesh; and Chuenpagdee and Juntarashote for Thailand, Chaps. 5 and 14, respectively (see also Mayer-Foulkes 2008). Thus, they get stuck. Families remain in poverty over generations. They can only continue to do what they already do, and children follow in the footsteps of their parents.

Another example is the idea of a resource dependency trap (see Christy 1986; Doulman 2004). In a study among Kenyan fishers, Cinner et al. (2008) found that poor fishers are less likely than wealthier fishers to exit a severely declining fishery. If fishing is the only livelihood available, resource degradation means less food and income, unless compensated by increased effort. But in the long run, they easily end up even worse. Fishing as a last resort occupation means that people are attracted to the activity because there are no other opportunities. If access is open, overfishing may easily occur. Garrett Hardin (1969) described this as the "tragedy of the commons." McEvoy (1986) named it "the fisherman's problem." Lack of alternatives to fishing is a problem that many of the small-scale fishers who were interviewed for this book also experience.

In yet another illustration, fishers in Norway were indeed "the poorest of the poor" during the international depression of the 1930s. But their problem was not a resource crisis; fish stocks were abundant and landings were big. When dockside prices fell, fishers responded by working longer hours, which meant increased supply and further fish price reduction. This vicious cycle was broken when new legislation allowed producer sales organizations to fix minimum prices (Hallenstvedt 1982). How to break the cycle and to prevent and relieve the poor from becoming entangled in such traps, is therefore an essential aim of poverty alleviation strategies. Policy interventions like the one employed in Norway would be one way to go, while developing alternatives to fishing would be another option, as discussed below.

4.3.2 Alternative and Supplementary Livelihoods

To be poor also means that all available resources are used for immediate necessities. Basic needs cannot be suspended; there has to be food on the table, and the costs of keeping the family business afloat must be covered (Bourdieu 1976). Further improvement requires a surplus for accumulation and saving; the ability to set something aside for later investment, but this requires some space or "room to maneuver" that the poor typically do not have (Clay and Schaffer 1984). Therefore, they are trapped. The provision of alternative livelihoods may help to increase this leeway. For this reason, it makes sense to look at fisheries development as the enhancement of freedom of choice. This perspective is in line with Sen (2000; see also Jentoft et al. 2010b). But, as Sen forcefully argues, enhancing freedom is not only something development initiatives should aim for; it is also a condition for development. Therefore, without freedom, there is likely no local initiative, self-induced growth, or innovation.

However, in fisheries-dependent communities, alternative livelihoods are not often readily available. Neither, as reported in several of the chapters that follow, do fishers always want to leave fisheries if such alternatives are provided. Partly, it is a question of having skills that can be converted into other income-generating activities. Partly, it is an issue of tradition and identity (Kraan, Chap. 8). As Onyango discusses (Chap. 6), fishing is not only what a person does. Fishing is also something that defines who he or she is. For these reasons, contrary to what regulatory schemes often aim to do, Kraan argues that it is both good resource and social policy to promote diversification instead of moving people out of the fishery altogether. Supplements to fishing can release pressure on resources, raise incomes, and reduce poverty without uprooting people from their heritage and way of life, rather than totally denying people access to fishing and transferring their labor to other occupations (see also González, Chap. 13; Allison and Ellis 2001; Hall et al. 2010).

In any case, alternative and supplementary sources of income cannot break the poverty trap unless people's capabilities to draw advantages from these new opportunities are developed. What capabilities people have, and how they can be strengthened is an obvious research issue, as well as a development challenge. The capability approach introduced by Amartya Sen in the 1980s has become very influential in poverty research, and it has inspired several of the authors in this volume (see Islam, Chap. 5; Onyango, Chap. 6; Knudsen and Koçak, Chap. 11). It has proven to be a flexible approach, and has been developed in a number of ways (Alkire 2002). For instance, in addition to poverty, the capability concept has been operationalized in the directions of inequality, social justice, and living standards (Clark 2005). It has also led to an emphasis on how institutions influence individual choices and liberties (Zimmerman 2006).

Some aspects of the capability approach have also met skepticism. A main critique focuses on what is seen as an inherent methodological individualism. This was, for instance, expressed in Townsend's (1985, p. 668) review of a paper by Sen (1983). Others argue that the capabilities approach tends to keep "an individualist hue that belies the radical implications of capabilities" (Jackson 2005, p. 102). Sen has responded to this criticism on several occasions (most recently in Sen 2009) by holding that collectives, like families, communities, and nations, can indeed be analyzed in terms of their capabilities to pursue valued goals. Furthermore, in much of what he has written, the social influences on both the formation and the "functionings" of the capabilities are emphasized. The critics, Sen (2009, p. 245) argues, do not "distinguish adequately between the individual characteristics that are used in the capability approach, and the social influences that operate on them."

But even if Sen views capabilities and functionings as culturally-dependent, there are few attempts in his work to discuss exactly how culture, community, and social structure influence them; particularly in terms of the interdependence between agency and the institutional and symbolic environments (Zimmerman 2006). Despite these shortcomings, Carpenter (2009) is of the opinion that the capability approach has the potential to more fully evaluate the influences of cultural and structural constraints on the production of inequalities and poverty. A focus on the spatial determinants of individual and collective capabilities and agency would be one way to go.

4.3.3 Space and Agency

The concept of "room to maneuver" discussed above, has similarly been seen as too individualistic. Webster and Engberg-Pedersen (2002, p. 7) therefore prefer the concept of "political space," which identifies the importance of "mobilization, organization, representation and empowerment"; i.e., the structures and processes that allow poor and vulnerable people to actively engage in changing the structural and institutional mechanisms and circumstances that hold them back. In a similar vein, in her case study from South Africa, Isaacs (Chap. 16) talks about the post-apartheid fisheries policy that aimed to expand the "action space" by reallocation of fisheries rights to previously disenfranchised groups.

However, since rights were primarily defined according to race and gender, but not allocated to the coastal poor *per se*, the policy ended up providing those already privileged with more space for action. Consequently, what was then left for the poor to do was to utilize their "informal" action space by employing what Isaacs refers to as "the weapons of the weak," i.e., poaching. Kraan (Chap. 8) is also interested in space and agency when describing how the Anlo-Ewe beach seine fishers of Ghana have expanded their fishing area by migrating up and down the coast (see also Salas et al., Chap. 10).

For small-scale fisher folk, the political action space is rarely empty to begin with. Rather, it is a competitive space where they must confront other stakeholders, who are often more powerful than they are. This is also why co-management arrangements often do not benefit the poor, but rather those who are already better-off; for instance, boat owners at the expense of the crew (Béné et al. 2004; Davis and Bailey 1996; Jentoft 2004). Still, several of the authors of this volume, such as

Andrade and Midré in the case of Guatemala (Chap. 19) and Hara for Lake Malawi (Chap. 12), think co-management is a viable option provided that one is conscious of the risks when setting it up.

The local co-management institutions in Tanzania (see Onyango, Chap. 6) and Mozambique (Menezes et al., Chap. 18) provide lessons that would be useful in other settings. Also, as illustrated in many of the chapters that follow, the analysis of the causes of poverty and what potential there is for alleviating poverty, requires a focus on how power works in fisheries (co-)management (Jentoft 2007). Co-management is about restructuring relationships between user-groups and government authorities to facilitate commitment and mobilize resources and interactive learning. Co-management arrangements are about creating action space and about providing the individual and collective capabilities that users need in order to fill it. The co-management concept also recognizes that the state has a role to play in defining, defending, and supporting this action space, as the state is sometimes the "only effective source of legitimacy for rule-making and rule-enforcing" (Wilson et al. 1999, p. 568).

Webster and Engberg-Pedersen (2002, p. 255) are critical of the tendency to view the poor only as passive victims. They hold that "[...] the poor are constantly seeking to maneuver within given conditions and to generate room for profitable activities." The political action space is not given once and for all. Rather, its boundaries are porous and fluid. As the space is filled, its boundaries are enlarged. How this works for poor small-scale fishers in actual settings is an obvious research issue. It is also one that has inspired the case studies presented in this volume; several authors point out that forming community organizations such as cooperatives is a way of equipping small-scale fishers with the means they need to operate more effectively and forcefully at the community level and beyond (see Amarasinghe and Bavinck, Chap. 17; González, Chap. 13; Nguyen and Flaaten, Chap. 15; Salas et al., Chap. 10).

Sen (2000) and Nussbaum (2000) talk about "the existential room" of human beings. They develop concepts to compare qualities of life, not primarily by asking about the command of resources or levels of satisfaction, but about "what people actually are able to do or to be" (Nussbaum 2000, p. 12). They both emphasize human life and human dignity as ends in themselves. They also discuss the characteristics of what it would mean to live a life that is essentially human, independent of varying preferences and cultural differences. Having minimum control over one's political and material environment has inherent value. Life without it, according to Nussbaum (2000), cannot be fully human.

The scholars who have been referred to above may differ on certain issues, but they all insist that widening people's action space and, hence, their level of freedom is essential for poverty alleviation. Poverty is characterized by a lack of entitlements and capabilities that provide a person with sufficient and secure action space, and therefore the freedom to choose the life that he or she wants. This idea also underpins the research by the authors of this volume in asking small-scale fishers what they appreciate and value. They are interested in hearing what made them become fishers in the first place, what entitlements and capacities they have, and what efforts they make in order to realize their goals. One important point is that small-scale fisheries, in many instances, *do not* "rhyme with poverty," to use Béné's (2003) often-quoted phrase. It is also important to recognize that, as Bavinck argues in Chap. 9 and Eide et al. stress in Chap. 2, in some parts of the world small-scale fisheries represent wealth and not poverty.

Still, for millions of people around the world, small-scale fisheries only provide a meager income and security. It is therefore not surprising when small-scale fishers prefer to leave if they could, as expressed by some of the fishers interviewed by Andrade and Midré (Chap. 19). But it should not be forgotten that, despite its many relative disadvantages (low income, lack of safety, political marginalization), small-scale fishing offers many a preferred lifestyle, exactly for the action space and subsequent freedoms that it involves. As one Thai fisher told Chuenpagdee and Juntarashote (Chap. 14): "Being a small-scale fisher is about freedom and independence. I can go fishing and come home every day as I like." This is a sentiment that is shared by many small-scale fishers interviewed by other authors of this volume. In fact, small-scale fishers do not necessarily perceive themselves as being poor, because they have something that is important to them that other people do not have (e.g., freedom, food, and pride).

In Onyango's case (Chap. 6), villagers reserve the concept of poverty for situations where a person completely lacks basic capabilities such as health. In most instances in the community he is studying, there is always something people can do to survive. In a crisis, they can also rely on other members of the community. The role of the community as provider of support and security is also described by other authors in this volume (for instance Gonzáles in the case of Nicaragua, Chap. 13).

But not all small-scale fishing communities are like this. Rather, they are characterized by fragmentation and conflict (see Bavink, Chap. 9). Lack of community support is felt by people engaged in the sea-snail fishery on the Turkish Black Sea coast described by Knudsen and Koçak (Chap. 11). These authors add political marginalization as another factor explaining the persistent poverty of small-scale fishers in their area. The same issue is emphasized in the Polish case study by Marciniak (Chap. 7), in the Mexican study by Salas and colleagues (Chap. 10), and by Amarasinghe and Bavinck for Sri Lanka (Chap. 17). In all these instances, marginalization may be understood as a process through which the political and social action space becomes restricted; it explains why people become poor and why they remain so, and what role the community may play.

4.3.4 Poverty as a "Wicked Problem"

Hardin's (1969) article on the tragedy of the commons is among the most frequent citations in fisheries social science research. (A *Google Scholar* search on fisheries and G. Hardin gave 18,800 hits on 28 December 2010). Hardin stated that poverty is among those problems that do not have a technical solution, but rather challenges our social values and sense of morality. Therefore, in order to avoid the tragedy of the commons, he argues that we must be willing to critically examine those values, and be ready to make hard choices that may possibly contradict them. Even if painful, restricting the freedoms of small-scale fishers in the commons may sometimes be beneficial for their long-term well-being.

How social values, morality, and even spirituality are challenged by fisheries management is illustrated by Chuenpagdee and Juntarashote (Chap. 14) for the case of Thailand. These Thai Muslim fishers said "the sea is God's fish pond" which is to be shared by all. Similarly, Hara (Chap. 12) reports that fishers strongly believe that "God will always provide." In the community Onyango studies (Chap. 6), the general view is that no one should be denied access to the lake since everyone has a God-given right to fish. Contrary to Hardin and most fisheries managers and academics, for these poor small-scale fishers, the commons is a "blessing" rather than a curse (Berkes et al. 1989; Kurien 2004). According to Hardin, such values and attitudes would unavoidably lead to the ruin of the commons. He therefore thinks that these values must be sacrificed, as to him "injustice is preferable to total ruin" (Hardin 1969, p. 1247).

Sen's notion of freedom leads in a different direction (Sen 2000; Jentoft et al. 2010b). His concept of freedom aims to provide users with the entitlements and capabilities they need to proactively cope with resource degradation and poverty, even imposing restrictions on themselves. His is not a freedom to overfish, but one that allows poor resource-users to make their own decisions on how the freedom of the commons should be addressed. Building the necessary competency and capability to do it rationally, ethically, and with reason, would be a means of realizing this freedom.

The rationale for a human rights-based, inclusive, participatory, and deliberative process of management that involves the poor also finds support in Rittel and Webber's (1973) idea of "wicked problems" (see Jentoft and Chuenpagdee 2009). For a wicked problem, no correct definition or definite solution exists, and no one can claim to be sure that they have the right answer. Rittel and Webber's prime example is in fact poverty. It matters how poverty is defined, be it low income, deficient cognitive, social, and occupational skills, or whether it is seen as caused by deficient physical and mental health. In fact, they argue that the image of the solution often comes before or simultaneously with the definition of the problem.

Menezes and co-authors (Chap. 18) portray poverty in the case of Mozambique in similar terms when they discuss the government's effort to determine which means are best suited to eradicate poverty in small-scale fisheries. Since knowledge regarding the health of the resource is scant in Mozambique, overfishing is not part of the definition of the poverty problem. Instead, poverty tends to be regarded as a distributional problem pertaining to fishing rights, lack of effective fishing gear for the poor, and inequitable infrastructure conditions. This view is also expressed by the authorities in the Vietnamese case study (see Nguyen and Flaaten, Chap. 15) which, despite the country's impressive economic growth in recent years, is still short of resources that can uplift small-scale fisheries in marginal coastal communities.

Nicaragua (González, Chap. 13) is yet another example of the wicked problem of poverty alleviation. Although there is general consensus about ecosystem degradation in Nicaragua's Pearl Lagoon fisheries, there is disagreement about two management plans with very different ideas regarding "*what* needs to be done, *with whom*, and *for what purpose*." This has caused delays in addressing ecosystem degradation, and has created frustration among local residents dependent on ecosystem services. Local inhabitants were not consulted in a consistent manner in either of these two planning processes, and definitions of the problem and ideas of solutions were largely imposed on the communities by external (foreign) experts.

Rittel and Webber (1973) argue that if there are different perceptions about what the problem is and what solutions are needed, as many voices as possible must be heard. The challenge confronting planners must be scrutinized from several angles. For this, Rittel and Webber prescribe an interactive, deliberative process among stakeholders. Value judgements are typically involved for each causal variable. This is particularly so when poverty is viewed as "unacceptable hardship," as this term has no objective criteria. Rather, it involves moral imperatives: When the hardship is unacceptable, something should be done about it (Gordon and Spicker 1999, p. 157). What to do exactly, has no correct or incorrect answer, only good or bad ones, and must be sorted out by those who are suffering these hardships. This is also because hardships and poverty are relative categories.

4.3.5 Relative and Absolute Poverty

An old and persistent controversy in poverty research concerns the "absolute" versus "relative" conceptualization of poverty. This distinction also triggered a discussion between Amartya Sen (1983, 1985) and Peter Townsend (1985). Poverty is often intuitively understood as the lack of fulfillment of basic needs. What these needs are, however, may vary from place to place, among groups, and over time. However, Sen (1983) insists that any definition of poverty must include a core of absolute deprivation. An individual or a household lacking sufficient food, without clothes to wear, and without shelter, is poor in an absolute sense. Also, Sen argues, when there is starvation and hunger there is clearly poverty, no matter what the relative picture looks like. But he also insists that poverty should be broadly defined. This is also the view of González for the Pearl Lagoon, Nicaragua (Chap. 13). He argues that policies aimed to improve the lot of the poor are likely to fail if they are based on too narrow a concept.

Townsend acknowledges that the distinction between conceptualizations of poverty as absolute or relative has some advantages if the former is linked exclusively to physiological needs. But he contends that an absolute concept of poverty "[...] imposes a minimalist as well as static view of the extent of human needs in poor as well as rich countries" (Townsend and Gordon 1991, p. 39). An absolute concept of poverty is thinner than the relative concept. Consequently, defined in terms of lack of basic needs, the poor part of a population will typically be smaller than when applying the alternative concept. With an absolute definition of poverty, the pressure on governments to implement pro-poor policies will be weaker.

Poverty from a relative perspective involves contextual analysis and comparisons between categories of people. Townsend (1979, p. 31) said that poverty should be defined as the lack of resources that excludes people "from ordinary living patterns and activities." Accordingly, he maintained that poverty must always be defined in the context of the material and cultural standards where the deprived are living (Townsend 1985). For example, small-scale fishers in Poland (Marciniac, Chap. 7) or Turkey (Knudsen and Koçak, Chap. 11) are not as poor in an absolute sense as small-scale fishers in Malawi (Hara, Chap. 12), Tanzania (Onyango, Chap. 6), Bangladesh (Islam, Chap. 5), or Mozambique (Menezes et al., Chap. 18), but they are still poor relative to other groups within their areas.

The Polish experience of poverty is also relative in the sense that people now contrast themselves with the situation as it used to be before the political and social transformation that has taken place since the fall of communism, and with entry into the European Union. Many have therefore left their home fishing communities to seek employment elsewhere, including abroad. Those who remain have become a marginalized group excluded from enjoying a living standard comparable to that of others. Starvation is not common, but is still a threat that looms in people's minds. Their frustrations often play out in excessive consumption of alcohol and domestic violence.

A relative concept of poverty would also be open for the reverse situation. Small-scale fishers in Thailand (Chuenpagdee and Juntarashote, Chap. 14) and Guatemala (Andrade and Midré, Chap. 19) do not see themselves as poor when compared with other people in their regions. This also resonates with the opinions expressed in the Mexican case study (Salas et al., Chap. 10). Fishers appreciate the fact that they always have enough to eat, which is more than many people in other professions and communities can say. A Lake Victorian fisher interviewed by Onyango (Chap. 6) stated that "when you are fishing, you are not worried about whether you will eat." This is also why people are often drawn to the fishery from the hinterland, where conditions for making a living are often worse than near the coast. The relative advantage of fisheries also explains why, in many parts of the world, the number of fishers has increased dramatically; in some instances by several 100% (see Chaps. 9, 12, and 18 by Bavinck, Hara, and Menezes et al., respectively). But even if small-scale fishers are not always among the poorest of the poor in terms of income, they are often so on all other poverty indicators, such as illiteracy, health, nutritional standards, social exclusion, and vulnerability, as Willmann (2004) points out.

According to Narayan et al. (2000), when asked about what poverty means to them, people in developing countries often speak of material resources like food, shelter, clothing, livestock, land, and other assets that are tied to their basic needs. The enumeration of these types of resources is also linked to a narrow concept of absolute poverty. One common feature in the different case studies reported in this volume is that informants tend to think about poverty in terms of absolute needs. The real poor, they say, are those who do not have enough to eat, who lack proper shelter, and require help from family members, neighbors, NGOs, churches or the government, in order to survive and be safe.

In the community, there are often also mechanisms for sharing and redistributing produce and providing mutual help. The needy are being cared for, as when they show up on the beach when boats land with their catch (see Salas et al., Chap. 10). In the case study by Amarasinghe and Bavinck for the case of Hambantota, Sri Lanka (Chap. 17), and that of Nguyen and Flaaten for Vietnam (Chap. 15), community cooperatives make sure that people are not destitute. Thus, cooperatives serve a welfare function which, in more developed countries such as Poland (Marciniak, Chap. 7) and Turkey (Knudsen and Koçak, Chap. 11), is a reponsibility of the government.

The difference between the relative and the absolute notions of poverty may, however, look greater than it actually is (Lister 2004, p. 20ff.). Sen (1983), for example, makes a distinction between poverty in terms of capabilities and commodities. For capabilities, he insists on an absolute notion of poverty. For instance, the capability of avoiding shame, a feeling of inferiority (cf. Marciniak, Chap. 7) because of not having goods like most people have, is absolute. Either you feel inferior or you don't. There is shame or no shame. The commodities needed to avoid these ills are, however, relative to historical and social circumstances. Adam Smith (in Sen 1983, p. 159), for instance, mentions the importance of wearing leather shoes, rather than footwear made of other materials, to avoid shame. Thus, the things required for filling absolute needs vary. As a consequence, there may be no conflict between the absolute and relativist positions. Gordon (2000, p. 51) argues that if "absolute poverty is defined in terms that are neither constant over time nor invariant between societies, then from an operational point of view, the concepts of absolute and relative poverty become virtually indistinguishable."

4.3.6 Poverty as a Relational Issue

As suggested by Townsend (1979) and others, a broader, relative concept of poverty (and affluence) includes social relationships. According to Room (2000), there are two types of relationships of particular relevance to the discussion of poverty: The relationship to the system of production, and the relationship to extended family, neighborhood, and informal help systems. The poor are deprived not only of material entitlements (production system) but also often of resources necessary to fulfill ordinary role obligations and to fully participate in community life (social connectedness). Conceptually, this notion of poverty is linked to social exclusion.

In a review of how poverty is discussed in Europe, Veit-Wilson (2000, p. 148) writes that "poverty is defined relationally, as the identifiable categorical or spatial characteristic of groups or individuals which hamper or prevent them from taking an adequate part in the dominant society." Social exclusion is an important issue since it may be both a cause and a consequence of poverty. It may be defined as a situation or condition where an individual "does not participate to a reasonable degree over time in certain key activities of his or her society; and (a) this is for reasons beyond his or her control, and (b) he or she would like to participate" (Burchardt 2000, p. 388).

Lack of income may be one factor contributing to social exclusion, but exclusion may also be among the factors that prevent people from having the opportunity to move out of poverty. Thus, if they are excluded from the labor market or from drawing a livelihood out of a particular resource like a fishery, poverty is what they will experience. Exclusion based on ethnicity, social class, or caste is reported by several chapter authors, such as in the Guatemalan, Nicaraguan, Turkish and Indian situations that Andrade and Midré, Chap. 19; González, Chap. 13; Knudsen and Koçak, Chap. 11; and Bavinck, Chap. 9 write about. Institutional mechanisms may also be at play.

The open access/common pool situation counteracts marginalization and social exclusion, as pointed out by Andrade and Midré (Chap. 19), but it may create more competition in a way that may eventually exclude poor fisher groups from the resource commons. Open access may have little to offer to small-scale fishers if their fishing grounds are encroached upon by industrial vessels, as is often described in this volume. Neither do management systems that aim to reduce or control access if rules are not enforced, as in the case of South Africa and Vietnam (Isaacs, Chap. 16; Nguyen and Flaaten, Chap. 15). In several of the case studies presented here (Turkey, Bangladesh, Nicaragua, Vietnam), poverty among small-scale fishers is also associated with lack of land ownership, which they need to sustain a diverse but secure livelihood of which fishing is only a part. This underscores the need to see small-scale fisheries as part of an employment system that often includes other sources of livelihood and which allows people to switch between activities according to a natural and seasonal cycle, or throughout their career. When fishers reach a certain age, they need to find other ways to earn a living that are less physically demanding.

The German sociologist Georg Simmel (1965 [1908]) discusses another aspect of the relationship between the poor and society at large. He suggests that the key characteristic, the distinguishing mark of the poor, occurs when they reach a specific marginal and degrading social status as recipients of public assistance. Hence, it is not economic or other problems *per se* that define poverty, but the social response to that situation (Coser 1965, p. 142). Simmel wrote on the basis of his observation of his country, Germany, an emerging welfare state at the time. In many developing countries, such programs are missing or are insufficient. Thus, one may think that his ideas are relevant only in countries where we find public welfare programs. But, as several of the authors in this volume describe, international NGO's as well as national development programs are operative in many cases. What kinds of relationships are established between these organizations and their clients is a research issue. Do such relationships add to the stigma or shame of being poor? Accepting help in order to survive often involves having to agree to the helper's definition of the situation, his or her ideas about the causes of poverty, images about the character of the poor, as well as his or her theories about how poverty can be alleviated (Eyben 2007).

The literature about failed development and management projects is immense, and there are several examples of such in the chapters that follow. There seems to be a growing concern about the ethical implications connected to the relationship between helpers and recipients of assistance. Although involving and listening to the poor as "the real experts" may have become more common, there are also examples in this volume that suggest that, as far as small-scale fishers are concerned, it does not always happen. The assistance to Sri-Lankan and Thai fishing communities after the 2004 tsunami are the cases in point (Amarasinghe and Bavinck, Chap. 17; Chuenpagdee and Juntarashote, Chap. 14). They are both examples of an external aid that had negative consequences at the community level.

4.3.7 Gendered Poverty

Women's roles, relationships, and agency are crucial for social and economic development (Okin 2003). Since the majority of children are dependant on their mothers' and grandmothers' well-being, gendered inequality and poverty also have vital negative effects on the situation of the children and the family. Women's poverty thereby affects the future generation's life chances and development possibilities.

As Sen (2000) and many others have noted, there are many cultural and other barriers against women's participation in labor markets outside households (Nussbaum 2000). Apart from leading to the loss of potential household incomes, women's positions relative to men's result in further injustice and impoverishment. Bennett (2005) argues that it is not necessarily lack of money, but the control of their earnings that is the problem for women; and makes the case for a gendered perspective where women's poverty must be analyzed in relation to that of men. Women are often disproportionally hurt by natural resource degradation (cf. Power and Harrison 2005). For all these reasons, a "feminization of poverty" is often detectable in fisheries-dependent communities (McCay 2005).

The role of women is therefore of central importance to human development in general. Okin (2003) argues that unless women's capabilities and voices are heard and fostered as much as those of men, improvement of the least-developed countries is doomed. These arguments are no less relevant as far as small-scale fisheries are concerned, where women are usually the primary caretakers (Weeratunge et al. 2010), running the household and caring for children. In many instances, they are key breadwinners in a diversified household economy, as described by Islam (Chap. 5) and Chuenpagdee and Juntarashote (Chap. 14; see also Béné et al. 2009). In India, women in fishing communities sometimes make their income from low-paid petty-trade (see Bavinck, Chap. 9). In addition to their employment outside the household, women are often indirectly and directly involved in their spouses' fishing business, and are responsible for supportive activities (making and mending gear, baiting and bait fishing, processing, sales). They sometimes partake in the actual fishing operations, as mentioned by Chuenpagdee and Juntarashote (Chap. 14). In some instances, they run their own fishing vessels, as illustrated by González in the case of Nicaragua (Chap. 13; cf. also Davis et al. 1992; Nadel-Klein and Davis 1998).

Isaacs (Chap. 16) emphasizes the need to create an action space that makes it possible for women in small-scale fisheries to thrive. For this to happen, gender issues must be mainstreamed in fisheries development and management, as Bennett (2005) holds. Gender neutrality, which often is legitimized from an equality point of view, means in practice that issues of gender are ignored, which easily leads to the reinforcement of women's marginalization and gender inequities, as when fishing rights are instituted (Munk-Madsen 1998). This is also why Sen (2009) insists on the "justice as fairness" principle. Targeting the poorest and most vulnerable through "positive discrimination" in fisheries development is warranted as social justice. But the lesson learned is that women's poverty and vulnerability are unlikely to be addressed without grassroots mobilization, as demonstrated in many parts of the world; for instance Maneschy and Álvares in the case of Brazil (2005), and Medard (2005) in Lake Victoria fisheries.

4.3.8 Poverty and Sufficiency

In several chapters of this volume, the people interviewed speak about their aspirations, what they hope to achieve in life, and why they find small-scale fisheries attractive both as an occupation and as a way of life. Although people may recognize that they are poor both in absolute and relative terms, they may still express that they have what they basically need and what they can expect to obtain. Fishing is what they know and what they do. Life is hard to imagine otherwise. As long as fishing provides enough for them to continue, they are satisfied. For many, fishing is not a choice but a life they are born into and grew up with. It is their identity, and has formed the history of their family and their community (Onyango and Kraan, Chaps. 6 and 8, respectively). On the same note, Cordell (1989, p. 2) argues: "Fishermen's livelihood claims often spring from deeper connections with the past and the community, from a special fraternity with the sea expressed in knowledge of marine localities [...] from commitment to a place, a group, and a way of life." The idea of having enough in the case of Guatemala (see Andrade and Midré, Chap. 19) is captured by the concept of making out, "*defenderse.*" The metaphor points to an external danger that the people seem to control. Their existence may be marked by risks, but they still manage to stay out of poverty. They do not have much, but enough to sustain themselves, which provides these fishers with great satisfaction.

In a similar vein, Thai small-scale fishers, according to Chuenpagdee and Juntarashote (Chap. 14), abide by a philosophy where having enough to get by is also good enough. Their level of aspiration may seem modest, but conforms to Buddhist values and norms. Even if external evaluators judge their living standards as low, the people interviewed say they are satisfied with what their fishing life may offer. Expressions of sufficiency should, we believe, be distinguished from utilitarian notions of "happiness." We would rather point to Sen's (1983) idea of capabilities and the possibility of having a life conforming to personal values. Sufficiency can thus be understood as an existence with some amount of freedom – from pursuing desires that are likely to lead to disappointment and disillusionment.

The idea of sufficiency implies a kind of moderation atypical in market economies (Baudot 2000). Sufficiency is based on the principle that the acquisition and consumption of something of value should be restrained at some point (Princen 2005). The question is: How much is enough? In an example of the Monhegan lobster fishery, Princen (2005, p. 274) finds that fishers' idea of sufficiency involves self-conscious decisions to impose harvest limits on themselves. When small-scale fishers in Thailand speak about sufficiency, the limits are again self-imposed. But it also involves adjusting aspiration levels to what can realistically be achieved, legitimized as a moral principle that is championed by the king.

Such moral principles are about social values, what people consider worth striving for but also what they believe they should do. As Sen (2009, p. 250) reminds us, "seeing people only in terms of needs may give a rather meager view of humanity. Human beings have needs, but they also have values and, in particular, cherish their ability to reason, appraise, choose, participate and act" (quoted in Onyango, Chap. 6). Onyango, therefore, concludes that fisheries management and poverty reduction strategies should be formulated within a framework that accounts for the social values and principles that small-scale fishers attach to their way of life, and the (moral) satisfaction, pride, and identity they derive from it.

4.3.9 Poverty as a Process

Poverty may seem to be an unchanging condition defining the fate of individuals and classes of people. For the poor, it is something to be endured. As Jensen (2000, pp. 130–131) argues: "They have no future, we say of those in direst need, not because they are going to die tomorrow, but because their lives are static." This view is consistent with the idea of poverty as a trap. It is also in harmony with the view of Oscar Lewis (1965, p. xliii) that poverty develops into a distinct culture "with its own structure and rationality, as a way of life which is passed down from generation to generation among family lines" and, once entrenched, it takes more than poverty eradication measures to effect change. Alternatively, poverty can be understood as a process, a situation that may be dynamic, spurred by the initiatives and actions of the poor themselves, and/or by circumstances in the natural and social environment (Davis 2010).

Barth (1966) argued that we may accurately and meticulously describe an observable, social fact, but we will still not have provided an explanation. Only a description of the processes generating the phenomenon will explain it. Regardless of the inherent methodological individualism in Barth's approach, his argument applies to any social phenomenon, poverty included. Thus, behind the global overview of poverty and well-being that was presented by Eide et al., Chap. 2, there are social, economic, and ecological processes for the researcher to discern.

We may register the entitlements and capabilities of poor, small-scale fishing people and assess their financial, social, and cultural capital, or what they lack of them. But in order to provide a full understanding of how people become poor and marginalized, we also need to understand how they acquired their skills and possessions in the first place, and how they may have been eroded or lost. We would note the social norms and cultural values that integrate and marginalize categories and classes of people (Sen 2004), but we also have to analyze how people respond to these constraints and stressors, particularly how they manage to confront and/or break them. From a process perspective, we therefore also need to study how small-scale fishers convert their assets into economic and social opportunities.

A focus would, in other words, be on how people cope with poverty and vulnerability, what they do or don't do, how they engage with others, and how they learn and adapt. From a poverty alleviation perspective, the focus would be on how poor fisher folk individually, collectively, or with the help of external supporters, manage to change the ramifications that hold them back or tend to exclude them from realizing the things that they value. Scoones (1998) distinguishes between three different coping strategies that poor people employ: Intensification, migration, and livelihood diversification. Bennett (2005) lists social, economic, and institutional coping mechanisms, whereas Perry et al. (2011) distinguish between coping strategies for the short-term and adapting strategies for the long-term.

How people cope with poverty and vulnerability and how they adapt to shifting socioeconomic and ecological circumstances are important research questions for all the authors of this volume. From a dynamic perspective, the investigation aims not only to understand how poor people accept and endure, but how they act upon their situation; how they adapt, change, and innovate. For Bangladesh, Islam (Chap. 5) generates his own list of coping strategies based on people's responses to environmental hazards. In the case of small-scale fishers in India, described by Bavinck (Chap. 9), coping involves collective uprising. Here, small-scale fishers took to the streets to protest against the newly developed trawl technology that threatened their livelihoods. Transformative change is also described by Isaacs (Chap. 16) for South Africa, where marginalized small-scale fishers were able to use their "informal action-space" to initiate a litigation process that resulted in a policy change in favor of the poor.

The argument is that it is not only important to learn by listening to what the poor say, but also to learn from investigating what they actually do. Thus, one would look at how small-scale fishers cope by acting alone and engaging with others, be they family members, other fishers, traders, processors, and government. Fishers may well be on the losing side in those transactions, and are thus increasingly marginalized (see Marciniak, Chap. 7 for Poland). But they may also end up stronger, as described by Andrade and Midré (Chap. 19, Guatemala) and González (Chap. 13, Nicaragua). In the case of Turkey (Knudsen and Koçak, Chap. 11), Ghana (Kraan, Chap. 8), and Bangladesh (Islam, Chap. 5), small-scale fishers have been able to advance through migration and/or resettlement. In comparing fishers with people in other occupations, Bavinck (2001, pp. 86–87) points out that: "The process of upward or downward mobility is often more pronounced in families engaged in fishing than in other professions." This, he partly explains by investment thresholds in artisanal fishing being "less absolute." Fishers are able to advance by building up their assets bit by bit.

4.3.10 Vulnerability

Small-scale fishers may be poor of means to sustain their livelihoods, but they are also vulnerable to environmental shocks and to social change (Macfadyen and Corcoran 2002; FAO 2005). How exposed and prepared small-scale fishers are to cope with environmental hazards is an issue addressed in several chapters of this volume. Agder (2006, p. 268) defines vulnerability as "the state of susceptibility to harm from exposure to stresses associated with environmental and social change, and from the absence of capacity to adapt." (For other definitions, see Kasperson et al. 2010). At the same time, he stresses that vulnerability is a dynamic phenomenon, "often in a continuous state of flux," shaped by biophysical and social processes that determine people's ability to manage (Agder 2006, p. 274). For the same reason, Kasperson and colleagues (2010, p. 266) argue that vulnerability is best perceived as a "set of cross-scale dynamics, and historical trajectories," and that learning and coping are therefore "essential ingredients in vulnerability." This perspective is also adopted by the authors of this volume.

Poverty is obviously among those factors that make people vulnerable, because it affects their coping capacity. Fishing people are also vulnerable because the ecosystems that they draw their livelihoods from are vulnerable; partly, but not exclusively, because of their own fishing practices. Often, fishing people are at the receiving end of ecosystem damage, as when the Pearl Lagoon in Nicaragua (González, Chap. 13) is polluted by fertilizer run-off and erosion from up-river logging. The link between poverty and vulnerability works both ways. In given circumstances, vulnerability causes deprivation. Still, as Béné (2009) argues, it is important to recognize that the two phenomena are not always systematically correlated. People may be poor but not necessarily vulnerable, and vice-versa. But in many instances, people are both very poor and extremely vulnerable, as with the small-scale fishers in the coastal communities of Bangladesh (Islam, Chap. 5).

Small-scale fishers are often vulnerable to a degree that other people are not. In his classic treatise, Raymond Firth (1946/1966, p. 3) compared Malay fishers with farmers and observed that: "Investment in agricultural land has a permanency not found in fishing enterprises; and fishing boats and gear, though perhaps as durable as agricultural implements and cattle, are, on the whole, more liable to sudden damage and loss." Small-scale fishers are also exposed to life threatening dangers when out on the sea to a degree that fishers operating from bigger and safer vessels are not. Living on the coast exposes them to natural disasters like cyclones or hurricanes. In some countries like Bangladesh and the Caribbean region of Nicaragua, Guatemala, and Mexico, this happens quite frequently (see Islam, Chap. 5; Gonzalez, Chap. 13; Andrade and Midré, Chap. 19; and Salas et al., Chap. 10). When such natural disasters occur, lives are lost, and livelihoods are devastated.

Fishers in Bangladesh also experience vulnerabilities that are social and institutional at their origin. Poor people are often defenseless against abuse, as when their gears are stolen, people are kidnapped for ransom, or when government officials or middlepersons take advantage of their weaker position. Thus, whereas poverty involves want or deprivation, vulnerability refers to what Sen (2000) calls lack of "protective security," which involves protection of poor, small-scale fishers from being excluded from the commons on which they often depend, to a larger degree than other people (cf. Beck and Nesmith 2000). In most instances, though, poor people are left to fend for themselves. The need for protective security calls for governance interventions. Both in the Polish and the Turkish cases (Marciniak, Chap. 7; Knudsen and Koçak, Chap. 11), there are social support mechanisms that provide fishers with a degree of security that those in other countries represented in this volume do not enjoy. In many situations, as mentioned by Islam and Onyango (Chaps. 5 and 6), people are able to get support within their communities or from moneylenders. But also in these situations, small-scale fisheries deserve more attention from policymakers, for instance, in the form of more supportive legislation (Macfadyen and Corcoran 2002; Thorpe et al. 2004; FAO 2005).

Government assistance in case of crisis would be helpful, as in the cases of Poland and Turkey, but there are several chapters in this volume which point to other measures that can reduce vulnerability at the local level. Salas and colleagues, and Amarasinghe and Bavinck (Chaps. 10 and 17), analyze how fishers cope with vulnerability by becoming members of community cooperatives; making them better prepared for natural hazards or household economy collapse. Cooperatives extend the action space needed to better cope with vulnerability. Although cooperatives seem to be out of fashion in fisheries development circles, due to their mixed success around the world (Hersoug et al. 2004), they can also fill a collective institutional vacuum that often exists at the local level, and thus enable communities to reduce vulnerability.

But from a governance perspective, cooperatives are complex organizations (Amarasinghe and Bavinck, Chap. 17; Jentoft 1985, 1986) that need careful design and effective leadership. Amarasinghe and Bavinck argue that, because of their welfare-centric approach, there is danger that cooperatives may end up enhancing fishing activity and pressure, unless they pay attention to ecosystem limitations and governance. Their success also relies on external support, such as resources and legislation that help them to fulfill some of their social functions, such as fisheries management and social security.

4.4 Conclusions

As demonstrated in several of the chapters of this volume (cf. also Stobutzki et al. 2006), overfishing and destructive use of fishing gear can degrade marine habitats and cause poverty and vulnerability. But poverty and vulnerability in small-scale fisheries are not uni-dimensional, and do not pertain only to income. Neither are poverty and vulnerability necessarily fisheries management or resource conservation problems, but structural and institutional problems where political reform is essential (Glavovic 2008; see also Bavinck, Chap. 9). Thus, overfishing can be a source of poverty, due to a Malthusian logic; and a tragedy of the commons mechanism.

But, as Béné (2003, 2004b) points out, the overfishing-poverty nexus is not the only reason why small-scale fishers are often poor. Small-scale fishers may be poor for precisely the same reason as other people are poor, because they lack access to basic natural, physical, or financial entitlements, or because they do not possess the social and human capital they need to be fully included, involved, and self-reliant (see Islam, Chap. 5). They may also be poor because they are trapped in exchange relationships that systematically work to their disadvantage, and because they do not have the alternatives, resources, power, and action space that it takes to break out (Isaacs, Chap. 16).

While providing a nuanced portrait of small-scale fisheries around the world, the authors of this book also paint a picture of fishing people in need of social and political change that will address their poverty and vulnerability more forcefully than has been the case in the past. This is as much an issue of the political will of decision-makers, as it is an issue of their capability. But addressing the same problem does not necessarily involve using the same remedies. The context varies, the system to be governed has unique characteristics from place to place that need to be taken into account, and the ecological and social stressors are not the same in all settings. These chapters give a deeper understanding of these conditions and mechanisms, which is a prerequisite for any initiative that will help to bring the change that is needed.

References

Agder WN (2006) Vulnerability. Glob Environ Change 16:268–281

Alkire S (2002) Valuing freedoms: Sen's capability approach and poverty reduction. Oxford University Press, Oxford

Allison EH, Ellis F (2001) The livelihoods approach and management of small-scale fisheries. Mar Policy 25:377–388

Barth F (1966) Models of social organization. Occasional paper, No. 23. Royal Anthropological Institute of Great Britain and Ireland, London

Baudot J (2000) The international build up: poverty and the spirit of the time. In: Gordon D, Townsend P (eds.) Breadline Europe: the measurement of poverty. Polity, Bristol, pp 25–34

Bavinck M (2001) Marine resource management: conflict and regulation in the fisheries of the Coromandel Coast. Sage, New Delhi

Beck T, Nesmith C (2000) Building on poor peoples' capacities: the case of common property resources in India and West-Africa. World Dev 29(1):119–133

Béné C (2003) When fishery rhymes with poverty: a first step beyond the old paradigm on poverty in small-scale fisheries. World Dev 31(6):949–975

Béné C (2004a) Small-scale fisheries: assessing their contribution to rural livelihoods in developing countries. United Nations Food and Agriculture Organization (FAO), Rome

Béné C (2004b) Poverty in small-scale fisheries: a review and some further thoughts. In: Neiland AE, Béné C (eds.) Poverty and small-scale fisheries in West-Africa. Food and Agriculture Organization of the United Nations/Kluwer Academic, Rome, pp 61–82

Béné C (2009) Are fishers poor or vulnerable? Assessing economic vulnerability in small-scale fishing communities. J Dev Stud 45(6):911–933

Béné C, Bennett L, Neiland A (2004) Managing small-scale fisheries with reference to poverty. In: Neiland AE, Béné C (eds.) Poverty and small-scale fisheries in West-Africa. Food and Agriculture Organization of the United Nations/Kluwer Academic, Rome, pp 83–102

Béné C, Steel E, Kambala Luadia B, Gordon A (2009) Fish as "bank in the water": evidence from chronic-poor communities in Congo. Food Policy 34:104–118

Béné C, Hersoug B, Allison EH (2010) Not by rent alone: analysing the pro-poor functions of small-scale fisheries in developing countries. Dev Policy Rev 28(3):325–358

Bennett E (2005) Gender, fisheries and development. Mar Policy 29:451–459

Berkes F, Feeny D, McCay BJ, Acheson JM (1989) The benefits of the commons. Nature 340:91–93

Bourdieu P (1976) The disenchantment of the world. Algeria 1960. Cambridge University Press/ Editions de la Maison des Sciences de l'Homme, Cambridge/Paris

Bourdieu P (1986) The forms of capital. In: Richardson JE (ed) Handbook of theory and research for the sociology of education. Greenwood Press, New York, pp 241–258

Burchardt T (2000) Social exclusion: concepts and evidence. In: Gordon D, Townsend P (eds.) Breadline Europe: the measurement of poverty. Polity, Bristol, pp 385–406

Carpenter M (2009) The capabilities approach and critical social policy: lessons from the majority world? Crit Soc Policy 29(3):351–373

Chambers R (1997) Whose reality counts? Putting the first last. Intermediate Technology Publications, London

Christy FT (1986) Special characteristics and problems in small-scale fisheries management in developing countries. In: Miles E, Pealy R, Stockes R (eds.) Natural resources. Economic and policy applications. University of Washington Press, Seattle, pp 118–151

Cinner JE, Daw T, McClanahan TR (2008) Socioeconomic factors that affect artisanal fishers' readiness to exit a declining fishery. Conserv Biol 23(1):124–130

Clark DA (2005) The capability approach: its development, critiques and recent advances. Working papers. ESRC Global Poverty Research Group, Oxford, pp 1–18

Clay EJ, Schaffer BB (1984) Conclusion: self awareness in policy practice. In: Clay EJ, Schaffer BB (eds) Room for manoeuvre: an exploration of public policy in agriculture and rural development. Heinemann, London, pp 191–193

Cordell J (ed) (1989) A sea of small boats. Cultural Survival, Inc., Cambridge

Cornwall A, Fujita M (2007) The politics of representing 'The Poor'. In: Moncrieffe J, Eyben R (eds.) The power of labelling: how people are categorized and why it matters. Earthscan, London, pp 48–63

Coser LA (1965) The sociology of poverty: to the memory of Georg Simmel. Soc Probl 1(3):140–148

Davis P (2010) Poverty in time: exploring poverty dynamics from life history interviews in Bangladesh. In: Addison T, Hulme D, Kanbur R (eds.) Poverty dynamics: interdisciplinary perspectives. Oxford University Press, Oxford, pp 154–182

Davis A, Bailey C (1996) Common in custom, uncommon in advantage: common property, local elites, and alternative approach to fisheries management. Soc Nat Resour 9:251–265

Davis A, Thiessen V, Jentoft S (1992) The veiled crew: an exploratory study of wives' reported and desired contributions to coastal fisheries enterprises in Northern Norway and Nova Scotia. Hum Organ 51(4):342–352

Dean H, Melrose M (1999) Poverty, riches and social citizenship. MacMillan, Houndsmill

Doulman D (2004) Fisheries management, poverty alleviation and the implementation of The Code of Conduct for Responsible Fisheries. In: Neiland AE, Béné C (eds.) Poverty and small-scale fisheries in West-Africa. Food and Agriculture Organization of the United Nations/ Kluwer Academic, Rome, pp 209–227

Eyben R (2007) Labelling people for aid. In: Moncrieffe J, Eyben R (eds.) The power of labelling: how people are categorized and why it matters. Earthscan, London, pp 33–47

FAO (2005) Increasing the contribution of small-scale fisheries to poverty alleviation and food security. FAO Technical Guidelines for Responsible Fisheries, 10. Food and Agricultural Organization of the United Nations, Rome

Firth R (1946/1966) Malay fishermen: their peasant economy. W.W. Norton and Company, New York

Glavovic B (2008) Poverty and inequity at sea: challenges for ecological economics. In: Patterson MG, Glavovic BC (eds.) Ecological economics of the oceans and coasts: current issues in ecological economics. Edward Elgar Publishing, Cheltenham, pp 244–265

Gordon D (2000) Measuring absolute and overall poverty. In: Gordon D, Townsend P (eds.) Breadline Europe: the measurement of poverty. Polity Press, Bristol, pp 49–78
Gordon D, Spicker P (eds.) (1999) The international glossary on poverty. Zed Books, London
Granovetter MS (1973) The strength of weak ties. Am J Sociol 78(6):1360–1380
Hall SJ, Dugan P, Allison EH, Andrew NL (2010) The end of the line: who is the most at risk from the crisis in global fisheries? Ambio 39:78–80
Hallenstvedt A (1982) Med Lov og Organisasjon. Organisering av Interesser og Markeder i Norsk Fiskerinæring. Universitetsforlaget, Oslo
Halpern D (2005) Social capital. Polity Press, Cambridge
Hardin G (1969) The tragedy of the commons. Science 162:1243–1248
Harriss J (2002) Depoliticizing development: the world bank and social capital. Anthem Press, London
Hersoug B, Jentoft S, Degnbol P (2004) Fisheries development: the institutional challenge. Eburon Publishers, Delft
Isaacs M, Hara M, Raakjær J (2007) Has reforming South African fisheries contributed to wealth redistribution and poverty alleviation? Ocean Coast Manage 50:301–313
Jackson WA (2005) Capabilities, culture and social structure. Rev Soc Econ 63(1):101–124
Jensen C (2000) I have seen the world begin: travels through China, Cambodia and Vietnam. Harcourt Inc., New York
Jentoft S (1985) Models of fishery development: the cooperative approach. Mar Policy 9(4):322–331
Jentoft S (1986) Fisheries co-operatives: lessons drawn from international experiences. Can J Dev Stud 7(2):197–209
Jentoft S (2004) Fisheries co-management as empowerment. Mar Policy 29:1–7
Jentoft S (2007) In the power of power: the understated aspect of fisheries management. Hum Organ 66(4):426–437
Jentoft S, Chuenpagdee R (2009) Fisheries and coastal governance as a wicked problem. Mar Policy 33:553–560
Jentoft S, Chuenpagdee R, Bundy A, Mahon R (2010a) Pyramids and roses: alternative images for the governance of fisheries systems. Mar Policy 4(6):1315–1321
Jentoft S, Onyango P, Islam MM (2010b) Freedom and poverty in the fishery commons. Int J Commons 4(1):345–366
Kasperson JX, Kasperson RE, Turner BL II (2010) Vulnerability of coupled human-ecological systems to global environmental change. In: Rosa EA, Diekman A, Dietz T, Jaeger C (eds.) Human footprints on the global environment: threats to sustainability. MIT Press, Cambridge, pp 231–293
Kooiman J, Bavinck M, Jentoft S, Pullin R (eds) (2005) Fish for life: interactive governance for fisheries. Amsterdam University Press, Amsterdam
Kurien J (2004) The blessing of the commons: small-scale fisheries, community property rights, and coastal natural assets. Working Paper Series, No. 72. Political Economy Research Institute. University of Massachusetts, Amherst
Lewis O (1965) La Vida: a Puerto Rican family in the culture of poverty – San Juan and New York. Random House, New York
Lister R (2004) Poverty. Polity Press, Cambridge
Macfadyen G, Corcoran E (2002) Literature review of studies on poverty in fishing communities and of lessons learned in using the sustainable livelihoods approach in poverty alleviation strategies and projects. FAO Fisheries Circular, No. 979. Food and Agriculture Organization of the United Nations, Rome
Maneschy MC, Álvares ML (2005) Identities in construction and in conflict: restructuring and the social roles of women in the fisheries communities of Pará State, Brazil. In: Neis B, Binkley M, Gerrard S, Maneschy MC (eds.) Changing tides: gender, fisheries and globalization. Fernwood Publishing, Halifax, pp 51–63
Matthews S (2002) Not often are the voices of the poor heard. Community Dev J 37(2):200–202. http://cdj.oxfordjournals.org/content/37/2/200.full.pdf+html. Accessed 6 July 2004

Mayer-Foulkes D (2008) The human development trap in Mexico. World Dev 36(5):775–796
McCay B (2005) Gender, globalization and a tragic choice on Fogo Island, Newfoundland. In: Neis B, Binkley M, Gerrard S, Maneschy MC (eds.) Changing tides: gender, fisheries and globalization. Fernwood Publishing, Halifax, pp 116–132
McEvoy AF (1986) The fisherman's problem: ecology, and law in the California fisheries, 1850–1980. Cambridge University Press, Cambridge
Medard M (2005) Women's strategies in the globalized Lake Victoria fisheries. In: Neis B, Binkley M, Gerrard S, Maneschy MC (eds.) Changing tides: gender, fisheries and globalization. Fernwood Publishing, Halifax, pp 78–91
Midré G (2005) Opresión, espacio para actuar y conciencia crítica. Líderes indígenas y percepción de la pobreza en Guatemala. Instituto de Estudios Interétnicos, Universidad de San Carlos de Guatemala, Guatemala
Midré G, Flores SA (2002) Élite ladina, políticas públicas y pobreza indígena. Instituto de Estudios Interétnicos, Universidad de San Carlos de Guatemala, Guatemala
Moser C, McIlwayne C (2001) Violence in a post-conflict context: urban poor perceptions from Guatemala, Conflict prevention and post-conflict reconstruction. The World Bank, Washington DC
Munk-Madsen E (1998) The Norwegian fishing quota system: another patriarchal construction? Soc Nat Resour Int J 11(3):229–240
Nadel-Klein J, Davis DL (1998) To work and to weep: women in fishing economies. ISER Press, St. John's
Narayan D, Chambers R, Shah M, Petesch K (2000) Voices of the poor: crying out for change. Oxford University Press/The World Bank, New York/Washington DC
Nussbaum MC (2000) Women and human development: the capabilities approach. Cambridge University Press, Cambridge
O'Connor A (2001) Poverty knowledge: social science, social policy, and the poor in the twentieth-century U.S. history. Princeton University Press, Princeton
Okin SM (2003) Poverty, well-being and gender: what counts, who's heard? Philos Public Aff 31(3):280–316. http://www.jstor.org/stable/3558008. Accessed 6 July 2004
Pender J (2002) Empowering the poorest? The world bank and the 'Voices of the Poor'. In: Chandler D (ed.) Rethinking human rights: critical approaches to international politics. Palgrave, Basingstoke, pp 97–114
Perry I, Ommer R, Barange M, Jentoft S, Neis B, Sumaila UR (2011) Marine social-ecological responses to environmental change and the impacts of globalization. Fish Fish. doi:10.1111/j.1467-2979.2010.00402.x
Pinker R (1999) Do poverty definitions matter? In: Gordon D, Spicker P (eds) The international glossary on poverty. Zed Books, London, pp 1–5
Power N, Harrison D (2005) The power of gender ideology in the face of resource decline in Newfoundland, Canada. In: Neis B, Binkley M, Gerrard S, Maneschy MC (eds.) Changing tides: gender, fisheries and globalization. Fernwood Publishing, Halifax, pp 229–242
Princen T (2005) The logic of sufficiency. MIT Press, Cambridge
Putnam RD (1995) Bowling alone: the collapse and revival of American community. Simon & Schuster, Toronto
Reis EP, Moore M (eds) (2005) Elite perceptions of poverty and equality. ZED Books, London
Rittel HWJ, Webber MM (1973) Dilemmas in a general theory of planning. Policy Sci 4:155–169
Room G (2000) Trajectories of social exclusion: the wider context for the third and first worlds. In: Gordon D, Townsend P (eds.) Breadline Europe: the measurement of poverty. Polity, Bristol, pp 407–440
Scoones I (1998) Sustainable rural livelihoods: a framework for analysis. Sustainable Development Programme, Institute of Development Studies, Working Paper. University of Sussex, Sussex
Sen A (1983) Poor, relatively speaking. Oxf Econ Paper 35(2):153–169
Sen A (1985) A sociological approach to the measurement of poverty: a reply to Professor Peter Townsend. Oxf Econ Paper 37(4):669–676

Sen A (2000) Development as freedom. Anchor Books, New York
Sen A (2004) How does culture matter? In: Rao V, Walton M (eds.) Culture and public action. Stanford University Press, Stanford, pp 37–58
Sen A (2009) The idea of justice. Allen Lane, London
Simmel G (1965 [1908]) The poor. Soc Probl 13:118–140
Stobutzki IC, Silvestre FT, Garces LR (2006) Key issues in coastal fisheries in South and Southeast Asia, outcomes of a regional initiative. Fish Res 78:109–118
Thorpe A (2004) The nature and causes of poverty. In: Neiland AE, Béné C (eds.) Poverty and small-scale fisheries in West-Africa. Food and Agriculture Organization of the United Nations/ Kluwer Academic, Rome, pp 9–36
Thorpe A, Reid C, van Anrooy R, Brugere C (2004) African poverty reduction strategy programmes and the fisheries sector: current situation and opportunities. Afr Dev Rev 16(2):328–362
Thorpe A, Andrew NL, Allison EH (2007) Fisheries and poverty reduction. CAB Rev Perspectives in Agriculture, Veterinary Science, Nutrition and Natural Resources 2(085):1–12. http://www.cabasbstractsplus.org/cabreviews. Accessed March 2008
Townsend P (1979) Poverty in the United Kingdom: a survey of household resources and standards of living. Penguin, Harmondsworth
Townsend P (1985) A sociological approach to the measurement of poverty: a rejoinder to Professor Amartya Sen. Oxf Econ Paper 37(4):659–668
Townsend P, Gordon D (1991) What is enough? New evidence on poverty allowing the definition of a minimum benefit. In: Adler M, Bell C, Clasen J, Sinfield A (eds.) The sociology of social security. Edinburgh University Press, Edinburgh, pp 35–69
Tvedten I, Hersoug B (1992) Fishing for development: small-scale fisheries in Africa. Scandinavian Institute of African Studies, Uppsala
Veit-Wilson J (2000) Horses for discourses: poverty, purpose and closure in minimum income standard policy. In: Gordon D, Townsend P (eds.) Breadline Europe: the measurement of poverty. Polity Press, Bristol, pp 141–164
Verba S, Kelman S, Orren G, Miyake I, Watanuki J, Kabashima I, Ferree GD (1987) Elites and the idea of equality: a comparison of Japan, Sweden, and the United States. Harvard University Press, Cambridge
Webster N, Engberg-Pedersen L (eds) (2002) In the name of the poor: contesting political space for poverty reduction. ZED Books, London
Weeratunge N, Snyder KA, Poh Sze C (2010) Gleaner, fisher, trader, processor: understanding gendered employment in fisheries and aquaculture. Fish Fish 11(4):405–420
Willmann R (2004) Poverty in coastal fishing communities. In: Neiland AE, Béné C (eds.) Poverty and small-scale fisheries in West-Africa. Food and Agriculture Organization of the United Nations and Kluwer Academic, Rome, pp 245–252
Wilson DC, Medard M, Harris CK, Wiley DS (1999) The implications for participatory fisheries management of intensified commercialization on Lake Victoria. Rural Sociol 64(4):554–572
Zimmerman B (2006) Pragmatism and the capability approach: challenges in social theory and empirical research. Eur J Soc Theory 9(4):467–484

Part II
Understanding

Poverty can mean different things to different people, and has multiple causes, which are not always characteristic of and unique to small-scale fisheries. In many instances, fisheries are an attractive alternative in comparison to what other sectors can offer. Small-scale fisheries are witnessing a pressure from newcomers, with the added burden placed on dwindling natural resources. But when poverty exists, it is not uniform. It is not necessarily felt the same by all. Poverty research has moved towards a broader perception of poverty, from a focus on income only, to that of other deprivations such as health, education, and vulnerability. It has led to a greater appreciation of the knowledge, experiences, and vulnerability of being poor; and the situational factors that limit options, as well as what might be done to improve the conditions. Understanding poverty from a high-up perspective is essential for broader policy frameworks and macro-economic strategies.

Poverty can and should be quantified and compared on a global scale. Understanding poverty from below and from the inside is a condition for a compassionate policy, where the poor and disenfranchised are seen and do matter. There is also need for a more precise identification of what in the concrete would make a difference in peoples' lives. Due to the nature of their environment, the locations in which they live and work, the technology they employ, and the resources they can drawn on, small-scale fisheries people are confronted with particular liabilities and risks. Small-scale fisheries are not only often shaky business, they can also be a perilous occupation, where fatalities have devastating effects on family and community.

People in this sector who are poor are often left to fend for themselves when crises hit. They have little support from government but would have to seek help among family and friends, who are often as poor as they are. Therefore, disasters are as much about poverty as about natural hazards. The five chapters of this section are all about what poverty and vulnerability mean to people in fishing communities. The challenges they face are both similar and different. But the chapters also illustrate why poverty and vulnerability are global issues that come with a particular reality, experience and definition, and why it is always important to understand poverty and vulnerability in concrete settings, whether we are in Bangladesh, Tanzania, Poland, Ghana, India, or somewhere else in this world.

Chapter 5
Living on the Margin: The Poverty-Vulnerability Nexus in the Small-Scale Fisheries of Bangladesh

Mohammad Mahmudul Islam

Abstract This chapter examines the relationship between poverty and vulnerability in small-scale fisheries of Bangladesh. For this purpose, data were collected in three coastal fishing communities. The results show that in small-scale fisheries, poverty is a complex issue, with a wide array of causal factors in effect. Small-scale fishers' livelihoods are threatened by: low productivity of fisheries and high dependency on certain species; seasonality in fishing; frequent natural disasters; heavy debt bondage; coastal piracy and other illegal rent seeking activities; mass illiteracy; and lack of participation in political processes and local institutions, to mention some of the problems. Thus in Bangladesh, small-scale fishers are forced to live on the margin of existence where they are extremely vulnerable to shocks such as environmental disasters. The study finds that a combination of different livelihood strategies is an important tool for escaping poverty in the fishing communities. I argue that to arrest poverty in small-scale fishing communities such as those of Bangladesh, addressing vulnerability is vital; and creating a buffer against crisis is urgent.

5.1 Introduction

Fish and fisheries have always been an inseparable part of the life and livelihoods of the people of Bangladesh. A widely known maxim *Maache-Bhate Bangali* meaning "Fish and rice make a Bengali" illustrates the importance of fish in the main diet of most Bangladeshis (cf. Alam and Thomson 2001). Fish alone supplies about 60% of animal protein intake by the population of Bangladesh. Fisheries also play a

M.M. Islam (✉)
Bremen International Graduate School for Marine Sciences (GLOMAR),
Research Center for Sustainability Studies (ARTEC), University of Bremen, D-28359
Bremen, Germany
e-mail: mislam@uni-bremen.de

S. Jentoft and A. Eide (eds.), *Poverty Mosaics: Realities and Prospects in Small-Scale Fisheries*, DOI 10.1007/978-94-007-1582-0_5,

major role in employment, foreign exchange earnings, and other aspects of the economy. The sector contributes about 4% of the national gross domestic product (GDP) and 7% to both the agricultural GDP and the country's total export earnings (Rahman et al. 2003). Moreover, fisheries provide full-time employment for roughly 1.5 million professional fishers, and another ten million people are part-time fishers; thus about one-eighth of the total population of the country is professionally related to fisheries. More than 90% of the total marine fish production in Bangladesh comes from small-scale coastal fisheries (BBS 2007; Bangladesh Economic Review 2008).

Though small-scale fisheries play a critical role in the livelihoods for a large percentage of the population in the country, this sector has failed to draw adequate attention to its poverty reduction potential – from academic research as well as from the policy arena. A substantial amount of academic literature reports on the significant poverty reduction that Bangladesh has achieved since the 1990s, but with a conspicuous absence of references to small-scale fisheries. Given that the small-scale fisheries sector is playing an important role in providing food security, income, and labor, it seems a paradox that so many of its participants are living in extreme poverty. The situation for the poor fishers is complex. Coastal small-scale fishers have the advantage of access to coastal resources (e.g., fish, forests) that are often not available for the poor living in inland districts. Moreover, the economic value of fish and other aquatic resources from wetlands has been found to be more than double the return from a single rice crop (Colavito 2002). Nevertheless, small-scale fishers of Bangladesh, as in many other parts of the world, are among those who are living in persisting poverty.

Fishers' poverty and vulnerability are intimately linked, as stated by Béné (2006): ". . . fishing activity may be seen as a source of vulnerability, where vulnerability becomes a source of poverty: People are more prone to poverty because they are more vulnerable; and they are more vulnerable because of the type of activities they pursue, namely fisheries" (p. 11). As will be demonstrated in what follows, the livelihoods of poor coastal small-scale fishers in Bangladesh are also enmeshed into a series of vulnerabilities. This chapter, therefore, aims at a deeper understanding of poverty and vulnerability, as they are common among coastal fishers in Bangladesh. To keep the analysis simple, but maintaining the essence of the problem, two main questions will be addressed. These are: (1) How do poverty and vulnerability manifest in fishing livelihoods? (2) How do fishers, individually and collectively, cope with poverty and resource crises and what are the conditions to break the vicious circle of their poverty?

The chapter is arranged as follows. The next section builds on theories on poverty and vulnerability which is used to analyze the empirical data. Section 5.3 describes the methods that were used to collect data, along with a brief description of the study areas. The empirical findings on small-scale fishing livelihoods of Bangladesh are presented in Sect. 5.4; whereas Sect. 5.5 shows how fishers cope with vulnerability and poverty, and Sect. 5.6 provides the conditions to break the vicious circle of poverty in fishing communities. The final Sect. 5.7 summarizes the main findings of the study and offers some concluding remarks.

5.2 Theoretical Perspectives

5.2.1 Poverty

The discourse on poverty and development has so far been dominated by the outsider perspectives and expertise of non-poor – professionals, politicians, civil servants, elected officials, academics; whereas, the perspectives of the poor themselves have been largely neglected (Narayan et al. 2000; Beck and Nesmith 2001). Questioning this practice, Beck and Nesmith (2001) argue that for eradicating poverty, poor people's knowledge and abilities must be understood within the socio-economic structure and context that reproduces poverty. This understanding must be incorporated into development planning. Narayan et al. (2000), in their *Voices of the poor: Crying out for change* formulated the key elements that contribute to poverty. As they state, according to the perspective of some of the 2.8 billion poverty experts, i.e., the poor themselves, poverty is multi-dimensional and complex; it is manifested in the lack of assets required for well-being. Notably, well-being is beyond material poverty; it has multiple, interlocking facets. Also, these dimensions mingle to create and sustain powerlessness, a lack of freedom of choice and action. Each dimension can further cause or compound the others. Caught in multiple deprivations, escapement is a greater struggle than many can handle. This trap is described by the poor by using metaphors like bondage, slavery, of being tied like "bundles of straw," leaving them unable to act (Narayan et al. 2000).

There is an overwhelming notion that fishers are among the poorest of the poor (Béné 2003); that small-scale fisheries are indeed equated with poverty. However, the widely accepted perception that fisheries "rhymes with poverty" experienced a recent shift from an old paradigm to a multi-dimensional model of poverty, which involves the idea that poverty in small-scale fisheries is not only the consequence of scarce resources, but of a number of other factors as well. Disaggregating the poor fishers into (socially) marginalized, (economically) excluded, (politically) disempowered, and (class-)exploited groups, the model reveals a more holistic range of different mechanisms that lead to impoverishment (Béné 2003).

5.2.2 Vulnerability

Vulnerability is conceived of as a key dimension of fisher's poverty (Béné 2009). When people are poor, they are also less resilient. They do not easily recover from shocks or crises. Any further shocks or crises such as a bad fish harvest, illness of the family head, financial asset loss etc. may easily push them into extreme poverty. Thorpe et al. (2007) find that fishing communities are vulnerable in heterogeneous ways, so that the poorest tend to be disadvantaged in receiving food and financial help to rebuild their livelihoods, as experienced in the case of the 2004 Asian Tsunami. However, though poor fishers are usually among the most vulnerable, vulnerability is not simply another word for poverty.

Chambers (2006) who distinguishes poverty from vulnerability describes the former as deprivation, lack or want; whereas, the latter is defined as "defencelessness, insecurity, and exposure to risk, shocks and stress" (p. 33). Again, vulnerability has two sides: "an external side of risks, shocks, and stress to which an individual or household is subjected; and an internal side which is defenceless, meaning a lack of means to cope without damaging loss" (p. 33). He argues that understanding vulnerability helps to disaggregate poverty. He further elaborates vulnerability as possessing inadequate capacities to cope with stress, crises, and shocks (which implicitly subsumes timely and effective external interventions), and the attendant's slow or limited recovery from crises.

The most vulnerable individuals, groups, classes, and regions are thus most exposed to potentially harmful perturbations, i.e., those who have limited coping capacity and suffer the most from the impacts of economic crises or environmental degradation such as natural disasters and climate change (Bohle et al. 1994). Bosher et al. (2007) assess four key determinants of vulnerability in rural India: people's access to assets, to facilities, to political networks, and to social networks. The most vulnerable people and communities typically have few capabilities, and therefore little choice about where and how they live (Sen 1981). They cannot therefore easily escape when disaster hits.

Yamin et al. (2005) argue that the concept of vulnerability in poverty studies is important because it draws attention to the multiple dimensions of deprivation, such as social exclusion and gender, as well as to poverty dynamics. It also helps to focus on the established patterns of coping and resilience used by those directly affected. Understanding vulnerability requires a deep understanding of the climatic, social, generational, geographic, economic, and political processes that generate poverty, particularly chronic poverty.

Bosher (2007) holds that in-built community level survival strategies (e.g., neighborly assistance, social capital) can provide a level of resilience that can reduce vulnerability. He argues that without support from civil society and social institutions, the conditions of the rural poor may never improve. Inequalities in vulnerability are also attributed to the institutions that, in some cases, have been accused of mismanagement and corruption (Kothari 1986; Kohli 1990; Narayanasamy et al. 2000; Robbins 2000 – as quoted in Bosher 2007).

Hence, it is necessary to understand the social and institutional dimensions of vulnerability that often come hand in hand with those that are related to natural disasters. Considering this, access to key socio-political institutions may enable people or whole communities to get hold of vulnerability-reducing resources, such as those that allow quick recovery. The types of social institutions and the strength of social networks may therefore influence an individual's survival strategies, in terms of "drawing upon communal resources," and "drawing on social relationships - patronage, kinship, friendship and informal credit networks" (Agarwal 1990). Such relationships may be important in providing vulnerability-reducing resources, particularly when government mechanisms are unavailable, or people have been deliberately or otherwise marginalized (Bosher et al. 2007).

5.3 Materials and Methods

This research on poverty in the small-scale fisheries in Bangladesh is mainly articulated with qualitative fieldwork data – collected through observations, individual and key informant interviews, and focus group discussions. Secondary data is drawn from documents and reports published by the government and different NGOs. The fieldwork was carried out in three fishing communities along the coast of Bangladesh from December 2007 to February 2008, and then updated in 2010 from January to April. For some variables, a household survey was conducted. In choosing the sampling area, diversity of fishing grounds, fishing gear, target species, and caste structure were considered. More specifically, the sampling areas cover: (1) One fishing community (Mothurapur) adjacent to the Sundarbans mangrove forest; (2) Two fishing communities (Selimpur in Chittagong District; and Thakurtala in Cox's Bazar district) from the South-eastern coast of Bangladesh (Fig. 5.1).

Focus group discussions lasting 45–60 min were conducted with seven to eight participants in each group. Ten key informants were interviewed. Several (8–10) key issues were covered by the in-depth interviews. The range of key informants spread through different backgrounds (e.g., fish harvester, crab collector, shrimp fry collector, fishers with boats, without boats, fish sellers, fishers' community leader, fisherwomen, NGO officials). The interviews were open-ended to allow some flexibility.

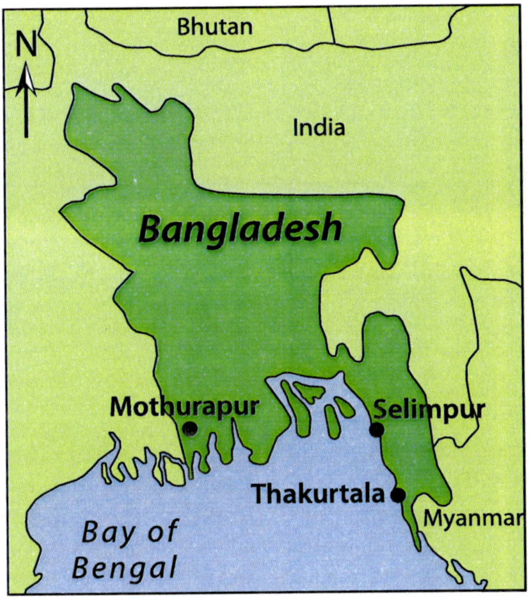

Fig. 5.1 Map of Bangladesh showing the three study areas. (*1*) The village of Mothurapur is situated in the Shymnagar *thana* (subdistrict) of Satkhira district which is located in the south-western corner of Bangladesh. (*2*) Selimpur is a caste-based Hindu fishing hamlet which is located approximately 20 km north of Chittagong, the main port city of Bangladesh. (*3*) The fishing village of Thakurtala is situated in Moheshkhali *thana* – a subdistrict of Cox's Bazar district

Secondary data was collected from Dhaka for public reports, and also from some local NGOs which are working in the coastal area with the small-scale fishers. Newspaper reports were also collected.

5.3.1 Study Areas

5.3.1.1 Mothurapur

The village of Mothurapur is situated in the Shymnagar *thana* (subdistrict) of Satkhira district which is located in the south-western corner of Bangladesh. Mothurapur is situated on a coastal embankment of a river bordering the Sundarbans Mangrove Forest. The community is comprised of both low caste Hindu and Muslim families who mostly rely on the extraction of common pool resources for their livelihoods. The access to resource extraction of the Sundarbans Mangrove Forest is *de jure* controlled by the Government. Fishers have to pay a certain amount of revenue to collect a pass for fishing in the forest for a certain period of time. In the village, the sanitation facility is poorly developed; only a few families have shared or own a sanitary toilet. Drinking water is in severe shortage; and people (mainly women and young girls) have to travel more than 1 km for collection of drinking water.

5.3.1.2 Selimpur

Selimpur is a caste-based Hindu fishing hamlet which is located approximately 20 km north of Chittagong, the main port city of Bangladesh. Most people of this community are involved in the fishery, and very few households are found to work in non-fishing related occupations, as young people have now to take up wage employment in the city. Proximity to the city and well connected communication pave the way for some NGO activities in the area. But still, sanitation facilities are poorly developed and most households have pit latrines. Drinking water access is secured for all by tube wells.

5.3.1.3 Thakurtala

The fishing village of Thakurtala is situated in Moheshkhali *thana* – a subdistrict of Cox's Bazar district. Moheshkhali is a coastal island connected to the mainland either by water way transportation to Cox's Bazar, or a bridge linked to Chakaria *thana* of Cox's Bazar district. Thakurtala is a traditional fishing village inhabited by low caste hereditary Hindu fishers who mainly fish in the Moheshkhali channel, the offshoot of the Bay of Bengal that separates the island from the mainland of Bangladesh. The village is situated in proximity to the *thana* headquarters of Moheshkhali. Land scarcity and increased population pressure make the community settlement very congested; and saline water intrusion is another problem. In general, the sanitation is poorly developed, though all have access to safe drinking water.

5.4 Small-Scale Fishing Livelihoods

5.4.1 Fisheries System and Fishing Assets

Along the Chittagong coast, where Selimpur is situated, traditional Hindu fishers catch mainly Bombay duck (*Harpodon nehereus*), Sergestid shrimp (*Acetes* species), and a few other species of estuarine fish using an Estuarine Set Bag Net (ESBN), using small engine-boats from mid-November to mid-April. The production in the sea during the remaining 3 months (i.e., mid-April to mid-July) is very low; partly due to non-availability of fish at that time as a result of high salinity in the coastal waters (during this period, fish move toward the deep sea), and partly for taking preparation (net mending or weaving, boat repairing, finance mobilization) for the ensuing peak season for Hilsha (*Tenuolosa ilisha*) (Kleith et al. 2003).

These fishers rely on the Hilsha fishery for their yearly income, which ranges from mid July to mid November. The Hilsha fishery constitutes the largest single fishery in Bangladesh, contributing about 1/6th of the country's total fish production. The Hilsha is a moderately sized fish; it may reach up to 60 cm in total length, and weight may reach up to 2.5 kg. This species obtains a high price in local and international markets. It is estimated that about two million fishers and traders directly and indirectly are engaged in this fishery (Kabir 2006).

Fishers from Thakurtala also use the ESBN for fishing in the Moheshkhali channel. Some fishers also fish offshore using a Marine Set Bag Net (MSBN). Those who use MSBNs cannot fish in the sea during the monsoon period of July to October, due to turbulent weather in the Bay of Bengal. Those fishers (our interviewed fishers) who fish in the Moheshkhali channel using ESBN, however, can fish all year. But nowadays productivity from the channel is seriously decreased, in part due to sedimentation in the channel. Many areas of the channel which were previously used for setting nets have now become unsuitable due to land accretion. Many species (mainly Hilsha species) have moved away from the channel. Consequently, the catches from ESBNs are drastically reduced, which take away direct income from the livelihoods of the fishers in the Thakurtala village.

In the Mothurapur study area of the Sundarbans mangrove forest, fishers use different gears to catch in the rivers, canals, and tributaries that criss-cross the forest. Most fishers do collective fishing in the forest by using beach seine on the river bank. Almost all women and young girls in the study area are involved with shrimp fry collection along the coastal embankment or in the vicinity of the forest. The boat, engine, and other fishing equipment are fishers' most productive assets. Social and economic differentiation is based on the ownership of these assets, as they dictate fishing strategies, access to fishing grounds, and influence economic benefits that the fishery brings to the fishing family. Access to productive assets (e.g., boats, nets) is, however, not secured for all coastal small-scale fishers.

5.4.2 Vulnerability and Risk

Risk and vulnerability are inseparable parts of fishing in the Bay of Bengal. The Bay of Bengal is one of the most disaster-prone regions in the world. Cyclones and tropical storms are yearly phenomena. In addition, the tidal activity is becoming increasingly turbulent, making fishing operations dangerous and limited. Rough seas, as well as frequent cyclones, often force coastal fishers to stay home or to abandon their fishing trip. Yet, due to very limited options for survival, many fishers defy warnings and continue fishing, which results in many fatalities every year. For instance, during cyclone Sidr in 2007, many fishers died due to ignoring the cautionary signal of a cyclone. During the last 50 years, about 0.7 million people were killed in Bangladesh due to cyclones and coastal storms.[1] Though there is no official record on how many fishers were killed or how many trawlers and nets were destroyed, the loss is certainly substantial.[2]

Fishing in the forest is also a risk, as tiger-human conflicts claim fishers' lives. During the study period, there were four cases of tiger-induced killings in the Shymnagar *thana*. In the study area, eight people who were the main earners in their families were killed by tigers over a period of about one decade. Income from risky fishing in the Sundarbans is further dissipated (mostly illegally) by rent-seeking activities of different levels from fishing to marketing.

> "Income from fishing in the Sundarbans by fishers goes to seven (*symbolic* number for many) families."
>
> – said one fisher from Mothurapur

He further explained,

> You have to give bribe to forest official to secure a pass for fishing. Extortion or ransom money goes to pirates; weightier and commission agent will exploit extra money in the fish landing centre, then *mohajon* or money lender will come to the scene to claim his money.

Fishers are exposed to piracy while at sea, which is particularly severe during the Hilsha fishing season, and in the case of the Sundarbans, all the year round. fishers are kidnapped for ransom. They are always afraid of being assaulted, and are concerned that dacoits will snatch their boat or nets. Ironically, the pirated property is then sold back to the fishers through brokers or through their own contact with the dacoits.

Moreover, in case of coastal fishing, a peak week (*joo*) of good harvest can be followed by a lean week (*dala*) of poor harvest due to spring tide and neap tide periods. Usually, fishers do not fish during lean weeks of the month; thus, fishing is virtually limited to 2 weeks in a month. Such seasonality is common in our study areas in the Sundarbans and along the south-eastern coast of Bangladesh. Fishers live near their workplace on the beach or coastal embankment. Living near the work place of the coast offers scope for many livelihood activities from common pool resources, but also makes them vulnerable to cyclones. When disasters strike,

---

[1]Daily Manabzamin (a Bengali Daily Newspaper) – 26 May 2009.

[2]The Mercantile Marine Department (MMD), the public authority for registration and licensing the industrial trawlers for fishing in the Bay of Bengal, does not keep track of fisher causalities. Small-scale fishing up to 30 m depth along the coast doesn't require a license.

limited finances make it difficult to restore homes and infrastructure. The loss of fishing gears, boats, livestock, and other household assets can wipe out entire livelihoods. Then families would have to rebuild their lives and livelihoods from scratch. Moreover, the accompanying loss of paddy fields and other food sources can worsen food insecurity along the coast, which often leads to health problems. The death of a household member capable of working can bring the whole family into extreme poverty and extended trauma.

In absence of safeguards against shocks, fisher households above the poverty threshold are pulled down; and those who are already below it experience further slippage (Hye 1996). Due to their distance from public facility hubs, fishing communities along the coast are usually the last to gain from economic development. There are critical food shortages, particularly after a cyclone; erratic production exhausts fishers' savings and entraps them into debt. Sometimes, in order to receive assistance such as cash or ration cards for food, they would even have to bribe the local government official concerned.

5.4.3 Access to Land and Infrastructure

Most of the surveyed households (more than 90% in the case of Mothurapur) are landless. They live on *khas* (government owned) land or on land owned privately by someone else. Thus, they are cut off from enjoying the insurance that ownership of land (which Sen (1981) would call an "exchange entitlement") can offer against a sudden loss of livelihood resources. The few, who do possess land, only have a marginal quantity, which would not allow them to use it for generating additional income, only for family settlement. Poor infrastructure, remoteness, and poor transport facilities are inhibiting factors for access to development and mobilization activities. Another entitlement deficiency is the poor transportation system that inhibits fishers from easy and expedient access to the markets. During the rush period of peak fishing, it is not feasible for fishers to sell their catch in the market directly (as it involves further time and labor from already tired fishers); so instead, they sell the fish at coastal landing sites. Notably, poor transportation facilities pave the way for the buyer, or the *dadondar,* to gain bargaining power over them.

5.4.4 Gender Perspective of Fishers' Poverty

The burden of poverty tends to bias toward women, especially female-headed households. Women from poor households and female-headed households are forced to involve in income-generating activities outside the home for their livelihoods. Almost all the interviewed fisherwomen in Mothurapur village were found to sell their labor in fishing activities. About one-fifth of the women did or do work outside fisheries where they often face the hardest conditions, such as hard physical work, e.g., digging ditches and maintenance of roads. Female-headed households tend to have a higher frequency of food insecurity, and their livelihoods and coping capacities are constrained by low education, poor skills, and low earning ability. Thus they usually earn less than their male counterparts. However, women participate in the fishery sector as processors, packers, and vendors. They are, in many instances, the financial mainstay of the fisher household. Although women in fishing households increasingly become active in income generation, they still face discrimination on the job. Although women claim that they do the same laborious job that men do, in the Sundarbans area, they are usually paid 70% of the wage from what men earn.

Another burden on the household is the substantial dowry required when a daughter marries.[3] The dowry exists from ancient times as a well-known matrimonial custom in traditional Hindu communities, which also includes Hindu fishing communities (Rahman et al. 2002). It is now also widespread in other communities. More than 90% of the households interviewed in three areas have said that they have to pay dowry either in cash money or in-kind. Most households consider dowry as something that has entrapped them into debt, as most of them had to take a loan either from an NGO or from a money lender.

5.4.5 *Political Processes, Local Institutions, and Social Networks*

Fisher households are less involved in political processes and local institutions, for instance, the Union Council.[4] Elected local government officials of the Union Council are perceived to be blind to the problem of fishing communities, just as they are to the welfare concerns of the fishing community in general. One fisher from the Chittagong district complained about a local government representative:

> They just come to us when they need our votes for election. After being elected, they just forget about us. Even when the police harass us in front of our commissioner (i.e. the local administration official), he doesn't protect us. We are unable to elect our own representative as we do not have power and money. We do not even have a suitable candidate from our own community because we are all illiterate.

In general, the small-scale fishers of the study areas are, for the most part, unorganized. The existing traditional organization (*samaj*) is not effective enough to promote their interests, and they have poor representation in the local administration of the Union Council. Well-off fishing households are often found to have strong social ties to rich people and influential persons involved in politics, while the poor have very few such networks. However, having such social networks is important for drawing benefits from state and private resources, and making better use of local opportunities (Rahman et al. 2009).

---

[3]*Dower* (paid by the husband to his wife) is an essential part of Muslim marriage as practiced in Bangladesh. However, dowry is prohibited by state law. The problem of dowry is now widespread. The practice of dowry increases the vulnerability of women in Bangladesh, turning them into liabilities for the families (Chowdhury 2010).

[4]The Union Council is the lowest layer of local government administration where the Chairman and members of the council are elected by public vote.

In the three communities, therefore, there is a clear correlation between possession of financial, physical, and social capital (see also Rahman et al. 2009). Social capital, which includes relationships with local level institutions such as the Union Council is important as a coping asset, as material and non-material benefits can be derived from these institutions. However, the poorer the households, the less connected they are to such networks, and the fewer assets they have to deal with their problematic livelihood situations.

5.4.6 *Human Capital and Capabilities*

A number of additional entitlement shortages limit the opportunities of poor small-scale fishers, and increase their vulnerability. Sen (2002) observes that, "health is one of the most important conditions of human life and a critically significant constituent of human capabilities which we have reason to value" (p. 660). In the study areas, frequent bouts of illness often impair fishers' capability to work. The cause of the most prevalent illnesses can usually be linked to the lack of knowledge about, and access to, proper sanitation. Though sanitation facilities are reaching around the countryside due to government and NGO assistance, fishing villages are among the few communities where most households still lack water-sealed sanitary latrines. Illness, especially among the earning members, is one of the major causes of families getting pushed into poverty. It often leads to family bankruptcy, which may even bring additional health problems because the cost of medicine is often paid by reducing the frequency of meals or, in extreme cases, with the starvation of family members. As expressed by a key informant in Chittagong:

> Fishers spend a large portion of their earnings buying medicine. It is unlikely that all the members of a family are in sound health at the same time. They are uneducated about general sanitation rules, and for that they suffer.

Fishers are considered to be more prone to alcoholism than other people in Bangladesh. They spend a lot of what they earn during the peak fishing season on alcohol. A number of key informants also blamed fishers for lacking the habit of saving. As a local proverb says, "*Jailla, dangai utle khhailla* - fishers earn money in the sea. When they come to land, they come empty handed."

As in many other rural communities in Bangladesh, the rate of illiteracy in fishing communities is very high. At the maximum, the children of fishing households go to school at least once in their life, but the drop-out rate is about 80% before completing 5 years of primary schooling. The present survey reveals that the literacy rate at 7+ years of age is around 30%, which is below the national average. Early drop-out of school-aged children from fishing households is common. One reason is that fishing is very labor intensive, and fishers cannot afford to take manpower from outside the household to allow the children to attend school. Seasonal and uneven income from fishing is not conducive to bear the regular costs for education. Failure to pay school fees regularly makes the students ashamed and many leave school.

5.4.7 *Market, Money Lender, and Dadon System*

There is the potential to avert poverty in small-scale fisheries, as fish is a highly priced product in comparison to other agricultural products in Bangladesh. However, a good catch does not ensure a good price of fish products for fishers. Hilsha fish enjoy lucrative domestic and foreign market prices, but their distribution disfavors small-scale fishers. Ali (2010), in a recent study, found that there are eight layers of middlemen from the fishers to the consumer table in the Hilsha marketing channel, and that fishers only get 1.5% of the final consumer price. However, in absence of a sound marketing system in general and effective state monitoring, this type of an exploitative fish marketing chain seems to persist.

Fishers' access to the formal credit markets (i.e., banks) is very limited due to lack of, or insufficient collateral assets like landed property. Therefore, they are dependent on informal credit mechanisms, like the *dadon* system. This informal arrangement is often blamed for exploiting the fishers. *Dadon* is a transaction built upon a verbal contract between the fisher and the money lender (called *dadondar)* - whereby the lender requires that the fisher sells the fish to him; or he gets a certain commission when fish is sold to a third person. Thus, the *dadon* system binds the fishers to the money lender in a debt cycle. Regardless of the amount of money owed, the borrowers must give all the fish they catch to the *dadondar* who determines the price of the produce (or a commission that ranges between 5% and 10% of sales revenue) (Habib 2001; Kleih et al. 2003).

One fisher from Chittagong complained during an interview that if he tries to bargain, the *dadondar* punishes him by reducing the previous bid. The *dadondar* usually allows at least three fishing seasons for the loan defaulter to repay loans. If there is a failure, he may confiscate the fisher's productive assets like boats, nets, home, or homestead land. Thus, the *dadondar* sometimes becomes the *de facto* owner of the family's productive assets and fish catches. A fisher interviewed by Alam (1996) said: "*Dadon* means selling everything to the Paiker (a word for money lender). It is not only your fish but also your freedom, boats, nets etc. But we have no way of being free from it" (p. 109). The exploitation within the *dadon* system occurs mainly within the Hilsha fishery; which requires huge investment to prepare nets, and boats suitable for fishing. Most fishers cannot afford such expenditures; so, therefore, they have to go to money lenders.

5.4.8 *Migration and Alternative Livelihoods*

Along the Chittagong coast, migrant fishers come mainly from nearby Noakhali or the Bhola and Barisal districts during the Hilsha fishing season. Most migrant fishers are either seasonal fishers, or work in agriculture in their place of origin. They migrate to the Chittagong coast during the rainy season when there are no agricultural activities where they live. Lack of productive assets is one of the factors pushing fishers (e.g., fishers of Thakurtala) to migrate elsewhere (e.g., to Chittagong) along the coast. In Chittagong, they are usually hired by local elites or *dadondars*. Migrant fishers go deep sea fishing for Hilsha, whereas local fishers usually catch Hilsha nearer to shore. Local boat owners also prefer to hire migrant fishers whom they perceive to be more experienced and skillful in deep sea fishing than locals.

This specialization of fishing between migrants and locals helps to avoid conflict between the groups, which may otherwise easily escalate. Most migrant fishers, who also work in agriculture, are found better off than the local fishers who rely only on fishing. Therefore, migrants are not always welcomed by local fishers, especially when migrant and local fishers are fishing in the same fishing ground. Still, as one migrant fisher in Selimpur said: "Migration is our right because the sea is for all and it is not the private property." Part of the problem is that locals see that migrant fishers contribute to their rent dissipation. Another issue is that they weaken their bargaining power with the *dadondars*. Furthermore, one local fisher from the Chittagong district complained that, "migrant fishers buy accessories that we need for boats and nets, and thus the price increases. Then we have to buy with extra price."

As a hereditary profession and habit, many fishers do not want to leave fishing. Some even think that it is their "holy duty" to feed mankind with fish. Fishing is not only a means for maintaining livelihoods. It is an interesting, challenging, and independent profession. However, for those who want to change and pursue alternative livelihoods, there are many obstacles. Fishing skills are not easily converted into other professions, and living in remote communities means little opportunity for developing other skills. The few existing alternate job opportunities (e.g., day labor, van puller) return less income than fishing. Habituated to large income during the peak season, fishers do not intend to work for less money, even during the lean season.

Summing up, vulnerability is the central theme when studying poverty and livelihoods in small-scale fisheries of Bangladesh. Vulnerability and poverty are closely but not the same. Natural disasters are frequent and hard to escape. When their assets are destroyed, small-scale fishers have to rebuild their lives. Most small-scale fishers also depend on the Hilsha species, which means risk of livelihood failure in the case of stock failure. The lack of alternative skills keeps fishers trapped in resource dependency. This can lead them to increase the pressure on the marine ecosystem, often by using destructive gear or targeting undersized species (to be further described in the following section). Fishers' access to the formal credit market (i.e., banks) is very limited due to lack of collateral assets like landed property; therefore, they are dependent on informal credit mechanisms, like the *dadon* system. The poverty averting potential that high market prices of fish products involve tends to be eroded, as middlemen control price setting. Small-scale fishers' vulnerability is also exacerbated by social mechanisms, such as underperforming institutions, and lack of the security that ownership to land provides. Without a functioning health care and welfare system, the death or illness of a family member may be disastrous. In many instances, fishers' livelihoods are also exposed to maltreatment by middlemen or government officials who often take advantage of them. Thus, all in all, small-scale fishers lack the "protective security" which Amartya Sen considers a basic instrumental entitlement (Sen 1999, p. 184–185).

Table 5.1 Factors that contributed to fishers' vulnerability, and the coping strategies to redress vulnerability

Factors contributing to vulnerability	Coping strategies to redress vulnerability
• Dependency on single species (e.g., Hilsha, shrimp or prawn larvae) • Seasonality and fluctuation of natural resources • Extreme weather conditions e.g. cyclones, and coastal storms • Human-tiger conflicts • Piracy and other unlawful activities • Landlessness and settlement in areas exposed to coastal disasters • Death of bread-winners and female-headed households • Dowry • Lack of access in political processes and local institutions • Market vagaries and fishers' limited capacity to bargain • Lack of access to the formal credit system	• Fishing undersized species and using illegal fishing practice • Family bonding – by sharing responsibilities, and jointly contributing to the family income • Social networking – borrowing money and taking help from other fishers and relatives • Women's role in income generation • Patron–client relationship with money lender, extortion and protection money to pirates • Alternative activities outside of fisheries • Migration and remittance from abroad

The factors that contribute to the vulnerability of fishers are summarized in Table 5.1. Table 5.1 also contains a summary of the coping strategies that fishers employ to redress vulnerability. These are described in detail in the following section.

5.5 Coping Strategies to Redress Vulnerability

Poor fishers cannot afford to be without income, or sit idle for long periods of time. Hence, they develop ways of coping with the changing conditions and vulnerabilities as they experience. The following are the coping strategies adopted by fishers of the study areas in order to enhance their livelihoods: (1) Enhancing resource exploitation; (2) Family bonding; (3) Social networking; (4) Women's contributions; (5) Patron-client relationships; (6) Job diversification; (7) Migration and remittance. Each of these coping strategies are addressed in more detail below.

5.5.1 Enhancing Resource Exploitation

In order to address the series of vulnerabilities and liabilities that fishers experience and the poverty that they face, they ultimately cope in a way they are capable of, that is by putting more pressure on the common pool of marine resources. Some of these

Fig. 5.2 Set Bag Nets (*SBN*) are extensively use in the rivers and canals of Sundarbans Mangrove Forest. This fine-meshed net is used to collect post larvae (*PL*) of shrimp and prawn

coping strategies only exacerbate vulnerability by depleting the resource base further. There are several signs that this is happening. For instance, the small meshed set bag net widely used by small-scale fishers along the coast of Bangladesh has been blamed for catching undersized fish. Fishers generally do not follow the restriction on mesh size of the cod end of ESBN imposed by law; thus mesh size of the cod end is getting smaller to capture more undersized fish. Many species (including Hilsha) are also targeted during the breeding season. Illegal fishing (e.g., catching of *Jatka* – the juvenile species of Hilsha), and the use of illegal gears (e.g., monofilament fishing nets) are also widespread. According to Khaled (2010), every year about 463 million *Jatka* (3,707 tons in weight) are caught using different types of nets including monofilament nets. By saving 20–30% of *Jatka,* an additional 0.10 million to 0.15 million tons of Hilsha production would be possible.

Fry collection (of very juvenile prawn and shrimp species) is the main occupation of women and young girls of Mothurapur, despite a ban on collection of wild fry. From a literature review, Ahmed and Troell (2010) find that, compared to any other fishery, it is assumed that prawn and shrimp fry collection have the highest by-catch rate (Fig. 5.2) and each year more than 98 thousand million juvenile fish and crustaceans are lost.

All in all, these coping strategies have detrimental impact on the marine ecosystem, and the resources that small-scale fishers depend on for their livelihoods. But fishers are also in a desperate situation where they need to survive and pay their bills on a daily basis.

5.5.2 *Family Bonding*

Family bonds are an important aspect of livelihood security. Therefore, fishers' households first cope with vulnerability through family cooperation. Reducing food consumption or simply skipping meals are the immediate strategy when there is a crisis. Especially women prefer to eat less rather than selling productive assets, which is a second strategy. In most cases, the fishing crew is from the same family (e.g., father and sons or brothers), as recruitment from outside the family involves payment, even if there is no catch. When the father gets too old to fish, the son usually takes over the skipper position. The father retreats to some easy job on land like net making or mending. In most cases, sons take care of parents when they grow old. Children also help out. Due to the high risk, fishers in the Sundarbans need to bring a crew of more than one person. Usually, children accompany their parents during fishing in the forest. Along the south-eastern coast, children do not go fishing in the sea. They do individual fishing along the coast to catch shrimp fry, or work in the fish landing center. Many children collect fish either by asking fishers, or they collect fish that are discarded, which allows them to earn around 50 cents to $1 per day.

5.5.3 Social Networking

Existing informal social networks and solidarity among fishing people are strengthened by working together. Fishers always live together in fishing helmets locally termed *samaj*. The *sarder* is the village head, whose role is to help settle disputes among fishers. He always tries to solve problems within the group, before going to the police. The latter costs money and often involves harassments. The fishers who live together usually also fish together in the same fishing zone. Partly, this is done for safety reasons. Most women replied that they first seek a loan from those who live next door in the form of daily necessities like rice, salt etc. The dowry needed for a girl's marriage from a poor family is often raised with the help of relatives. In the case of a widow or a woman-headed household involved in fish marketing, she can get fish on credit from a kin fisher and pay back after the fish is sold; thus she may not need to invest cash.

5.5.4 Women's Contributions

Fisherwomen also have some reputation for preserving the savings culture more than fishermen. They often save money in secret as insurance against misfortune. The women's supplemental income is gained through, for instance, rooftop or yard gardening. Some have jobs in marketing of fish or fish drying, whereas others provide funds for their family by net making and mending. In the Sundarbans area, almost all women (and most young girls) in fishing families are active in income generating activities, mainly shrimp and prawn fry collection. Thus, women have, in some instances, become the financial mainstay of the fisher households. Many young girls from fishing households in the Chittagong area work in readymade garment factories, or in the Export Processing Zone. Some women also raise poultry and other livestock as an investment. This comes into use to meet daily needs during the lean periods, or to overcome sudden shocks like illness that involve extra expenditures.

5.5.5 *Patron–Client Relationships*

The patron-client relationships that fishers have with *dadondar* (or money lenders) are generally perceived as exploiting the fishers. Still, it also makes a vital contribution to fisher's security. In the Chittagong district, the peak fishing period for Hilsha starts after 2 months of want and scarcity. During this period, fishers need money for preparations, net mending, the maintenance of boats and engines, as well as for subsistence (Habib 2001). During this period, *dadondar* supply fishers with the money they need to maintain their livelihoods. Money is provided on trust rather than by written document or for collateral property. Social security is also provided by *dadondar*. During any mishap, they help fishers, for instance, to get legal aid. For a fisher to have a relationship with the *dadondar,* means proper and timely marketing and payments for his products (even though the price is low). If a fisher has no permanent *dadondar*, his product may be targeted by several *dadondars* to buy, which may result in mishandling, improper, and uncertain payments (Alam 1996).

5.5.6 *Job Diversification*

Most fishers want to cling to their hereditary profession. Those who want to diversify their income-base usually allow a son or daughter to find alternative jobs and encourage their wife to do homestead income generating activities (such as gardening) or some job related to fishing like net mending or net preparations). Such job diversification by family members also helps to redress vulnerability, as loss of one member's income can be compensated by another's. About 70% of ascending households have at least one earning member who does not fish. It appears that occupational diversification, especially the capacity to switch from lower-productivity fishing activities to higher productivity non-fisheries activities, plays a crucial part in the process of escaping from poverty and vulnerability. Income from alternative jobs or income from household savings may serve as working capital for another family member.

5.5.7 *Migration and Remittance*

In some cases, like for fishing families in the Chittagong, remittances sent by family members from abroad are found to help families getting out of poverty. However, migration (for doing the same fishing job in other areas) is less common for traditional Hindu fishing communities of the study areas (except Moheshkhali). However, there are some cases in Selimpur where fishers also migrate for work outside of fishing, thus contributing with remittances to the family. However, fishers from other areas usually migrate to Chittagong during the Hilsha fishing season. In Moheshkhali, Hindu fishers who do not have their own boat and net are found to migrate to the Chittagong coast to work as hired fishers in the Hilsha fishery. For migrant fishers along the Chittagong coast, income from the Hilsha fishery is a means to cope with family expenses (e.g., earned money used to repay loans taken during crisis periods).

Although there is also some inward migration in coastal areas by poor people from other parts of the country who are in search of a job, outward migration by young members from fishing households in Selimpur into the city of Chittagong is very prevalent. This outward migration is largely facilitated by the close proximity (20 km in distance), alternative job opportunities in the city, as well as a willingness to leave because of dwindling fish resources. In the Sundarbans area, permanent outward migration is limited, as most fishers find it difficult to find alternative jobs in the city, or because they lack "migration capital" (e.g., costs related to travel, settlement in the city, subsistence cost for the family for the period of absence).

5.5.8 Summary of Coping Strategies

To sum up, small-scale fishers of Bangladesh develop a range of coping strategies to shield themselves from a host of vulnerabilities. Some of these coping strategies only exacerbate their vulnerability by depleting the resource base they depend on. This includes exploitation of fry and juvenile species with high by-catch, and use of destructive fishing gears. Some other coping strategies create buffers against crises. Fishers create a family network by sharing their responsibilities and joint contributions to family income; providing non-material or moral support during the periods of crises. Women are not passive beneficiaries of men's income in a family. Rather, they are active with diverse alternative income generating activities. In many cases, fishing alone is not sufficient to sustain livelihoods; so, fishers do ancillary or alternate jobs. Women are, however, more active in the search for alternative jobs than fishermen who tend to cling to their occupation. In the absence of "protective security" (Sen 1999), many fishers take refuge in money lenders when they face difficulties. Money lenders can also provide security during livelihood crises. Even though migration is considered as an ultimate coping strategy, outward permanent migration in the study areas seems not to be very pronounced, as fishers do not migrate permanently. Seasonal inward and outward migration occurs to reap the benefits of peak fisheries production of other areas.

5.6 Reducing Vulnerability: Breaking the Circle of Poverty

In a study of rural poverty in Bangladesh, Sen (2003) found that the households who escape poverty tend to accumulate natural, human, physical, and financial assets faster than non-ascending households. This, he attributes to their ability to diversify into other income and sustenance sources. They allocated more investment into non-agricultural activities on land. In general, they displayed strong non-agrarian orientations by engaging in alternative activities such as trade, service, migration (remittance), and non-agricultural labor.

In the present study, it was also observed that to avert poverty, small-scale fishers tend to better utilize the natural, human, physical, social, and financial capitals that they have access to (Table 5.2).

Table 5.2 Fishers' perceptions of the drivers of poverty

Assets	Factors averting household poverty	Reasons for deteriorating household income
Natural assets	• Good income from peak season fishing • Income supplement from other sources than fishing (e.g., forest resources) of nearby areas	• Low productivity of fisheries • Natural disasters
Human assets	• Family head skilled and industrious • Women's mobility for work • Development of skills through education • Family members work outside of fishing jobs • Small household size	• Low level of education or no formal education • Death of earning family members • Women without work • Family illness • Alcoholism among family head or other members • Big household's size • Lack of motivation • Fatalism
Financial assets	• Self insurance through savings • Frugality, endowment to land • Remittance	• Lack of financial assets (e.g., owningboats, nets) • Exploitation by *dadon* system • Erratic income, habit of lack of savings
Physical assets	• Income diversification through gardening, agriculture • Direct marketing facility of good time catch • Well connected by road	• Lack of direct fish marketing due to poor infrastructure • Unavailability and/or limited alternative activities
Social assets	• Good family network • Cooperation between relatives • Cooperation between communities	• Dowry problem during girl's marriage • Unlawful elements (e.g., piracy, extortion etc.)

5.6.1 Changes in Natural Assets

For most coastal communities, who depend on natural resources, changes in the natural capital particularly impinge on their vulnerability (Kleith et al. 2003). So productivity in the fishing grounds have direct bearing on the poverty status of fishing communities as primary sources of income. As fish and shrimp are always commodities of higher market price in comparison to most of the other agricultural products, availability of fish and getting appropriate price always holds potential for poverty alleviation in fishing communities. Good income from Hilsha fishing is a key factor for many families in Chittagong for bringing them out of poverty. In the Sundarbans area, those fishers who are involved in crab collection can earn a good income; fishers can also supplement their income from wood and non-wood forest resources. Natural disasters are found to be the most prominent factors that drive fishers to abject poverty by wiping out the livelihoods base.

5.6.2 Changes in Human Assets

According to Sen (2003), a declining demographic dependency ratio[5] had positive implications for rural income growth and poverty reduction. In the present study, a considerable number of fishers are aware and adopt a family planning program to curb family size. Despite the significant progress in curbing the growing number of members in poor fishing households, the dependency ratio is considerably higher compared to other groups or classes. In the present study, it was found that the burden of children and elderly were greater in poor fishing households than in more well-to-do families.

The present study also confirms the findings of Ellis (1998) that education and development of human skills are among the factors that help people to cope with social and ecological change. It also helps people to find alternate livelihood options more easily. As Ellis argues: "Since poverty is closely associated with low levels of education and lack of skills, education is also a key factor contributing to the greater ability of better off families to diversify compared to poorer families" (Ellis 1998, p. 27).

As noted in Table 5.2, human capital also plays an important role in the transition of ascending fishing households. Fishers escaping poverty have at least primary education of a few years, or some other family members have it. There is often the opportunity to take part in awareness raising or extension programs by the government or by an NGO. An educated skilled family member can serve as a buffer against vulnerability by qualifying for a job outside of fishing. In the Chittagong district, women work in the garment industry, which helps the family to raise their income. This is particularly helpful during the lean fishing periods. In terms of human assets,

[5]The dependency ratio is calculated by dividing the total number of dependent members by the total number of earning members of each family (Rahman et al. 2002).

descending or chronically-poor fishing households suffer different levels of illiteracy, death of earning members, and widowhood, leaving them without viable sources of income. Family illness and alcoholism lead to further slippage into poverty.

Individual attributes make a crucial difference in poverty levels in fishing households in the same fishing communities. The level of poverty may differ among fishers working on the same boat, like among crew members or more significantly between a crew and skipper. The crew who becomes a skipper can, for the most part, change their fate and get out of poverty; whereas the other crew usually remain poor. But moving from crew to skipper is rather demanding for most of them. One key informant from Chittagong said:

> Those crews who will become skipper in the future depend on entrepreneurial skills, hard labour, and high aspirations.

5.6.3 Changes in Financial Assets

To most of the fishers, lack of financial assets, for instance, cash money is synonymous with poverty. For fishers who have very few collateral assets, other financial assets serve as informal insurance against sudden shocks or risk. Similarly, Sen (2003) found that households of rural Bangladesh who ascended from poverty had higher access to institutional credit than both the chronic poor and the descending households. His finding suggests that access to financial capital is an important element in the process of moving out of poverty. The poor fishers mainly lack capital needed for investment in productive assets, e.g., boats or nets, as their income always falls short of meeting running costs and investment needs of their fishing operations, as well as of their daily household consumption. They do not have access to formal institutional credit, and have to rely on the informal *dadon* system which binds them in a relationship of dependency and poverty, which they have problems to free themselves from. It is evident that households who have some agricultural land ("investment capital") are better diversifiers than landless fishing households.

The few fishing households who have both homestead and agricultural land are mostly found to graduate from poverty. Moreover, to address the seasonality of fishing and the vulnerability of livelihoods, ascending households were able to invest in various fishing gears, which allowed them to spread the risk. They tended to have several types of nets for different fishing seasons and fish species, and this makes them able to switch target species easily, thus utilizing the full productivity of the productive assets. Moreover, once one fishing gear is lost or destroyed, they can sustain their livelihoods by utilizing another one.

5.6.4 Access to Social Network, Market and Information

Many fishers feel that existing informal social network and solidarity strengthened by working together are the most instrumental for livelihoods and can act as a buffer against vulnerability. Those fishers who go fishing jointly with other fishers (e.g., shore seining in the case of the Sundarbans) are found to be in good condition. Fishing together with the same fishing gear and craft helps to improve welfare conditions by risk spreading, sharing knowledge and skills for income generation. Fishers tend to help any new entrepreneur in the marketing channel by selling fish on loan. They help new entrepreneurs, because an inclusion of a new person in the marketing channel will increase competition among the middlemen; thus fishers may get better prices in a competitive market. Furthermore, close proximity to market and other ancillary services such as fish landing and processing facilities, tools and facilities allow fishers to bargain for proper price of their products.

There are sufficient anecdotal and replicated examples that demonstrate that access to information is one of the factors that can help to improve incomes and help to empower the poor (Greenberg 2005). The present study also provides the evidence that use of mobile phones may inform of rich fishing grounds and markets. For instance, news of a bumper catch in certain areas can be relayed to other fishers to come and reap the fish available.

5.7 Conclusions

The meaning of poverty may be instinctively clear, but its measurement is complex (Adams et al. 2004). Monetary income is not the only significant variable that explains economic and social deprivation in a fishing community. Hence, it is vital to develop a framework to guide poverty research to understand both the complexity and drivers of poverty that will help in developing the adequate policy response at scales appropriate to the context-specificity of many drivers of both poverty and resources degradation, as poverty is a changing phenomenon (Thorpe et al. 2007).

An FAO (2007) policy document claims that both sustainable fisheries and goals related to poverty reduction can be more readily achieved by reducing fishers' vulnerability and strengthening their basic human rights. This claim is also consistent with the findings presented in this chapter. In my study of small-scale fishing in three coastal communities in Bangladesh, people are indeed vulnerable; and several conditioning variables are identified as affecting how people are able to deal with it. The policies needed to reduce poverty in small-scale fisheries must also address these factors. Different livelihood crises and natural hazards put small-scale fishers in Bangladesh on the margin of destitution, where any further shocks or crises can lead to livelihood failure. Thus, for sustainable livelihoods of poor fishers of Bangladesh, it is urgent to create buffers against vulnerability and crises. A resource management system that works to sustain the resource base is essential, but is not the only necessary remedy.

One mechanism for building such buffers is the development of human capital through investing in education. In the present study, fishers' literacy rate lags behind the national average. By providing global scale evidence, Psacharopoulos (1994) shows that education has a strong positive return. He observes that primary education will continue as the number one investment priority in developing countries, and that investment in women's education is, in general, more profitable than that for men. Thus, he concludes that, "investment in education continues to be a very attractive investment opportunity in the world today - both from the private and the social point of view" (p. 1325).

This evidence is clearly supported by the empirical findings of the present study. Fishers who have some basic education prove to be more resilient in times of crisis by having more capability to avail alternative livelihood options. Exploring the benefits of education, Bebbington (1999, p. 2034) thus argues that:

> The individual's ability to read and write not only enhances likelihood to secure better jobs and more efficient performance - it also enhances his or her ability to engage in discussion, to debate, to negotiate, to add their voice to the multitude of voices influencing households, local and national discourses on development and other issues.

In the same way, fishers with basic education can improve their position in marketing channels or as a skipper, where most of the profits from fishing activities accrue.

The same is the situation if family members like daughters go to primary school. Women who receive primary education will more easily find a job outside the household (like in a garment factory) than women without education. Sen (1999) also explores the crucial role of female education, female employment, and female ownership rights for the economic fortune, well-being, and freedoms of family members. Thus, poverty alleviation strategies and policies for fisheries communities should not only target the fisher, but aim at uplifting the women and the contributions they make.

Poor fishers' lack of capital for initial investments in acquiring any productive assets is another cause of vulnerability. Most small-scale fishers of Bangladesh live on the margin of survival; rarely do they have any money to save after meeting consumption costs. The existing informal *dadon* credit system severely limits the economic freedom of fishers by binding them into long-term exploitative debt bondage. Thus, access to the formal credit systems will create economic opportunities for them, and is therefore something development policies should aim for.

In Bangladesh, as in most other settings, social bonding is traditionally embedded in the local community. The effective utilization of this social capital can play an important role in reducing vulnerability. As Snowden (2005) argues, social capital reduces community distress. However, the existing *dadon* system in fishing communities has shifted the traditional kinship arrangement into one of patronage, and has thus shrunk considerably the scope of social capital. The patterns of traditional social bonding can only, to some extent, help people to survive in times of crisis; and are not effective in lifting them out of poverty. Most fishing communities suffer from lack of organization; they have no cooperative organizations. Such organizations may provide an effective buffer against crisis, as organizations serve as cushions to shocks and stress, and would therefore be an obvious thing to create in order to reduce vulnerability (see Amarasinghe and Bavinck, Chap. 17 and Salas et al., Chap. 10).

Good communication in terms of information and social networks, and access to markets not only help to ensure a proper price of products, but also determines access to education, information regarding the labor market and other social services. Building of human capital through access to health service and training or extension services also has a very positive impact on livelihood security. However, there is no single route that is sufficient to achieve such development goals. In a similar vein, large stocks of one single asset are also of little use, as it is rather the right mix of different assets that creates an effective buffer against crises (Kasperson et al. 2010). Given the multiple dimensions of poverty and vulnerability, a combination of different strategies as those mentioned here is essential, and should therefore be supported also by outside interventions.

Acknowledgments The author wishes to thank Svein Jentoft, Michael Flitner, Maarten Bavinck and Maria-Victoria Gunnarsdottir for helpful comments and critical review on earlier drafts of this article. Thanks also to the Norwegian Research Council for funding the PovFish project, and GLOMAR for funding this research work.

References

Adams WM, Aveling R, Brockington D, Dickson B, Elliott J, Hutton J, Roe D, Vira B, Wolmer W et al (2004) Biodiversity conservation and the eradication of poverty. Science 306:1146–1149

Agarwal B (1990) Social security and the family: coping with seasonality and calamity in rural India. J Peasant Stud 17(3):341–412

Ahmed N, Troell M (2010) Fishing for prawn larvae in Bangladesh: an important coastal livelihood causing negative effects on the environment. Ambio 39:20–29

Alam QK (1996) Two fishing villages of Bangladesh: a community study. Ph.D. thesis, Aalborg University, Aalborg

Alam MF, Thomson KJ (2001) Current constraints and future possibilities for Bangladesh fisheries. Food Policy 26:297–313

Ali SMZ (2010) Supply chain and prices of Hilsa fish in Bangladesh. Bangladesh Institute of Development Studies (BIDS), Dhaka. http://www.bdnews24.com/details.php?id=155054&cid=2&aoth=1. Accessed 4 Mar 2010

Bangladesh Economic Review (2008) Agriculture. Finance Division, Ministry of Finance, Government of the People's Republic of Bangladesh, chap 7

BBS (2007) Statistical yearbook of Bangladesh. Bangladesh Bureau of Statistics, Statistics Division, Ministry of Planning, Government of the Peoples Republic of Bangladesh, Dhaka

Bebbington A (1999) Capitals and capabilities: a framework for analyzing peasant viability, rural livelihoods and poverty. World Dev 27(12):2021–2044

Beck T, Nesmith C (2001) Building poor people's capacities: the case of common property resources in India and West Africa. World Dev 29(2):119–133

Béné C (2003) When fishery rhymes with poverty: a first step beyond the old paradigm on poverty in small-scale fisheries. World Dev 31(6):949–975

Béné C (2006) Small-scale fisheries: assessing their contribution to rural livelihoods in developing countries. FAO Fisheries Circular No. 1008. FAO, Rome

Béné C (2009) Are fishers poor or vulnerable? Assessing economic vulnerability in small-scale fishing communities. J Dev Stud 45(6):911–933

Bohle H-G, Downing TE, Watts MJ (1994) Climate change and social vulnerability: toward a sociology and geography of food insecurity. Glob Environ Change 4(1):37–48

Bosher L (2007) A case of inappropriately targeted vulnerability reduction initiatives in Andhra Pradesh, India? Int J Soc Econ 34(10):754–771

Bosher L, Penning-Rowsell E, Tapsell S (2007) Resource accessibility and vulnerability in Andhra Pradesh: caste and non-caste influences. Dev Change 38(4):615–640

Chambers R (2006) Vulnerability, coping and policy (Editorial Introduction). IDS Bull 37(4):33–40

Chowdhury FD (2010) Dowry, women, and law in Bangladesh. Int J Law Policy Family 24(2):98–221

Colavito L (2002) Wetland economic valuation using a bio-economic model: the case of Hail Haor, Bangladesh. Paper presented at a workshop on conservation and sustainable use of wetlands: learning from the world. IUCN-The World Conservation Union, Kathmandu

Ellis F (1998) Household strategies and rural livelihood diversification. J Dev Stud 35(1):1–38

FAO (2007) Social issues in small-scale fisheries. COFI/2007/6, Committee on Fisheries Twenty-Seventh Session, Rome, 5–9 Mar 2007. ftp://ftp.fao.org/

Greenberg A (2005) ICTs for poverty alleviation: basic tool and enabling sector. ICT for development secretariat, Department for Infrastructure and Economic Cooperation Sida. http://www.eldis.org/fulltext/sidaictpoverty.pdf. Accessed 22 April 2008

Habib A (2001) Delipara: an obscure fishing village of Bangladesh. CODEC, Chittagong, Bangladesh

Hye HA (1996) Below the line: rural poverty in Bangladesh. University Press Limited, Dhaka

Kabir SMH (2006) Hilsha. In: Islam S (ed.) Banglapedia: national encyclopaedia of Bangladesh. Asiatic Society of Bangladesh, Dhaka. http://www.banglapedia.org/httpdocs/HT/H_0123.HTM. Accessed 21 Mar 2010

Kasperson JX, Kasperson RE, Turner BL II (2010) Vulnerability of coupled human-ecological systems to global environmental change. In: Rosa EA, Diekmann A, Dietz T, Jaeger C (eds.) Human footprints on the global environment threats to sustainability. MIT Press, Cambridge, pp 231–294

Khaled MH (2010) NGOs-stakeholders collaboration needed to raise Hilsa production. http://www.thefinancialexpress-bd.com/more.php?news_id=96762. Accessed 24 Apr 2010

Kleith U, Alam K, Dastidar R, Dutta U, Oudwater N, Ward A (2003) Livelihoods in coastal fishing communities and the marine fish marketing system of Bangladesh. Natural Resources Institute (NRI) Report No 2712. University of Greenwich, London

Narayan D, Chambers R, Shah MK, Petesch P (2000) Voices of the poor: crying out for change. Oxford University Press, New York

Psacharopoulos G (1994) Returns to investment in education: a global update. World Dev 22(9):1325–1343

Rahman MM, Haque MM, Akhteruzzaman M, Khan S (2002) Socio-economic features of a traditional fishing community beside the Old Brahmaputra River, Mymensingh, Bangladesh. Asian Fish Sci 15(4):371–386

Rahman MM, Chowdhury ZA, Sada MNU (2003) Coastal resources management, policy and planning in Bangladesh. In: Silvestre G, Garces L , Stobutzki I, Ahmed M, Valmonte-Santos RA, Luna C, Lachica-Aliño L, Munro P, Christensen V, Pauly D (eds.) Assessment, management and future directions for coastal fisheries in Asian countries. WorldFish Center Conference Proceeding 67, pp 689–756

Rahman PMM, Matsui N, Ikemoto Y (2009) The chronically poor in rural Bangladesh: livelihoods constraints and capabilities. Routledge, Oxon

Sen A (1981) Poverty and famines: an essay on entitlement and deprivation. Oxford University Press, Oxford

Sen A (1999) Development as freedom. Oxford University Press, Oxford

Sen A (2002) Why health equity? Health Econ 11:659–666

Sen B (2003) Drivers of escape and descent: changing household fortunes in rural Bangladesh. World Dev 31(3):513–534

Snowden LR (2005) Racial, cultural and ethnic disparities in health and mental health: toward theory and research at community levels. Am J Community Psychol 35(1/2):1–8

Thorpe A, Andrew NL, Allison EH (2007) Fisheries and poverty reduction. CAB Rev Perspect Agric Vet Sci Nutr Nat Resour 2(085):1–12

Yamin F, Rahman A, Huq S (2005) Vulnerability, adaptation, and climate disasters: a conceptual overview. IDS Bull 36(4):1–14

Chapter 6
Occupation of Last Resort? Small-Scale Fishing in Lake Victoria, Tanzania

Paul O. Onyango

Abstract Small-scale fisheries have been conceptualized as a "safety valve" – the last reliable livelihood when no other exists for fishers, who are considered poor. This perception appears to be the grounds upon which poverty alleviation and resource management policies are defined. This chapter looks at this notion and questions whether small-scale fisheries are really an "occupation of last resort." Based on an ethnographic study on a Lake Victoria fishing community in Tanzania, data indicate that regardless of their poverty status, small-scale fisheries are perceived as offering a rich way of life that fishers join by choice. By discussing what fishers consider as the underlying issues in their choices, this chapter argues that fisheries management (in technical terms) should shift to governance that supports opportunities and processes for fishers to pursue the kind of life they want, and create an environment in which they can pursue that life, respectively. Such a shift would also benefit from a set of management-relevant social variables and indicators that focus on peoples' judgments of their well-being, capabilities, and satisfaction to aim toward sustainable fisheries management and poverty reduction. The chapter therefore emphasizes that if managers and policy makers/governors do not understand the full meaning and satisfaction that small-scale fishers attach to their occupation, policies instituted to curb overfishing risk not only misfire but also backfire.

6.1 Introduction

"There is considerable evidence that coastal fishing is an economic activity of last resort" (Panayotou 1982, p. 30). Fisheries are seen as the only viable livelihood option that poor people have, and not something they would otherwise prefer. Being

P.O. Onyango (✉)
Norwegian College of Fishery Science,
University of Tromsø, 9037 Tromsø, Norway
e-mail: paul.onyango@uit.no; onyango_paul@udsm.ac.tz

S. Jentoft and A. Eide (eds.), *Poverty Mosaics: Realities and Prospects in Small-Scale Fisheries*, DOI 10.1007/978-94-007-1582-0_6,

an occupation of last resort implies that fishers are compelled to enter the fisheries, and adopt practices that enable them to earn their livelihood. They do not enjoy any bit of being in the fisheries but are mostly in it for survival.

This is a perception highly rooted in the debate on poverty in fisheries, especially small-scale fisheries, as well as the method of management of these fisheries (Bailey and Jentoft 1990; FAO 2000; Payne 2000; Smith et al. 2005). The whole idea here is that the open access, common pool nature of fisheries allows an increasing number of fishers to enter and overexploit the fish resource, thereby impoverishing themselves even further (Hardin 1968; World Bank 1992; Jalal 1993). Consequently, from a management perspective, this therefore requires a response which checks on such practices so as to avoid the collapse of fish stocks.

In this chapter, I argue that perceiving fishers exclusively from a dimension of survivors misusing the resources fails to fully recognize the human dimension of fisheries. Fisheries are generally a human phenomenon in which the changing fish stocks in the water are simply, to a large extent, a reflection of what happens in the society (Jentoft 1999; McGoodwin 2001).

This chapter examines whether small-scale fisheries are really an occupation of last resort for those who are poor. It explores this by focusing on the Nyakasenge fishing community located in Lake Victoria, Tanzania. Fishers here are poor and no strangers to hardships, risks, and suffering. However, fishing is much more than the methods and types of fishing used, the species targeted, the natural fish habitat, and/or monetary gains (how much is being fished and how it is being fished). Deviating from the sustainable livelihood concept (Allison and Ellis 2001), I argue in this chapter that fishing in this community is also a way of life that yields satisfaction, provides identity, and is a window through which fishers see the world. This chapter aims to demonstrate that fishing is not only a means of ensuring livelihood but it is also a desirable, rewarding, and meaningful way for people to live their lives (way of life).

The chapter argues that fishing is a valuable occupation to fishers (as an *opportunity*). It also points out that fishing provides an environment in which fishers choose to live the way they prefer (a condition best described as a *process*). The meaning and value they attribute to their occupation have great significance for understanding the implications of changes in the fisheries, including those that occur as a consequence of resource management measures. Therefore, any fisheries management and poverty reduction strategy commencing from the image of small-scale fishing as a last resort, and which focuses on effort reduction, risks violating peoples' perceptions of what constitutes a preferred life and the values they cherish (Carothers 2008; see also Kraan, Chap. 8). The chapter further proposes the need to develop management-relevant social variables and indicators focusing on people's judgments of their well-being, capabilities, and satisfaction that are essential to their understanding of fishing as their "way of life," so as to guide fisheries management to achieve sustainability and poverty reduction. The chapter begins with a discussion on the culture of small-scale fishing, followed by a brief background on the fisheries of the lake. The methods used are also briefly discussed before presenting fishing life and operations in this fishing village.

6.2 The Culture of Small-Scale Fishing

Fishing is among the world's oldest means of livelihood (Thompson et al. 1983; van Ginkel 2009). It was also among the ways in which life was lived in Africa. To a large extent, it defined migration patterns, manner of life, and above all the culture of those involved in it. Fishing also defined relationships both among and within community members, as well as shaping the world view of fishers (Thompson et al. 1983; Iliffe 2000).

In a seminal article, Acheson (1981) pointed out that despite the existence of anthropological studies of fishing communities, such studies have been seen to belong to archaeology, physical anthropology, or subareas of social anthropology. Consequently, what has been witnessed is a body of literature and sets of concepts focused on how fishers have adapted to earning a living from fisheries. This body of literature comprises studies that have highlighted the difficulties and risks fishers face in the natural environment (Graaf et al. 2006); the techniques through which fishers extract fish resources (Tzanatos et al. 2006); the uncertainty with which fishing markets and prices fluctuate; and the physical and psychological problems associated with fishers having to work away from home for long periods of time.

Fishing is seen as an activity and basically a means of support or subsistence (livelihood or occupation leading to a culture of fishing), and not so much as fishing peoples' behaviors habits, and customs that are typical of the way of life as lived in fishing communities – that is, their total way of life which is a fishing culture. These two perspectives are what have been referred to as "anthropology of fishing" (denoting the former – livelihoods or occupational perspective) and "maritime anthropology" (denoting the latter – total way of life perspective) (Andersen and Wadel 1972; Gatewood and McCay 1988). Acheson therefore argues that "those studying fishing societies have … made an important contribution by documenting the ways man has adapted to earning a living from a highly alien and dangerous environment" (Acheson 1981, p. 307). He further asserts, as can be implied from some later writers (Smith 1988; Hersoug 2004; Jentoft 2004; Johnsen et al. 2009; van Ginkel 2009), that the focus in managing fisheries has therefore been concentrated on how to spread risk awareness by use of norms, institutions, and networks.

The livelihood perspective has since then become very influential among those who write about small-scale fishers (Allison and Ellis 2001; Béné 2003; Neiland and Béné 2004). As livelihoods, fishing and fish resources are seen as assets, access to which is mediated by institutions, organizations, and social relations and is affected by trends and shocks that impact livelihood security and environmental stability. This idea of livelihoods originated from the need to understand different capabilities of rural households to withstand crises (Allison and Ellis 2001). The livelihood concept is thus well linked to concepts such as vulnerability, sustainability, resilience, social exclusion, and sensitivity, some of which are quite common in the context of poverty (Béné 2003; Du Toit 2004). It therefore appears that understanding of poverty in small-scale fisheries draws considerably from this perception of fishing as a livelihood (Béné 2003).

The livelihood argument has not only gained momentum among academics but has also shaped and driven development (Mathie and Cunningham 2003) and, by extension, management of fisheries in rural settings where small-scale fisheries are located. Macfadyen and Corcoran (2002) argue on how the livelihoods framework is pro-poor and people-centered. The framework therefore allows scientists and development agencies to see impoverished communities from the point of view of an observer or an *etic* (outsider) perspective. The framework enables objective identification of the livelihood platform, how access is modified, and the resulting outcomes. Consequently, this has resulted in a number of actors – researchers, donor agencies, government, and the media – to be focused on the deficiencies, problems, and needs of poor communities, such as small-scale fishers within a framework of problem-solving missions (Mathie and Cunningham 2003).

From a livelihood perspective, fishing is primarily an economic and subsistence activity (Schumann and Macinko 2007; Wartena 2006 – as cited in Kraan 2009). Smith et al. (2005) indicate that the livelihood notion as used in the context of resisting poverty by rural communities, such as those of small-scale fishers, largely depends on opportunities offered by the natural resource-based production system which is affected by the wider economic, institutional, and political environment. This economic dimension provides an exemplary explanation for perceiving fishing as an occupation, hence the argument of occupation of last resort (Cunningham 1993; Béné 2003; Smith et al. 2005). It has, however, been noted that non-economic considerations, in most cases, play an important role with regard to livelihood choices (World Bank 2000; Sievanen et al. 2005; Cinner et al. 2008; Wartena 2006 – as quoted in Kraan 2009). Bebbington (1999) also noted earlier that rural societies stuck to "nonviable" livelihood activities to maintain their cultural and social practices that accompany rural residence. Thus, small-scale fishing provides an avenue for people to practice and maintain their culture and social life. These observations, in a way, remind of what Sir Walter Scott observed way back in 1816: "It's no fish ye're buying – its men's lives" (Sir Walter Scott, *The Antiquary*, 1816 – cited in Thompson et al. 1983, p. 3). This, therefore, implies that reducing fishing to a livelihood narrows the understanding of it and may possibly explain, in part, the difficulty in managing this resource, as management would not only interfere with fishing activities but have a broader impact on people's lives as well.

6.2.1 *Fishing as a Way of Life*

Although distinguishing fishing as a livelihood, as opposed to a way of life, is quite challenging if not problematic, this chapter perceives the former to be part of the latter. There is much more to fishing than it being a career where income is the principle benefit (Gatewood and McCay 1988). Fishers would be reduced to *homo economicus*: rational and broadly self-interested actors,[1] who have the ability to make judgments regarding their subjectively defined ends (Kraan 2009). The alternative perception is that human behaviors such as fishing is a social activity embedded in human communities. As Seligman (1993) argues, humans are not always led by their logic but by their emotions, which are often irrational from a means–end perspective. Neither are humans necessarily self-interested. They perform altruistic acts like charity, volunteerism, lending a helping hand, parenting, and even in some situations giving one's life for a higher cause. They also perform self-destructive acts like substance abuse, negative addiction, negative risk-taking, procrastination, inability to complete projects, masochism, and suicide (Elliot 2007). Factors that lead to satisfaction/happiness/pleasure are not only confined to economic values, but include non-materialistic and non-economic values (van Ginkel 2009).

Fishing, therefore, needs to be perceived not only as a means people resort to in order to address their impoverishment in a materialistic sense, but as an activity that provides satisfaction and meaning to life (Bebbington 1999; Gudeman 2001). This calls for an understanding of fishing from a cultural perspective – as a way of life (Williams 1981), not just a strategy for material survival (Béné 2003, 2004). Although it can be argued differently, behaviors and actions that go well with cultural values of the community would need to generate favourable approval by community members. Such approval, in most cases, becomes a source of happiness and/or joy for the one being appraised, but also for the community as a whole. Then a fisher's behavior is turned into habits that he wants to be identified with. These habits become deeply embedded in a fishing culture to the extent that fishers see them as their way of life, and the only way life is worth living. As Berger and Luckmann (1971) argue, such values are constructed into realities in which societies develop procedures for maintenance.

Thus a people's culture is a means through which one can understand a community and its members' behaviors. Keesing (1981) puts it more clearly when he argues that culture is an ideational design for how people live and behave. It is their shared knowledge about language, history, myths, religious beliefs, world view, values, normative behavior, means of subsistence, and customary modes of social, economic, and religious organization (McGoodwin 2001). This chapter, therefore, perceives fishing culture in line with Geertz's (2006) perspective on culture, which to him constitutes webs of significance that man is suspended in and spins – the set of shared attitudes, values, goals, and practices that characterizes an institution, organization, or group. Culture is a whole way of life which defines, and is manifested in social and cultural activities such as language, styles of art, and thinking (Williams 1981).

---

[1]The issue of rationality has been questioned. However, it is implicit in the livelihoods framework, especially in assuming that rationality of choice could lead to prediction of actual actions – the livelihood strategy (Sen 2009).

Culture is learned and internalized by its members and then passed from one generation to the next (McGoodwin 2001; Kroeber 2006). McGoodwin (2001) argues that a collapse in fish stocks may have a symbolic importance in community traditions, mythology, religion, and cultural identity by leaving these components severely impoverished and incapable of being quickly revitalized. This does not rule out less dramatic events such as a change in the fisheries from subsistence to commercial which can also affect community members' self-esteem, patterns of household consumption, modes of kinship, and gender roles. Culture is also adaptive and able to change with changing circumstances. Given this possibility to adapt and/or change, culture should not be contrasted with modernization (McGoodwin 2001).

Therefore, as a way of life, fishing becomes interwoven throughout the fabric of community and culture, and is central to the individual and collective identity of fishers, expressed through popular myths, folktales, and local history (McGoodwin 1990; Gudeman 2001). Fishers become proud of their identity as fishers (Apostle et al. 1985; Gatewood and McCay 1988; Pollnac and Poggie 1979 – as cited in McGoodwin 1990). It is the case, as McGoodwin (1990, 2001) argues, that people living in one place share the same heritage and identity, similar lifestyles, and similar feelings about the world and how it ought to be (Layder 1994).

Fishing as an activity requires certain adaptations and behaviors which necessitate the development of certain cultural characteristics. Thus, a fisher community's fishing activities, the gears and methods they use, and the organization of fisheries' activities all result from trials undertaken over a long period of time.

6.3 Methodology

The study involved 8 months of fieldwork in the Nyakasenge fishing community. This was in addition to the contact with this community for the better part of over 10 years of research on Lake Victoria fishing communities. Data were collected through participant observations, structured and unstructured interviews, as well as focus group discussions (FGDs). In total, 30 FGDs and 60 interviews were carried out through the duration of the fieldwork. The data collected focused on fishers' reasons for joining the fisheries, their daily activities, the type and use of their fishing equipment, and their behavior.

The research team consisted of the author and two research assistants. The team lived in this community during the fieldwork period. The research team adopted a program of daily discussions, which focused on issues that emerged from the field. From these initial interviews, topics were identified for more targeted questioning on the following days. Issues were discussed to a saturation point. The team also discussed with the fishers some of the preliminary analyses of data collected before moving to other topics. Respondents were sampled based on the category of fishing activity to which they belonged, thus the following categories of fishers were interviewed: crew members fishing Nile perch (*Lates niloticus*), Tilapia (*Oreochromis niloticus*), Dagaa (*Rastrineobola argentea*), and/or Haplochromines. Additional

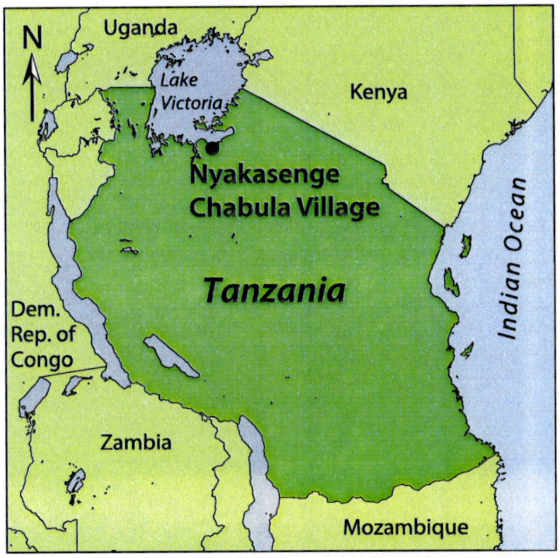

Fig. 6.1 Lake Victoria is the world's second largest freshwater lake, with a surface area of 68,800 km^2 and an average depth of 40 m. It is located between latitudes 00 2′N and 30 0′S and longitudes 310 41′E and 340 52′E, and it lies at an altitude of 1,134 m above sea level in East Africa – Kenya, Tanzania, and Uganda. Its catchment area stretches further across Rwanda and Burundi and covers an area of about 194,000 km^2. Nyakasenge village is located in a remote area of the Magu district, in the Mwanza region. It is one of the six hamlets of Chabula village on the southern part of the lake some 40 km east of Mwanza city

categories were gear owners and fish factory agents. In addition, fisheries personnel at the division, district, and regional levels were interviewed. Other relevant government officers such as the Regional Cultural Officer and the Village Executive Officer were also interviewed. The data presented here draw from all these categories, but the focus is mainly on Nile perch fisheries.

6.3.1 Site Selection and Description

Nyakasenge was systematically and randomly selected from over 500 beaches/fishing communities in Lake Victoria, Tanzania. The selection process considered the following criteria: a fishing community where three types of fisheries were present (i.e. Nile perch, Tilapia, and Dagaa); a fishing community recognized by the fisheries authorities; a community having a functional beach management unit (BMU); a fishing beach where there are indigenous as well as migrant fishers; a beach with more than 100 fishing boats; and finally, a beach accessible by road. From over 200 beaches that fulfilled these conditions, Nyakasenge was randomly picked.

Located in a remote area of the Magu district in the Mwanza region, Nyakasenge is one of the six hamlets of Chabula village (Fig. 6.1). Although the exact period of its establishment is not known, residents claim that the beach is older than the Nile perch fishery introduced sometime during the 1950s (Pringle 2005). The beach was a landing site for some of the indigenous species (see Verschuren et al. 2002 on indigenous species) fished from parts of the Speke Gulf. However, with the proliferation of Nile perch in Lake Victoria during the 1980s (Abila and Jansen 1997), the beach grew in terms of human population, the number and type of houses, and accessibility (Verschuren et al. 2002). The demand for Nile perch in the European countries and other places also led to a management system changing from a more traditional one in which local community chiefs were the de facto fisheries authorities to a central system in which riparian states and governments took responsibility in managing the fisheries (Owino 1999; URT 1997).

6.4 Fishing Life in Nyakasenge

To an outsider, Nyakasenge fishing village appears as a poor community. A first-time visit only leaves one touched by the deplorable conditions in which fishers and their families live. It feels even worse when one has to live in the village for an extended period. Indeed, one then has to come to terms with having to either bathe in the lake just as the fishers do or simply collect water from the lake (there is no piped water system) and take a bath in makeshift bathtubs that are smaller than those who use them. Additionally, one has to reconcile oneself to using unrefined water directly from the lake for all domestic purposes. There are no health facilities nearby, such as a hospital, dispensary, or clinic (including health personnel) in the village, so when people get sick, they must travel 5 km. Despite these conditions in Nyakasenge, fishers there do not perceive themselves as poor. They have access to basic needs such as water and food, which they get from the lake. Poverty as they understand is attributed to not being able to use one's hands, head, and legs to engage in an activity. In other words, to be poor, one must be disabled, and have no relatives who are morally obliged to feed, clothe, and house those who do not have these needs. Moreover, these relatives should also ensure accessibility to health facilities.

In October 2009, Nyakasenge fishing beach had 184 fishers (crew, boat, and gear owners). There were 87 small-scale traders including shopkeepers, tailors, and restaurants. There were 33 boat and gear makers as well as repairers, and 31 fish traders. These are among other unregistered fishers living on the beach. These people have come from different ethnic riparian communities surrounding Lake Victoria. However, since the beach (which is part of Nyakasenge) is located right in the midst of Sukumaland, the Sukuma people are the dominant ethnic group. The beach is simply part of the Nyakasenge hamlet, which includes the adjacent Shoka island and a 2 km radius from the beach. The portion of the hamlet outside the beach area is occupied by other people involved directly or indirectly in fishing and agricultural activities. Almost everybody in the hamlet has a relationship with fishing activities. There are farmers, livestock keepers, and government employees such as teachers in the local primary and secondary schools who live in the hamlet and who are also fishing, in partnership with fishers, or in the fish trade. It is also at the beach where most activities and services – for instance shopping, posho meals, and selling food stuffs – happen.

A tour of the village revealed to us that each home had at least fishing gear or an item related to fisheries. We also observed that whatever they did was greatly influenced by fishing activity; for example those who go to the gardens would leave the gardens in good time to look for baits for their evening fishing. The daily life of fishers followed a strict routine, which involved preparation for fishing, fishing, recreation, eating, and sleeping. Community meetings were convened to mobilize community efforts toward issues such as preparing for visitors or contributions toward funeral expenses for a community member. These gatherings were occasional, and were not (in most cases) planned for in advance. Thus, fishers appeared to spend their time thinking and planning their fishing activity.

During our stay in Nyakasenge for the period of our fieldwork, we observed a pattern of various activities. We noted that life on the beach was oriented toward the lake. That is to say that people residing here had a tendency of moving from their houses to the area in which boats anchor. Moreover, in the mornings, there was a fish market between the edge of the lake waters and the houses – an open sandy space that cannot be used for building houses. It served the function of a public square and marketplace in town. Many activities such as community meetings, celebrations, football training, children playing, drying of fresh fish, drying of nets, and makeshift shades for selling fruits are undertaken in the open area between the water's edge and the houses. This area is cool due to cool winds/breezes blowing from the lake on the days that, for the most part, are sunny and very warm.

From our previous visits to this and other fishing communities at the lake,[2] our perception of Lake Victoria fishers was a group of poor people who entered the fisheries as an occupation of last resort, had no other opportunities in the place where they came from or in other sectors, and therefore moved to the lake to earn their livelihoods. To us, their life was still poor despite being in the fisheries. They also seemed to have a lifestyle of intensive fishing, without regard to stock status. They appeared as people who saw life as only important today and not tomorrow, and were therefore extravagant in expenditure, especially on entertainment (drinking), living lives void of savings. We saw them as morally uncontrolled and irresponsible, lacking a developmental mind, and insensitive about their own future. We were greatly influenced by our earlier observations during the Nile perch proliferation in Lake Victoria (the period between the 1980s and 1990s). Fishing during this time, in our understanding, was an occupation for making easy money for those who did not have anything else to do.

However, our attention was drawn to several issues that made us understand these fishers differently during our stay in this village in November 2009. Although

[2]The author has been doing research at this lake for over 10 years before doing fieldwork related to information collected for this chapter.

Fig. 6.2 A group of fishers from Shoka Island, Chabula village in Lake Victoria, Tanzania, out in the lake setting their long lines (Photo by Paul Onyango 2009)

our objective was to generate information regarding how poverty is understood and experienced among these fishers,[3] we could not avoid seeing how fishers were motivated in their fishing activities. We noted why they understood poverty differently from being in a state of low wages, labor exploitation, inequality, political disfranchisement, social exclusion, powerlessness, and lack of freedom. While they agreed that one could lack some basic needs, they believed that so long as a person has the ability (hands, head, and legs) to get what he/she lacks, that person is not poor.

Additionally, poverty in this community meant inability to respond to emergency. For instance, responding to a crocodile attack that is a frequent occurrence in this village.

6.4.1 *Fishing Activities*

Life in Nyakasenge is about fisheries from morning to evening, all year round (Fig. 6.2). We have gone back to this fishing village several times, and every time we visit, fishing activities have occupied the time of most fishers. As we observed,

[3]The meaning of poverty in this community is published elsewhere by the author (see Onyango 2010).

fishers of Nyakasenge include Nile perch (*Lates niloticus*), Tilapia (*Oreochromis niloticus*), and Dagaa (*Rastrineobola argentea*) fishers. Dagaa fishers also catch other species. Each type of fishery has got its special demands and time schedules.

6.4.1.1 Nile Perch Fishers

The Nile perch fishers were once in majority in the lake but they are now replaced by Dagaa fishers, which include those who fish with longlines and those who use gill nets. Gill nets are also slowly being phased out here, and fishers are increasingly using longlines. The longliners either fish during the day or the night. Fishing with longlines also requires baits. Fishers therefore set times to look for baits. We observed that, on average, fishers fish only three times a week, and this schedule depends on the availability of baits. Looking for baits can take as long as 2–3 days to be able to get enough for the average of 1,500 hooks per boat. Those who cannot fish the baits by themselves have no option but to buy from bait fishers – one live bait costs T.shs 100 (US$0.08).

Fishing during the day (locally known as *tega zibua*) involves leaving in the morning (6:00–7:00 a.m.) to set the nets. Fishers stay in the lake, haul the hooks, and come back in the evening (5:00–7:00 p.m.). We also noted that gill-netters prefer *tega zibua* because of gear theft. Gears are stolen when left in the lake when no one is seen attending to them. The other group of fishers prefer to fish during the night – *tega mlazo*. This group always sets off at 3:00–4:00 p.m. to the lake to set their hooks, and come back at 7:00–8:00 p.m. They spend the night, like everybody else, on the beach and then leave at 6:00 a.m. to haul their hooks (if their gears are not stolen), and then land the fish from 11:00 a.m. to 1:00 p.m. Gill-netters who fish during the night do not, however, come back to the beach after setting their nets; they stay by the lake. Some, at other beaches nowadays, stay for about 3–4 days before they come back. They chill their fish in containers.

6.4.1.2 Dagaa Fishers

The Dagaa fishers, on the other hand, follow the lunar cycle – fishing especially during the dark and half-moon periods. This makes them fish effectively for 2 weeks in a month, while fishing less for the remaining 2 weeks. The dark or half-moon implies that the night is either completely or partially dark for about 2 weeks. Such periods are very conducive for fishing Dagaa. During these periods, fishers set off between 5:00 and 8:00 p.m., only to come back in the morning between 5:00 and 7:00 a.m. Dagaa fishers try to fish every day during this period unless conditions such as weather, availability of crew and boat, and gear conditions require attention. These Dagaa fishers are mostly seen on the beach during the day. They enjoy sleeping in the shade close to the lake during the day as they wait to set off for fishing. They also like spending their time playing pool, cards, and checkers.

6.4.1.3 Tilapia Fishers

The Tilapia fishers are a different group from the Dagaa and Nile perch fishers. The latter two groups comprise mostly people who have migrated from different places and converged at Nyakasenge for the purpose of fishing. Tilapia fishers, on the other hand, are mainly people who are indigenous to the area; they live outside the beach area, but within a 2 km radius from the beach. This group comprises mainly young boys aged between 13 and 23 years. They are rarely seen on the beach in the morning because they have to go to school, or those who are married go to their gardens. This group combines fishing and farming as part of their daily activities. They fish with single hooks tied on a wetland reed. They could have as many as ten of these hooks. They are actually not registered with the local BMU as fishers. Their fishing is undertaken mostly in the evenings from 5:00 to 7:00 p.m. On average, each one lands about 5 kg, with the highest being 10 kg and lowest 0.5 kg or nothing. A tilapia fisher does not go fishing every day, but occasionally and depending on the availability of baits and the weather. They fish in rocky areas. When they return, they sell any catch which they think is over and above the family needs. As is typical of fishers, the money they get is used to purchase flour, paraffin, cooking oil, and other kitchen stuff depending on each one's needs.

6.4.2 Motivational Factors

After about 4 months' stay in the community, new issues started emerging. Although we had discussed several issues with fishers, we started looking at their responses with a new lens. For instance, their responses to the questions regarding why they were in fisheries and not other higher income-generating activities led us to perceive these as either pull or push factors, or both factors in many instances. Pull factors refer to attributes of fisheries that attract them to join, while push factors refer to those that drive them away from where they were before they decided to become fishers.

The reasons fishers gave and were attributable to push and pull factors were identified from the statements that they made.

> My father wanted me to tend to our livestock and do nothing else … I left home so that I could come to fish instead of looking after livestock.

His father had 40 cows, which is by all standards a sign of being rich in this community. In fact, ownership of only four cows is an indication of richness. He had to look for something else to do and not depend exclusively on his father's wealth.

Another fisher tells his story. He says that his home is located in another fishing community where his uncle fishes. In this community reaching class seven is considered as good enough for someone to start looking for a job.

> I completed class seven and decided to get involved in the fishing activity in my uncle's fishing crew as a means of engaging in an income-generating activity for myself.

Here is a story from another fisher who comes from an area where they grow rain-fed maize and rice.

> Back in my village, we are farmers depending on rain fed crops … conditions are not very good during the dry periods because incomes from farming are low.

This implies that they can only plant twice a year, with only one good harvest expected during the year. Indeed, we observed that there is extensive growing of rain-fed rice, which is a popular staple food in this area.

One fisher explained why he was attracted to the fisheries.

> My mother and father were living in extreme hardship. … We fished for about one week and I managed to save T.shs 6500 (USD 6 in June 2005). … I took the money home and on that day the whole family was able to get good food, which made me receive blessings from my parents. I returned to the lake and joined fishing forthwith. … I therefore joined fishing because I felt that I could get fish easily.

We learned that the parents were old and inactive, and they could not undertake agricultural activities effectively or involve in other activities. They had to be taken care of by their children. The blessings and joy he received inspired him to like fishing. He said that at times his parents did not have enough to eat. They lived in a grass-thatched house with a leaking roof and holes in the walls. He went to a friend of his who had a beach seine to ask for help to address his parents' situation. He was promised that the only help he could get was to take him to the lake to fish.

Another fisher said: "I grew up in a fishing community – not Nyakasenge – and therefore liked the activity." There were other fishers who also told us that they were born in fisher families and grew up seeing their parents and grandparents fish, so fishing became more admirable as they matured.

Another fisher explains:

> My father was a civil servant in addition to owning a fish boat and fishing gear. … He asked me to take charge of supervising the fishing activity, thus I decided to join fishing as a means of engaging in an income-generating activity.

Fishing to this person is a family activity. He decided to be engaged in it because of benefits it had brought to his family.

Another fisher had this to say: "I observed a neighbour fisher come home well dressed; he brought enough fish and was able to build a burnt brick house." He explained how he felt fishing was a means of improving his welfare and how he had seen his neighbor live from it for so long. To him, fishing seemed to be something that was enjoyable and fulfilling.

We did conduct a follow-up on these reasons given, by asking whether there are other income-generating activities that they could have chosen other than fishing. We learned during our discussions that most of them had not finished primary education, and so they could not get jobs or pursue income avenues that required schooling. Still, we learned that there are several income-generating activities in the areas that they come from, or they could access besides fisheries. They had opportunities of joining small-scale mining, which attracted young people with similar backgrounds and who were these fishers' neighbors in their home villages.

There was also the business of transporting people and/or luggage on bicycles (locally known as *bodaboda*). This was an activity that had steadily gained momentum since the late 1990s, and some of those who started this activity had moved from bicycles to motorcycles, an indication that the transport business was developing fast as an attractive employment alternative. Then there was small-scale horticultural farming for items such as onions, tomatoes, and watermelon. We did observe that young people of fisher age have equally ventured into this activity. Thus, during an in-depth discussion, one of the fishers hit the point that was suggested by others in different ways by stating:

> The reality that our education cannot guarantee us a job like yours (he was referring to us) or a specialized training, we are able to do other things. ... Anyone does what he likes according to his heart and mind.

The group discussion revealed that these fishers preferred fishing to other activities. Fishers were generally happy as long as they had the opportunity to fish and live in the fishing community.

6.4.3 Happiness and Satisfaction

Fishing is a risky activity. For instance, the winds and currents that erupt while out fishing can cause capsizing. This risky nature has made fishers adapt to these conditions while out fishing. Being able to maneuver the winds and currents is a delight to fishers because besides receiving a favorable appraisal from fellow fishers at the beach, as we noted, one fisher also told us: "I feel that I have some level of control." Indeed, for the first 6 months that we were in Nyakasenge, there were two occasions on which the local BMU office had to organize a search for fishers who went missing, and only their boat was spotted on the lake. There are also unexplained occurrences in the lake – which fishers refer to locally as *mashetani* (dangerous spirits). A number of fishers told us that when they are out fishing, at times they see things emerging from the water and then disappearing. They believe that such things can bring bad luck in fishing, make the boat capsize, and even lead to death. Fishers also tell of how they are often attacked on the lake by people who either want the fish they have caught or are simply in need of their fishing gears. We witnessed at least five conflicts that the BMU addressed of fishers who fought for fish while on the lake.

Fishers also agree that fishing provides an avenue for whoever is in it to feel that their contribution is useful and valued. They feel that "everybody has a responsibility when you go out fishing and every person gets equal pay."[4] One day, when we went out to fish with one of the fishing groups (we had hired our boat while the fisher group was in their usual boat), we observed that one boat using longlines had five fishers. The whole group was highly dependent on each other such that without

[4]Incomes differ depending on catch but on average each crew gets between 2,000 and 15,000 (US$1.5 and 11.5) per fishing trip. This is besides an income of over US$30 earned by fish agents.

even one of them, the whole fishing activity would be impaired. We noted that in the boat there is no specific leader, no one tells another what to do, or reprimands the others, or gives commands. There is a driver (the fisher who pilots the boat) who has to ensure that the boat goes in the correct direction. The driver ensures that the boat does not collide with other boats, and that longlines are set in a manner that they do not entangle with other fishers' longlines. There are two paddlers who ensure that the boat moves, while the remaining two are involved in the actual hook setting. These last two cannot set hooks if the boat is not directed and paddled to the fishing grounds, creating a very strong dependency among themselves. Such dependence has encouraged fishers to cooperate both during fishing and at the beach. For instance, when a boat lands, fishers will gather around the boat and together pull it onshore. We also observed how they cooperate during capsizing, and how they search for those who go missing.

We observed that when one fisher is bereaved, all fishers feel obliged to contribute some money to the relatives or loved ones as a means of assisting them to meet the funeral expenses. Fishers are happy that they are not alone during the bereavement period, and in facing the emotional and physical difficulties of life. There is also a by-law in the community according to which all adults contribute T.shs 1500 (US$1.15) whenever a death occurs. Fishers do not want to be seen going against this by-law. In fact, one fisher summarized this by stating: "We have to support one another when we live here because our relatives are far away." He was talking about a cultural practice where the relatives of a person who dies take full responsibility of ensuring that the family members of the deceased are taken care of during and after the funeral.

There was a general consensus among fishers that fishing provides easy food and water. The fish from the lake and the access to water (accessible to all in the village) are important things that fishers typically do not have easy access to in their home villages. Water, for instance, was a major concern for fishers. It was noted by fishers that during dry periods, access to water in their home villages was sometimes very difficult. One of the beach leaders told us that in his village, which is within the district, there are times when they have to wake up very early in the morning to go and line up for water coming from a spring. "You can line up only to miss water when you are just three people away from the water."

In other villages, people buy water from those who have dug wells. But when they come to the lake, he claims: "Water is there for free and in plenty, nobody refuses you from collecting and using water in the way you want." Indeed, we observed that people use water without any restrictions. With regard to fish, we actually observed a practice where fishers give fish for free to fellow fishers, especially if the one being given the fish did not go fishing on that day. We inquired of this practice from several fishers including those who received fish from others. The main reason we were given was that the fish was "the food for that day." However, fishers also claimed that "you cannot refuse fish to a fellow fisher." This is so, because by giving out fish, the fisher is actually saving for a day when he may also not go to fish. Additionally, fish given to non-fishers, who will not pay back, appeared to be based on generosity. Fishers are generally generous despite giving fish for other reasons.

Another day when we accompanied another fishing group on a fishing trip, we noted the joy with which fishers at times do their work. During this trip, the gear and boat owner who is respected for his knowledge in hook fishing accompanied us. He informed us of how they do the actual hook setting, and how it requires knowledge in calculating the depth to lower the hooks. What caught our attention during this trip was the manner in which this group was accommodating themselves on the lake. The gear owner was responsible for setting the hooks; he appeared to do this quickly and accurately. He set 1,500 hooks in a record of 1 h 25 min, whereas, if it were done by his other crew, they would have taken 3–4 h to set the same hooks. His accuracy was in hooking live baits, and throwing them in the lake. Dead baits do not attract Nile perch, so the baits must be able to swim when thrown in the water.

Besides such a level of accuracy and a sign of experience, the group was fishing with extreme enthusiasm, singing cheerfully and looking quite involved. They looked very relaxed like people who were doing something they were best at. Most of the times they would sing, tell stories to each other, and be cheerful. Although they also admitted that not all fishing days are the same, "going into the lake to fish is pleasurable," as one of them put it and others seemed to agree. It is something they said that they always longed for, were excited to do at anytime, uplifted their spirits, and was pleasing to do.

6.4.4 *Identity and Meaning*

When you meet Nyakasenge fishers you may be led to conclude that they are people who are withdrawn, and do not want to talk openly about what they do at the beach. They want to know what has brought you to their village, thinking more about you in terms of an investigator who is not interested in their fishing, but is only there to look for reasons to deter them from fishing. Or maybe you are looking for some fishers with the intention of arresting them.

But when we lived and interacted with them more often, we saw a totally different group of people. They are people who look at themselves as strong, independent, self-reliant, autonomous, and aggressive, which helps them to confront the risks while out fishing.

> You cannot go to the lake to fish if you are fearful or shy; otherwise you may end up not coming back. … When you are out on the lake, every person needs to make independent decisions.

He was explaining the idea that fishers should be as independent as possible, in as much as they depend upon one another in their fishing activity. One fisher noted: "Those people who some of you think are fishers are actually not fishers." He was referring to the group of gear and boat owners who have advanced loans to others to be able to fish. "We are the fishers!"

"A fisher must be able to confront difficulties, not just sit somewhere and wait for money," he said, gesturing with pride regarding what he had said about who he is.

He went ahead and compared himself to us in a manner suggesting that our physical looks do not show any signs of hardiness or brawn: "Look at how soft your bodies are. You cannot do physical work for a long time like a fisherman can." His fellow fishers responded in agreement by laughing and smiling at such a comparison. What these fishers seem to agree on is that fishing is more than the income that it gives; it is more than employment or taking up a job. Fishing has to do with the characteristics it produces, for which they are pleased to be identified.

One of the areas that we found challenging to discuss with fishers was what fishing means to them. It was challenging to us because fishers could not understand our question, given their assumption that the meaning of fishing should be obvious and not something to be asked about. One fisher said that "fishing is the only activity that I grew up thinking about." He actually stated how fishing is simply his life. "This is what I live from. Fishing gives me the things I need to live as everybody else." He was comparing his life with other fishers he interacts with. Another fisher stated: "It is through fishing that I was able to know what life really is about." He was referring to having matured as a person because he was in the fishery, an opinion that was shared by several fishers we interviewed. Fishing, as others agreed, made them understand what masculinity is all about; it made them realize that they were "real men." One fisher told us:

> When you go to fish and are able to control or subdue the forces in the water/lake to attain what you want, then you feel you are a real man.

They told us that to understand what life means, one only needed to go into fishing. We therefore took time to understand what they meant when they talked about life by going to sea with them and to observe for ourselves what they were referring to.

Life obviously also means being able to eat, have shelter, and dress. The fisher was referring to the guarantee of getting food, water, collaboration, and independence. One fisher explained: "When you are fishing, you are not worried about whether you will eat." He explained how fishing also had enabled him to buy 18 goats, which he exchanged for three cows. With the value that this community places on cows, this fisher feels that by owning cows, he is perceived as being somebody among the community members. He feels that he is highly valued, if not perceived as rich. Fishing has not only given him enjoyment, food on the table, and money in the pocket. It has also made him what he considers himself to be – that is, a person of respect.

6.4.4.1 Remuneration

We did note that the remuneration system in fishing (especially among the crew members and between the crew and boat and gear owners) also makes fishers perceive fishing as ensuring and/or providing equity. While remuneration differs from one fishing group to another, it is basically done through a share system in which boat and gear owners on one side equally share the proceeds for each sale with the crew. Other modes of payment are through allocating days for crew and for gear and boat owners. For instance, 3 days could be allocated for the crew, and then the next two fishing days are allocated to the boat and gear owners. For those who divide each day's sales, although the share is equal, the boat and gear owners end up getting more than the crew. This is because the crew has to share among themselves 50% of the total income, while the boat and gear owners nest the other 50%.

This remuneration system has guaranteed fishers' equity. They see that each one gets paid according to the agreement, and there is no undercutting when it comes to payment. Thus they perceive fishing as an activity that they can participate in without much fear when it comes to modes of payment among themselves. At this level, they do not fear being cheated by their fellow crew. Moreover, they see each other as equals, sharing what they get, and perceiving themselves to be equally valued.

6.5 Discussion

To an outsider, Nyakasenge fishers could be seen as concentrating on activities that are necessary to meet their short-term, basic needs, and less focused on opportunities that could lift them up from their deprived/poor state characterized by vulnerability because of storms, drought, and robbery, dependency on middlemen, lack of basic facilities such as health care, and income poverty (Béné 2004; Thorpe 2004). These conditions are not unique to Nyakasenge fishers. Non-fishers also suffer from these conditions, as do people in adjacent lake communities (RAWG 2004).

Notably, poverty in small-scale fisheries, like in Nyakasenge, has been understood in different ways. A general perspective has been that poor fishers face conditions in which they lack certain facilities and services that enable them to meet their basic needs (Spicker 1999; Béné 2004; Thorpe 2004). This understanding of small-scale fishers' conditions creates a view about them as experiencing a number of problems and unfulfilled needs. This again raises questions as to how they are able to cope with such conditions (Allison and Ellis 2001). In addition, it creates a perception about small-scale fishers as people who are in need of problem solvers (Mathie and Cunningham 2003).

The needs perspective illustrates how people living in dire conditions, and who are no strangers to adversity and suffering, find their only hope in joining the fisheries – fishing being an occupation of last resort. It also explains how once they are in fisheries, they cannot leave because the available alternative opportunities are less attractive (in terms of incomes) – hence the argument that "they are fishermen because they are poor" (Béné 2004, p. 26; Brookfield et al. 2005).

Although the needs perspective has much to offer, I argue that needs and problems are not the only means through which we can see fishers who are poor. Amartya Sen puts it very explicitly when he says: "Seeing people only in terms of their needs may give a rather meagre view of humanity. … Human beings have needs, but they also have values and, in particular, cherish their ability to reason, appraise, choose, participate and act" (Sen 2009, p. 250).

Indeed, this study has shown that despite the needs of Nyakasenge fishers, the value and meaning of fishing to them both play an important role that defines how they live their lives, interact with fish resources, and what they cherish in it. This is to say that fishing is more than just looking for a means of sustenance, or simply generating income, fulfilling basic material needs, or finding refuge for their poverty (Béné et al. 2010). Fishing is not even necessarily related to their survival needs. Fishing offers an opportunity for those with hands, head, and legs to explore their potential. Fishers are inspired by factors beyond their survival needs. They join fisheries out of choice rather than simple necessity (Cinner et al. 2008).

Also, the decision to join fisheries indicates that they are not essentially self-interested and short-term maximizers. On the contrary, they seek to achieve a sense of personal power, to act with a degree of freedom (opportunity and process) and autonomy, and to experience the joy and self-fulfillment that comes with it. They appreciate the fisheries for the income they bring. In addition, there are also other very important reasons: satisfaction, happiness, and above all identity and meaning. These reasons are important in shaping their perspective on fisheries, and how they interact with it. For instance, as Pollnac and Poggie argue: "If we expect to reduce fishing pressure by convincing fishers to shift to alternative occupations through provision of training programs, these jobs must provide some of the same non-monetary benefits as fishing" (2008, p. 198). This is also where management strategies and poverty reduction mechanisms should focus. Management strategies and poverty reduction mechanisms should create an opportunity and a process for fishers to do what generates value and meaning to them.

To be connected to others and to be part of a community are other elements that count. This also comes with obligations and responsibility (van Ginkel 2009). Fishing is embedded in the way a community has defined their life (McGoodwin 2001). It is therefore difficult to separate it from the community and its members. The members "are" the community, and the community comes alive in them. Although a fisher is objectively different from his fishing, his fishing also defines the fisher. Thus they cannot be easily separated. Fishing is part of the culture of the community to which the fisher belongs. This bondage between a fisher and fishing makes it hard for a fisher to abandon his activity, because he would have to leave something of himself.

The underlying difference between fishing as an occupation and fishing as culture lies in what the fisher gets from it. As an occupation, the fisher evaluates the advantages and disadvantages on the basis of costs and benefits: the idea of thinking rationally within a framework of a logic of consequence (Elster 1983; March and Olsen 1995; Sen 2002). As a rational thinker, a person is assumed to be thinking more about the outcome, what he expects to achieve, and guiding what he wants to pursue.

Thus, a poor person would evaluate his condition and realize that he would maximize benefits if he joint fishing. Rationality in this case is used as a predictor of actual choice[5] (see Sen 2002, 2009). But when fishing is the only option

[5] Amartya Sen (2009) questions whether rationality can actually be a determinant of actual choice. That rationality is the "disciple of subjecting one's choices to reasoned scrutiny." That reasoning is likely to favor rationality of choice, and consequently favor maximization of what people pursue. Choices are not made by rational, self-interested motives alone. Choices are also defined by fulfillment of social obligations, cultural conventions, and the enactment of routines (Jentoft et al. 1998).

available – sitting idle is not considered an option because the fisher has "capabilities" in terms of hands, legs, and head which he can use – then the issue of choice does not apply.

A fisher will do what is before him and what he has to do. He will do what is expected of him, and what he has committed himself to do and what is therefore his responsibility. He may not even be conscious of the consequences that will follow. He will not follow the logic of consequence. Instead, he will follow what March and Olsen (1995) is termed as the "logic of appropriateness." Fishing will not be an occupation chosen for some particular gain, but something that feels like a "natural" way of life, a life that, given the circumstances of being part of a community where he feels at home, represents the obvious thing to do and one that should be lived. If Nyakasenge fishers did follow the logic of consequence only, then they would have been inclined to join an occupation that has the highest returns, such as mining. Moreover, they would have followed a daily fish-catching strategy that is most profitable to them such as fishing longer hours and improving their gears to make them more effective, but they are not.

On the contrary, the fishers' strategic behavior is both tempered and inspired by cultural norms and values that are prevalent in the community and among peers, in which being a fisher is seen as a person who is able to handle risks, be independent, and maneuver his way while out on the lake fishing, and deciding whether should go fishing or not. A fisher is perceived to be self-reliant, selfless, courageous, and aggressive, and also cooperative, that is concerned about other community members, and is willing to work together with others and to look out for them (McCay 1989; Pollnac and Poggie 2008).

6.5.1 Freedom of Choice as a Central Issue in Assessing Human Life

Fishing as a way of life also demands an understanding of life's opportunities. As an occupation of last resort, fishing is simply an opportunity that a fisher would end up taking out of lack of not only better alternatives but also other alternatives. The poverty condition in which people find themselves, as in the case of Nyakasenge fishers who are considered poor due to their living conditions, compels them to join the fisheries because they cannot take any other decision; it is either fishing or not being able to survive.

This, however, is not the case in Nyakasenge. Despite their deprivation, they have several opportunities accessible to them such as mining, bicycle transport, horticulture, and even fishing. Out of these alternatives, fishing seems to have more attractive pull factors than all the rest. For instance, in fishers' conclusions about the various reasons for joining fishing, they indicated that joining was more related to doing what they like according to their heart and mind given that they have hands, head, and legs (implying that they do not see themselves as poor). In addition, because of those alternatives that are available to them, they also have the freedom to choose what they want to do and how to live. This is why they have left those other accessible and equally good opportunities to join the fisheries.

The ability to choose "is a valued aspect of living that we have reason to treasure" (Sen 2009, p. 227). Being poor with no other alternative than fishing is a very different situation from being poor but with a variety of occupational opportunities within reach. This also relates to Sen's argument that "[i]n assessing our lives, we have reason to be interested not only in the kind of lives we manage to lead, but also in the freedom that we actually have to choose between different styles and ways of living" (2009, p. 227).

Sen's argument on the ability to make choices is more relevant here because the freedom to make choices is an important factor in explaining actions/behavior. If fishers have the freedom and ability/capability to make choices with regard to how they want to live, they could use this freedom to choose a sustainable fishing life. Sustainable fisheries management demands to be understood not only as involving managing people rather than the fish, as is often the point that social scientists make (Jentoft 1999; Berkes et al. 2001). The issue is also *how* people are managed, whether they are managed in a command and control fashion, or whether they are involved in an interactive process of participation and mutual learning (Jentoft et al. 1998; Kooiman et al. 2005).

Managing people is not as simple as a matter of effort control as argued by biologists and economists. Neither can one manage people as if they are objects and tools. Managing people involves the appreciation that people are also their own governors and stewards of their own lives, families, and communities. Therefore, there are ethical and moral issues involved which cannot be ignored but require a broader understanding of, and respect for, what fishing mean to fishers and the freedom they have in the way they relate to fish (Jentoft et al. 2010). Therefore, fisheries management and even development (Wilson 2003) in this perspective would require a broader governance perspective and approach that is also based on social and cultural variables and indicators beyond those of biology and economics.

Two aspects appear to be very important here in the discussion on choice: opportunity (i.e. those things that we value) and process (not being forced into some state because of imposed constraints). Nyakasenge fishers appear to have both opportunity and process with respect to their choosing how they would want to live their lives. Thus, we can see these fishers not merely from the point of view of their means of living and their livelihood strategies of fishing as an occupation, but through their lives, that is fishing as a preferred way of life. Choices influence behavior and actions (March and Olsen 1995). It is argued here that when choices are made, they are often based on the influence of the community to which the person making the decision belongs. The community sets the rules, duties, and mannerisms for its members. These rules, duties, and mannerisms are morally accepted and perceived as the means through which every community member should behave. Once internalized, the community members view them as good practice, something that individual community members should adhere to.

Second, the idea of decisions being influenced by consequences or appropriateness brings into discussion the issue of responsibility (Sen 2009). When a fisher considers the consequences of a choice (consequence-sensitive reasoning), he acts according to that choice. When he makes a choice regardless of the consequences, then it can be argued that this fisher is essentially acting in a responsible manner: being able to make moral or rational decisions on one's own, and therefore being answerable for one's behavior.

As Sen argues: "Responsible choices are based on the chooser's evaluation of states of affairs, including consideration of all the relevant consequences viewed in the light of the choices and the comprehensive outcomes associated with what happens as a result" (2009, p. 218). But decisions that people reach and their actions take into considerations not only the good consequences but also what the person *has* to do, regardless of the consequences. That is: "A person not only has good reason to note the consequences that would follow from a particular choice, but also to take an adequately broad view of the realizations that would result, including the nature of the agencies involved, the processes used and the relationships of people" (Sen 2009, p. 219).

This is what we see with Nyakasenge fishers. Fishers feel obligated to the people they are related to, especially their parents, siblings, and significant others. It is not only good but also right to maintain these relationships, comply with what parents are saying, and not be simply idle. This aspect of responsibility is also evidenced in their interaction with the fish resources. To them, their fishing practices are simply the way fishing is supposed to be undertaken. Their choice of gears and fishing times reflects the standard norm for how fishing is supposed to be done and how it is actually performed among Nyakasenge fishers. They assess their fishing lives based on the choices they make, on how to fish but not on the fishing gears they use or fishing times and frequency.

Indeed, the means through which fishers assess their lives could explain why they do not consider themselves poor. They asses their lives, including their poverty status, not only on the basis of their material and/or tangible possessions, but also on the basis of their observations, expectations, and aspirations related to what they can possibly do.

6.6 Conclusions

Management of fisheries, whether economic or biological, has persistently remained an issue focusing on two main questions: (1) How much should be fished? (2) How should it be fished? (Kolding and van Zweiten 2010). The focus on these questions can only be beneficial to those who get involved in fisheries as an occupation or livelihood. Correspondingly, the management of Lake Victoria fisheries draws much from this "how and how much" perspective. Indeed, the management of Lake Victoria has been, and continues to be, focused on effort levels as an important driver for fisheries. Moreover, increasing fishing pressure has convinced fisheries authorities and researchers that overfishing is an important threat to the lake fisheries (Njiru et al. 2005; Matsuishi et al. 2006; LVFO 2008). The argument behind these two questions – "how and how much" – rests on the concept of sustainable fisheries. This is generally understood to mean a level of fishing that does not result in the loss of a potential yield, or stock components erosion to a point where the stock structure loses diversity and resilience to environmental fluctuations.

This line of thinking follows Gordon (1954), Beverton and Holt (1957), and Hardin (1968), whose arguments in relation to fishing sustainably were built on the exploitation rate of fish resources. But judging from the story of Nyakasenge fishers, the ideas of Gordon, Beverton, and Holt, and even Hardin, are problematic in achieving an effective management of the lake fisheries. This is because the relationship between the fishers and the fish seems not to be driven by factors directly related to the "how and how much" questions (van Ginkel 2009). Rather, the relationship is driven by a desire to live lives that fishers prefer. Such lives are built on fishers' values, norms, and morals, not only concerning the "how and how much" questions, but most importantly the *why* question. Thus, the relationship between the fisher and the fish must be understood from a broader community perspective. This understanding has to relate to how fishers make choices (not to be poor or live in poverty) regarding the way they want to live their lives in the environment that they find themselves and what they aspire to.

Fisheries management should therefore not only focus on the stock status, species diversity, and exploitation pattern and rate but also on human life. For instance, if one would ask about the value of fish, then it would be reasonable to recognize not only the fish as a resource, a food item, and a commercial commodity but also the opportunities that it offers to fishers to build a life for themselves and their significant others in the community. The impact of fishing on fisher's lives should be among the principle considerations in assessing the value of fish and the way fisheries are managed. Fishing is valuable if it also enhances the quality of human life. In a way, this is what Sen argues when he states: "It is, therefore, not surprising that environmental sustainability has typically been defined in terms of the preservation and enhancement of the quality of human life" (see Sen 2009, p. 248). Human life, here, is understood in the context of Nyakasenge fishers, who view values of cooperation, risk taking, autonomy, and being responsible as most crucial in enabling them to live life the way it is supposed to be lived.

Poverty reduction strategies and fisheries management mechanisms therefore need to be re-examined (Berkes et al. 2001; Béné 2003; Béné et al. 2010). Such a re-examination should begin by focusing on the "why" question, rather than the "how and how much." There are other ways in which these strategies and mechanisms can be re-examined. The "why" question would primarily lead to two issues: opportunity and process (freedom) (Sen 2009). Formulation of poverty alleviation strategies and fisheries management mechanisms should be focused on whether poor fishers have an *opportunity* to pursue the kind of life they want, and whether they have the environment in which they can choose to live such a life (*process*) (Sen 2009). Poor, small-scale fishers should have opportunities, such as fishing, in which they can make decisions to be engaged in a manner that allows them to generate meaning, derive satisfaction, and be happy. It is not an occupation which comes with stigma, but also with a policy of making people free of the burden of being a fisher. This is the policy of providing alternatives to fishing, which is a good thing because it gives people more choices.

Management and poverty reduction should, therefore, be viewed in the context of providing an environment (a process) in which human beings can live their lives. There should be a move from management (in its technical terms) to governance[6] (Bavinck et al. 2005; Kooiman et al. 2005; Jentoft et al. 2007; Jentoft and Chuenpagdee 2009). Such a movement to governance requires development of management-relevant social variables and indicators (Smith 1978) beyond employment, migration, age, and population (Hamilton and Butler 2001). Variables and indicators focus on people's judgments, perceptions, and meanings in relation to their well-being, capabilities, and satisfaction. Such social variables and indicators should capture fisher communities' sociocultural values and qualitative life aspects such as challenge, adventure, cooperation/collaboration, independence, and belonging. The indicators should complement the economic and biological indicators and should be able to guide the management of fishing activities and provide feedback on the extent to which management and poverty eradication objectives are being met.

It is important to recognize that the poor have various capabilities, including the capability to make choices. It is worth noting that it is one thing to choose what one wants to pursue in life, but it is another to be forced into doing something – not forced in the direct way (like by force) but because there is nothing else to do. When forced to pursue fishing, as is argued by the proponents of fishing as an occupation of last resort, fishers are seen to pursue material things of convenience such as incomes, possessions, and livelihoods. This dimension of thinking does not recognize the totality or fullness of human life, which is more focused not only on the means of living, such as fishing, mining, or agriculture, but more so on freedom or particularly opportunities of living. This would be what someone wants, values, and decides to choose. Human life is grounded on preservation and possibly expansion of human freedoms and capabilities, opportunity, and process, on which management mechanisms and poverty reduction strategies should be built.

Fisheries management and poverty reduction strategies should therefore be formulated within a broader framework where values and principles such as the meanings that fishers attach to their fishing, the satisfaction they generate, and the identity they receive are central. Policies should be formulated in a manner that will enable fishers to make sustainable decisions on their own and not force it on them. If management is not built on these fundamental issues about what fishing means to those who fish but on purely technical assumptions with regard to the "how and how much" question as is the case with most fisheries' management strategies, then such management mechanisms are likely to misfire and/or backfire. Fishers may adopt strategies directed at enabling them to live their lives the way they prefer. Such strategies may not be congruent with the management mechanisms at a macro level, therefore leading to misfiring or backfiring of management mechanisms.

---

[6]Governance is used here in its broader sense, beyond accountability, transparency, rule of law, and vibrant civil society, to include integrated processes of governing actors, and emphasizes on principles, values, and goals that underlie problem solving and building of institutions.

Acknowledgments I would like to thank several people for their constructive comments to this and earlier versions of this chapter. The list includes Svein Jentoft, Jahn Petter Johnsen, Eirik Jansen, Fiona Nunan, other anonymous reviewers, and all those who provided comments and ideas during the writing period. I would also like to thank Maria-Victoria Gunnarsdottir for the editing services. Above all, I wish to acknowledge funding from the Norwegian Research Council that has enabled the production of this chapter.

References

Abila RO, Jansen EG (1997) From local to global markets. The fish exporting and fishmeal industries of Lake Victoria: structure, strategies and socio-economic impacts in Kenya. Socio-economics of the Lake Victoria Fisheries, Report No. 2. IUCN Eastern Africa Program 1997. IUCN-EARO, Nairobi

Acheson JM (1981) Anthropology of fishing. Annu Rev Anthropol 10:275–316

Allison EH, Ellis F (2001) The livelihood approach and management of small-scale fisheries. Mar Policy 25:377–388

Andersen R, Wadel C (1972) North Atlantic fishermen: anthropological essays on modern fishing. University of Toronto Press, Toronto

Apostle RL, Kasdan L, Hanson A (1985) Work satisfaction and community attachment among fishermen in southwest Nova Scotia. Can J Fish Aquat Sci 42:256–267

Bailey C, Jentoft S (1990) Hard choices in fisheries development. Mar Policy 14(4):333–344

Bavinck M, Chuenpagdee R, Diallo M, Kooiman J, Mahon R, Williams S (2005) Interactive fisheries governance: a guide to better practice. Eburon Publishers, Delft

Bebbington A (1999) Capitals and capabilities: a framework for analyzing peasant viability, rural livelihoods and poverty. World Dev 27(2):2021–2044

Béné C (2003) When fishery rhymes with poverty: a first step beyond the old paradigm on poverty in small-scale fisheries. World Dev 31(6):949–975

Béné C (2004) Poverty in small-scale fisheries: a review and some further thoughts. In: Neiland E, Bènè C (eds.) Poverty and small-scale fisheries in West Africa. FAO/Kluwer, Rome/Dordrecht, pp 61–82

Béné C, Hersoug B, Allison EH (2010) Not by rent alone: analysing the pro-poor functions of small-scale fisheries in developing countries. Dev Policy Rev 28(3):325–358

Berger PL, Luckmann T (1971) The social construction of reality: a treatise in the sociology of knowledge. Penguin, Harmondsworth

Berkes F, Mahon R, McConney P, Pollnac R, Pomeroy R (2001) Managing small-scale fisheries: alternative directions and methods. International Development Research Centre, Ottawa

Beverton RJH, Holt SJ (1957) On the dynamics of exploited fish populations. Fish. Inv, ser. 2, Sea Fisher, 19, 1–533

Brookfield F, Gray T, Hatchard J (2005) The concept of fisheries-dependent communities. A comparative analysis of four UK case studies: Shetland, Peterhead, North Shields and Lowestoft. Fish Res 72:55–69

Carothers C (2008) "Rationalized out": discourses and realities of fisheries privatization in Kodiak, Alaska. In: Lowe ME, Carothers C (eds.) Enclosing the fisheries: people, places and power. American Fisheries Society Symposium 68, Bethesda, pp 55–74

Cinner JE, Daw T, McClanahan TR (2008) Socio-economic factors that affect artisanal fishers' readiness to exit a declining fishery. Conserv Biol 23(1):124–130

Cunningham S (1993) Fishermen's incomes and fisheries management. Research Papers No. 61. CEMARE, University of Portsmouth, Portsmouth

Du Toit AD (2004) Social exclusion discourse and chronic poverty: a South African case study. Dev Change 35(5):987–1010

Elliot A (2007) The social animal. Worth Publishers, New York

Elster J (1983) Sour grapes: studies in the subversion of rationality. Cambridge University Press, Cambridge

FAO (2000) Poverty in coastal fishing communities. Advisory Committee on Fishery Research Third Session 5–8 Dec. FAO, Rome

Gatewood JB, McCay BJ (1988) Job satisfaction and the culture of fishing. MAST 1(2):103–128

Geertz C (2006) Thick description: toward an interpretive theory of culture. In: Moore HL, Sanders T (eds.) Anthropology in theory: issues in epistemology. Blackwell, Malden, Oxford and Carlton, pp 236–243

Gordon SH (1954) The economic theory of a common property resource: the fishery. J Polit Econ 62:124–142

Graaf M, van Zwieten PAM, Machiels MAM, Lemma E, Wudneh T, Dejen E, Sibbing FA (2006) Vulnerability of a small-scale commercial fishery of Lake Tana's (Ethiopia) endemic *Labeobarbus* compared with African catfish and Nile tilapia: an example of recruitment-overfishing. Fish Res 82:304–318

Gudeman S (2001) The anthropology of economy. Blackwell, Oxford

Hamilton LC, Butler MJ (2001) Outport adaptations: social indicators through Newfoundland's cod crisis. Hum Ecol Rev 8(2):1–11

Hardin G (1968) The tragedy of the commons. Science 162:1243–1248

Hersoug B (2004) Exporting fish, importing institutions – fisheries development in the third world. In: Hersoug B, Jentoft S, Degnbol P (eds.) Fisheries development: the institutional challenge. Eburon Publishers, Amsterdam, pp 21–92

Iliffe J (2000) Africans: the history of a continent. Cambridge University Press, Cambridge

Jalal KF (1993) Sustainable development, environment and poverty nexus. Occasional papers No. 7. Asian Development Bank, Manila

Jentoft S (1999) Healthy fishing communities: an important component of healthy fishing stocks. Fish Manage Essays 24(5):28–29

Jentoft S (2004) The community in fisheries management: experiences, opportunities and risks. In: Hersoug B, Jentoft S, Degnbol P (eds.) Fisheries development: the institutional challenge. Eburon Publishers, Amsterdam, pp 93–129

Jentoft S, Chuenpagdee R (2009) Fisheries and coastal governance as a wicked problem. Mar Policy 33:553–560

Jentoft S, Bonnie JM, Wilson DC (1998) Social theory and fisheries co-management. Mar Policy 22(4–5):423–436

Jentoft S, Son TC, Bjørkan M (2007) Marine protected areas: a governance system analysis. Hum Ecol 35:611–622

Jentoft S, Onyango PO, Islam MM (2010) Freedom and poverty in the fishery commons. Int J Commons 4(1):345–366

Johnsen JP, Murray GD, Neis B (2009) North Atlantic fisheries change: from organic associations to cybernetic organizations. MAST 7(2):55–82

Keesing RM (1981) Cultural anthropology: a contemporary perspective. Holt, Rinehart and Winston, New York

Kolding Z, van Zweiten P (2010) The tragedy of our legacy: how do global management discourses affect small-scale fisheries in the south? Public lecture given at the University of Tromsø, Tromsø, March 2010

Kooiman J, Bavinck M, Jentoft S, Pullin R (eds.) (2005) Fish for life: interactive governance for fisheries. Amsterdam University Press, Amsterdam

Kraan M (2009) Creating space for fishermen's livelihood: Anlo-Ewe beach seine fishermen's negotiations for livelihood space within multiple structures in Ghana. Ph.D. thesis, Amsterdam University, Amsterdam

Kroeber AL (2006) The concept of culture in science. In: Moore HL, Sanders T (eds.) Anthropology in theory: issues in epistemology. Blackwell, Malden, Oxford and Carlton, pp 36–40

Layder D (1994) Understanding social theory. Sage, London

LVFO (2008) The fisheries management plan for Lake Victoria 2009–2014. LVFO, Jinja

Macfadyen G, Corcoran E (2002) Literature review of studies on poverty in fishing communities and lessons learnt in using livelihoods approach in poverty alleviation strategies and projects. FAO Fisheries Circular No. 979, FIPP/C979. FAO, Rome

March JG, Olsen JP (1995) Democratic governance. The Free Press, New York

Mathie A, Cunningham G (2003) Who is driving development? Reflections on the transformative potential of asset-based community development. Occasional paper series No. 5. Coady International Institute, St. Francis Xavier University, Antigonish. http://www.sed.man.ac.uk/research/events/conferences/documents/participation%20papers/mathie.pdf. Accessed 10 Mar 2010

Matsuishi T, Muhoozi L, Mkumbo O, Budeba Y, Njiru M, Asilla A, Othina A, Cowx IG (2006) Are the exploitation pressures on the Nile perch fisheries resources on Lake Victoria a cause for concern? Fish Manage Ecol 13:53–71

McCay BJ (1989) Sea tenure and the culture of the commoners. In: Cordell J (ed.) A sea of small boats. Cultural Survival Report No. 26. Transcript Printing Company, Cambridge, pp 203–227

McGoodwin JR (1990) Crisis in the world's fisheries: people, problems and policies. Stanford University Press, Stanford

McGoodwin JR (2001) Understanding the cultures of fishing communities: a key to fisheries management and food security. FAO Fisheries Technical Paper 401. FAO, Rome

Neiland A, Bènè C (2004) Poverty and small-scale fisheries in West Africa. FAO/Kluwer Academic, Rome/Dordrecht

Njiru M, Waithaka E, Muchiri M, van Knaap M, Cowx IG (2005) Exotic introductions to the fishery of Lake Victoria: what are the management options? Lakes Reservoirs Res Manage 10:147–155

Onyango PO (2010) Re-configuring poverty: the wickedness perspective. Afr Trop Hydrobiol Fish 12:37–46

Owino JP (1999) Traditional and central management systems of the Lake Victoria fisheries in Kenya. Socio-economics of the Lake Victoria Fisheries Report No. 4. IUCN Eastern Africa Program, Nairobi

Panayotou T (1982) Management concept for small-scale fisheries. FAO Fish Technical Paper 228. FAO, Rome

Payne I (2000) The changing role of fisheries development in policy. Natural Resource Perspective 59. DFID, London

Pollnac RB, Poggie JJ (1979) The structure of job satisfaction among new England fishermen. Anthropological Working Paper no. 31. Kingston, RI: International Center for Marine Resource Development

Pollnac BP, Poggie JJ (2008) Happiness, well-being and psycho cultural adaptation to the stress associated with marine fishing. Hum Ecol 15(2):194–200

Pringle RM (2005) The origins of Nile perch in Lake Victoria. Bioscience 55(9):780–787

RAWG (2004) Tackling vulnerability: an approach to poverty reduction. A plain language guide to the 2002/03 Tanzanian participatory poverty assessment. Poverty Eradication Unit and Research on Poverty Alleviation (REPOA), Research and Analysis Working Group, Tanzania

Schumann S, Macinko S (2007) Subsistence in coastal fisheries policy: what's in a word? Mar Policy 31:706–718

Seligman M (1993) What you can change and what you can't: the complete guide to successful self-improvement. Vintage publishers, New York

Sen A (2002) Rationality and freedom. Belknap, Cambridge

Sen A (2009) The idea of justice. The Belknap Press of Harvard University Press, Cambridge

Sievanen L, Crawford B, Poolnac RB, Lowe C (2005) Weeding through assumptions of livelihood approaches in ICM: seaweed farming in the Philippines' and Indonesia. Ocean Coast Manage 48:297–313

Smith EV (1978) Four issues unique to socio-cultural indicators. Soc Indic Res 5(1):111–120

Smith ME (1988) Fisheries risk in the modern context. MAST 1(1):29–48

Smith LED, Khoa SN, Lorenzen L (2005) Livelihood functions of inland fisheries: policy implications in developing countries. Water Policy 7:359–383

Spicker P (1999) Definitions of poverty: eleven clusters of meaning. In Gordon D, Spicker P (eds.) The international glossary of poverty. ZED Books. CROP International Series on Poverty. Dhaka University Press, Dhaka

Thompson P, Wailey T, Lummis T (1983) Living the fishing. Routledge and Kegan Paul, London, Boston, Melbourne and Henley

Thorpe A (2004) The nature and causes of poverty: an overview. In: Neiland E, Bènè C (eds.) Poverty and small-scale fisheries in West Africa. FAO/Kluwer, Rome/Dordrecht, pp 7–36

Tzanatos E, Dimitrious E, Papaharisis L, Roussi A, Somarakis S, Koutsikopoulos C (2006) Principle socio-economic characteristics of the Greek small-scale fishermen. Ocean Coast Manage 49:511–527

United Republic of Tanzania (URT) (1997) National fisheries policy and strategy statement. Government Printers, Dar es Salaam

van Ginkel R (2009) Braving troubled waters: sea change in a Dutch fishing community, vol 4, MARE Publication Series. Amsterdam University Press, Amsterdam

Verschuren D, Thomas CJ, Kling HJ, Edgington DN, Leavitts PR, Brown ET, Talbot MR, Hecky RE (2002) History and timing of human impact on Lake Victoria, East Africa. Proc R Soc Lond B 269:289–294

Williams R (1981) Culture. Fontana Press, Glasgow

Wilson DC (2003) The community development tradition and fisheries co-management. In: Wilson DC, Nielsen JR, Degnbol P (eds.) The fisheries co-management experience: accomplishments, challenges and prospects. Kluwer, Dordrecht/Boston/London, pp 7–30

World Bank (2000) Voices from the village: a comparative study on coastal resource management in the Pacific Islands. Pacific Island Discussion paper series No. 9. World Bank, Washington DC

World Bank (1992) World development report. Oxford University Press, Oxford

Chapter 7
Vanished Prosperity: Poverty and Marginalization in a Small Polish Fishing Community

Boguslaw Marciniak

Abstract This chapter presents results of a study that was conducted in a small fishing community located on the shore of the Vistula Lagoon, where fisheries have played a significant role in the socio-economic and cultural life of the inhabitants of this area. After World War II, the populations of Vistula Lagoon communities and the surrounding areas were almost completely changed. Fisheries became a very important source of income for the local people, and a very important element in their social integration. Fishers were part of the social elite in their communities. Recently, a new political and economic reality (especially the introduction of the free market economy), aquatic ecosystem degradation, decline in fish stocks, and few employment alternatives have resulted in the diminishing economic power of fishers and other inhabitants of coastal communities. Fisheries can no longer guarantee sufficient income to live on. This chapter discusses strategies and activities that ought to be considered to improve socio-economic conditions in such communities; finds answers to questions on what kind of governance may prevent overuse and degradation of natural resources; and provides remedies that are suited to reduce the risk of poverty. Unstructured interviews and focus group discussions with local inhabitants, many of whom have been employed in the fishing industry, were the main research methods.

7.1 Introduction

In the 1980s and 1990s, Poland underwent major political change. First, the communist system, which functioned primarily through central planning and state-owned enterprises, was replaced with a free market economy and private ownership.

B. Marciniak (✉)
Sea Fisheries Institute in Gdynia, ul. Kollataja 1, 81-332 Gdynia, Poland
e-mail: bmarciniak@mir.gdynia.pl

S. Jentoft and A. Eide (eds.), *Poverty Mosaics: Realities and Prospects in Small-Scale Fisheries*, DOI 10.1007/978-94-007-1582-0_7,

Especially for those living in the larger cities, the new situation created previously unknown economic and social possibilities. For others, mainly residents of small towns and villages, the new system was often manifested in unpredictable changes for the worse in terms of unemployment, and social and economic instability (Milic-Czerniak 1993; Beskid 1993a, b; Jarosz 2005). Such changes also impacted on coastal fishing communities. Faced with limited employment prospects, and simultaneously, the progressive degradation of the natural environment, many residents found themselves threatened with poverty and starvation.

The second major change occurred when Poland, on 1 May 2004, acceded to the European Union. With this, European Union law and directives began to be introduced in Poland, which also included fisheries directives. One of the most controversial European Union directives was regarding the significant and rapid downsizing of the Polish fishing fleet, the aim of which was to adapt the fishing capacity to the resources available in the Baltic Sea (Horbowy and Kuzebski 2006).

The outcome of these two events can be seen in the fact that each year in Poland there is an increase in the number of small, coastal localities where no resident is employed in fisheries, and where there are no longer active fishing boats (Rakowski 2008). Now, the seaside locations of these communities are exploited solely as tourist attractions. In other coastal localities, the number of fishers, fish cutters, and boats based in these ports are decreasing systematically.

An important fact is the degradation of the natural environment of the Baltic Sea and the errors in fisheries resource management. Overfishing has reduced fish stocks in the Baltic Sea and has caused economic decline. Aging of the fishing population, lack of young people willing to work in this industry, and very limited possibilities for fishers to participate in decision-making processes regarding the fisheries and the marine natural environment have become the most serious barriers in recruitment of new fishers. Marine fisheries are increasingly unable to guarantee its practitioners adequate incomes to support families, and poverty is becoming a reality in the lives of fishing families.

The degradation of the natural environment (e.g., the overfishing of eel stocks, and contamination from agricultural activities which reach the Baltic Sea via rivers (Igras and Pastuszak (2009))) has had a direct impact on the quality of life of the residents of coastal communities. As the size and value of catches decline dramatically, threats to the economic well-being of fishers and their families become a reality. Such a situation is also creating an inverse co-dependency; poverty among the residents of these regions often leads to the further degradation of the natural marine environment. While the residents of these localities are fully aware that continuing to fish endangered species will lead to the further degradation of the natural environment, they have no alternative but to continue pursuing such a destructive activity, as this is the only source of income. Questions that demand answers include: (1) To what extent will initiatives to alleviate poverty and the threat of starvation lead to the protection of the entire ecosystem? (2) How will fishers and coastal residents survive the current crisis situation?

In this chapter, I argue that the optimum solution appears to be to establish permanent links between the fight against poverty and the preservation of the natural environment. But how should this be done? In order to answer these questions, one must first consider how poverty should be defined so that the description and analysis takes into consideration the socio-economic situation of the residents of small, coastal fishing localities. Who exactly are the impoverished members of these communities? Why are they impoverished? Are fishers among them?

Based on the analysis of the socio-economic situation of the Vistula Lagoon fishers, the state of the natural environment of the lagoon, and one particular community, I hold that solving the environmental problem of resource degradation is significantly more complex than just preventing illegal catches or reducing excess numbers of fishing vessels. The degradation of the natural environment and the impoverishment of fishers are mutually dependent phenomena which must be resolved simultaneously. I also argue that solving the problem is only possible with the active participation of fishers, other inhabitants of coastal communities, local administrative authorities, as well as the central authorities of the nation. Integrated actions must form the basis of creating long-term strategies that take into consideration the needs for environmental protection and the socio-economic situation of the fishers.

In this chapter, the issues of poverty, resource degradation, and governance implications are discussed through a case study undertaken in the Tolkmicko district, a small coastal fishing community situated by the Vistula Lagoon. The case study is preceded by a presentation of the characteristics of the community and its fisheries. Here, I draw on the opinions of local people as to how these problems should be addressed. The final section discusses what kind of governance might achieve the aim of alleviating poverty, while protecting the environment.

7.2 Study Methods

The study used various research methods and techniques that allowed multi-thematic analysis of the social situation. The study was performed based on the triangulation procedure, which is the application of a combination of research methods to collect materials as well as the combination of qualitative and quantitative analyses. This permitted testing the same hypotheses without the burden of error stemming from the limitations and flaws of individual methods (Neuman 2003).

Data were collected in 2008/2009 using unstructured interviews with local fishers and other inhabitants of towns and administrative districts (25 people) and unstructured interviews with people currently or previously serving in important community positions, who had relevant information regarding the lives and work of local residents. These included representatives from local administrations, and non-governmental organizations (NGOs) (10 people). Informative focus group discussions were held with fishers (9 people), fishers' spouses (12 people), and the oldest residents of the administrative districts (8 people). Additionally, available official documents were analyzed, and private talks were held with inhabitants of the region in order to discover underlying dimensions of the poverty problem in this community.

The statistical data presented in this chapter illustrate the scale of impoverishment, while the catch results presented are official data that do not always reflect reality.

It was impossible to determine the actual levels or sources of income of the members of the community or of individual households. It was also impossible to establish the actual scale of unemployment, the employment level, or the magnitude of migrations. The reason for this was that many people are employed either informally or seasonally – primarily in construction, agriculture, and fisheries.

Pilot studies that focused on a few fishers preceded the study proper. The fishers chosen had decided to liquidate their vessels, or were those who had lost their jobs as a result of decisions made by owners to scrap vessels. Within the scope of pilot studies, conversations were held with fishers who were continuing to work and with representatives of local government.

Tolkmicko, the fishing community that was the subject of the study, was chosen because, until recently, fishing was extraordinarily important as a source of income (although not the only one) for many residents. The term "community" is defined as a group of people whose lives are played out in a defined area; or those who identify with this area and hold it in particular esteem, even if they reside outside of it (Jałowiecki 1988; Wodz 1989).

7.3 Tolkmicko District

The study was conducted in the administrative district of Tolkmicko, which comprises the small fishing town of Tolkmicko and the neighboring localities of Suchacz and Kamienica Elblaska. Tolkmicko is a fishing and agricultural community in which the former has always played the leading role in the socio-economic and cultural lives of the residents.

The Tolkmicko administrative district is located in northern Poland on the southern coast of the Vistula Lagoon, not far from the border with Russia (Fig. 7.1). The surface area of the lagoon is 838 km^2, of which 328 km^2 are located within the borders of Poland.

The length of the Vistula Lagoon is 90.7 km; of this 35.1 km are located within the borders of Poland. The width of the lagoon ranges from 6.8 to 13 km, and the mean depth is 2.7 m. The Vistula Lagoon is the natural aquatic environment for species such as eel (*Anguilla anguilla, Anguillidae*), pikeperch (*Sander lucioperca, Percidae*), herring (*Clupea harengus membras, Clupeidae*), perch (*Perca fluviatilis, Percidae*), bream (*Abramis brama, Cyprinidae*), ziege (*Pelecus cultratus, Cyprinidae*), and pike (*Esox lucius, Esocidae*). The Vistula Spit, which was formed by westerly winds and marine currents carrying sand masses, separates the lagoon from the Baltic Sea.

Thanks to the advantageous position on the shore of the Vistula Lagoon, the village and district developed progressively and became an important port and fishing center in this area many years ago (Wielopolski 1946). Many inhabitants of Tolkmicko still have old, edited before World War II, postcards presenting a local fishing harbor full of fishing and cargo boats. The population, however, has undergone fundamental change.

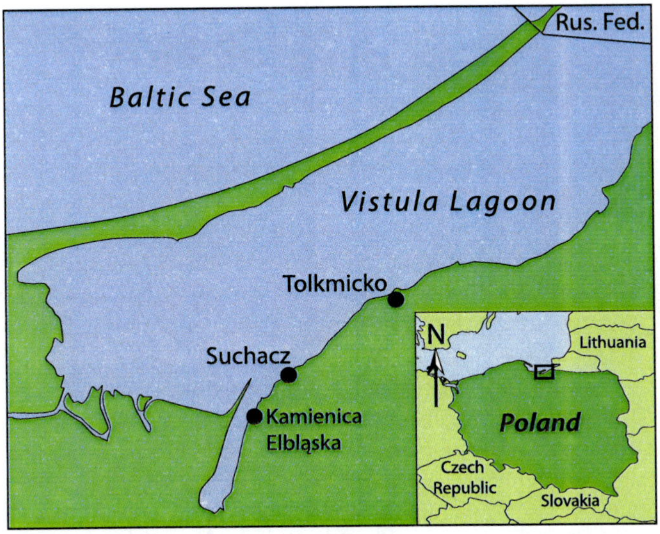

Fig. 7.1 Map of the Vistula Lagoon and the Tolkmicko district (Source: Sea Fisheries Institute in Gdynia)

As a result of long-term German colonization, the area was inhabited nearly exclusively by Germans prior to World War II. After the conclusion of World War II and the annexation of Tolkmicko to Poland, most Germans left this area in 1945–1946. The current residents of Tolkmicko are exclusively Polish, according to information provided by representatives of local administration.

According to the information provided by the local government, the population of the Tolkmicko district in 2008 was 6,663 residents, of which 2,793 live in the town of Tolkmicko and 3,870 live in the remaining areas of the district. The residents are employed in fisheries, vegetable processing, various services, the local administration, and education. For approximately 140 people, work in fisheries is one of the most important sources of income.

In Poland, education is traditionally one of the most important indicators of social status. It is not surprising that the majority of young Tolkmicko residents who graduate from vocational high schools and universities decide against returning to the Tolkmicko district. The peripheral location of the town and the district, combined with severely limited employment opportunities, has meant that only those with guaranteed employment or those with very low levels of education decide to stay in the area. The district has five schools, a kindergarten, a local cultural center, a public library, two community homes, a post office, a health center, a police station, five churches, and centers of village and district state administrations. The schools, the Roman Catholic Church, and the Public Library are strongly engaged in a campaign to protect the local natural environment and restore fish stocks in the Vistula Lagoon waters. They organize meetings with authorities and hold public lectures on the protection of the natural environment, as well as the cultural heritage of the community.

7.4 Fishing in the Tolkmicko District

The Tolkmicko district has three fishing ports: Tolkmicko, Suchacz, and Kamionek Wielki. There were also two large enterprises linked to the fisheries in the district. These were the Zalew Fishing Cooperative and its subsidiary fish-processing plant. The cooperative also owned a small repair shop where boats, nets, and other fishing gear could be repaired. Community ties among the new residents of the district, who had come from different parts of the country, were first established through fishing and the processing industry. The teamwork required by fishing quickly promoted social and cultural integration of the new arrivals and helped to create a new community in this area.

The socio-economic changes in Poland in the 1990s, combined with significant decreases in the size and profitability of catches, led to the liquidation of the fishing cooperative, the repair workshop, and then later the fish-processing plant. The fishers who owned fishing boats received an offer to buy the cooperative's fishing nets, working clothes, and other fishing equipment. The entire fishing fleet, including all equipment and fittings, became the private property of the fishers. In subsequent years, the number of fishing vessels operating out of the ports in the Tolkmicko district more than halved during the period from 2004 to 2008. The number of boat fishers in the Tolkmicko Commune has steadily dropped to 43 in 2008 compared to around 100 in 2004 and 150 in 2000. The decline in the numbers of fishing boats is represented in Table 7.1.

Sweeping changes also occurred in the size of the Polish fishing fleet during the period from 2004 to 2008. Until the end of 2008, the fleet reduction had resulted in a 34% withdrawal of fishing vessels that had conducted fishing activities in the Baltic Sea (Kuzebski 2009). This was in response to the decreasing economic effectiveness of catches, and the implementation of the fishing capacity reduction program under the auspices of the Common Fisheries Policy of the European Union. This took on great proportions when the owners of fishing boats learned that substantial financial compensation was being offered to scrap vessels (Ministerstwo Rolnictwa i Rozwoju Wsi 2004).

Owners of fishing vessels were free to make the decision to withdraw their vessels. According to the fishers, the financial compensation received far exceeded the market value of the vessel, in many instances. It was awarded, although to widely varying degrees, to both vessel owners and crew members who lost their jobs as a result of the decision to scrap their vessels, and who were then barred

Table 7.1 Fishing boats in the Tolkmicko district ports during 2004–2008

Port/year	2004	2005	2006	2007	2008
Tolkmicko	23	14	9	7	6
Suchacz	9	6	6	6	6
Kamienica Elbląska	3	2	1	1	2
Total	35	22	16	14	14

Source: Fisheries Economics Department, Sea Fisheries Institute in Gdynia from returning to work in fisheries for 1 year (Ministerstwo Rolnictwa i Rozwoju Wsi 2004).

Most owners of scrapped boats expressed positive opinions about this offer. Some of them invested the received money in boats that remained. Due to advanced age, some planned to end their work and the compensation money made their future more secure. Others could pay their debts. According to a boat fisherman, aged 60+, with more than 21 years work experience:

> Fishing isn't as attractive a job as it was many years ago. With the sizes of catches and the new restrictions that are constantly being introduced, it's impossible to provide for your family or to have any job security. This is the main reason that fishers are no longer among the social elite of our community. Because of my age, and the age of my fishing partner, we would have had to retire soon anyway. Neither his children nor mine wanted to work in the fisheries since it is becoming harder and harder to support a family and there's no job security. If I had sold the old boat, I would never have got as much as I did by scrapping it. This is an excellent solution for older fishers with nobody to take over their businesses.

Younger fishers, especially crew members, had a different opinion. Many of them were not satisfied with such an option. They received financial compensation of €10,000 each when they declared no fishing for the following 1 year, but this was not a long-lasting solution for them and their families. A boat fisherman, aged 40–49, with more than 6 years work experience stated:

> This is a terrible decision. Why in the world are people getting money to destroy boats that could have provided jobs for many of us? There is no other work here, and that money won't last long. I don't want to leave this area because of my family. I don't have any education that would allow me to work in any other job. They threw us out into the street. And what about all the people who were working under the table fishing and repairing gear? These people didn't get any money, and there are far more of them than there are of us fishers.

The negative effects of the fleet-restructuring program included the irreversible loss of many jobs for people in the community, and the liquidation of many fishing boats that were in good condition.

The realization of the program to reduce the fishing capacity of the fleet and protect natural marine resources effected far-reaching changes in the lives of a considerable segment of the coastal population. Some of these inhabitants did not know how, or could not adapt to the new reality. A fisherman, aged 50–59, with more than 21 years work experience said:

> I never dreamed that after the political transformation things would get so bad. The fishing co-operative guaranteed that the catches made by its fishers would be paid for at a good price. Fishers had paid vacation, insurance, and access to state health care. The state systematically stocked lagoon waters with eel fry. The women had work in the processing plants or in agriculture. Today's youth can't believe that it was ever like this. The unsuccessful privatization of everything made a very few people rich, but left most with no way of earning a living. The fish weren't exploited – they were plundered, and this has ruined stocks. Maybe the situation will improve and the boats will again set sail for fishing grounds, but this is becoming more and more difficult to believe.

The boats from Tolkmicko district ports operate only on the lagoon waters and only in the Polish part. Fishing cruises are short, and usually do not exceed 7 h. Formally, crews on each boat comprise 2–4 fishers depending on the fishing season

Fig. 7.2 Fishers at work. Formally, two or three and very seldom four fishers are employed on each boat and 5–10 persons cooperate as an informal land crew, depending on the type of fish caught, the fishing season, and the gear used (Photo: B. Marciniak / Sea Fisheries Institute in Gdynia)

(Fig. 7.2), the type of fish caught, and the gear used. In reality, the number of crew working in connection with one boat is much larger and comprises 6–12 people. These additional people are known as the land crew (Marciniak 2008). They are employed informally, meaning that they are not registered as fishers, do not pay tax, and have no unemployment and pension benefits. They work as needed to repair gear, sort catches, repair boats, and take part in fishing sporadically. These people are usually the spouses, children, and relatives of the formal crew, but local, unemployed people, and retired fishers are also involved. It is a local practice that formal crew members are usually hired based on family ties; however, recently, these rules have not been strictly followed.

In an effort to counteract overfishing, several short-term catch limitations have been imposed by the state administration. These include quotas, closed seasons for particular fish species, closed fishing in spawning grounds, and regulations regarding permissible gear mesh size. The increasing number of catch limitations has meant that the fisheries are no longer a very attractive job choice, and it has become necessary to look for new crew recruits outside of family circles.

Within the framework of the Common Fisheries Policy, the European Union has also implemented catch restrictions on boat fishing. Fishers cannot register new fishing boats until they withdraw the old boats from fishing. It is also now impossible to upgrade engines aboard fishing boats to more powerful models (Ministerstwo Rolnictwa i Rozwoju Wsi 2004). Periodical closed seasons for individual fish species are also in force.

In recent years, the equipment on board fishing boats has improved substantially, including much more powerful engines, more durable fishing gear, and better rescue equipment. All of this is exceedingly expensive, and acquiring it is a financial burden for fishers and their families. Fishers have to do it if they would like to successfully compete with other fishers, and improve the safety of their work. However, fishers are faced with problems such as a lack of employment stability and decreasing incomes. The young residents of the district do not accept this, and are increasingly relocating either temporarily or permanently to other parts of Poland or even abroad.

Fishing no longer guarantees a reliable and sufficient income. Consequently, the significance of the fisheries as a source of living is decreasing quickly. Working in fishing is, in many cases, accompanied by working in other areas such as construction, agriculture, and tourism, but the opportunities for diversifying into other sources of income are temporary, and are highly restricted by the number of options available, which cannot guarantee sufficient incomes. The community has been, and is now, particularly sensitive to changes in the economic situation and the state of the coastal ecosystem. Additionally, the fisheries management system is very unstable, and there is no possibility to increase the variety of species caught.

Consequently, fishers experience socio-economic degradation: unemployment has increased in coastal districts; unemployed fishers are threatened by risk of poverty; previous social standing and prestige is lost; and feelings of helplessness in confrontations with local and central state administrations have become common.

7.5 Absolute Versus Relative Poverty

Poverty is often defined and understood as the situation in which an individual does not have sufficient income or the means to earn sufficient income to meet his or her basic biological needs. This is referred to as "absolute poverty" (Brym 2001). For many people, however, poverty is defined as economic, social, and political marginalization resulting in individuals or entire social groups being deprived of the possibilities and rights that are available to other members of the same social group. This is referred to as "relative poverty" (Galbraith 1970; Townsend 1976; Gerber and Macionis 2002, p. 283). The causes of poverty thus defined, which are consequences of marginalization, include the lack of education and the lack of acceptance by others based on ethnicity, race, gender, or health issues.

Duffy and Mandell (1994) present a similar definition of poverty, according to which "the poor [are] those who have too little to get by and who are unable to participate in any meaningful fashion in the social, political, educational, or spiritual life of the nation. While these individuals are not (necessarily) starving or homeless, they are "relatively deprived" in the nation and community in which they live."

In Tolkmicko, we could include in this category: elderly, youth, Euro-orphans (children left by their parents working abroad (Strybel 2009)), single parents, victims of addiction, disabled people, unemployed, underpaid "working poor" (Gerber and Macionis 2002, p. 287), and many former fishers, especially crew members.

In writing about poverty among Polish fishers, it is very risky and indeed also somewhat unethical to use the definition of "absolute poverty." The standard of living and the income level, as well as the state health care and social security systems available, are markedly superior to those in many countries substantially poorer than Poland. This is not to infer, however, that so-called rich countries do not have poor citizens whose standards of living are significantly lower than the majority of the citizens of these countries. Although cases of "absolute poverty" do occur in these countries, the number of them is smaller and the duration is usually shorter than those in developing countries. Thus, it is more appropriate to use the definition of relative poverty to describe the poverty experienced by Polish fishers.

Both "absolute poverty" and "relative poverty" can result in social marginalization and social exclusion. But there are other causes of marginalization and exclusion that are not related to poverty (e.g., self-imposed marginalization and exclusion for improper behavior). Marginalization is a process through which individuals or social groups assume a peripheral position in relation to the other members of a given community (Mahler 1996). Although poverty might be one of the causes of marginalization, it does happen that marginalization is one of the causes of poverty. Severe, long-term marginalization can lead to social exclusion.

According to Narodowa Strategia Integracji Spolecznej (The National Strategy for Social Integration) (2003), social exclusion is defined as having limited or no possibilities for participating in, influencing, or using the basic public and market institutions that should be available to everyone, especially the poor. The document also designates social groups that are susceptible to social exclusion and those that are seriously threatened with social exclusion. This list includes, among others, the long-term unemployed, single mothers, people addicted to alcohol or drugs, the homeless, and illegal workers. According to those who study this issue, many of the marginalized and excluded can be classified as belonging to the so-called underclass (Myrdal 1962; Duffy et al. 2007).

For this study, key information concerns how the local community defines impoverishment and who is considered to be poor and/or marginalized. The majority of those who participated in the study spoke of poverty in terms of "absolute poverty," the index of which comprises the level of current income and the risk of starvation. They designated the impoverished as those who are unemployed or who have earned so little that they are unable to meet the basic material needs of their families. These included the so-called working poor. This category also included former fishers, fish-processing employees, and those who earned extra income illegally and are known as the "land crew" of fishing vessels. The reduction of the fishing fleet in Tolkmicko, the old age of most of the fishers, the relatively low level of education, the lack of other professional qualifications, the lack of sufficient financial means necessary to relocate, and the lack of alternative employment have caused rapid deterioration in the material and social status of these people and their families, making them relatively poor in comparison with other people of the district.

In reality, most of the residents of the Tolkmicko district are experiencing a feeling of deprivation because of their material situation. Residents who are still formally employed in the fisheries or are receiving assistance payments from the years worked in fishing cooperatives are not included in this group. Although fisheries remain the main source of income, their current living conditions are worse than before. The situation of former private fishers is worse since retirement benefits are not as good as those paid to former cooperative employees. Poverty is not an issue for residents who have moved to the district recently, after the political system turnover. Most of them arrive with capital earned in other regions of Poland or abroad, and they choose to build their homes in the most picturesque areas of the district that are far away from the older residents and their problems. This is yet another source of stress for those who have never been, are not, and will never be able to afford a comparable standard of living.

7.6 Causes of Poverty

There are many reasons why the situation of Tolkmicko residents, including local fishers, has taken such a radical turn for the worse. Mismanagement of the lagoon ecosystem is certainly one of them. This is especially true for eel fisheries, as there were no limitations on catches of this resource for years (toward the end of the 1970s and the beginning of the 1980s). Eel was an especially attractive catch because of the income it earned. Fishers who exceeded the planned eel catch were rewarded by the buyers and cooperative administration with cash bonuses. Fishing was, therefore, an extraordinarily attractive job for all the inhabitants of the district.

This continued for as long as the lagoon waters were stocked annually, especially between 1973 and 1984 (Borowski 1994). Funding for stocking programs came largely from income taxes levied against the fishers (4% of fisher's income). After the fishing cooperative was liquidated (taken off from the register list in 2003) and the local fisheries were fully privatized, the stocking program was discontinued since there was no institution to administer it. Catches of eel declined in subsequent years (from approximately 28,000 kg in year 2000 to 3,781.7 kg in 2004 and 628.5 kg in 2008 according to the Sea Fisheries Institute in Gdynia).

The erosion of ethical standards in the community is manifested in illegal fishing and informal employment, which are other reasons why poverty now affects the residents of the district. Poaching in the Vistula Lagoon leads to significant decreases in fish resources. Informal employment in the fisheries also leads to the excessive exploitation of the lagoon's fish resources, and those who worked informally currently have no right to unemployment benefits, health insurance, or retirement pensions. Many of these people are impoverished and must turn to various types of social services.

Alcohol addiction is another cause of poverty among a significant segment of the local population. Excess alcohol consumption is traditionally part of the social relations and the culture of fishers and the residents of many small, coastal communities (Peace 1991). The lack of alternatives for spending free time and the inherited patterns of this cultural behavior mean that the social lives of some inhabitants of these communities center around alcohol consumption. Fishers in particular are known for their ability to consume alcohol frequently, and for some this is confirmation of their professional success and their competence in fishing (Peace 1991). A fisher's daughter, aged 50–59, who is also married to a fisher, and has more than 6 years work experience in fishing services said:

> My father was a drinker and a fisher. My husband and his brothers were drinkers, and they were also fishers. Right now, because of a lack of money, most fishers can't afford to buy alcohol often. Years ago, the fishers worked much more and under much worse conditions than did any of the other local residents. They also drank a lot more alcohol. Good catches meant that the fishers had excellent incomes, and this is something they would show off in the local restaurants. There was a time when fishers and their guests were really the only customers in these places. Drinking vodka in restaurants and being able to buy rounds for others was a way of proving that you were rich. It gave you great social status and was a way of proving your "manhood".
>
> It was accepted as the norm that most of the fishers were alcoholics, and drinking together was an important way of maintaining friendships among the fishers. Alcoholism ruined the lives of many of these men; it drained their pockets of money, destroyed their families, and, in time, it affected their health. Today, there are no more rich fishers, and the alcoholics around are usually unemployed fishers. Alcohol provides an escape from hopelessness and the lack of perspectives for a better future. In order to get money to buy alcohol, they work illegally, beg, or even steal. Today, the problem of alcoholism is no longer identified with belonging to the fishing elite; it has become a symbol of total social marginalization and the lack of will to fight for a better future.

As indicated in this quote, excessive alcohol consumption is perceived as an indication of the erosion of moral values, and is linked to the degradation of the natural environment. Alcohol-dependent fishers are often willing to fish illegally to earn money to purchase alcohol.

Low-paying jobs, which are supposed to be alternatives to fisheries work, have been attributed a far less significant role in the fight against poverty. With no alternative employment opportunities, the residents of the district are forced to accept work that does not guarantee a sufficient wage for supporting the household. Some of the residents earn extra money from illegal fishing or illegal forest exploitation.

The fishers themselves have contributed to the degradation of the ecosystem. During the period of no restrictions on eel catches, the number of fishers and those willing to work informally in the fisheries increased rapidly each year. Many of the new fishers, often without training, were concerned only with increasing their earning in the shortest possible period. Catches of eel, the most valuable species, were decreasing, and it was increasingly difficult to find buyers for herring. Low prices and the poor diversity of fish species inhabiting the lagoon meant that catches of other species were not as profitable for fishers as were catches of eel. After the full privatization of the fisheries, fishers ceased paying for stocking programs. In subsequent years, the magnitude of stocking decreased and then stopped altogether. For a short period of time, companies from Germany and Holland participated in stocking programs, and in exchange for providing stocking materials, they purchased adult eel from the fishing cooperative (information provided by former members of the cooperative).

A free market economy that occurred as a result of the rapid political changes forced the implementation of new, previously unknown operational methods.

Among the former members of fishing cooperatives and state employees, nobody had any experience or knowledge of how to work in the new economic reality. Literally overnight, these people had to adapt to operating independently on the economic market and to organizing catches and the repair of equipment. They also had to take over the sale of the catches, the management of the fishing grounds, and the privatized ports. Not everyone knew what to do, and not everyone understood that "top to bottom" management on a local scale had ceased to exist. It had been replaced by a "self-management" model that required immense independence and the ability to make appropriate decisions. The erosion of social connections resulting from rapid change did not permit organizing cooperative efforts to manage the fishing grounds or the sale of fish. Cut-throat competition among fishers from the Vistula Lagoon eliminated many of them from the fisheries. The threat of poverty became a reality, and their positions in local communities became much weaker.

During the system transformations, all of the production enterprises operating in the district were purchased by private owners from other areas. For various reasons, these new owners decided against continuing production, and most of the enterprises ceased production within a short period of time. Many of the district residents became unemployed, especially those of middle and older age. Those who managed to keep their jobs had to accept the grounds for employment established by the new owners. Wages fell considerably and social benefits were reduced significantly. Employees were expected to increase yield and show discipline without any guarantee of increased earnings or permanent employment.

Herring catches which were meant to replace eel catches could not guarantee the profitability. The same was true for catches of pikeperch. In both fisheries and vegetable processing, very few employees could count on stable, full-time employment. Such guarantees were only offered by positions in the state administration and the services sector, but there were few such jobs in the district.

The peripheral location of the district, far from the main transportation routes, and the liquidation of the railroad connection to the nearest large city both significantly limited possibilities of finding employment outside the district. The numbers of tourists who visited the district were limited for the same reasons. The income of residents, many of them former fishers and members of fishers' families, employed in seasonal jobs serving the tourist trade (food sales, accommodations, souvenir sales), decreased to such a degree that most quit these jobs. Unemployment benefits, informal employment, and illegal fishing became the most important sources of income for many families.

Paradoxically, the position of older residents improved since they had earned the legal right to retirement or disability pensions. In many families, the retirement or disability pensions of the older members were the only regular, guaranteed source of income. The social position of the older members of the community was strengthened considerably, and their membership in families became the best financial guarantee for the family's well-being. The situation was particularly dire for older people who had not earned the right to a retirement or disability pension and who had no relatives who were willing to help. These people (among them were former informal

Table 7.2 Number of residents of the Tolkmicko district who relied on social assistance from the District Social Services Center during 2006–2008

2006		2007		2008	
Families	Family members	Families	Family members	Families	Family members
721	1,988	625	1,748	570	1,527

Source: Information provided by District Social Services Center (GOPS)

fishers) ended up in the care of the social services (Table 7.2) and relied on the daily assistance of social workers and free meals provided by the District Social Services Center (GOPS).

In 2008, in the Tolkmicko district, more than 1,500 residents relied on this assistance. According to information provided by the local GOPS, 80 took free meals; others required daily assistance and various services. The situation was the worst in 2006, when very few local people decided not to migrate to Western Europe to look for job opportunities, and when many West European countries did not allow Polish citizens to work there legally. Moreover, in 2006, many more people were able to fulfill the formal criteria to qualify for social assistance which later expired and could not be renewed.

The fishers who decided to continue working in the fisheries contend that the threat of poverty facing their families is increasingly real. The fishers attribute this to a number of factors:

- continually changing directives and management systems set forth by the central state administration;
- poor species diversity of commercial fish;
- too many fishing vessels and fishers operating on the lagoon in relation to fish resources;
- limitations imposed by fishing seasons for particular species;
- closed areas of the lagoon;
- administrative limitations on the modernization of fishing boats; and
- illegal fishing or the deployment of illegal gear.

The fishers also declared that catches are made in a basin that is too small, as fishing is not permitted on the Russian side of the lagoon. The multiple causes of experienced and potential poverty suggest a broad poverty alleviation approach and a complex formula aimed at governance structures, management mechanisms, and ecosystem damage repair.

An increasingly urgent problem, and partly related to what the fishers have listed above, is the lack of young people willing to work in the fisheries. The young, who are aware of all these difficulties, do not want to take up the fishing profession. The lack of heirs is increasingly at least part of the reason why vessel owners decide to scrap their boats or hire older people or the unemployed as crew. Because of a lack of qualifications and formal work permit (fisher license), the majority of these people are employed for only short periods of time.

7.7 The Face of Poverty

According to information provided by the Regional Office of Employment in Elblag, the rate of unemployment in Tolkmicko Commune during 2000–2007 exceeded 30%. In Tolkmicko district, people attempt to hide their poverty, but this is not always possible. High unemployment is visible in queues for unemployment benefits at the employment office. Although still relatively rare, there are cases where people start begging from relatives and neighbors. This would suggest that for many poverty is moving from being "relative" to becoming "absolute." The small catches of fish only confirm that it is increasingly difficult for fishers to support themselves and their families through fisheries. This is also a confirmation of the advanced state of the degradation of the marine resource base.

Many residents describe themselves as working poor, which means that their incomes only provide the basic needs of their families. They report that household consumption is significantly limited and that no purchases are planned for means of transport such as cars, motorcycles, or bicycles. They also report that purchases of large household items such as furniture, televisions, washing machines, or refrigerators are impossible. Fewer people are able to leave their places of residence to go on vacation, and more of them are spending their vacation working in agriculture or construction, which in most instances is informal work, as defined above. Purchases of clothing and food are also curtailed. Children and youth from poor fishing families are deprived of access to expensive educational aids such as computers and books, which puts them at a disadvantage in comparison to their peers from other parts of the country. Private lessons, which help children to perform better in school, thus increasing their chances of continuing education, are also too expensive.

Feelings of inferiority are experienced by an increasingly large group of people. If impoverishment lasts for any length of time, it begins to be noticed by others. These people are stigmatized as poor, which leads to a loss of confidence in their ability to get out of their dismal situation. They deal with this stress by taking it out on their closest family members, friends, or various material possessions found in their places of residence. The very strong systems of social control that used to function in small communities are no longer as effective as they once were. Acts of aggression and violence committed within families, and vandalism in one's own community are a common affirmation that those who are suffering from poverty are not dealing with it well. This is also confirmed by numerous instances of petty theft, illegal taking of fish from set gear, and theft of wood from forests.

Migration abroad in search of work is now common. Those who depart refuse to give in to the local hopelessness caused by a lack of real prospects for any kind of change. Approximately 200 people, many fishers among them, from Tolkmicko district have emigrated to Great Britain, Germany, Norway, and Sweden. Those who left were either young or middle-aged. To date, generally one family member has gone, most of them for a relatively short period of time. Language difficulties, the lack of qualifications, and no understanding of the job market are the reasons immigrants settle for informal jobs at much lower wages than are paid to local workers. Currently, those emigrating realize that, as residents of the European Union, they have a right to legal employment that guarantees an adequate salary and social benefits. An increasing number of these people, who see no real prospects for finding permanent, well-paid work at home, are emigrating with their whole families. These emigrations, especially when the children begin to go to school in the foreign country, might turn out to be permanent. The passive attitude of the local authorities toward this issue signals their acceptance of this substantial reduction in the human capital of the local community (Amarasinghe 2009) – something that does not bode well for the future of Tolkmicko.

A small group of district residents conduct trans-boundary trade at the border with Russia. This occupation is semi-legal and entails crossing and recrossing the border several times during the same day carrying the permitted quantity of alcohol and cigarettes each time. The magnitude and scale of this activity is not known, but for some it is a significant source of income.

Fishers, especially former crew members of fishing vessels that were scrapped, often find work on foreign fishing vessels or in the fish-processing industry mainly in Great Britain, Ireland, and Scandinavian countries. Thanks to their professional skills and ability to perform arduous work, they do not have a very difficult time finding work.

Impoverishment is also apparent in the local fisheries. Most of the fishers (especially those who did not receive any financial compensation for scrapping their boats) are unable to accumulate sufficient funds to purchase new gear, to upgrade vessels, or to invest in port infrastructure (refrigeration equipment, means of transport).

One consequence of the declining significance of fisheries to local economies is that the opinions of fishers were ignored regarding the rational management of the local ecosystem. For example, fishers were not consulted when Natura 2000 (an ecological network of protected areas, set up to save Europe's most valuable species and habitats) (Natura 2000 2004) was being developed, even though the aim of the program was to exclude some fishing grounds from continued fishing.

7.8 Poverty Alleviation

The fishers and other residents of the Tolkmicko district, who were interviewed, declared their willingness to combat the threats of poverty and the processes of social marginalization. They believed that the government is responsible for creating a management model for the natural marine environment, for developing strategies, and for implementation that aims to improve the state of the lagoon ecosystem. This would guarantee the economic effectiveness of the proposed solutions, and also take into consideration the social consequences of their implementation. People interviewed also insisted that there should be a dialogue between scientists, representatives of government, as well as the local users.

Actions taken should include the monitoring of catches, and sanctions for damage to the natural environment. Regulations formulated by the government should set forth the number of fishers entitled to fish the lagoon, the periods in which catches of particular species of fish are permitted, the types of permissible gear, technical parameters of the vessels licensed to fish, and the amount of financial compensation awarded during closed seasons as stipulated by regulations. People expect the government to identify categories of people which are especially susceptible to poverty. These include the disabled, older people, single parents, children without adequate supervision (since their parents are working abroad (Euro-orphans)), and the working poor, who should receive financial assistance and social services when they are in difficult material circumstances.

Concrete strategies for reducing the threat of poverty are developed by the local state administration, local NGOs, local resident ad hoc groups, and task groups. The GOPS assists residents who qualify for material assistance, and these are usually older residents, single mothers, people with disabilities, the unemployed, and people with alcoholism. More than 260 families take advantage of this kind of assistance. The Catholic Church assists those who, for various reasons, are ineligible for assistance from the GOPS. They provide material assistance, can help in finding work, and organize sports and cultural events for youth and older residents. The church participates actively in all activities aimed at rebuilding the social ties that have been destroyed by the financial crisis and provides an equal chance for the impoverished. A priest from the Roman Catholic Church stated:

> We try to help the members of the community who are living in poverty because of unemployment and low wages and have no hope of living like others in our country. It is also difficult for older people, and those who are either ill or handicapped. Most of these people are too proud to seek assistance from Social Services. When we visit them in their homes, we can see what their standard of living is. Times are such that the church must become involved in the fight against poverty so that these people don't lose their sense of self-worth or their faith in life. Together with the Public Library, the sports club, and the Community Culture Center, we organize various cultural and sporting events, meetings for community groups, excursions, and after-school programs and vacations for children. We also have a meals program. All of these are free of charge. When the situation improves, we'll withdraw from these types of activities, but, for the time being, we have to work to minimize the social marginalization of many people and prevent them from feeling helpless.

The village cultural center and the public library both organize cultural events, make computers available for residents, and organize excursions for those who would otherwise not be able to afford them in an effort to limit social marginalization among district inhabitants. The local schools also make an effort to build local human capital, so that children from poor families receive the same education as their peers from wealthy families or urban areas. All of these organizations work to increase awareness for the necessity of protecting the local ecosystem and to build a sense of belonging within the district.

Local administrative authorities see better opportunities for combating poverty through the development of tourism, rather than through maintaining the fisheries at the current level or even developing them. In their opinion, tourism creates new employment opportunities for local inhabitants, while the profits earned will contribute to the district economy. Since the size of the local fishing fleet has decreased considerably, the fisheries are no longer a significant element of the local economy.

However, the negative sentiment of government officials with regard to the potential of the fishery seems to be slowly changing recently. This is based on the publication of the Operational Program "Sustainable Development of the Fisheries Sector and Coastal Fishing Areas 2007–2013" (Republic of Poland, Ministry of Agriculture and Rural Development 2008), which presents directions on how to use funds allocated in Poland under the European Fisheries Fund 2007–2013. They earmark large sums of money for the development of infrastructure in fishing communities (Kuzebski 2009). The condition for funding is that there are active fishers and fishing in the district.

Fishers have therefore become more interesting to district authorities. Thanks to the presence of fishers and fisheries, the district can obtain large sums of money for the development of local infrastructure. Thus, the fight to preserve the fisheries in the district has become a pawn in the play for large sums of money. Consequently, many local administrative leaders have understood that even a small group of fishers can play an important role in helping Tolkmicko transform itself into an attractive place for both tourists and investors.

According to the Operational Program "Sustainable Development of the Fisheries Sector and Coastal Fishing Areas 2007–2013," the money obtained under Axis 4 of this program will be used to rebuild local tourism and fishing infrastructure like wharfs, roads to ports, and parking places (own elaboration on the basis of the Operational Program). There is a great expectation among residents of the Tolkmicko district that it might create new, well-paid jobs for local residents who will rebuild local tourism, fisheries, and other community infrastructure, and thus curb informal unemployment significantly. Their hope is also that it will help reinforce the socio-economic position of women by offering better education and new professional qualifications that will permit them to work in the district or its vicinity.

Recently, fishers, local administrations, and NGOs operating in the district made the decision to cooperate and coordinate their activities. The aim is to include as many residents as possible in the discussion about plans for district development and their realization. It is anticipated that representatives of all the organizations and institutions will participate in decision-making processes for the development of the district, the fisheries, and the protection of the lagoon environment. Local fisheries management can be used as a way to protect local fish resources, the lagoon ecosystem, and the interests of the local fishers and other residents.

The fishers interviewed hope that Common Fisheries Policy funding will permit them to participate in the decision-making process with district authorities on what the money will be used for. They see the money as providing an opportunity for starting up a local purchasing and sales market for fish and agricultural products, which would allow them to eliminate fish brokers from the market. They also expect that the funding could be used for local businesses and to build material infrastructure that is essential to the functioning of the fisheries. The funds might also facilitate diversification in the fisheries sector into fish farming in the waters of the lagoon.

The fishers also think that they should be granted the exclusive right to fish in certain areas, and that they should be included in a system to monitor the exploitation of the resources. Another proposal mentioned in the interviews is to increase the power of local fisheries organizations to negotiate with local and state administrations. The fishers would like to rebuild the prestige of the profession of fishers among the residents of local coastal communities by illustrating the contribution of fishers and fisheries to the development of these communities. They also propose searching for new possibilities of rationally exploiting marine natural resources in such a way as to halt further degradation, while simultaneously permitting their further exploitation (e.g., restocking and linking fishing with tourism). Another proposal is to organize assistance for former fishers and their families. This should be available to the oldest fishers who are on disability pensions and the widows of fishers. Special importance is ascribed to encourage the young who are beginning their careers in fisheries to stay with it and to continue working in this field and to make all district residents aware that local coastal fisheries are not just where the fishers work; they are also the main tourist attraction in the district.

7.9 Conclusions

Many Polish coastal fishing communities are faced with issues of poverty and the progressive degradation of the natural environment. The result has been that these communities have fallen behind those in other areas of Poland ("relative poverty"). Most residents of many coastal fishing communities, just like those of the Tolkmicko district, live in small, spatially isolated communities. In comparison to urban dwellers, the residents of such communities have severely restricted access to educational and cultural institutions, shopping centers, and health-care institutions. Possibilities of finding alternative employment are also substantially limited. The consequences of territorial marginalization are distinctly evident in the marginalization of the residents of these communities in their economic, cultural, and political lives. Both individuals and whole social categories are kept away from many opportunities that are available for other Poles, but in most cases, local residents refer their feeling of being deprived first of all to their poor material situation.

Social marginalization and economic insecurity are consequences of fishing community living conditions being hugely dependent on the state of the natural environment, and highly variable governmental regulations that determine the principles used to manage the fisheries. The number of people gainfully employed as fishers has decreased systematically. In most instances, the owners of the fishing vessels are not threatened with poverty. Many of the remaining owners belong to the economic elite of their respective communities. In addition, the boat owners who decommissioned their vessels received as compensation a sum of money much higher than when selling their boat on the market. The money was spent on equipment for boats that remained in operation, and into tourist infrastructure belonging to their families. Thus, all of them are well secured and still maintain their status in the community. Poverty affects mainly those who have worked for many years as crew members aboard fishing boats. Very often, such employment is seasonal and informal, and thus does not guarantee retirement, disability, or unemployment benefits. Poverty also affects the spouses and children of these fishers.

Poverty in Polish fishing communities and the degradation of the marine natural environment in the lagoon and in other coastal areas are multidimensional phenomena, but they differ in how they manifest themselves and the intensity at which they occur. Each is a complex configuration of elements that are mutually dependent, and these relationships are dynamic and constantly changing. These two phenomena are related, and almost always happen simultaneously. Limiting the occurrence and reducing the impact of these phenomena can only be done when they are both addressed simultaneously. This is possible only if the battles against poverty in coastal communities and marine ecosystem degradation are fought together, and together they become a joint goal for local and central government administrations, as well as for formal and informal social groups that are part of the coastal communities.

When Poland acceded to the European Union, a variety of issues regarding the fisheries were still unresolved. One fundamental issue was the lack of a concept for the sustainable development of the fisheries that would permit the rational exploitation of resources. Responsible coastal fisheries (cf. Code of Conduct of Responsible Fisheries 1995) can play an important role in alleviating the impacts of poverty in small fishing communities, but this idea is still not well accepted by all fishers. Creating strategies for implementing responsible fisheries at the local, regional, and state levels that include the active participation of the fishers in the management of local resources could lead to significant reductions in the level of poverty and social marginalization.

In their writings about the role of communities in fisheries management, Pinkerton (1989) and Jentoft (2000) contend that the community is frequently ignored in the decision-making process. Because of this, opportunities to reinforce community strength and to render management systems more effective are lost. The same is true for the role of the community in the formulation of strategies for counteracting the threat of poverty or the occurrence of poverty itself. Community-based resource management and poverty alleviation can be founded on many of the same legislative acts, similar social norms, and aim for similar goals. Both can also be realized by the same people, social groups, institutions, and community organizations.

The governance model (Kooiman et al. 2005), which takes into consideration the full spectrum of interactions necessary for resolving problems, appears to be a method for identifying problems and formulating the most effective action strategies. These must take into consideration the ecological, economic, cultural, and social contexts in which these problems occurred. The multivariate analysis of the problem and the participation of many parties should be part of the solution. The issue of the degradation of the marine ecosystem, the poverty of the inhabitants of coastal communities, and the role of coastal fisheries in creating the basis for economic and social life in these areas should be an inseparable part of an integrated governance system for coastal areas.

Acknowledgments I am obliged to all residents of Tolkmicko district, representatives from local administration, local authorities, representatives of the church and NGOs, and Tolkmicko district fishers for their constructive comments and patience through our collaboration. I would also like to acknowledge the comments and assistance of Svein Jentoft and Maria-Victoria Gunnarsdottir. Notwithstanding these contributions, I am solely responsible for the conclusions stated here.

References

Amarasinghe O (2009) Social capital to alleviate poverty. Fisheries cooperatives in southern Sri Lanka. The Newsletter No. 51, Summer

Beskid L (1993a) Nowa rzeczywistość ekonomiczna. In: Rychard A, Federowicz M (eds.) Społeczenstwo w Transformacji. Ekspertyzy i Studia. Instytut Filozofii i Socjologii Polska Akademia Nauk, Warszawa, pp 23–25

Beskid L (1993b) Nastroje spoleczne wobec zmian sytuacji materialnej. In: Rychard A, Federowicz M (eds.) Społeczenstwo w Transformacji. Ekspertyzy i Studia. Instytut Filozofii i Socjologii Polska Akademia Nauk, Warszawa, pp 42–44

Borowski W (1994) Koncepcja poprawy rybolowstwa na Zalewie Wislanym na tle uwarunkowan społeczno-gospodarczych i przyrodniczych. Morski Instytut Rybacki w Gdyni, Zakład Biologii i Ochrony Zasobow, Gdynia

Brym RJ (2001) New society: sociology for the 21st century, 3rd edn. Harcourt, Toronto, pp 166–167

Code of Conduct of Responsible Fisheries (1995) FAO, Rome

Duffy A, Mandell N (1994) The widening gap: social inequality and poverty. In: Glenday D, Duffy A (eds.) Canadian society: understanding and surviving in the 1990s. McClelland and Stewart, Toronto

Duffy A, Glenday D, Henslin MJ, Pupo N (2007) Sociology: a down to earth approach. Fourth Canadian edition. Pearson Allyn and Bacon, Toronto

Galbraith JK (1970) The affluent society. The New American Library Inc, New York

Gerber LM, Macionis LJ (2002) Sociology. Fourth Canadian edition. Prentice Hall, Toronto

Horbowy J, Kuzebski E (2006) Impact of the EU structural funds on the fleet and fish resources in the Baltic fisheries sector. WWF Poland, Sea Fisheries Institute, Gdynia, pp 8–10

Igras J, Pastuszak M (2009) Udzial polskiego rolnictwa w emisji zwiazkow azotu i fosforu do Bałtyku. IUNG-PIB, Pulawy

Jałowiecki B (1988) Spoleczne wytwarzanie przestrzeni. Ksiazka i Wiedza, Warszawa

Jarosz M (2005) Wygrani i przegrani polskiej transformacji. Instytut Studiow Politycznych PAN, Oficyna Naukowa, Warszawa

Jentoft S (2000) The community: a missing link of fisheries management. Mar Policy 24:53–60

Kooiman J, Bavinck M, Jentoft S, Pullin R (eds.) (2005) Fish for life: interactive governance for fisheries. Amsterdam University Press, Amsterdam

Kuzebski E (2009) Description of the vessels withdrawal plan in the framework of the European Fisheries Fund 2007–2013. In: Kuzebski E, Marciniak B (eds.) Less boats - more fish? Socio-economic consequences of fleet capacity reduction programmes in the Baltic Sea. WWF Poland, Warsaw

Mahler F (1996) Maldevelopment and marginality. In: Danecki J (ed.) Insights into maldevelopment. Uniwersytet Warszawski, Warszawa

Marciniak B (2008) Kobiety: matki, zony i corki rybakow – niewidoczni członkowie załog. Wiadomosci Rybackie, 3–4 (162)

Milic-Czerniak R (1993) Zroznicowanie sytuacji ekonomicznej gospodarstw domowych. In: Rychard A, Federowicz M (eds.) Społeczenstwo w Transformacji. Ekspertyzy i Studia. Instytut Filozofii i Socjologii Polska Akademia Nauk, Warszawa, pp 25–27

Ministerstwo Rolnictwa i Rozwoju Wsi. Departament Rybolowstwa (2004) Instrukcja uzytkowa, Instrument Finansowy Wspierania Rybolowstwa, Warszawa
Myrdal G (1962) Challenge to affluence. Pantheon Books, New York
Narodowa Strategia Integracji Spolecznej (2003) Dokument przygotowany przez Zespol Zadaniowy do Spraw Reintegracji Spolecznej, ktoremu przewodniczył Minister Gospodarki, Pracy i Polityki Spolecznej – Jerzy Hausner. Zespol zostal powolany 14.04.2003 r. przez Prezesa Rady Ministrów. www.mpips.gov.pl/userfiles/File/mps/NSIS.pdf
Natura 2000 (2004) Ustawa z dnia 16 kwietnia 2004 o ochronie przyrody. Dziennik Ustawa z 2004, no 92, position 880
Neuman L (2003) Social research methods: qualitative and quantitative approaches, 5th edn. Pearson Education Inc, Toronto, pp 137–139
Peace A (1991) Fishing, drinking and the construction of identity in rural Ireland. Marit Anthropol Stud 4(2):3–16
Pinkerton E (ed.) (1989) Cooperative management of local fisheries: new directions for improved management and community development. University of British Columbia Press, Vancouver
Rakowski M (2008) Charakterystyka społeczno-ekonomiczna regionu nadmorskiego. In: Marciniak B Spoleczno-ekonomiczne Skutki Realizacji Wspólnej Polityki Rybackiej (WPR). Zakład Ekonomiki Rybackiej. Morski Instytut Rybacki, Gdynia
Republic of Poland, Ministry of Agriculture and Rural Development (2008) Operational programme "Sustainable Development of the Fisheries Sector and Coastal Fishing Areas 2007-2013", Warsaw, October 2008. http://ec.europa.eu/fisheries/cfp/structural_measures/op/poland_en.pdf
Strybel R (2009) Poland's new underprivileged: Euro-orphans. Polish Times, 3/2009. www.mypolishtimes.com
Townsend P (1976) Sociology and social policy. Penguin, Harmondsworth
Wielopolski A (1946) Tolkmicko nad Zalewem Wislanym. Wydawnictwo Związku Gospodarczego Miast Morskich, p 8–9
Wodz J (1989) Poczucie przynaleznosci lokalnej – niektore czynniki konstytutywne. In: Wodz J, Jalowiecki B, Sowa K, Dudkiewicz P (eds.) Społecznosci lokalne. Terazniejszosc i przyszłosc. Instytut Geografii i Studiów Regionalnych UW, Instytut Gospodarki Przestrzennej, Warszawa, pp 50–60

Chapter 8
More Than Income Alone: The Anlo-Ewe Beach Seine Fishery in Ghana

Marloes Kraan

Abstract Ghanaian artisanal fisheries have dominated the West African coastal region for over 100 years. Due to natural conditions (upwelling) in the Gulf of Guinea, Ghanaian fishers have long been migrating to follow the fish. While migrating, they spread their technical knowledge of boat building and fishing, as well as knowledge of management institutions to other coastal communities. Fish stocks in West Africa – and in Ghana – are now in crisis. Due to declining catches, the contributions that fisheries make to poverty reduction are becoming threatened. This chapter describes the history and current situation of the Anlo-Ewe beach seine fishers, one of the coastal ethnic groups involved in fishing. This chapter presents four main findings: (1) fishing in West Africa is not always a last resort activity – which has often been suggested; (2) artisanal fisheries have been very profitable; (3) fisheries mean more to fishers than earning money – it is a way of life; and (4) policies aimed at providing "alternative" livelihoods for fishers to solve problems of resource scarcity are likely to be unsuccessful. This chapter concludes by pointing out how the inclusion of strong artisanal fisheries in fisheries governance is crucial for preventing stock depletion and growing poverty.

8.1 Introduction

Reports on the rapid depletion of the world's ocean resources underline the importance of fisheries management. According to the Food and Agriculture Organization (FAO), 80% of the world's major fish stocks are either fully exploited, over-exploited, rebuilding, or depleted (FAO 2009). These findings are worrying

M. Kraan (✉)
Centre for Maritime Research (MARE), Nieuwe Prinsengracht 130,
1018 VZ Amsterdam, The Netherlands
e-mail: marloes@marloeskraan.eu

S. Jentoft and A. Eide (eds.), *Poverty Mosaics: Realities and Prospects in Small-Scale Fisheries*, DOI 10.1007/978-94-007-1582-0_8,

from a global environmental perspective, and echo the findings on the rapid depletion of, for example, tropical forests. Yet, they are even more alarming in terms of food security in developing regions. About 90% of fisheries worldwide are small-scale, producing 50% of the fish, and providing livelihoods to millions of people in poor fishing communities (FAO 2005).

In Africa, any depletion in fish stocks will have serious implications in terms of food security and livelihoods. In many African countries, fish is generally still considered a cheap source of animal protein, affordable to poorer population groups (Heinbuch 1994; Feidi 2001). Moreover, fish has quite a long shelf life (up to 6 months) due to processing techniques like smoking and drying. This facilitates distribution and consumption in inland areas (DoF 2003; Mensah et al. 2006). Fish supplies in Africa are in crisis (see for instance Christensen et al. 2004). The per capita fish consumption is declining due to a growing population and a decline in fish production. The contribution that fisheries make to poverty reduction is becoming threatened due to increasing scarcity (WorldFish Center 2005). Small-scale fisheries are becoming increasingly marginalized, with the scarce resources being concentrated in fewer and fewer hands (Berkes et al. 2001; Hauck 2008). The importance of sustaining small-scale fisheries, and thereby enhancing development, is being increasingly recognized (Allison and Ellis 2001).

Fisheries in Ghana are enormously important in relation to livelihoods, with an estimated 10% of the population directly (as fishers and processors) or indirectly (as traders, canoe carvers, or petrol sellers) dependent on fishing (Interview with G. Hutchfull, 13 October 2005; Mensah et al. 2006; Akyeampong 2007). This is because fisheries in Ghana are largely rural in character, being dominated by artisanal fisheries[1] responsible for 75% of marine catches. Fisheries, therefore, play a major role in poverty alleviation (Mensah et al. 2006). Together with the (semi-) industrial marine and inland sector, fishing contributed US$380 million to the national economy in 1996[2] (Atta-Mills et al. 2004), and was US$94 million worth of exports in 2002 (FAO Ghana Country profile[3]). The fishing sector is therefore very important for Ghana in terms of livelihoods and food security.

The link between poverty and small-scale fisheries is often strongly emphasized in the literature (De Vries 2003; Pauly 2006), yet should be met with some caution. The almost universally accepted perception that "fishery rhymes with poverty" (Béné 2003) has been challenged by empirical data as being at least more complex than often presented. In this chapter, I will demonstrate how Anlo-Ewe net owners in Ghanaian coastal communities have been considered to be rich rather than poor.

---

[1] I will use the concepts of artisanal and small-scale fisheries interchangeably. Both concepts have their own connotations (see Johnson 2006). With these concepts, I refer to the sub-sector in Ghana using canoes (or even only nets such as cast nets) for their fishing operations. The artisanal or small-scale subsector stands opposed to the (semi-)industrial sector.

[2] In 1996, Ghana had a GDP of US$6.9 billion. http://devdata.worldbank.org/AAG/gha_aag.pdf [Accessed date: September 2008].

[3] http://www.fao.org/fishery/countrysector/FI-CP_GH/en [Accessed date: March 2009].

However, the existing perception has still led to the idea that the root of the problem lies in the economic and biological aspects of the activity (Béné 2004): "Poverty in fisheries has been explained through a linear relationship between the low incomes of fishers (due to low catch), and the over-exploited resources (created and/or maintained by the open access nature of the fisheries)."

This chapter describes the importance of fisheries for the Anlo-Ewe beach seine fishers, the history of this fishery, how it is currently conducted, and the importance of artisanal fisheries in light of a long history of scarce livelihood alternatives. By describing the Anlo-Ewe case, it will become clear that artisanal fisheries in West Africa have not always been associated with poverty – on the contrary. The sector has developed itself into a thriving business, and has proven to be a sector worthwhile protecting due to – amongst others – its importance in terms of food safety and livelihoods.

8.2 Theoretical Debate on Fisheries and Poverty

The complex relationship between poverty and fisheries can be better understood by making use of concepts and debates in the social sciences. First of all, the environmental entitlements debate has shown that it is not so much the scarcity of a resource or lack of production that increases poverty, but lack of access to it (Sen 1981). As the livelihoods approach has emphasized, income is not a true assessment of poverty; one may lack financial resources but have access to natural resources, and therefore not be poor (Béné 2004). As Béné assesses, fishing communities often reflect the general lack of development of the rural areas in which they exist. However, fishers are not always the poorest of the poor (Odotei 1991; Béné 2004; Mensah et al. 2006; see also Hoorweg et al. 2009).

The focus on income and over-exploited resources can also be attributed to the dominance of economists and biologists in fisheries governance (both in theory as in practice). There has been insufficient input from the social sciences, yet their contribution is imperative. First of all, social scientists have highlighted the dynamics and multidimensionality of poverty by using the livelihoods approach. The usefulness of the livelihoods approach for fisheries research lies in the fact that it shows how *all* assets and capabilities, and the relationships between them, are potentially important. This is particularly necessary as fisheries research has, to a large extent, been performed under the so-called paradigm halieutique (Chauveau et al. 2000). This refers to a primary focus on access to the natural resource and related assets (such as the fishing gear), while disregarding other assets such as family labor, physical strength, skills, political influence, identity, and infrastructure (to name a few).

Also, a more historical perspective often misses in the "fisheries rhymes with poverty" debate. Although many artisanal fishers around the world are currently confronted with declining catches and thus struggle for their living, in the past artisanal sectors have also been thriving. The Ghanaian case is a good example hereof.

The Ghanaian case shows how fishing in West Africa is not always a last resort activity (which has often been suggested), and how artisanal fisheries have actually been very profitable. Finally, it is important to understand, in the fisheries–poverty debate, that fishers fish not only to earn an income, but that being a fisher is a way of life. Fishing is their profession in which they have specialized as the generations before them. Being a fisher is part of one's identity, as for instance expressed in the songs sung during fishing and in the decoration of the canoes.

8.3 Existing Fisheries Governance Structures

At the beginning of the 1990s, new approaches to fisheries governance were sought after, when clear signs of over-exploitation of the world's fish stocks became apparent. These approaches had to be directed to conservation and related to environmental, social, and economic concerns. The FAO developed the Code of Conduct of Responsible Fisheries, which was adopted on 31 October 1995 (FAO 2005). The second general principle of the Code of Conduct (article 6.2) states:

> Fisheries management should promote the maintenance of the quality, diversity and availability of fishery resources in sufficient quantities for present and future generations in the context of food security, poverty alleviation and sustainable development.

Article 6.18 recognizes the important contributions of small-scale fisheries to employment, income, and food security:

> ...protect the rights of fishers and fish workers, particularly those engaged in subsistence, small-scale and artisanal fisheries, to a secure and just livelihood, as well as preferential access, where appropriate, to traditional fishing grounds and resources in the waters under their national jurisdiction.

In response to the growing awareness of poverty in fishing communities, and the lack of attention these communities often receive, the FAO developed the Sustainable Fisheries Livelihood Program (SFLP),[4] together with the Department for International Development (DfID) and 25 participating countries in West Africa. This program, which commenced in 1999, is a "regional development project whose overall objective is to reduce poverty in inland and coastal fisheries communities through the sustainable improvement of their livelihoods."[5] Its two main working and reference tools are the Sustainable Livelihoods Approach, and the FAO Code of Conduct. The result was a lot of research that put the issue of poverty and small-scale fisheries firmly on the political map.

Allison and Ellis (2001, p. 387) applied the sustainable livelihoods approach to small-scale fisheries research and found that:

> Fisheries sector development analyses have tended to focus on what small-scale fisher folk do not have access to: infrastructure, finance and technology – rather than what they do have – adaptable and flexible income-generating strategies, resilient resource management institutions, knowledge, skill and social capital.

---

[4]http://www.sflp.org/briefs/eng/flyer.pdf.

[5]For more information, see www.sflp.org.

Small-scale fisheries have long been ignored by national governments due to their preoccupation with modernising the fisheries, and therefore on developing a (semi-)industrial sector. When small-scale fisheries were included in policy, they were encouraged to "develop" or "modernize" with a view to increase the efficiency of fishing effort (Allison and Ellis 2001, p. 382) by "supplying artisanal fishers with improved boat designs, or subsidising credit for the purchase of outboard motors, promoting nets made of more durable materials."

A lot of modernization programs have, in effect, undermined the adaptive capability of small-scale fisheries due to a lack of understanding of small-scale fishers' livelihoods (see also Platteau 1989). As Lewins (2004, p. 44) formulates it: "Well-meaning policy interventions have so often failed to produce change because the social and political realities faced by the poor are rarely understood or considered." The livelihoods approach provides a means by which to better understand the nature of small-scale fishery production systems, in which flexibility, geographical mobility and livelihood diversification are characteristic adaptive responses (Allison and Ellis 2001).

In my research in Ghana (2003–2005), I tried to understand what constitutes the livelihood space of Anlo-Ewe fishers, one of the questions being: How have the Anlo-Ewe fishers organized their livelihood? I studied the Anlo-Ewe beach seine fishers, who historically took up fishing in the 1850s, and for whom this fishery quickly became the main driver of economic prosperity. Although fishing villages are not paved in gold, it was obvious that fishing had been, and perhaps still is an attractive livelihood activity. Net owners in the past had been the wealthiest inhabitants in Anlo.

The societal problem of declining catches should be addressed by improved fisheries governance. To create a system of fisheries governance in which social scientists as well as artisanal fishers play a larger role. As fishing is more than a means to earn an income, it is understandable that policies aimed at providing "alternative" livelihoods for fishers to solve problems of resource scarcity will therefore not be successful. In this chapter, I will present the case of the Anlo-Ewe beach seine fishers in Ghana – a case of fishers within the larger, lively and historically dominant artisanal fishing sector. This is a sector which has been, and still is, attractive to the many professionals working in it. Understanding the strength and working of the sector is imperative for developing good governance for the fisheries. The chapter will therefore conclude by pointing out how the improved inclusion of a strong artisanal fishing sector in fisheries governance is crucial for preventing both stock depletion as well as growing poverty.

8.4 Research Methods

My research took place between October 2003 and December 2005, when I spent almost one and a half years in Ghana. The research was intended to contribute to a better understanding of the practice of fisheries governance. The combination of an important artisanal fishing sector with specialist fishers and declining catches is, in fact, a societal problem – a problem which can only be addressed with a better understanding of the sector.

My research was oriented around two important questions: (1) What makes up the livelihoods of Anlo-Ewe beach seine fishers? (2) How can they and do they negotiate their livelihood space (within multiple governance structures)? Answering these questions means gathering data at lower levels, whereby one keeps an eye open for heterogeneity within the case unit (Anlo-Ewe beach seine fishers in Ghana). For my research questions, I had already made an internal differentiation between fishers at home, within their own social system, and fishers on migration, based on the hypotheses that it influences their negotiation practices and outcomes.

As a consequence, I differentiated within the case unit at the village level – my second unit of analysis. I compared Anlo-Ewe beach seine fishers who fish from their hometown (Woe) with Anlo-Ewe beach seine fishers who fish from villages along the Ghanaian coast outside their home area in places they have migrated to (Akosua Village, Half Assini). For an overview of the characteristics of the three research villages, see Table 8.1. Gathering data at the village level would not, however, provide all the answers I wanted. I had to gather the data at lower levels and then aggregate the data at a higher level (village) so I could answer the main questions. The levels at which I had to gather most of the data would create three additional units of analysis: the company level; the household level; and the individual level. As such, I performed a multiple embedded case study (Yin 1994) in three research locations (Woe, Akosua Village, Half Assini), and within the cases through the use of subsets (companies, households, and individuals).

Most of the data were collected in Woe and Akosua Village, with Half Assini added as a control migration community. Woe lies in the home territory of the Anlo-Ewe; Akosua Village is a migrant village for the Anlo-Ewe in the Central Region in Ghana; and Half Assini, a town in the Western Region, is an Anlo-Ewe migrant

Table 8.1 Main features of the three research locations

Villages	Region	Traditional state	Population size	Main fishing groups	Characteristics
Woe	Volta	Anlo-Ewe	8,545	Anlo-Ewe	Rural town, mixed agriculture and fisheries nearby Togo. Nearby Keta lagoon
Akosua Village	Central	Effutu	630	Effutu in Winneba, Anlo-Ewe migrants in AV	An Anlo-Ewe migrant fisher settlement, close to Winneba, a major urban town. Nearby Muni lagoon
Half Assini	Western	Nzema	11,734	Fante and Anlo-Ewe migrants	Capital of the district, quite large with separate neighborhoods for the migrant fishers. Nearby a couple of small lagoons

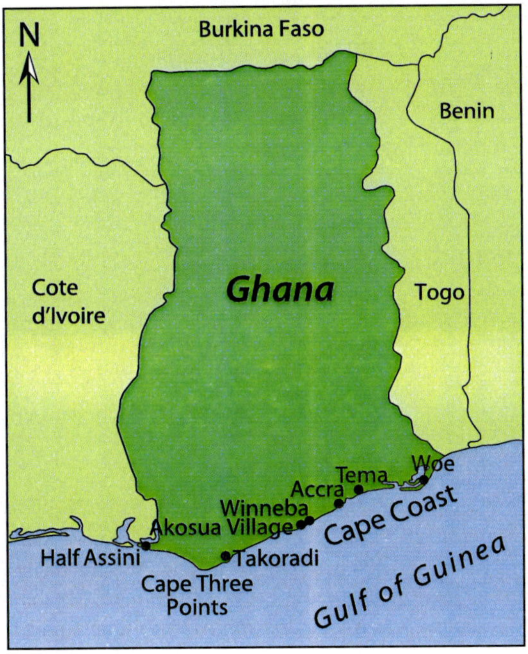

Fig. 8.1 Map showing the research locations (Half Assini in the west, Akosua Village in the center, and Woe in the east) along the coast of Ghana, and some other major coastal towns (Source: Kraan 2009)

fisher community. The three research locations were thereby spread evenly along the Ghanaian coast (Fig. 8.1; Table 8.1).

A few different research methods were used for this study: participant observations; photo and film; interviews, questionnaires and household surveys; focused group discussions; mapping; as well as collection of documents and fishing records.

Different approaches were used for the interviews: a combination of informal talks, formal interviews, open interviews, standardized interviews, and topical interviews. The interviews were held with representatives of the Ghanaian government, representatives of traditional governments, and fisher folk. In total, 98 interviews were held. In addition, I distributed a number of questionnaires among net owners ($N=31$) and crew members ($N=116$).

The household surveys achieved two goals. The first goal was to gather basic data related to assets, and access to certain services (such as water and electricity) of fisher households which would give a better picture of the fisher households' livelihoods. The second goal was to get some idea of the size and composition of households, the demographic composition, and the link to fisheries and/or other income-deriving activities. However, the household surveys also provided an opportunity to observe (to look behind the fences of the compounds), to meet people and explain to them why I was there. In Woe, we met with 107 households; in Akosua Village 105; and in Half Assini 41. N total=254, resulting in data on 1,411 individuals.

Finally, beach seine fishing companies keep quite detailed records such as account books (recording the catches and the sharing of the catch), sales books, fine books, expense books, and loan books. In all three research locations, I managed to acquire insights into some of these records.

Other documents that I collected were court cases, company contracts, payment receipts of migrant net owners (relating to the land leased from the Chief), and government documents related to fishing.

8.5 Fisheries in Ghana

Ghana is located in the central part of the Eastern Central Atlantic, along the Gulf of Guinea, between Côte d'Ivoire and Togo (Fig. 8.1). It had a population of over 22 million in 2008, with an annual growth rate of 2.6%. Ghana lies in the tropical equatorial belt where climatic conditions change mainly due to the amount and distribution of rainfall. There are two distinct wet seasons each year – a major one in May–June and a minor one in August–September (Mensah et al. 2006). Ghana's oceanography is subject to two upwelling periods – a major one from July to September and a minor one of 3 weeks between December and January. In these upwelling periods, fish production increases sharply (MFRD 2004).

Ghana has a long fishing history and, together with Senegal, has the largest fishing industry in West Africa. Old European travel reports provide evidence that Ghanaians (Fante in particular) were already fishing at sea before 1471 (Odotei 1991; Haakonsen 1992). Fishing is one of the most important economic activities in the country (Ferrais et al. 1997). The contemporary Ghanaian fishing sector consists of marine fisheries, inland fisheries (with most of the fishing taking place on Lake Volta), and aquaculture. The marine sector is the most important, providing 80% of domestic supply.[6] Artisanal fishing is the most important sub-sector within marine fisheries, contributing 60–70% of the marine fish output (Mensah et al. 2006).

The artisanal fishers in Ghana belong to different ethnic groups, each of which have specialized fishing techniques. The Anlo-Ewe specialize in beach seine fishing; the Fante, Effutu, and Ga in purse seine and drifting gill net (*ali*) fishing; and the Dangbe and some Ga in line fishing with ice (*lagas*) (Ferrais et al. 1997; Mensah et al. 2006). This specialization can be explained by the environment that fishers live in. For example, fishers living nearby rocky areas are more likely to use line gear. This ethnic-technical divide of the artisanal fishing sector has allowed for internal fishers' migration (Kraan 2009). The fishers in my research also come out of fishing families; their parents and grandparents were also involved in fishing before them. I never met a former farmer coming from the inland – as described by Pauly (2006) – who sought opportunities on the coast. Thus, the theory of Malthusian overfishing, as has been shown in Senegal,[7] appears not to be the case in Ghana.

[6] http://www.fao.org/fishery/countrysector/FI-CP_GH/en [Accessed date: July 2008].

[7] From the literature, Malthusian overfishing (as Pauly (2006) has called it) can be recognized in Senegal (De Vries 2003; Pinnegar and Engelhard 2008).

The common fishing craft is a dugout canoe carved out of a single trunk of wood, symmetrical in shape, double-ended, and ranging in size from 3 to 18 m in length and 0.5 to 1.8 m in width (Ferrais et al. 1997). The targeted species are small pelagic (most important are the highly variable sardinella and anchovies); large pelagic (mainly tuna); and demersal species. The demersal species show clear signs of stress – landings exceed the potential yield (Mensah et al. 2006). Traditional processing methods such as smoking, salting, and drying are used all along the coastline to preserve most of the fish caught – both by the artisanal and inshore fleets (MFRD 2004).

Fish is Ghana's most important non-traditional export commodity.[8] In total, the fisheries sector accounts for 5% of the agricultural gross domestic profit (GDP), worth a total of US$96 million in 2002 (MFRD 2004). Fish and seafood exports from Ghana consist mainly of tuna (76% caught by the industrial sector); frozen fish (mostly demersals); and shrimps, lobsters, cuttlefish, and dried/smoked fish (MFRD 2004). Ghana's marine waters are home to small and large pelagic and demersal species.

In 2009, the total quantity of fish caught by Ghanaian vessels in marine waters was 317,446 t (personal communication with Paul Bannerman, head of research, Marine Fisheries Research Division (MFRD)). The total production (including production from inland waters) is not enough to sustain Ghana's demand for fish, which has been estimated at 600,000 t per year (Mensah et al. 2006). Ghana therefore imports fish from Europe and other West African countries, mainly from Morocco, Mauritania, Namibia, Norway, the Netherlands, Belgium, Senegal, and Gambia (FAO Ghana profile[9]).

8.5.1 Migration

Migration is a characteristic of artisanal fisheries in Ghana, as it is in other countries in the West African region. Ghanaian migrations have been recorded from the beginning of the twentieth century (Chauveau 1991). Most Ghanaian migrant fishers can be found in Togo, Benin, and Côte d'Ivoire (Odotei (1995, 2002); Mensah et al. 2006), as well as in the Congo, Cameroon, Guinea, and Sierra Leone (Hendrix 1985; Haakonsen and Diaw 1991; Wagner 1991; Jul-Larsen 1994; Odotei 1995; Ferrais et al. 1997; Solie 2006).

In Liberia, Côte d'Ivoire, Benin, and Togo, Ghanaian fishers have a strong foothold in the artisanal fishing sector, being responsible for a large part of the catches.

[8] Since the 1980s, Ghana has been pursuing an export-diversification strategy of development with a greater emphasis on the non-traditional export sector. Non-traditional exports are agricultural, processed and semi-processed, and handicraft products. In Ghana, fish is one of these, and then mostly tuna, shrimps, lobsters, and prawns (Addo and Marshal 2000).

[9] http://www.fao.org/fishery/countrysector/FI-CP_GH/en [Accessed date: March 2009].

Furthermore, Ghanaians have had a strong technical influence on the fishing sector in West Africa. They have taught a lot of West Africans to fish using their techniques and equipment, and the Ghanaian canoe is used in many countries in West Africa. Fishers in Ghana are also internally mobile, whereby fishers move temporarily or permanently to other regions dominated by different ethnic groups.

The mobility of artisanal Ghanaian fishers has, for the most part, been explained by the movement of fish species due to upwelling. The sardinella start migrating from the west toward the east, with the fishers in their wake. Yet, it has also been recognized that fishers left their homes due to coastal erosion, population pressure, or overfishing, as well as being positively triggered to move to new places.

In addition, migration to neighboring countries has been explained by the possibility to earn the stronger FCFA[10] (in the context of economic hardship in Ghana), and that it was possible to save money away from family. Other explanatory factors have been that migration is seen as an adventure whereby the experience adds to one's social status. The availability of existing migration networks has been an important factor which explains early fisher migration. This finding has underscored the fact that mobility is neither a recent nor an exceptional phenomenon, and that it should be understood in conjunction with wider social economic and political developments.

8.5.2 *Livelihood Space*

In addition to these more push-and-pull characterizations of migration, I have argued (Kraan 2009) that the fisher migration can only be understood when seen as a livelihood strategy linked to the fishing activity. I thereby make use of the concept "livelihoods space." Livelihood space refers to three elements of space: (1) space to work (fish and market the fish), live safely, and make use of facilities and services; (2) space within the fishing sector (niche creation); and (3) space where one is accepted – where one has a place in society (Kraan 2009).

Previous fisher migration research explains how fishers manage to find and maintain the first (space to work and live) and third (space where one is accepted) elements of livelihood space. The second element of livelihood space – niche creation and maintenance – however is equally important for understanding fisher migration. The ethnic-technological divide in the Ghanaian fishing sector corresponds to a spatial divide at sea which allows for internal migration. As the different ethnic groups have specialized in different techniques and thereby use different spaces, they create room for other ethnic groups to come and fish in their territory. The niche differences also explain the differences in type and duration of their migrations.

[10] CFA Franc – CFA stands for Communauté Financière Africaine (African Financial Community). Currency used in (amongst others) the neighboring countries of Ghana – Benin, Togo, and the Ivory Coast.

8.6 Anlo-Ewe Beach Seine Fisheries

8.6.1 Introduction of the Beach Seine – Yevudor

The beach seine is a commonly used technique all along the coast of Ghana (Fig. 8.2). The Anlo-Ewe have specialized in beach seine fishing. They learned sea-fishing from Fante fishers who are very mobile along the Ghanaian coast (Overå 2001). Marine fishing only really became central to the economic life of the Anlo with the introduction of the beach seine. This net is called "*yevudor*," which means "the white man's net" (Nukunya 1991, p. 209). It was introduced around 1850 by a European slave trader and his Anlo-Ewe wife (Greene 1996; Akyeampong 2001). It soon became the dominant fishing technique used by the Anlo-Ewe (Nukunya 1999, p. 6): "In the great majority of towns in this area, almost the whole population is engaged in sea-fishing."

8.6.2 An Expensive Net

The *yevudor* was (and still is) an expensive net due to its expensive European twine and its considerable size (Akyeampong 2001). Not all Anlo men had access to enough

Fig. 8.2 A beach seine being pulled in at Woe beach, Ghana (Source: author)

Table 8.2 Cost of fishing equipment of a beach seine company in cedis[a] in 2004

Net			
1. 3/8 × 9″ net	200 yards	(2 bundles @ 20,000,000)	40,000,000
2. ¾ × 9″ net	50 yards	(1/2 bundle)	10,000,000
3. ½ × 9″ net	50 yards	(1/2 bundle)	7,000,000
4. 1 × 9″ net	200 yards	(2 bundles @ 7,500,000)	15,000,000
5. 1 ½ × 9″ net	100 yards	(1 bundle)	7,000,000
6. 2 × 9″ net	400 yards	(4 bundles @ 3,00,000)	12,000,000
Subtotal			91,000,000
Subtotal in Euros			8,272
Floating			
	2,000 pieces	(40 sets @ 750,000)	30,000,000
Leads			
	1,500 pieces	(30 sets @ 250.000)	7,500,000
Ropes			
20 mm	4 coils	@ 600,000 (*kahehe*)	2,400,000
18 mm	30 coils	@ 500,000	15,000,000
14 mm	20 coils	@ 500,000 (*doblaka*)	10,000,000
Canoe			
			35,000,000
Outboard motor			
			25,000,000
Total			215,900,000
Total in Euros			19,627

Source: Kraan (2009, p. 108)

[a]In this chapter, the currency value from 2004 is used. In 2007, the cedi was re-denominated, and 10,000 cedis became 1 cedi

financial resources to buy a *yevudor*, and many therefore chose to work for the wealthier *yevudor* owners (Greene 1996). Moreover, a lot of people were needed to handle the large nets and all of these factors led to the formation of fishing companies (Akyeampong 2001). According to Akyeampong (2001), the origin of the term "company" appears to be "a conscious emulation of the commercial companies that proliferated in the Gold Coast from the late nineteenth century" and "reflects the heavy infusion of capital by net owners, who were not necessarily fishers." The *yevudor* owners, a new class of accumulators, acquired a reputation for their wealth in Anlo (Akyeampong 2001).

In Woe, one of the three research communities, a net owner made a list for me to show what it would cost to put together a new net, with detailed information on how much would be needed of each netting material with a particular mesh size, how many ropes, their sizes, and how many floats and lead (see Table 8.2). All in all, the net would cost (excluding lead and floats) about €8,000. Another €11,000 would be needed for lead, floats, ropes, canoe, and outboard motor. Beach seines are not always newly constructed, and often fishers inherit the net from their family. A lot of net owners started out with small nets and added to them over time, when they had more money to invest. These new net owners often started fishing in a lagoon or in the surf before fishing at sea.

8.6.3 The Beach Seine Technique

A beach seine is a net operated from the shore, after it has been set by the crew in a canoe at sea (Fig. 8.3).

Once the net has been set, it will be hauled in from the shore. A large net may need almost 30 or 40 people pulling on each side. The two groups will gradually walk toward each other while pulling. This can take 3–7 h, depending on the number of ropes used (which also depends on the size of the net). The people pulling on the beach (and also the fishers in the canoe when they set the net) often sing while working. The singing is an important element; crew members get paid extra when they have a leading role in the singing. The lyrics express norms and values or can be funny; they can be part of a repertoire or improvized (Kraan 2009). When the cod end arrives to the shore, it is dragged up onto the beach where it is left for a bit while the fish die. Some crew start collecting the net while others, including a lot of women

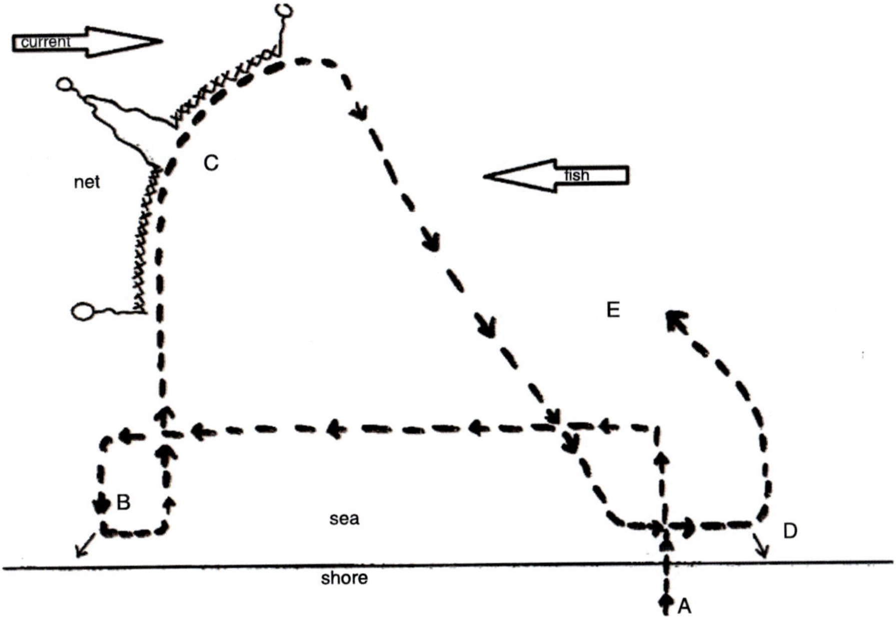

Fig. 8.3 The canoe sets off from the shore at point (**a**), and follows the path visualized by the dotted line in the direction of the arrows moving against the current. The net is cast in such a way that the fish, also swimming against the current, swim into the net. The first rope is let out of the boat and brought to the shore by a swimmer at point (**b**). The net is cast against the current (coming from the east – *left* in this image), whereby the fish, swimming against the current, can swim into the net at point (**c**) before it is drawn toward the coast and closed. After casting, the canoe either returns to the shore with the crew from where the second rope will start to be pulled, or – as depicted at (**d**) – all the crew except one or two will jump out of the canoe to swim through the surf to the shore. The boat then returns (**e**) to the sac of the net (Source: author)

and children, then empty the cod end, sort the fish, and remove the garbage.[11] After that, the sharing and selling can start.

Beach seines, operated from the shore, can differ in size (between 150 and 1,800 m in length and between 6 and 22 m in depth). A beach seine net needs a lot of maintenance and these nets are continuously being repaired – in fact, after every day's fishing. The beach seine is used throughout the year, although catches fluctuate. Peak catches occur in July, and higher catches in the minor upwelling season in January, as compared to the low season (March, April). It is the second most important gear with regard to the total catch (73,848 t in 2004) after the Ali-Poli-Watsa (APW) canoe (154,946 t in 2004). The beach seine contributed 28% of the total value of the artisanal marine catch in 2004. It is most efficient in terms of catch per unit effort (CPUE), which means that it catches a fairly large amount of fish (catch) based on a fairly low amount of time (effort in days).

8.6.4 Impact of the Fishing Gear

The beach seine technique has been criticized for its negative impact on fish stocks (Hosch 2002). The non-selectiveness of the gear, which catches almost everything within the scope of the net due to the small mesh sizes used, is a major point of concern. As near-shore waters (especially in the vicinity of lagoons and estuaries) play an important nursery role, beach seines with small mesh sizes are said to have quite negative biological effects. This has been shown in research performed by Nunoo et al. (2006). They therefore recommend "a co-managed (fishers and government) 3-month ban on beach-seining (between May and July) as the most appropriate control measure toward the sustainability of Ghanaian fish stocks."

In some West African countries (Gabon and Gambia), beach seines have been banned all together.[12] As discussed above, the beach seine is important in Ghana (as it is in other countries such as Togo and Benin), so an outright ban would be politically difficult to implement.[13] In addition, fisheries officers and politicians are aware of the social function of the beach seine, providing work, income, and fish for many fishing communities where alternatives are often not available. The beach seine is in that sense a controversial technique.

---

[11] Beach seines in Ghana catch quite a lot of garbage, consisting mainly of plastic bags that have been blown into the sea or washed into the sea via the open gutters and sewage system.

[12] www.fao.org/DOCREP/005/Y3274E/y3274e09.htm [Accessed date: January 2009].

[13] A ban was suggested when a new Demersal Fisheries Management Plan was drawn up (Yeboah 2002). See Chap. 6 in Kraan (2009) on how difficult it is to implement the by-laws in the Keta district (Volta Region) due to the power of net owners at the local level, let alone ban the beach seine.

8.6.5 *Beach Seine Fishing Companies*

The larger the company, the better – the sooner the net comes in, the fewer the fish that can escape. Crew members often work under a contract. They receive an advance payment at the beginning of the fishing season (which lasts from 9 months up to 5 years), and are paid their full salary at the end of the contract period. In between, they receive food, money for food, or fish; their medical bills are paid; shelter is provided (especially on migration); and they are able to borrow money from their net owners in case of other costs. The loans and advance payments are subtracted at the end of the contract period. Whatever is left over is their salary for the fishing season. The contract system has advantages for both net owners and crew. The workers have access to money, and the net owner has the assurance of available workers.

Net owners often have crew members who are related to them. However, even if they are not related, the relationship between the net owner and the crew is often referred to in terms of "children" or "boys," and "master" or "owner." The oldest members of the crew are referred to as "uncle." These words reflect the expectations associated with the relationship, namely that the "sons" obey the net owner, and the net owner takes care of his "sons." Elders in the community are addressed as "our father," and people refer to co-companies as "our brothers."

Going out to sea with the canoe and casting the net is a truly skillful job, whereas pulling in the net, which on average takes between 3 and 7 h, mainly requires strength. The last part of the pulling, when the net is almost ashore again, is more crucial and entails skilled crew members making the right decisions and doing the right thing in the water and on land (Kraan 2009). Often more people than only the crew (including for instance students, elderly, mentally challenged) assist in the pulling, and these helpers are always given some fish. That is an important social side to beach seine fisheries.

When the net has almost been landed, more skillful work needs to be done such as diving, carrying the net in the water, deciding when the net should close, and how fast it should be pulled. Most of the special skills can be learned on the job except that of the clerk, the one who keeps the company's records. This person needs to be able to read, write, and calculate. The community function and the fact that women and children show up and help, the singing while setting the net, and community pulling, all show how deeply embedded beach seine fishing is in village life.

8.6.6 *Fishing for a Living*

Beach seine fishing is a business, and seldom done only for subsistence or because of a lack of alternatives (as has been suggested by Jorion (1988)). This will be addressed in more detail in the next section. Net owners do intensive bookkeeping (recording catches, loans, sales, fines, and expenses). Those working with a contract system need to maintain these records because the money is shared at the end of the contract period.

Table 8.3 Example of how a catch is shared in Woe in cedis (GHS) in 2004, given to me during fieldwork in 2004. The company catches five pans of fish, with a value of 1,000,000 cedis together. The leaders of the company first take off all the costs leaving 750,000 to be divided amongst owner and crew. The miscellaneous post and the remainder are kept by the net owner

Catch	Cedis	Euros
Five pans, each 200,000 cedis	1,000,000	90.9
Chop money	60,000	5.45
Transportation fee	10,000	0.91
Drinks (bottle of Schnapps)	16,000	1.45
Outboard motor (petrol)	40,000	3.64
Carrier of outboard motor	5,000	0.45
Divers	20,000	1.82
Carriers of net (on shoulder)	10,000	0.91
Singers	10,000	0.91
Tying of the net	20,000	1.82
Swim to sac	5,000	0.45
Total expenses	196,000	17.82
Income	804,000	73.09
Miscellaneous	54,000	4.91
Left	750,000	68.18
Divided in five parts		
One part is net owner's money	150,000	13.64
One part for the leaders (*bozu*, paddlers, menders)	150,000	13.64
One part is also for net owner (transport for buying net)	150,000	13.64
Two parts for the company (crew 40 people)	300,000	27.27
Thirty strong (all get 8,000)	240,000	21.81
Three children	15,000	1.36
Seven women (get half of men)	28,000	2.55
Remainder	17,000	1.55

Source: Kraan (2009, p. 136)

In my research villages, I was allowed to look into the books of some of the companies. The books show how first expenses are deducted, including the part for the crew members with special roles who earn extra. The remainder is shared between net owner and crew. This is shared either in three parts or in five parts. From all sharing systems, it can be deduced that the net owners receive a fairly large amount, often with a separate part earmarked for investments in the canoe and net, and always with the cost of the expedition already deducted (Table 8.3). Another aspect of fishing that can be seen perfectly in the bookkeeping is that catches are seasonal; the prices fishers get for their catches vary depending on the season, and on the catches of other companies.

A net owner in Woe gave me an example of how companies share the catch (Table 8.3). The expenses are first deducted from the proceeds; what is left is shared in five parts.[14]

[14]Sharing systems in fishing companies differ. Some share in three parts, others in five parts. The way these parts are then divided among the workers differs. In general, the effect of sharing in three or in five parts is that, if shared in five parts, the crew gets a relatively larger part.

Two parts go to the net owner,[15] one part is shared among the leaders in the company, and the remaining two parts are shared among the crew members. In the end, if you are a strong crew member, you will earn 8,000 cedis based on this example. If you also perform a special task, you can earn even more. For instance, if you carry the outboard motor, you will earn an additional 5,000 cedis (€0.45). On average, we could conclude that fishers earn between 5,000 and 40,000 cedis (between €0.45 and €3.63) per fishing expedition.

To get an impression on how poor or rich these data make fishers in Ghana is not easy. Research performed in 1999 showed that the average daily income per capita in Ghana was €0.50 per day, with the average daily household income being €2.16 (Trades Union Congress 2004).[16] These figures give us an indication. They show that if catches are good and fishermen have two expeditions a day, a fisherman can earn about €7 (three times the daily *household* income of 1999). Yet if catches are bad and income is minimal, a fisherman can go home with only €0.45 (still almost average per capita income). Incomes in fisheries are highly variable; it depends on the weather, the luck, the timing, and the season.

Thus when catches are good and many fishing expeditions can be carried out, it shows that fishers can earn quite a lot of money. Fishing in Anlo has been lucrative for quite some time. In an article by Nukunya (1989, p. 162) we read that:

> Company owners are easily among the most richest men, not only in Woe but also in all the settlements along the littoral (…).

However, with regard to the crew members, Nukunya (1989, p. 166) found that:

> Their incomes compare favourably with those of their counterparts in farming and other occupations. The large fish content of their diet means also that they eat better than their counterparts.

Furthermore, Nukunya (1989, p. 170) states:

> As for the ordinary fishers, the least that can be said about them is that, in terms of achievement, measured in buildings, clothes and food habits, they can hold their own against the average farmer.

In my research, I tried to gain a better picture of the importance of fishing as a source of income for fisher households, and asked whether people engaged in other income-generating activities in addition to their fisher or processor profession (in which most of them worked). The majority of the households indicated that their income comes only from fishing-related activities (85% in Woe; 82% in Akosua Village; and 93% in Half Assini). It became clear that when other income-deriving activities were carried out, farming was one of them in the home village of Woe, whereas the fishers on migration seldom had that alternative available to them.

[15] The financial returns are different for net owners and crew members. The net owners get a large part of the catch, but they also have to reinvest part of it in the business. The canoe, net, and motor all need to be maintained continuously, and crew members should always be able to arrange a loan with net owners, although net owners also use the financial returns as income.

[16] In the publication the dollar is used; US$0.60 per capita per day and US$2.59 per household per day. For comparison reasons in this chapter I have converted it to Euros, which had a conversion rate of 1.2.

Overall, fishers are mainly active as specialized fishers. Only 11% of the fishers answered that their household members carried out an activity outside fishing. The combination of fishing and farming was quite common, as is the combination (in the households) of fishing with petty trading.

Income from fishing is important for Anlo-Ewe beach seine fishers as they are quite specialized and do not have many other income-generating activities. Declining catches can impact strongly on the income of the fishers (depending on how market prices react). The crew members, due to the catch-dividing system, are more affected than the net owners. Net owners have the possibility of placing a large part of the risk on the crew. This is also reflected in the job satisfaction of crew members, which is much lower than that of the net owners (Kraan 2009).

8.6.7 *Fishing as a Way of Life*

Jorion states, in a much cited article about the Xwla and Anlo-Ewe in 1988, that no one would ever become a full-time marine fisher by choice because it is too dangerous and economically risky (Jorion 1988). Nukunya (1989, p. 159) replies in a reaction to this article that the fishers do choose their profession because of the financial returns and because of their love for it[17]:

> One simply has to listen to these fishers recalling some of their famous exploits at sea, the challenges they faced and the heroism required to meet them, their big catches, the big monies they earned etc. etc. Again during the off season when they appear idle, they look forward to the onset of the next season; no doubt with some financial motive, but also for the joy the activities bring, because fishing is their life.

During my fieldwork, I also heard accounts of how beautiful the fishing profession is and how lucrative it can be, for instance when I met the small (12 people) company in Togo in January 2004, who had managed to buy a new boat after 1 year of fishing. When I saw them in Togo, they were painting it in bright colors. Some months later, I met them again in Ghana and I could see the result. The canoe was named "We are also coming," which refers to their successful migration.

The singing in the boat and while pulling the net, the decorated canoes, the decorated houses of the net owners, the completely new attire and happiness in the community which accompanies the return home of a migrated company are also evidence of how fishing is much more than just a job: "It is a culture, a way of life" (Akyeampong 2007, p. 180). Dr. Dovlo, an Anlo-Ewe professor, originating from a fishing community, explained to me how the home-coming of fishing companies on migration took place (Interview, 12 June 2004):

> As a child I remember how the family fishing company returned from migration, in trucks – all of them wearing their new company uniform. We welcomed them as heroes! Women ran outside, happy as they were to see their men and sons again. I, as a schoolboy, felt jealous, seeing my age-old nephews returning as men. They then also started treating me as boy instead as someone of the same age. Migration made a man out of a boy.

[17] See also the critique of Jul-Larsen (1994, pp. 13–18).

Many Anlo-Ewe even refused to move away from the coast when they were faced with severe coastal erosion (Akyeampong 2001, 2007). Fishing is an integral aspect of their lives. They grew up at the seaside, on a coastal littoral sometimes only a few kilometres wide where they, as small boys, learned to swim and fish (Nukunya 1989); and the sea plays an important role in their religious and ritual life (Akyeampong 2001).

The importance of beach seine fishing for the coastal fishing communities also implies more than it just being a source of income to individuals; it has communal importance as well. Beach seine companies are often used to earn money for the community. Every now and then the net owners and the fishers are asked by the town council or the traditional council to fish a day extra for the community. In Woe, community fishing often takes place on Saturdays, a day when people often do not fish because of the number of funerals that take place on that day. One reason why community fishing days are held is, for example, to generate money for school furniture (Interview with migrant fisherman in Togo, 28 January 2004). School children often help with the fishing if organized on Saturdays or on holidays.

Nukunya (1989, p. 159) claims that the fishing industry is important to the Anlo-Ewe homeland because it has positive effects not only on the fishers, but also on the whole community:

> The quality of their housing alone is sufficient to tell him he is in an area which is, at least, less poor than most parts of Ghana and Togo. … The traditional mud and thatch houses have been completely replaced, all within the last 40 years or so. … In terms of education, the Anlo coast has more than its fair share of facilities … most of these schools have been built by local rather than government initiative.

I was able to observe the social importance of beach seines many times in the research locations, given the participation by everyone in pulling in the nets and, in that way, earning some fish. You often see old men and physically or mentally challenged people joining in, usually at the back of the rope, near the women. Members of shrines can also come to the beach and ask for fish, and they do not even have to help pull in the net. Their requests are rarely refused and it has become the norm to give fish to these people when they ask for it.

8.7 Alternative Versus Supplemental Livelihood Programs

Alternative livelihood programs are often suggested as a way of addressing the problem of declining catches due to other causes, which are often interpreted as necessitating a reduction in fishing effort. A reduction in effort will unavoidably lead to fewer fishers being able to live from fisheries or to fishers earning less, but depending on how the reduction is organized, meaning people or time. Ideally, this loss in income must be compensated for and in that context alternative livelihood options are often considered. See the following passage in the study of the National Coordination Unit (NCU) Ghana of the SFLP (NCU Ghana 2001, p. 27):

> Are there too many people fishing and too many gears in the water? The answer in the marine fishery is clearly yes. …Given that more than 500,000 people are involved directly or indirectly in marine artisanal fisheries, the task of limiting access and banning gears will inevitably mean that people will be displaced from the fishery. If this happens, for many the

obvious choice will be to do as others have done in the past, and move their seining gear to the Lake [Lake Volta]. This clearly cannot be part of any fisheries rationalization process, and alternatives must be sought. Government will be required to develop and fund a coherent strategy for assisting the transfer of fisheries community members into other income-generating activities, or into different kinds of involvement within the sub-sector. Inherent in such a process would be a policy decision to favour the expansion of artisanal communities into areas of the fishery currently occupied by industrial vessels.

At the local level, a few initiatives have been undertaken. In Half Assini, we were told by the Director of the Agriculture Department about their experience with these programs and fishers (Interview, 20 October 2005):

Director: We have been training fishers in alternative livelihoods. Catches are going down, so we trained them how they can go in pig farming or cassava farming. But the fishers are not used to waiting, so they haven't taken it up. Waiting is a problem, you first have to raise the animal, grow the crop before you can cash it. Also learning something new is difficult for them. Fishing is all they do, all they know. They have been fishing from childhood. No schooling, no alternatives to easily catch up with.

MK: Were they interested?

Director: We invited them, first we sensitized them. Then we asked them what they were interested in; they mentioned pigs, grass cutters, small ruminants. Then they came for the training. But since then there has not been any progress.

MK: How did you train them? By giving them the animals to start with?

Director: No, we showed pictures; invited them to a meeting room like this [the interview was taking place in a meeting room of the District Assembly in Half Assini]. Then visited a farm so they could see, informed them about the costs, the waiting time and when they might expect results.

MK: Did you give them money, for them to start?

Director: No we didn't give money. There are no funds and also there was no demand. Nowadays, we work on a demand-driven basis. They have to ask for it, they have to apply. Like the coconut farmers, they write letters.

In a later conversation, we came back to this issue. He then explained to me that these programs were not such a success with their fishers, since they were not willing to invest in the enterprise. So they came to the meetings, they followed the courses, but it was not taken any further (Interview, 17 December 2005).

From the literature, we do not get the impression that alternative activities are a great success either. A recent study on selected successful Marine Protected Areas (MPAs) in South-East Asia showed that many alternative livelihood programs proved to be unsustainable (Leisure et al. 2007). Dropping commodity prices, rising costs of inputs, and lower quality of the produce than of competitors meant that many programs lasted only for a short period of time. Leisure et al. (2007) state: "Most alternative income-generating activities are better suited for offsetting income initially lost due to establishment of no-fishing areas rather than as long-term tools to improve incomes or move people away from fishing." The new job opportunities in tourism, related to the established MPAs, turned out to lead to longer-term gains in non-fishing income.

Crawford (2002) looks into seaweed farming as an alternative livelihood for small-scale fishers in Asia. He points out that alternative livelihood programs have been suggested for over two decades now and that they generally have two objectives: first to raise the economic standard of living of fishers and coastal communities;

and second to reduce fishing effort. Three assumptions underlie these programs: (1) small-scale fishers are poor, and this is related to the over-exploited nature of the resources upon which they depend; (2) fishers are willing to give up fishing in favour of more lucrative economic opportunities; and (3) as fishers take up alternative livelihoods, this will reduce pressure on the fisheries (Crawford 2002).

Crawford rightly shows, and we have seen it before (Nukunya 1989; Odotei 2002; Béné 2004; Akyeampong 2007), that these assumptions do not always prove to be true. Crawford refers to a study on job satisfaction by Pollnac et al. (2001) in which they showed that fishers do not always wish to leave fisheries because fishing earns them more money than other jobs. Fishers are not always the poorest of the poor – as my research has also shown. It is also important to differentiate between fishers – in the case of beach seine fishing between net owners and crew. Fishers are not only interested in fishing because of its economical returns, but also because they are fishers – it is part of their identity. Many anthropologists successfully explain how fishers see their fishing as more than just what they do for a living; fishing is part of their identity (Acheson 1981; Van Ginkel 2007) whereby they share certain characteristics throughout cultures.

Thus it might be better to think of *supplemental* livelihoods instead of alternative livelihoods. It should, however, be kept in mind that this means that the fishers are limited with regard to space for taking up additional livelihood activities, since they need to be near the fishing grounds. Taking up alternative livelihoods, by some fishers, will not automatically lead to a reduction of fishing effort (Crawford 2002, p. 16): "Unless livelihood strategies are combined with resource management strategies that address the open access nature of coastal fisheries, progress toward improved fisheries management will be limited." In the end, Crawford goes one step further and suggests that economic diversification – and then seen from the household level – might even be a better goal than alternative or supplemental livelihood for fishers.

The suggestion made by the National Coordination Unit of the SFLP program of the Ghanaian government is indeed important (NCU Ghana 2001, p. 27): "*A policy decision* [is inherent in such a process – mk] *to favour the expansion of artisanal communities into areas of the fishery currently occupied by industrial vessels.*"[18] The Government would have to make some hard choices as to which is preferred in its fishing strategy: a (semi-)industrial sector important for revenue and export earnings or a thriving artisan sector important for livelihoods and food security.

The whole idea of alternative livelihood strategies seems to come from the way the sustainable livelihoods approach is used. Wartena's research on styles of

---

[18]This viewpoint is shared by the international NGO International Collective in Support of Fishworkers (ICSF), which was erected in 1984 out of concern for too much attention for the commercial, industrial, scientific, and fishery resource aspects at the expense of "the actual real-world, life-and-blood people involved in fishing worldwide fishworkers who are often sections of the population marginalized from mainstream society" (http://icsf.net/icsf2006/jspFiles/icsfMain/about/english/aboutIcsf.jsp[Accessed date: Jan 2009]).

making a living (instead of using the term livelihood) indicates that "non-economic considerations often play a dominant role in people's livelihood choices" (Wartena 2006). She has therefore chosen to use the term styles as a way out because it draws more attention to socio-cultural values which she misses in the way other academics have interpreted livelihood in their research. The image of actors acting as a *homo economicus* is still dominant in a lot of research, as reflected in the usage of the term capitals, and by the fact that a number of researchers use the concept of livelihood as a synonym for income (see, for example, Akyeampong (2007): "Marine fishing is more than a livelihood; it is a way of life"). Researchers are often confused between livelihood activities and livelihood "outcomes" (Wartena 2006). Combined with the idea that livelihood "outcomes" should be solely understood as income, the idea of *alternative* livelihood activities as a solution to loss of income is easily created.

For the Anlo-Ewe (migrant) fishers, alternative livelihood options are probably limited. The educational level of fisher households is not high. Data from the questionnaires showed that in Woe 63% of the fisher population was illiterate or had reached elementary level (1–6 years of primary school); in Akosua Village this was 57%; and in Half Assini, 75%. The crew and net owners' responses to the questionnaire underscored these findings. Seventy percent of the fishers were illiterate (19%) or had reached primary school. Thus, alternative livelihood options will not be plentiful, since the majority of the fishers are not well educated. Also from our study, it can be seen that many fisher households have already diversified to a certain level.

8.8 Improving Ghana's Fisheries Management

Considering the importance of the fisheries sector in Ghana, and considering the size of the artisanal sector therein, it is imperative to include the artisanal fishing sector in fisheries governance at the national level. Only then the problem of declining catches and the possible effect on poverty can be appropriately addressed. The FAO Code of Conduct clearly points in that direction (see the beginning of this chapter).

Including fishers in management is not as strange as some might find it. Fishing is a risky business with many uncertainties, but it also offers many opportunities. It is obvious that fishers try to organize their business collectively in such a way that risks are reduced and opportunities can be pursued. The assumption, thus, that fishers will try to manage their fishing well if they depend on it for their livelihood has proven to be right in my research (Kraan 2009, Chap. 7).

Ghanaian beach seine fisheries are indeed managed quite extensively, by the fishers themselves. Fishers have regulated the access to the fishing grounds, the interaction (including conflicts) between the fishers both within fishing companies and between different groups (and gears), and the extraction of fish and access to the market. My study (Kraan 2009) has thereby shown that the success of self-management depends to a large extent on local leadership.

At the national level, however, there is not much attention for the artisanal sector. Rules and regulations at the national level, often with a relation to preventing further stock depletion (such as mesh size), are seldom actively controlled nor explained to the fishers, and thus often not complied with. Yet only with compliance and/or control can management be effective. As compliance is the preferred situation, management is much more about managing people than it is about managing fish (stocks). However, fisheries management literature has often focused mainly on either managing fish stocks (which elsewhere has been called the *paradigm halieutique*) or on understanding the market. This focus is based on the assumption that the human activity is the main cause of resource decline, a vision which has been questioned in the New Ecology thinking. Moreover, the state has often been ascribed a central role which has obscured the fact that fishers themselves are largely active in managing activities.

The government of Ghana's regulating activities with regard to the artisanal sector is oriented much more around modernising the sector and conserving fish stocks. However, the artisanal sector is, for the most part, a self-regulated system in which the government of Ghana does not play a major role. It has also been recognized that the fishers are primarily concerned with managing their livelihoods. This is becoming more difficult as catches are declining. The topic of declining catches can be one on which government and fishers may share views. Although views on the decline may differ regarding the causes (see Kraan 2009), both the government and fishers are concerned about the occurrence.

8.9 Conclusions

Ghanaian fishers are specialized professionals, and being a fisher is often not only a means to earn an income but is part of one's identity. Fisheries are enormously important for the Ghanaian economy with almost 10% of the population being dependent on it. Fisheries have also proven to be crucial for coastal communities, being firmly rooted in the culture and engaging women, children, and people with little options to sustain themselves otherwise. Poverty in fishing communities can definitely be seen in Ghana, yet this is much more a reflection of a general lack of development in rural areas.

This chapter has shown how the link between poverty and small-scale fisheries in Ghana is, at the very least, more complex than at first glance. Fisheries are often one of the few options to sustain a livelihood in coastal communities. The declines in catches are, therefore, a serious threat and will undermine the strength of the sector. Thus far, the artisanal sector has been a self-regulated system in which the government of Ghana has not played a major role. More attention from the Ghanaian government for this sector, in light of its importance for livelihoods, and in light of declining catches, is imperative. It is, however, important for the government – by doing so – to connect to the world view and knowledge of fishers, and connect to their interest in social and economic outcomes of regulations. By improved inclusion of the artisanal sector, it is also important to connect to the existing institutions and strengthen them instead of crafting new institutions. The latter is often suggested and done, yet is doomed to fail as institutions are not embedded in the reality of plural normative orders, and if they ignore heterogeneous local needs (see Kraan 2009). Finally, the path of alternative livelihood programs turns out to not be very successful, as the options are limited and often prove not to be sustainable. Typically, the assets of fisher households need to be broadened, in order to really enhance resiliency. The education of the children is an important point to focus on, and points to a longer term perspective.

Acknowledgments Many thanks to Svein Jentoft and Maria-Victoria Gunnarsdottir for their constructive comments. Thanks also to the Amsterdam Institute for Social Science Research of the University of Amsterdam for funding the research. Gratitude to the Norwegian Research Council for funding the PovFish project.

References

Acheson J (1981) Anthropology of fishing. Annu Rev Anthropol 10:275–316

Addo E, Marshal R (2000) Ghana's non-traditional export sector: expectations, achievements and policy issues. Geoforum 31(3):355–370

Akyeampong EA (2001) Between the sea and the lagoon: an eco-social history of the Anlo of southeastern Ghana c. 1850 to recent times. James Currey, Oxford

Akyeampong EA (2007) Indigenous knowledge and maritime fishing in West Africa: the case of Ghana. Special Volume Indigenous knowledge systems and sustainable development: relevance for Africa. In: Boon E, Hens L (eds.) Tribes and tribals, Kamla-Raj Enterprises, Delhi, India No. 1, pp 173–182

Allison E, Ellis A (2001) The livelihoods approach and management of small-scale fisheries. Mar Policy 25:377–388

Atta-Mills J, Alder J, Sumaila R (2004) Decline of a regional fishing nation: the case of Ghana and West Africa. Nat Resour Forum 28:13–21

Béné C (2003) When fishery rhymes with poverty: a first step beyond the old paradigm on poverty in small-scale fisheries. World Dev 31(6):949–975

Béné C (2004) Poverty in small-scale fisheries: a review and some further thoughts. In: Neiland A, Béné C (eds.) Poverty and small-scale fisheries in West Africa. Kluwer, Dordrecht

Berkes F, Mahon R, McConney P, Pollnac P et al (2001) Managing small-scale fisheries: alternative directions and methods. IDRC, Ottawa

Chauveau J (1991) The historical geography of fisheries migrations in the CECAF region (end of the 19th century – 1980s). In: Haakonsen J, Diaw C (eds.) Fishermen's migrations in West Africa. FAO/DANIDA/NORWAY: IDAF/WP/36

Chauveau J-P, Jul-Larsen E, Chaboud C (eds.) (2000) Les Pêches Piroguières en Afrique de l'Ouest. Pouvoirs, Mobilités, Marchés. Karthala, Paris

Christensen V, Amorim P, Diallo I, Diouf et al. T (2004) Trends in fish biomass off northwest Africa, 1960–2000. In: Palomares M, Pauly D (eds.) West African marine ecosystems: models and fisheries impacts. Fisheries Centre UBC,Vancouver. Fish Cent Res Rep 12(7):215–220

Crawford B (2002) Seaweed farming: an alternative livelihood for small-scale fishers? Coastal Resources Centre, University of Rhode Island, Working Paper. http://imcafs.org/download/Alt_Livelihood.pdf. Accessed 26 Apr 2008

De Vries L (2003) 'La Mer est Pour Nous Tous'- Local Dynamics in the Senegalese Artisanal Fisheries. M.Sc. thesis, Wageningen University, Wageningen (unpublished)

DoF (2003) Ghana: post-harvest fisheries overview. Directorate of Fisheries, Accra. Draft version 14 Jan 2003

FAO (2005) Increasing the contribution of small-scale fisheries to poverty alleviation and food security. Technical guidelines for responsible fisheries. FAO, Rome

FAO (2009) The state of world fisheries and aquaculture 2008. FAO, Rome

Feidi I (2001) Trading amidst constraints. Trade in fish and fishery products in most states of the organization of the Islamic conference faces varied constraints. Samudra 30:11–17

Ferrais J, Koranteng K, Samba A (1997) Comparative study of the dynamics of small-scale marine fisheries in Senegal and Ghana. Paper presented at the First International CEOS Symposium, 6–8 Sept 1994, Monterey Colloques et Séminaires. ORSTOM, Paris

Greene S (1996) Gender, ethnicity, and social change on the upper slave coast: a history of the Anlo-Ewe. Heinemann, Portsmouth

Haakonsen J (1992) Conference report: a regional policy on fishermen's migration in West Africa. Seminar on artisanal fishermen's migrations, Cotonou, Benin, 3–5 Dec. Mar Policy 16(3): 237–238

Haakonsen J, Diaw MC (1991) Fishermen's migrations in West Africa. IDAF Project, DIPA/WP/36, Cotonou

Hauck M (2008) Rethinking small-scale fisheries compliance. Mar Policy 32:635–642

Heinbuch U (1994) Animal protein sources for rural and urban populations in Ghana. IDAF Technical Report 58. FAO, Rome

Hendrix MK (1985) Technology and maritime fisheries on the Sierra Leone Peninsula, c. 1600–1980. In: Stone J (ed.) Africa and the sea. Aberdeen African Studies Group, Aberdeen, pp 64–79

Hoorweg J, Wangila B, Degen A (2009) Artisanal fishers on the Kenyan Coast: household livelihoods and marine resource management. Brill, Leiden

Hosch G (2002) Don't jump the fence. Fisheries development and management must choose an integrated approach. Samudra, March 2002

Johnson D (2006) Category, narrative, and value in the governance of small-scale fisheries. Mar Policy 30:747–756

Jorion P (1988) Going out or staying home: seasonal movements and migration strategies among Xwla and Anlo-Ewe fishermen. MAST 1(2):129–155

Jul-Larsen E (1994) Migrant fishermen in Congo: tradition and modernity. CMI Report R 1994:6. Chr. Michelsen Institute, Bergen

Kraan ML (2009) Creating space for fishermen's livelihoods. Anlo-Ewe beach seine fishermen's negotiations for livelihood space within multiple governance structures in Ghana. Ph.D. thesis, African Studies Centre, Leiden

Leisure C, Van Beukering P, Scherl LM (2007) Nature's investment bank: How marine protected areas contribute to poverty reduction. http://www.prem-online.org/index.php?p=projects&cid=19. Accessed Mar 2008

Lewins R (2004) The sustainable livelihoods approach: the importance of policies, institutions and processes. In: Neiland A, Béné C (eds.) Poverty and small-scale fisheries in West Africa. Kluwer, Dordrecht

Mensah M, Koranteng K, Bortey A, Yeboah D (2006) The state of world fisheries from a fish-worker perspective: the Ghanaian situation, Samudra Monograph. ICSF, Chennai

MFRD (2004) Information on fisheries in Ghana. Marine Fisheries Research Division MFRD, Accra

NCU Ghana (2001) Study on the impact of policies, institutions and processes on fisheries livelihoods in Ghana. National Coordination Unit. FAO, SFLP, Rome

Nukunya G (1989) The Anlo-Ewe and full-time maritime fishing: another view. MAST 2(2): 154–173

Nukunya G (1991) The Anlo-Ewe fishermen's migration. In: Haakonsen J, Diaw C (eds.) Fishermen's migrations in West Africa. FAO/DANIDA/NORWAY: IDAF/WP/36

Nukunya G (1999) Kinship and marriage among the Anlo-Ewe. The Athlone Press, London

Nunoo F, Eggleston D, Vanderpuye C (2006) Abundance, biomass and species composition of nearshore fish assemblages in Ghana, West Africa. Afr J Mar Sci 28(3–4):689–696

Odotei I (1991) Migrations of Fanti fishermen. In: Haakonsen J, Diaw C (eds.) Fishermen's migrations in West Africa. FAO/DANIDA/NORWAY: IDAF/WP/36

Odotei I (1995) Migration of Ghanaian fishermen within the ECOWAS sub-region. In: Twum-Baah K (ed.) Migration research study in Ghana, vol II. Ghana Statistical Service, Accra

Odotei I (2002) There is money in the sea. Ghanaian migrant fishermen and women in the Ivory coast. University of Ghana, Institute of African Studies and Accra: Royal Gold Publishers, Legon

Overå R (2001) Institutions, mobility and resilience in the Fante migratory fisheries of West Africa. CMI Working Paper. Chr. Michelsen Institute, Bergen

Pauly D (2006) Major trends in small-scale marine fisheries, with emphasis on developing countries, and some implications for the social sciences. Marit Stud 4(2):7–22

Pinnegar J, Engelhard G (2008) The 'shifting baseline' phenomenon: a global perspective. Rev Fish Biol Fish 18:1–16

Platteau J (1989) The dynamics of fisheries development in developing countries: a general overview. Dev Change 20:565–597

Pollnac R, Pomeroy R, Harkes I (2001) Fishery policy and job satisfaction in three southeast Asian fisheries. Ocean Coast Manage 44(7–8):531–544

Sen A (1981) Poverty and famines: an essay on entitlement and deprivation. Clarendon, Oxford

Solie K (2006) Migrations de pêcheurs artisans marins en Guinée. SFLP: Programme pour les moyens d'existence durables dans la pêche en Afrique de l'Ouest (PMEDP). Projet Pilote 2 Aménagement participatif des pêches en zone côtière (unpublished working document)

Trades Union Congress (2004) Incomes in Ghana. Policy discussion paper. Global Policy Network. http://www.gpn.org/data/ghana/ghana-analysis.pdf

Van Ginkel R (2007) Coastal cultures: an anthropology of fishing and whaling traditions. Het Spinhuis Publishers, Apeldoorn/Antwerpen

Wagner K (1991) Fishermen's migrations in Sierra Leone: a case study of Tombo fishing village. In: Haakonsen J, Diaw C (eds.) Fishermen's migrations in West Africa. FAO/DANIDA/NORWAY: IDAF/WP/36

Wartena D (2006) Styles of making a living and ecological change on the Fon and Adja plateaux in South Benin, ca. 1600–1990. Ph.D. thesis, Wageningen University, Wageningen

Worldfish Center (2005) Fish and food security in Africa. WorldFish Center, Penang, Malaysia

Yeboah D (2002) Beach seining in West Africa: to ban or not to ban? SFLP Bulletin, PMEDP, Nr. 7

Yin R (1994) Case study research: design and methods, 2nd edn. Sage, Beverly Hills

Chapter 9
Wealth, Poverty, and Immigration: The Role of Institutions in the Fisheries of Tamil Nadu, India

Maarten Bavinck

Abstract This chapter explores two concurrent processes in the fisheries of Tamil Nadu, India, over the past century: technological modernization and demographic growth. The first process is closely connected to the Blue Revolution instigated by the Government of India after Independence, as well as to the globalization of markets. It has resulted in substantial increases in sectoral wealth. The second process is the increasing size of the fishing population through natural growth and immigration. I situate the poverty that still occurs in Indian fisheries in the confluence of these two processes, arguing that varying institutional arrangements which structure participation have an important effect on poverty's availability and location. The chapter centers on one particular district – Ramnathapuram – which has witnessed particularly dramatic increases in its fishing population compared to other parts of the South Indian coastline. This has resulted in specific patterns of poverty and riches.

9.1 Introduction

An earlier chapter in this volume (Eide, Bavinck, and Raakjær, Chap. 2) pointed out that wealth has characterized marine fisheries in the twentieth century as much as poverty has. As a result, many of the world's fisheries have witnessed demographic expansion, also through immigration, and a changing divide between rich and poor. In the present chapter, the details of this process are explored in the context of Tamil Nadu, India. The focus lies on one particular coastal district called Ramnathapuram,

M. Bavinck (✉)
Department of Human Geography, Planning, and International Development Studies, University of Amsterdam, Nieuwe Prinsengracht 130, 1018VZ Amsterdam, The Netherlands
e-mail: j.m.bavinck@uva.nl

S. Jentoft and A. Eide (eds.), *Poverty Mosaics: Realities and Prospects in Small-Scale Fisheries*, DOI 10.1007/978-94-007-1582-0_9,

or Ramnad, which stands out for its particularly high rate of immigration into fishing, and on the question of institutional control. I argue that the institutional configuration in Ramnad, compared, for example, to the Coromandel Coast, which lies to the north, was relatively weak, and therefore conducive to population inflow. This has contributed to severe competition on the fishing grounds, social conflicts on shore, and overfishing of marine resources.

Schlager and Ostrom (1993) distinguish two requisites in the fishing profession: rights of access and rights of extraction. The former pertains to the ability of establishing preconditions for the act of fishing: possessing equipment and beach rights, and actually being allowed to set out to the fishing grounds in question. The second requisite is the right to actually set the fishing gear, catch fish, and market the fish through available channels. Institutional controls over access and extraction thus constitute a core aspect of fisheries governance the world over (Charles 2001).

In many cases, such governance is exercised under conditions of legal pluralism (Von Benda-Beckmann 2002; Bavinck 2005). Legal pluralism implies the application of different legal ideas, principles, and systems to the same situation (Vanderlinden 1971). In many cases, one finds state law defining fishing rights juxtaposed over customary law. The former is implemented through governmental agencies and courts, whereas customary law relies on a range of traditional authorities. I argue that while state law has unambiguously striven to open doors into the fishing profession in India, customary law has frequently created informal barriers. The latter, however, is not distributed evenly along the coastline. Varying expressions and potencies of customary law have contributed, in interaction with state law, to diverse patterns of immigration and distributions of wealth (Bavinck 2003).

The following two sections describe the modernization of fisheries as it occurred in Tamil Nadu since Independence (1947), and the available distribution of wealth and poverty among the fishing population. I then delve into the change process in Ramnad District, highlighting its demographic features. These are linked, in comparison with other coastal regions, to institutional controls prevailing in the fisheries.

9.2 Blue Revolution in Tamil Nadu Fisheries

Like most new nations in the developing world, India took up the cause of industrializing capture fisheries after Independence. The so-called blue revolution was engendered by the Indian government in parallel to the green (agriculture) and white (dairy) revolutions. It commenced in the 1950s, and resulted in the establishment of a modern fishery sector, next to an old and widespread small-scale fishery (Bavinck and Johnson 2008). This modern fishery sector, which consisted – and still consists – largely of a fleet of small trawling vessels, developed rapidly. This was mainly because of the successful connection that was made in the 1960s to the international seafood market (Salagrama 2004). The differential between national and international prices, particularly for shrimp, resulted in what John Kurien (1978) has appropriately called a "pink gold rush."

Table 9.1 Emergence of trawler fishing in Tamil Nadu: vessel numbers and production compared (1948 and 2000)

Year	Small-scale vessels	Trawler vessels	Trawler catches (t)	Total fish catch (t)
1948	13,204	0	0	27,135
2000	41,770	8,009	200,468	377,483

Source: Department of Fisheries 2000

Although the fishing profession generally has a low status in Indian society, and is associated with low castes, the opportunities for making a fast buck drew in investors from fields as diverse as big business, the film industry, politics, and the professions (Bavinck 2001a). Throughout the subcontinent, a wave of conflict between modern and small-scale fishers subsequently emerged, as trawler fishers were operating on the same fishing grounds as small-scale fishers (Bavinck 2005).

The continuation of conflict ultimately resulted in state governments enacting legislation to spatially separate the two categories of fishers. On paper, trawler fishers were thereby relegated to the offshore fishing zone (beyond 3 nautical miles). In practice, however, these rules were scarcely effectuated and conflicts continued (Bavinck 2001a). Industrialization also impacted the small-scale fisheries, however, introducing new vessel designs (often engine-propelled), gear types, and fishing practices. As a consequence of these developments, some observers (e.g. Johnson 2006) suggest that there are now three categories of fishers in the country: semi-industrial and industrial, small-scale, and intermediate.

Tamil Nadu was at the forefront of the industrialization process of capture fisheries in India, closely following the states of Kerala and Goa, which had taken the lead. The Fisheries Department played a highly proactive role, establishing boat-building yards and refrigeration facilities, distributing new gears and fuels at high subsidy rates, and generally encouraging technological innovation. Its officers had two concerns in mind: (1) increasing total output and contributing to the generation of foreign exchange and (2) uplifting the fisher population from its condition of "backwardness." The latter was of more than nominal importance. Government officials held the fishing population to be poorer and more backward than average, and in great need of social and economic development.

In accordance with the general trend in India, the Tamil Nadu government focused its efforts on establishing a trawl fishery along its 1,000 km coastline. Looking back, it was more than successful. After an initial phase of reticence and trial and error, there came a genuine rush to invest in this new technology. The trawl fleet thus expanded in leaps and bounds, and fisheries production boomed (Table 9.1). As a consequence, the government was able to terminate most of its primary involvement in trawl fisheries (such as boat-building) by the 1970s, henceforth leaving expansion to the private sector to carry out. Now the Fisheries Department, largely in reaction to demands from the fishing population, began to involve itself in regulation.

The Tamil Nadu Marine Fishing Regulation Act (1983) and the accompanying Rules brought about a licensing regime and successfully tied trawlers to specific ports, thereby creating conditions for further governmental control. And although the attempts to spatially separate trawlers from small-scale fishers failed miserably, an effectual system of time-zoning was indeed established, at least in the three central coastal districts, including Ramnad (cf. Bavinck 2003).

Meanwhile, the small-scale fisheries in Tamil Nadu too were changing. Although government policy had tended to neglect this segment of the fishing population, particularly in the initial period of industrialization, technical innovation nonetheless took place (Bavinck 1997; Bavinck and Karunaharan 2006). By 2000, most small-scale fishers in Tamil Nadu were making use of light-weight and strong, synthetic fishing gears of various – frequently new – designs, and many of them had motorized their craft. Their range of operation had increased, they were fishing more days in every year, and catch per unit of effort had gone up.

Table 9.1 provides primary statistics on Tamil Nadu capture fisheries, as gleaned from departmental statistics.

The first point to note is the absolute growth of fish production. Catches in 2000 are no less than 14 times higher than five decades earlier. Much of this increase of production is caused by the modern fishery sector, which developed from zero vessels at Independence to approximately 8,000 vessels in 2000. These vessels are based in 13 harbor locations along the coast, and contribute more than 50% of annual production.

But small-scale fisheries too have grown, in terms of both the number of fishing units and average production. Thus, there were more than three times as many fishing units in 2000 than in 1948, and each unit caught on average about twice as much annually (from 2 to 4 t/unit). Together, the small-scale fishers of Tamil Nadu still account for almost half of the annual fish catch.[1]

The growth in fish production in Tamil Nadu, and in India, is linked to the development of the export market – first for shrimp, and then for a range of other seafood products. Separate figures for export of seafood from Tamil Nadu are not available. Salagrama (2004, p. 15), however, notes that the export of seafood from India increased from 15,732 t in 1961–1962 to a phenomenal 343,041 t in 1999–2000. In terms of value, exports went up from Rs 40 million to Rs 50,117 million (US$1,189 million) in the same period.

Discussing global trends, Delgado et al. (2003, p. 37) note that "fresh and frozen fish have shown a long-term increase in their real prices since the second world war." This conclusion also seems to apply to the domestic market in India. Salagrama (2004, pp. 14, 78) points out that, on the basis of wholesale price indexes for the period 1981–1996, "the increase in real value of fish is much faster than that of other food items." This is indeed the experience of consumers, for example in the urban conglomerate of Chennai, who complain about the regular rise of fish prices on the local market (field notes from author).

The long-term development of seafood prices, which obviously provides incentives for fishers to intensify their harvesting efforts, does not, however, quite capture the jolt experienced by fishers in the South, faced by the opening of the export market. Although Kurien's (1978) account of the almost miraculous development of the

---

[1]Compare Salagrama's (2004, p. 13) analysis of all-India trends which argues that catch per unit effort (CPUE) of artisanal fishers has declined since 1980.

Table 9.2 Population growth in Tamil Nadu 1951–2001: general figures and fisheries compared

	1951	2001	Growth (%)
Population in Tamil Nadu	30,119,047	62,405,679	200
Fisher population in Tamil Nadu	95,735[a]	679,711[b]	700

Source: Department of Fisheries 2000; Chacko et al. 1957; Govt of India Census 1952, 2001

[a]Data 1948

[b]Data 2000

demand for shrimp (from fertilizer for coconut plantations to prized export product) does not apply to all other seafood products in India (for many, a local demand does exist), the interval between domestic and international prices has been large enough to excite everyone engaged in marine fisheries.

What consequences has this had for the fishing population in Tamil Nadu? In the time period under consideration, the fishing population (men, women, and children) too has increased manifold. Available sources (Department of Fisheries 2000) suggest that the marine fisher population increased from 95,735 in 1948 to 679,711 in 2000 (an increase of 700%). This also resulted in a multiplication of fisher settlements. While the Tamil Nadu coastline in 1948 counted 233 villages and neighborhoods, this number had increased to 591 in 2000 (an increase of 250%).

Compare the above with the general demographic trend in Tamil Nadu. At the time of the 1951 census, Tamil Nadu counted 30 million inhabitants. This figure had more than doubled to 62 million by the time of the 2001 census.

Table 9.2 points out that the growth of the fisher population outstripped general population growth by far. The logical explanation for this trend is that a net movement occurred into fisheries during the period under consideration, triggered by the enormous economic potential of marine fisheries.

The sweeping process of modernization, or, to follow Smith's (2000) terminology, "industrialization," of the Indian oceans has had important consequences for the marine ecosystem, many of which are yet to be documented. Overall production data, however, point to a leveling of catches since the late 1990s, and scientists have recorded widespread evidence of "fishing down the food web" (Bathal 2005; Vivekanandan et al. 2005). These scientific analyses corroborate the observations fishers in Tamil Nadu have been making over a much longer time period of diminishing catches and sizes of fish, and vanishing species (Bavinck 2001a).

9.3 Location and Nature of Wealth and Poverty in Tamil Nadu Fisheries

There is evidence of extensive material poverty in the coastal fisheries of Tamil Nadu in the period before Independence. In the first decades of the twentieth century, the Madras Fisheries Bureau, established by the colonial government for the development of fisheries in southern India, thus concluded that the fishing population of the eastern coast was "backward" and in dire need of upliftment. To substantiate its claims, it described the living conditions of fishers of Chingleput and South Arcot districts, in the heart of the Coromandel Coast, as follows (Madras Fisheries Bureau 1916, p. 135):

> Their huts and surroundings are dirty and they are illiterate without any desire to improve their condition. Intemperance is the curse of the community.

More to the south of the Coromandel Coast, the population of Akkaraipettai was viewed as being (Madras Fisheries Bureau 1916, p. 136) "very backward and [leading] a hand-to-mouth life without paying any attention to the sanitary condition of their hamlets, education of their children, etc." And in the Catholic belt, in the southern districts, the situation was not much different; with the exception of merchants monopolizing the processing and trade of fish, fishers are described as "mostly poor."

Anugraham (1940, pp. 16–17) provides details on the housing situation of fishers in Chennai at the inception of World War II:

> Most of the fishers live in huts. Normally a hut is about 8 feet by 10 feet with low walls and low roof of palm matting. There is only one doorway which is hardly 5 feet high. In some cases, a bamboo tatty which serves as a provisional "door" is placed at the entrance and fastened by ropes to either door post. Most of the huts have no windows, as the need for ventilation is neither valued nor even felt by the fisherfolk. The ventilators, if any, are nothing more than holes to send out smoke. Incidentally, they let in a little light. ... It is remarkable that a hovel hardly 8′ by 10′ serves as a store room, a kitchen and dormitory, all in one.

These observations by outside academics and bureaucrats do not mean, however, that the situation of fishers at the time was undifferentiated. Certain categories, such as the owners of the capital-intensive beach seine companies, which operated along at least some shorelines, were most certainly quite wealthy (Hopewell 2004; cf. Salagrama 2008). Oral history suggests that within hamlets of small-scale fishers too there were distinctions between have and have-nots. The dividing lines between such categories, however, were relatively fluid, and poverty tended to affect the whole fishing population, to one extent or another.

The blue revolution brought about massive rearrangements of wealth and poverty in the fishery sector. As statistical evidence on income distribution within the fisheries sector is still limited, we will have to make use mainly of qualitative sources of information.

Today wealth is no longer concentrated with beach seine owners, who have beaten a quiet retreat. Instead, it is situated in the trawler ports situated along the Tamil Nadu coastline, and in a class of trawler owners. Here again, one finds a great deal of variation. Reviewing the dynamics of mechanized boat fishing in one of these ports, in Chennai, Bavinck (2001a) documents the existence of a successful group of trawler owners, who possess multiple craft and have diversified investments into other fields outside of fishing. Their children have pursued higher education and are moving into other societal domains. These families congregate in new, middle-class neighborhoods on the perimeters of the harbor area. In contrast stands a category of marginal trawler owners with old equipment, high debts, and uncertain returns. Their societal position is shaky, and they may as easily move down as up.

But the trawl fishery has also generated a new male and female working class, employed on the boats and in the extensive service sector surrounding each harbor. Their work is dirty, insecure, and tiring, and they work long hours for low pay. Residing in (rehabilitated or non-rehabilitated) slums that spread out in the city, alcohol abuse, violence, and prostitution are common, and upward social mobility limited. Still, in comparison with their predecessors in the pre-Independence era, these workers will agree that absolute poverty has gone down. For them, this is a consequence perhaps less of the blue revolution than of the numerous welfare schemes initiated by the Fisheries Department and other governmental agencies particularly since the 1970s. These include the supply of essential food items at subsidized rates (ration schemes), subsidized housing, life insurance, and a saving-cum-relief scheme to provide for the annual off-season; to this can be added the benefits of the public medical system in Tamil Nadu. Despite its many deficiencies, this has increased average life expectancy from 40 to 63 years, and the general decline of poverty (Government of Tamil Nadu 2003). As we shall see later, many members of the fisher working class currently derive from outside the fishing sector.

Wealth and poverty in the 500-plus fishing villages of Tamil Nadu are more subtly arranged. Although many village populations still operate social "safety nets" (Kurien 1995) to offset individual or collective mishap, this does not prevent the accumulation of wealth and poverty in certain echelons. Salagrama (2006) has made an extensive analysis of fisher livelihoods along the east coast of India, focusing on the state of Orissa, and his conclusions are relevant for Tamil Nadu as well. Salagrama (2006, pp. 101–103) uses composite wealth-ranking methods to divide the coastal fishing population into four categories, ranging from "very well-off" to "extremely poor" or "destitute."

The occupations in the top- and bottom-most wealth categories are primarily located in the urbanized, trawler ports, whereas village fishers seem to congregate in the middle range. Thus, the owners of (motorized and non-motorized) beach-landing craft figure among the wealthier in rural fisheries, whereas petty traders (often women) and fishers with limited assets rank among the poorer. However, Salagrama (2006, p. 101) points out that there are frequent fluctuations in these middle ranks, with many people constantly moving up and down the ladder "as a result of the dynamic nature of their access (i.e. entitlements) to various resources."

9.4 Fisheries and Immigration in Ramnad District

I argued above that, parallel to an explosion of wealth in the capture fisheries of Tamil Nadu, the fishing population of the state has grown substantially, with a large proportion of this increase deriving from the movement of non-fishers into the sector. For simplicity sake, I define "non-fishers" as people who do not belong to a

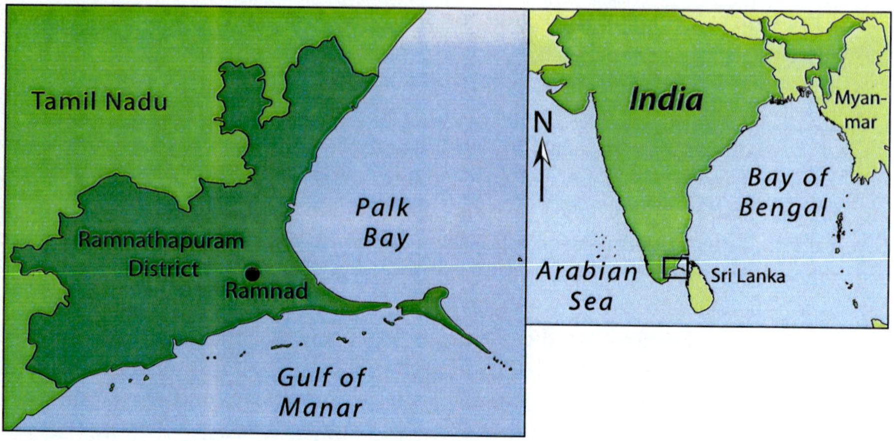

Fig. 9.1 The location of Ramnathapuram District within the state of Tamil Nadu as well as India

traditional fishing caste and who have memories of ancestral activity in economic sectors other than fishing.[2]

I now turn to examining the process of migration more closely, the argument being that immigration into Tamil Nadu fisheries has not been distributed evenly throughout the coastline. Instead, it has been differentiated geographically and socio-technically, concentrating in sub-sectors and regions which are characterized by high opportunity as well as low entry thresholds. High opportunity follows from the combination of natural, economic, and technological features, such as the availability of adequate aquatic resources, high market values, and the presence of suitable techniques for extraction. Low thresholds are the result of imperfect institutional barriers, created by state or non-state authorities. In the Tamil Nadu context, this has had two implications. First of all, immigration was strong in the new trawl fisheries, which government – making use of judicial and extra-judicial means (Bavinck 2001a) – shielded against protest and control by traditional fisher authorities. Second, it concentrated in geographical regions where traditional institutional barriers are weaker than elsewhere. Ramnad District illustrates both trends (Bavinck and Karunaharan 2006; Jentoft et al. 2009).

Ramnad is one of the largest administrative districts in Tamil Nadu, and its coastline of 237 km is by far the longest (Fig. 9.1). Within Tamil Nadu, Ramnad is known for its aridity, isolation, and limited level of development, and fisheries constitutes one of the major economic activities. Of the 1.2 million inhabitants, almost 10% depend on fishing for a livelihood. Bavinck and Karunaharan (2006) count 141

[2]Debates on the correlation between caste and occupation in India are heated and inconclusive as to the details. Observers will agree, however, on the main marine fishing castes of Tamil Nadu – the Pattinavar, Paravar, and Mukkuvar – who dominate the Coromandel Coast, the Gulf of Mannar, and the southern reaches of Tamil Nadu, respectively.

Fig. 9.2 Small-scale "vallam" fishers returning from fishing in the Palk Bay, Ramnad District

fishing settlements spread more or less evenly along the shoreline, with the trawler industry concentrated on or close to Pamban Island.[3]

The northern shore adjoins the Palk Straits, which is a shallow, sheltered sea area. The civil war, which took place in neighboring Sri Lanka from the early 1980s until 2009, has radically affected the fisheries in this region. As fishing activity in Sri Lanka has declined in lieu of the violence, many of the trawl fishers of Ramnad District relocated their operations to Sri Lankan waters, meanwhile risking a confrontation with border authorities (Suryanarayan 2005). Other small-scale fishers ply the near-shore region, in search of mud crabs, squid, and other commercial species (Fig. 9.2).

The southern shoreline of Ramnad District faces the Gulf of Mannar, which is recognized as one of the richest biodiversity regions of India. Because of its ecological significance, the Government of India in 1986 declared the 18 islands off the coast a no-take national marine park; the international community subsequently recognized the larger region as a biosphere reserve. Meanwhile, the Government of India (1976) ratified the Convention on International Trade in Endangered Species (CITES), and actively prosecutes those engaged in clandestine trade of the marine species on the CITES lists (Bavinck and Vivekanandan in press; Rajagopalan 2008). These institutional proclamations are slowly beginning to affect fishing practice, and fishers are complaining. There is no evidence, however, of significant numbers

[3]The 2000 census of the Fisheries Department of Tamil Nadu mentions 184 fishing settlements in Ramnad District. This includes, however, a large number of interior villages in which marine fishing is not the dominant, but a supplementary, activity.

Table 9.3 Religious composition of fishing settlements in Ramnad District (2005)

Religion	Number of settlements	Percentage of total
Hindu	92	65.2
Muslim	7	5.0
Christian	21	14.9
Mix of all	21	14.9
Total	141	100.0

Source: Bavinck and Karunaharan 2006

Table 9.4 Caste composition of fishing settlements in Ramnad District (2005)

Name of the caste	Number of settlements	Percentage of total
Paravas	17	12.1
Pattamkatti	4	2.8
Ambalar	59	41.8
Vanniyar	7	5.0
Pillai	3	2.1
Muslim castes	7	5.0
Mixed caste	44	31.2
Total	141	100.0

Source: Bavinck and Karunaharan 2006

of fishers shifting to alternative occupations. On the contrary, fishing intensity seems to be maintained and even increasing within the park as a result of demographic trends and new gear types.

Other than many fishing regions of India, the contemporary fishing population of Ramnad District displays an extraordinary diversity. There are Hindu, Muslim, and Christian (Roman Catholic and Protestant) settlements and settlements of mixed composition. Caste-wise, the district fisheries are highly differentiated too, with eight major castes being represented. Only two of these castes have a tradition of marine fishing, with one other consisting of former inland fishers. The others lack a long-time historical connection to the fishing occupation and have moved in relatively recently. There is no evidence of castes or religious groups coinciding with, or controlling, larger geographical areas. Muslim villages are alternated with Christian or Hindu settlements, and the same is true for castes (Bavinck and Karunaharan 2006). Tables 9.3 and 9.4 present data on the contemporary religious and caste diversity of fishing settlements in Ramnad.

Evidence of the incidence of large-scale immigration into Ramnad fisheries is of three types: cartographic, statistical, and observational. Cartographic evidence derives from a comparison of contemporary information on the Ramnad shoreline with a detailed map of the district drawn by the Madras Survey in the year 1892. The commissioning of this map, which measures approximately 1.5 by 2 m and is kept in the British Library, paralleled the publication of the so-called Ramnad Manual – an elaborate inventory of the physical, social, and political layout of the region (Raja Ram Rao c. 1889), intended for the support of the colonial administration.

Table 9.5 Fishing population of Ramnad District in historical perspective (1957–2000)

	Census 1957	Census 1978	Census 2000 (corrected)
Number of settlements	45	79	141
Number of fishers	11,250		
Fishing population	30,304	40,152	117,291
Average pop/settlement	673	508	832

Source: Chacko et al. 1957; Director of Fisheries 1982; Department of Fisheries 2000; Bavinck and Karunaharan 2006

The 1892 map shows a smattering of approximately 65 hamlets within 1 km of the coastline, and a total of only six villages or towns with a population of over 2,000 people. Not all of these habitations will also have served fisher people, however. Some of them were obviously more related to salt production, coconut cultivation, religious pilgrimage, and coastal trade than to fishing. Raja Ram Rao (c. 1889, p. 12) makes a mention of a diverse and what can be interpreted as an ecologically rich fishery, but concludes that "the condition of the fishermen is not very encouraging; they are able to obtain only a hand to mouth livelihood."

Statistical evidence derives from census reports which the Fisheries Department of Tamil Nadu has composed at various points of time, the most important being the census figures of 1957, 1978, and 2000 (Table 9.5). The latter figures were corrected by Bavinck and Karunaharan (2006).

The first point to note is that the number of fishing settlements in 1957 is not very different from the number indicated by the 1892 map (Table 9.5). This suggests that, aside from natural population growth, the fishing population remained at relatively similar levels during the first half of the twentieth century. The next point, however, is the threefold increase in the number of fisher settlements that took place in the period 1957–2000. Bifurcation and resettlement of a section of the population is a regular process in the fishing villages of South India, which takes place in response to population growth, carrying capacity of local fishing grounds, and social and political frictions. Re-zoning of administrative units and more relaxed criteria for inclusion in the census constitute other reasons for an absolute increase of settlements, at least on paper. Together, however, these factors cannot explain the extreme increase in the rate of fishing settlements in Ramnad District, which is far higher than the average for Tamil Nadu in the same time period. Thus, Tamil Nadu counted 242 fishing villages in 1957, and 591 in 2000 – an increase of 240%. In Ramnad, the rate of increase was 310%, far above average.

Travels along the coastline point out another aspect of population growth: the geographical expansion of certain settlements over others.[4] Although in 1892, Ramnad District possessed only four towns – the district capital Ramnad, Keelakarai,

[4]Interestingly, this expansion has not resulted in a significant growth of the average size of the settlements in Ramnad (see Table 9.5). Instead, it seems as if expansion has been complemented by a process of administrative subdivision. At specific locations, I therefore observe a clustering of fisher settlements into smaller and larger towns.

Devipattinam, and Rameswaram – of some size, there is now a whole range of towns along the coastline. All these settlements have expanded primarily because of suitable fishing conditions.

9.5 Immigration and the State

A major impetus for immigration in Ramnad District clearly derived from the state. I have argued elsewhere that Article 19(1g) of the Constitution of the Republic of India has had significant effect on the fisheries of the country (Bavinck 2001a). This Article, which is a part of a set of provisions defining fundamental rights, defines the right of every citizen "to practice any profession, or to carry on any occupation, trade or business." Trawler fishers, who often have had a non-fishing background, have frequently invoked this Article to justify their participation in the profession, and also to question the government's right to impose restrictions on their business (Bavinck 2001a, p. 229). Indeed the government of India – which was striving to implement a blue revolution in the country – defended and furthered the rights of the trawl industry, such as by constructing fishing harbors and removing them from the control of traditional authorities (Thomson 1989; Bavinck 2001a).

Another clause in the same Article – Article 19(1e) – provides citizens with the fundamental right "to reside and settle in any part of the territory of India." The combination of these clauses legalizes professional migration, such as in fisheries. It is limited only by the possibility afforded to the state – Art. 19(5) – to impose "reasonable restrictions" on the above rights "either in the interests of the general public or for the protection of the interests of any Scheduled Tribe." The government of Tamil Nadu, which the Constitution has entrusted responsibility for inshore fisheries in a zone up to 12 NM from the shore, has made some use of this opportunity to impose restrictions especially on the trawl fisheries. Such action generally followed large protests by small-scale fishers.

The Tamil Nadu Marine Fishing Regulation Act of 1983 thus authorized the registration and licensing of trawl vessels, and imposed conditions hereto. Although this legislation has taken many years to be implemented, and is still flawed,[5] it is one of the state's more successful interventions in the fishing sector. More recently, the state government has also implemented registration and licensing for small-scale fisheries. This effort has been more successful in Ramnad District than in other parts of the state.

Registration and licensing provides the state government with instruments to control fishing activity in different ways, including time and space zoning. It has also provided the government with opportunities to address the negative effects of trawler migration, which was a regular phenomenon in the period between the 1960s and 1980s.

[5] In 2007, the Tamil Nadu government announced an effort to re-count and re-register trawler vessels in the state. This exercise was considered necessary in lieu of evidence of gross over-registration.

Thomson (1989) made a thorough study of a village fishing economy in southern Tamil Nadu, describing the way in which trawler fishers obtained landing rights in an area in which small-scale fishing previously dominated. He distinguishes three methods, which differ as to the actors involved and their geographical origins. Two of them are relevant for our topic. The first method centers on the state which creates "private shore rights" by evicting small-scale fishers from places where harbors and jetties are constructed (Thomson 1989, p. 125, 137). The state's legal stamp on these locations provided the trawler sector with a measure of independence vis-à-vis small-scale fisher settlements. Although firm evidence is still lacking, this mode has probably played a role in the growth of several trawler fishing centers of Ramnad District, such as Rameswaram, Mandapam, and Tondi.

The second method consisted of migrant trawl fishers making agreements with the leadership of host villages. In exchange for landing rights, the trawl fishers thus paid weekly taxes to the local village fund. Thomson (1989, p. 138) notes that such contracts were quite common, until small-scale fishers' opposition to trawl fisheries increased. In my own work on the Coromandel Coast, I have noted (Bavinck 2001a, pp. 218–220) that contracts of this type probably also existed in this geographical region in the early days of trawl fishing, terminating – as in Thomson's case – in the 1980s. It also seems to have played an important role in Ramnad District, with one major variation. Contrary to other coastal regions of Tamil Nadu, Ramnad District has a history of large landowners running coconut plantations in the coastal area. These plantations are private property (*paddaa* title). Several cases where migrant trawl fishers have bypassed objections of small-scale fishing settlements by making landing site arrangements with coconut plantation owners have come to my attention. In other cases, there are serious disputes over the land on which trawl fishers are settled.

9.6 Non-State Regulation of Immigration

Thomson's (1989) description of trawler fishers' second strategy for accessing rural landing sites leads to the question of local leadership: who are these "leaders" who negotiate with outsiders over fishing rights? And to what extent does their authority extend above the local, to supra-local levels? Studies in other parts of Tamil Nadu have demonstrated varying patterns of non-state authority connected to particular caste groups. My own work among Pattinavar fishers demonstrates the prevalence of strong village councils, called *panchayats*[6] along the Coromandel Coast (see also Bharathi 1999; Bavinck 2001a, b). These non-state organizations – which possess headmen called *chettiyar* or *naaddaar* – take charge of a variety of collective needs, such as the regulation of fishing. Their authority is based on notions of territorial

[6]These non-state institutions are found throughout rural India, and are to be distinguished from the Gram Panchayats instituted by the government of India as the lowest tier of administration and political representation.

rights to adjacent waters, and on a prerogative to control the fishing that occurs there. Panchayat control extends to the landing site and includes the right to regulate its use, such as through taxation on landed produce. As territorial rights of neighboring villages are contiguous, and one village's rights commence where another's end, the entire coastline is thus nominally under non-state panchayat control.

In addition to regulating space and access to resources, these village panchayats also exert control over people. A clear system of village membership defines which males are part of the community, and are charged with responsibilities and rights. This same system allows, such as in the case of marriage, for transfers of membership from one location to another. It also allows for migrant fishermen – such as in Thomson's (1989) case of trawl fishers seeking local access – to request permission for the use of village resources and facilities. In principle, therefore, the non-state panchayats along the Coromandel Coast have substantial power over population movement into village fisheries.

So-called *panchayat circles* (Mandelbaum 1970), which coordinate decision-making for larger groups of villages and stretches of coastline, were traditionally in charge of supra-local affairs along the Coromandel Coast. Although such circles have mostly largely fallen into disarray among the Pattinavar caste in this region in modern times, coordination mechanisms between villages are generally still in place. Studies of local fishing regulation demonstrate that adjacent villages, through processes of regular communication and imitation, frequently adopt similar measures (Bavinck 1996, 1998).

Fishers of the Mukkuvar caste in southern Tamil Nadu appear to have a very similar set of arrangements (Ram 1991; Sundar 1999; Subramanian 2009), but with one big difference: being Roman Catholic and not Hindu, the village council in that region has been replaced by a parish council, and the headman by a parish priest. For regulation at the supra-local level, the Church appears to provide the necessary organizational infrastructure.

A conspicuous similarity between the two regions mentioned above is that the fishing populations are dominated by single castes, and that the organizations that play important regulatory roles are, in an important sense, caste institutions. Mandelbaum (1970, pp. 269ff) describes the role of panchayats and their leaders in India as "maintaining the jati" (or caste) as a social grouping. Although his analysis concentrates on cultural and social aspects, and less on aspects of work and profession, he does point out (Mandelbaum 1970, p. 316) that caste members "commonly carry on joint enterprises; they usually work together to advance and defend the jati's status in its local order." This is reminiscent of Kraan's (Chap. 8) analysis of how ethnic groups in Ghana occupy, and defend, various technologically defined livelihood spaces in fisheries. It suggests that, in the context of the India fisheries, caste identity may under certain circumstances constitute a powerful rallying point against immigration of "outsiders." The circumstance that, in the northern and southern reaches of Tamil Nadu, single castes continue to dominate village fisheries, and immigration is limited, lends support to the hypothesis.

Although our information on the history of the fishing population in Ramnad is limited, written sources suggest that even in the late nineteenth century it already had a mixed composition. Raja Ram Rao (c. 1889, p. 12) notes that those actually engaged in fishing at the time of writing were Muslims, Paravars, or other "low-caste Hindus" (without specification). Three decades later, descriptions provided by the Madras Fisheries Bureau (1916) of numerous landing centers along the coast confirm the impression of social diversity. Since that time, variation appears only to have increased. As pointed out in Tables 9.3 and 9.4, not only are many settlements of different religious and caste composition in Ramnad interspersed, almost a third of all fishing settlements are internally mixed.

A study of six fisher settlements in different parts of Ramnad District (Bavinck and Karunaharan 2006) pointed out that although every site has some form of non-state organization for the regulation of fishing, these institutions are generally not as comprehensive or as powerful as in other parts of the state. Moreover, these authors argue that "the many social differences (religion and caste) that prevail in the district, and the many differences in fishing style [...] prevent the easy translation of local law into regional law. Each village tends to have its own arrangements" (Bavinck and Karunaharan 2006, pp. 47–48). Other than in the north and south of Tamil Nadu, control over entry into fishing is therefore relatively weak, both at the local and the supra-local levels. Immigrants – single or in groups – thus have a greater number of potential entry points, and tend to face less effective opposition to their participation in fishing.

9.7 Are Migrants Poorer Than Non-Migrants?

One important question, which brings the argument of this chapter full circle, remains: are immigrant fishers poorer than people with hereditary connections to the profession? The evidence from Tamil Nadu is ambiguous. I noted earlier that the blue revolution provided opportunities for outside investors to partake in trawler fishing, and that these people belong to the current fishing elite. On the other hand, many of the laborers employed in and around the trawler fishery also derive from outside the sector, and, as Salagrama (2006) points out, many of them can clearly be categorized as poor. Within the urbanized trawl fishery, a case therefore seems to be made for connecting immigration and poverty (but also, at a numerically smaller scale level, for linking immigration with riches).

For rural, small-scale fisheries, which generate lower net incomes per unit of effort than the trawl fishery does, and therefore generally make up a lower economic level, the evidence is less clear. Fieldwork in Ramnad District does suggest that recent immigrants into the fishery possess fewer fishing skills and less financial capital, and tend to congregate in simpler fishing métiers such as crab fishing and squid jigging. This might result in lesser income levels. Added to this is the fact that older fishing communities sometimes succeed in marginalizing newcomers, for example, by denying them rights to housing program, piped water, and electricity. However, the extent to which this has taken place, and the consequences for well-being, are yet to be studied. The same holds true for the question as to whether immigrant fishing has significantly increased competition on the fishing grounds, as well as in the market, and what the consequences hereof have been for the poverty level of old-time fishers.

9.8 Conclusions

The main argument of this chapter has been that, even if one is mainly interested in the phenomenon of poverty in fisheries in India, it is worthwhile paying attention to the enormous wealth that fisheries have come to represent. Processes of industrialization and globalization, which have enveloped and driven fisheries in the course of the past century, have contributed to a steep rise in catch levels, economic returns, and fisher well-being. This has resulted not only in retention of employment in fisheries, but in demographic growth levels which suggest that there has been a significant movement into the sector.

To argue that fisheries in India is synonymous with poverty therefore amounts to a misrepresentation of the facts. This does not mean to say, however, that no hardship is to be found in the fisheries. Rather, I argue that whatever poverty exists is connected, directly or indirectly, to the ongoing process of wealth generation. By widening the scope to include riches, attention is redirected toward processes of social mobility on the one hand and exclusion on the other.

This chapter has paid special attention to processes of immigration into fisheries, emphasizing the fact that the institutional landscape that facilitates or hinders migration varies from region to region. While some geographical regions and hotspots have been characterized by high migration rates into fisheries, both at the bottom and the top of the economic hierarchy, this process is less pronounced in other regions. Ramnad District is a good example of a coastal area in which, over the past century, there has been very significant movement from the agricultural hinterland into both the rural and the urban fisheries. I argue that immigration in this case was facilitated by a relative lack of institutional defenses among the existing fishing population, which in turn relates to social heterogeneity and a limited capacity for collective action. It was also furthered by governmental action, which broke down existing social barriers and created new fishing sectors. Finally, I noted that there are linkages between immigration and the poorer parts of the fishery, with immigrants frequently collecting in less fortunate locations.

At a general level, my analysis confirms the argument made by Midré and Jentoft (Chap. 4) that poverty in fisheries is not an independent but rather a relational phenomenon – it is about how people, as individuals and in groups or categories, associate with one another within the dynamic framework of the larger economy. This viewpoint has important implications for the research agenda on poverty. As Harriss (2007, p. 10) points out:

> Instead of subjecting international poverty research to attempts to refine measurement and to test hypotheses for establishing predictive theory, it will be more productive to redirect greater attention to the analysis of the social processes, structures and relationships that give rise to poverty, recognizing that the creation and re-creation of poverty is inherent within the dynamics of capitalism.

Institutions are known to play an important role in creating, guiding, or reinforcing "social processes, structures and relationships" (cf. North 1990; Jentoft 2004; Bromley 2006). It is therefore reasonable to assume that, in the context of the wealth generated by a century of industrialization and globalization in the world's fisheries (Eide, Bavinck and Raakjær, Chap. 2), institutions also exert great influence on the state and distribution of riches and poverty. This chapter has demonstrated, however, that such institutions are often far from coherent. Instead, one frequently finds institutional arrangements that differ from region to region, displaying contradictions and gaps, and possessing varying levels of effectiveness (Von Benda-Beckmann 2002). In many cases, such arrangements are characterized by legal pluralism. It is to the dynamics of such legal pluralist frameworks that, if we want to know more about poverty in fisheries, more attention is to be devoted.

Acknowledgments In writing this chapter, I have benefited greatly from the constructive comments of Svein Jentoft, Ratana Chuenpagdee, and other members of the PovFish group. I am also grateful to the Netherlands Organisation for Scientific Research (NWO), and particularly to the IDPAD program which funded my Ramnad research under Project Nr 5.2.110 entitled: *Cooperation in a Context of Crisis: Public-Private Management of Marine Fisheries in South Asia.* Appreciation is extended to the Norwegian Research Council for funding the PovFish project.

References

Anugraham DD (1940) The fisherfolk of Madras – an economic survey. Ph.D. dissertation, University of Madras, Madras (unpublished)

Bathal B (2005) Historical reconstruction of Indian marine fisheries catches, 1950–2000, as a basis for testing the marine trophic index. Fisheries Centre Research Report, Fisheries Centre. University of British Colombia, Vancouver

Bavinck M (1996) Fisher regulations along the Coromandel Coast: a case of collective control of common pool resources. Mar Policy 20(6):475–482

Bavinck M (1997) Changing balance of power at sea; motorization of artisanal fishing craft. Econ Polit Wkly 32(5):198–200

Bavinck M (1998) 'A matter of maintaining the peace' – state accommodation to subordinate legal systems: the case of fisheries along the Coromandel Coast of Tamil Nadu, India. J Leg Plur 40:151–170

Bavinck M (2001a) Marine resource management: conflict and regulation in the fisheries of the Coromandel Coast. Sage, New Delhi

Bavinck M (2001b) Caste panchayats and the regulation of fisheries along Tamil Nadu's Coromandel Coast. Econ Polit Wkly 36(13):1088–1094

Bavinck M (2003) The spatially splintered state: myths and realities in the regulation of marine fisheries in Tamil Nadu, India. Dev Change 34(4):633–657

Bavinck M (2005) Understanding fisheries conflicts in the South – a legal pluralist perspective. Soc Nat Resour 18(9):805–820

Bavinck M, Johnson D (2008) Handling the legacy of the blue revolution in India – social justice and small-scale fisheries in a negative growth scenario. Am Fish Soc Symp 49:585–599

Bavinck M, Karunaharan K (2006) Legal pluralism in the marine fisheries of Ramnad District, Tamil Nadu, India. IDPAD Report 2. Indo-Dutch Programme for Alternative Development, New Delhi

Bavinck M, Vivekanandan V (in press) Conservation, conflict and the governance of fisher wellbeing – analysis of the establishment of the Gulf of Mannar national park and biosphere reserve. Environ Manage 47:593–602

Bharathi SB (1999) Coromandel fishermen: an ethnography of Pattanavar subcaste. Pondicherry Institute of Linguistics and Culture, Pondicherry

Bromley DW (2006) Sufficient reason: volitional pragmatism and the meaning of economic institutions. Princeton University Press, Princeton

Charles A (2001) Sustainable fishery systems. Blackwell Science, Oxford

Chacko PI, George S, Krishnaswamy PP (1957) Census of the sea fisherfolk and fishing crafts and gear in Madras state, 1957. Madras Fisheries Statistics Report No. 53. Government Press, Madras

Delgado CL, Wada N, Rosegrant MW, Meijer S, Ahmed M (2003) Fish to 2020: supply and demand in changing global markets. In association with WorldFish Centre. International Food Policy Research Institute, Washington, DC

Department of Fisheries (2000) Tamil Nadu marine fisherfolk census year 2000. Government of Tamil Nadu, Chennai

Director of Fisheries (1982) Census report on marine fisheries. Government of Tamil Nadu, Madras

Government of India (1952) Census of India 1951. Census Commissioner for India, New Delhi

Government of India (2001) Census of India 2001. Registrar General of India, New Delhi

Government of Tamil Nadu (2003) Human development report. Social Science Press, New Delhi

Harriss J (2007) Bringing politics back into poverty analysis: why understanding social relations matters more for policy on chronic poverty than measurement. CPRC Working Paper 77. Simon Fraser University, Vancouver

Hopewell J (2004) 'When the Shore Seine is Shot, the Whole Village Eats!' The change in shore seine organization in Valinokkam village and the decline of shore seining in southern Ramnad District, Tamil Nadu, India. Masters thesis, University of Amsterdam, Amsterdam (unpublished)

Jentoft S (2004) Institutions in fisheries: what they are, what they do, and how they change. Mar Policy 28:137–149

Jentoft S, Bavinck M, Johnson DS, Thomson KT (2009) Fisheries co-management and legal pluralism: how an analytical problem becomes an institutional one. Hum Organ 68(1):27–38

Johnson DS (2006) Category, narrative, and value in the governance of small-scale fisheries. Mar Policy 30:747–756

Kurien J (1978) Entry of big business into fishing, its impact on fish economy. Econ Polit Wkly 13(36):1557–1565

Kurien J (1995) Income spreading mechanisms in common property resource – Karanila system in Kerala's fishery. Economic and Political Weekly, July 15, pp 1780–1785

Madras Fisheries Bureau (1916) Madras fisheries statistics and information, West and East coasts. Bulletin 9, Madras

Mandelbaum DG (1970) Society in India, vol 2. University of California Press, Berkeley

North DC (1990) Institutions, institutional change and economic performance. Cambridge University Press, New York

Raja Ram Rao T (c. 1889) Ramnad manual, [title page missing]. Madras Presidency, Madras

Rajagopalan R (2008) Marine protected areas in India, SAMUDRA Monograph. International Collective for Support of Fishworkers, Chennai

Ram K (1991) Mukkuvar women: gender, hegemony and capitalist transformation in a South Indian fishing community. Zed Press, London and New Jersey

Salagrama V (2004) Policy research: implications of liberalization of fish trade for developing countries – a case study for India. Report Project PR 26109. Food and Agriculture Organization, Rome

Salagrama V (2006) Trends in poverty and livelihoods in coastal fishing communities of Orissa state, India. Fisheries Technical Paper 490. Food and Agriculture Organization, Rome

Salagrama V (2008) The social, economic, technological and environmental aspects of beachseining in India. ICM Integrated Coastal Management, Kakinada (unpublished report)

Schlager E, Ostrom E (1993) Property rights regimes and coastal fisheries – an empirical analysis. In: Anderson TL, Simmons RT (eds.) The political economy of customs and culture: informal solutions to the commons problem. Rowman and Littlefield, Lanham, pp 13–41

Smith HD (2000) The industrialization of the world ocean. Ocean Coast Manage 43:11–28

Subramanian S (2009) Shorelines: space and rights in South India. Stanford University Press, Stanford

Sundar A (1999) Sea changes: organizing around the fishery in a South Indian community. In: Barker J (ed.) Street-level democracy: political settings at the margins of global power. Kumarian Press, Toronto, pp 79–114

Suryanarayan V (2005) Conflict over fisheries in the Palk Bay region. Lancer Publishers & Distributors, New Delhi

Thomson KT (1989) Political economy of fishing: a study of an indigenous social system in Tamilnadu. Ph.D. dissertation, University of Madras, Madras (unpublished)

Vanderlinden J (1971) Le Pluralism Juridique: Essai de Synthèse. In: Gilissen J (ed.) Le Pluralisme Juridique. L'Université Bruxelles, Brussels, pp 19–56

Vivekanandan E, Srinath M, Kuriakose S (2005) Fishing the marine food web along the Indian coast. Fish Res 72(2–3):241–252

Von Benda-Beckmann F (2002) Who's afraid of legal pluralism? J Leg Plur 47:37–82

Part III
Coping

Poverty and vulnerability may have a crippling effect on people and communities, and this requires a response. People are getting up in the morning to begin their day. Before dawn they bring out their boat and head towards the fishing grounds. Every day they work, and every day they provide food on the table and care for their loved ones, while thinking of and preparing for the future. Thus, even when trapped in poverty, people are not idle, their lives are not static. They need to be as resilient as they can be. They must cope and adapt to circumstances that shift by the day. What is brought ashore today is not guaranteed tomorrow. There are seasons for things. People grow, families multiply, and with age comes new challenges. They must be looking out for new opportunities. Are there other ways to bring food and income? Are there possibilities that are still untapped, not tried, not looked into? Are there things yet to be experienced and learned?

The chapters in this section are all about coping and adapting. They come with a positive perspective on what small-scale fisheries can possibly be, based on in-depth studies of what people actually do, whether they are living on the Yucatan Peninsula of Mexico, the Black Sea Coast of Turkey, along the shores of Lake Malombe in Malawi, the Pearl Lagoon of Nicaragua, or are fishing off the coast of Thailand. The stories told are about how small-scale fishers diversify, both within and outside the fishery in order to broaden the basis for their livelihoods, for instance by combining fishing with farming and tourism, but which in some instances creates conflict with other resource users within or outside the sector.

In these chapters we also see how people empower themselves by organizing, by forming cooperatives, that help them become more resilient. But the chapters also illustrate the fact that poverty and vulnerability are problems that require structural reforms and government initiatives that help people cope more effectively. People are flexible and inventive, but within limits that require collective action and support, for instance in the form of secure rights to land and marine resources.

Chapter 10
Addressing Vulnerability: Coping Strategies of Fishing Communities in Yucatan, Mexico

Silvia Salas, Maiken Bjørkan, Felipe Bobadilla, and Miguel A. Cabrera

Abstract In this chapter, we present a case study from Yucatan, Mexico. The main hazards that fisher groups are confronted with in coastal areas are explored, as well as the coping strategies fishers have developed to face them. We also investigate the sense of well-being according to fishers' perceptions, and contrast with the level of marginalization reported in official records. Our findings suggest that fishers do not consider themselves poor, as long as they have access to fishing. Fishing gives them food security, but declining catches and other factors beyond their control, such as increase in the frequency of hurricanes and red tides, also expose them to risk and vulnerability. Several social and political issues generate concern among fishers as well. They employ proactive and reactive strategies at the individual and community levels to face those challenges. However, our research discovered that there are differences between communities and groups of fishers regarding those strategies. We contend that socio-economic conditions and levels of organization influence the ways fishers develop coping strategies. We discuss our findings in light of strategies that can be promoted to improve adaptive capacity of fishers in coastal communities, averting them from vulnerable conditions.

S. Salas (✉) • F. Bobadilla • M.A. Cabrera
CINVESTAV, Instituto Politécnico Nacional, Km 6 Antigua Carretera a Progreso, C.P. 97310 Mérida Yucatán, Mexico
e-mail: ssalas@mda.cinvestav.mx; maculato@gmail.com; mcabrera@mda.cinvestav

M. Bjørkan
Norwegian College of Fishery Science, University of Tromsø, 9037 Tromsø, Norway
e-mail: maiken.bjorkan@uit.no

S. Jentoft and A. Eide (eds.), *Poverty Mosaics: Realities and Prospects in Small-Scale Fisheries*, DOI 10.1007/978-94-007-1582-0_10,

10.1 Introduction

In Mexico, 90% of the national fishing fleet is comprised of small-scale boats between 8 and 12 m long, and close to 300,000 people depend on small-scale fisheries (Fernández et al. in press). In Yucatan, this sector generates about 15,000 direct jobs. The relevance of the fishing activity lies on its contribution as a source of jobs, food and foreign currency. However, as in other regions, fishing also carries its own risks, given the uncertainty of resource availability, hostile environmental conditions and increase in market demands. The latter has been an incentive to remain in fishing (and sometimes violate regulations) as long as there is a buyer. This condition also brings newcomers into the activity, increasing competition in the coastal areas (Sethi et al. 2005; Seijo et al. 2009).

In the last decade, declining catches in Mexico, and in Yucatan in particular, have brought concerns to those who depend on fishing. This situation has been associated with different factors, as suggested by several authors (Fernández et al. in press; Fraga et al. 2008; Mexicano-Címtora et al. 2009). Such factors include increasing fishing pressure, habitat deterioration and ineffective management practices. In addition, fishers argue that a surge in frequency and intensity of phenomena like hurricanes and red tides in recent years has contributed to augment the stress of fishers. For instance, hurricanes can have an impact on their assets, their activities and their personal life. As in the case of red tides, fishing operations are reduced as fish die or move away during such a phenomenon, reducing caches. The demand for sea food also decreases when this phenomenon occurs. The situation of stress has also been exacerbated by the entry of newcomers who, on a temporary or permanent basis, increase the pressure on the already limited resources. The results are rent dissipation and local conflicts, which affect the livelihoods of fishers (Salas et al. 2007).

All these conditions can increase the sense of vulnerability, and threaten people with poverty conditions (OECD 2001; Béné 2009). Fishers are highly dependent on fishing resources and the health of the marine environment. Consequently, any change in these conditions may affect their livelihoods in ways that are detrimental to them. How people cope and recover from stress and shocks varies, and is partly context dependent (Cinner and Pollnac 2004). Thus, coping strategies are framed by people's circumstances and the options available to them within the communities they live in. One factor to be reckoned with is how fishing communities are organized and the extent to which fishers cooperate, especially under conditions of stress.

In Yucatan, there are three forms of organizational strategies which are available for fishers to utilize, and which may influence their reactions to conditions of crises. First, fishers can operate independently and individually; that is, they do not have any organizational affiliation and are therefore basically on their own when a crisis hits. Second, in some communities fishers belong to producer cooperatives, which may provide support in times of crisis. Third, some fishers run their own business enterprises; that is, they have people employed either in processing and/or as fishers, and have therefore some kind of responsibility for the welfare of others during a crisis.

In light of this, we were interested to investigate how these different strategies affect how fishers cope with the above noted challenges. In order to do this, we studied two neighboring communities: San Felipe and Dzilam de Bravo. These are relevant case studies, since fishers have opted for quite different organizational strategies in each case. In San Felipe, the majority of fishers are mainly cooperative fishers, while independent fishers dominate the organizational landscape in Dzilam de Bravo. We explored these issues by asking the following questions: (1) How do people perceive poverty conditions and which conditions generate for them a sense of vulnerability? (2) How do the different groups of targeted fishers perceive the impact of hazards that can affect their livelihoods? (3) How does each group deal with vulnerability?

In this chapter, we first introduce the theoretical framework used for our analysis associated with poverty, vulnerability and coping strategies. Next, we describe the research methods and the characteristics of the two communities and fisher groups targeted. We report our results and discuss our findings in light of what kind of strategies can be promoted to improve the adaptive capacity of fishers – to improve the livelihood conditions for people from coastal communities, averting them from vulnerable conditions and hence from poverty.

10.2 Theoretical Framework

Poverty is a complex matter as underlined by a large body of literature, and particularly in the context of small-scale fisheries several authors have addressed issues on this matter (OECD 2001; Macfadyen and Corcoran 2002; Béné 2003, 2009; Fafchamps 2003; Béné et al. 2007; Fisher and Christopher 2007; Thorpe et al. 2007; Olmos et al. 2008; Seijo et al. 2009). One of the main issues when dealing with poverty alleviation has been to define indicators that can account for poverty reduction policies; however, "changing the numbers" of some of those indicators could give a false idea of improving conditions. In many cases, the indicators related mainly to income, and gross enrolments are somewhat the basis to define short-term policy changes. However, as stated by Thorpe et al. (2007), fishers can be very vulnerable given their lifestyle (which could involve several activities in addition to fishing) and the pressure that exogenous shocks (storms, hurricanes, tsunamis) can impose on them. Hence, economic evaluation of fishers' households alone cannot properly portray actual conditions regarding poverty.

According to the human development index (PNUD 2009), Mexico ranks 53rd, way below other Latin American and Asian countries. At a national level, the coastal and rural areas are considered to have limited livelihood standards; Yucatan ranks 19th out of the 32 Mexican states. In Yucatan, 51% of the population is suffering from income poverty, 26% of them have limited skills to get out of poverty, and 44% do not have access to health services (Pérez 2009). Only 3 out of 106 municipalities scored high on the human development index by 2005 (Pérez 2009). These indicators give an idea of the poverty situation in the region. However, this index does not necessarily apply equally to all sectors. Even within its own sector, fishers are a heterogeneous group, and some are poorer or more vulnerable than others. This circumstance brings up questions about how different groups of fishers perceive poverty, and which factors make them more vulnerable and at risk of becoming poor. It is also relevant to understand what kind of coping strategies people develop in order to face hazards that engender vulnerability. Those strategies are complex and diverse, and they can vary by social group, household, gender and the way they are organized (Bærenholdt and Aarsæther 1998; Béné 2009). In this context, the concepts of vulnerability, coping strategies and resiliency become relevant to understand how people from coastal areas have been facing different types of disturbances that may expose them to poverty.

Chambers (1989) defines vulnerability as the condition under which people feel helpless, insecure and exposed to risk, shocks and stress. Several authors (cited by Béné 2009) define this concept as a factor that is comprised of three components: exposure to risk, susceptibility and adaptive capacity. Under circumstances where people feel vulnerable, they develop strategies that make them master, tolerate or minimize stress (Chambers 1989; Jóhannesson et al. 2003; Clay and Olson 2008; Jiménez-Badillo 2008). According to several authors, poverty and vulnerability may constrain a fisher's ability to engage with state policies oriented to promote resource conservation (i.e. Adams et al. 2004; Allison et al. 2006; Fisher and Christopher 2007; Olmos et al. 2008).

In this context, it is important to stress that, even if poverty and vulnerability are linked, they are not necessarily the same. As stated by Béné (2009), not always the poorer are the most vulnerable. Furthermore, a program that aims at reducing poverty would not necessarily reduce vulnerability, but the contrary may apply (Macfadyen and Corcoran 2002; Béné and Friend 2011). Hence, in order to ensure sustainable fisheries, it is important to find under which circumstances people can best solve these matters in a way that promotes sustainability for their socio-ecological system. How such a socio-ecological system responds to different sources of stress, i.e. social, political or environmental change, and how such stress may affect fishers' situations are key issues. In addition, it is important to ask to what extent and in what form the two issues may be linked.

When a system can respond without suffering long-term damage or modification, it can be defined as resilient (Adger 2000; Gunderson 2000, 2002; Janssen et al. 2007; Gibbs 2009). Resilience has to do with the system's adaptive capacity, which is the ability to deal with change and disturbance (internal and exogenous) through learning, knowledge sharing and responding to feedbacks (Gunderson 2000; Fabricius et al. 2007). The higher the system's adaptive capacity, the more resilient it will be (Smith and Wandel 2006). When facing hazards, people tend to modify their behavior by generating strategies to cope not only with immediate problems; sometimes this can also generate a more robust socio-ecological system (Adger 2000, 2006; Folke 2006; Gallopín 2006; Janssen et al. 2007). However, one risk with adaptations is the possibility of accepting things "the way they are" as normal.

In this chapter, we will lean on Fafchamps (2003) classification of coping strategies as ex ante (proactive) and ex post (reactive) strategies, depending on how the system responds to the stressors. The former is in place before a shock has occurred, while the latter takes place afterward. According to Macfadyen and Corcoran (2002), proactive strategies include a set of preventative actions in order to reduce vulnerability. Reactive strategies include changes in livelihood activities that can solve the problem in the short term; however, they often increase vulnerability in the medium and long term. Reactive strategies are generally the first response when facing hazards, and through a learning process the adaptation can lead to proactive actions. The process is dynamic and both strategies are not exclusive. Under those conditions, people can also be creative and innovative in their choice of strategies (Bærenholdt and Aarsæther 1998). This innovation can be based on building social networks and cultural identity, which in turn can build adaptive capacity.

Networking can be developed over time, but can also be strengthened by cooperative processes. In this context, the generation of fishing cooperatives throughout the world has had different outcomes with mixed success (Jentoft and Davis 1993; Jentoft and Sandersen 1996; Råkjær Nielsen et al. 2004). However, there are also examples of how they can improve the development of social capital, in addition to facilitating members' access to material goods (Bjørkan 2005; Halpern 2005). Here, we understand social capital as an abstract property of relationship which facilitates cooperative action (Bourdieu and Wacquant 1992; Halpern 2005). Accordingly, social capital plays an important role in people's coping strategies, especially with regard to networking and cooperation (Bærenholdt and Aarsæther 1998).

10.3 Methodological Approach

From previous work in Yucatan, we are aware of the heterogeneity that characterizes the fishing communities in the region. Despite the fact that people have fishing as the main activity, target the same fishing resources and operate under the same fishing regulations, they still display significant differences in the way they deal with environmental challenges and undertake their fishing activities. We chose two neighboring communities – Dzilam de Bravo (DB) and San Felipe (SF) (Fig. 10.1) – which share many features, including the same fishing grounds. Still, they differ in the level of social and community organization (Salas and Pitcher 1999; Fraga et al. 2005), which makes an interesting contrast for this study.

A key difference between the two communities is the organizational landscape. In SF, the dominant organizational alternative for fishers is the cooperative. This is a political force with strong ties to the local government. Two fishing cooperatives operate in the SF community. In DB, the permit holders are the strongest political group, and the majority of fishers that work in the area are independent fishers who work on their own or get hired by permit holders. Only a small fishing cooperative exists in this area. Many of the independent fishers are migrants from other regions who mainly participate in the octopus fishery. This is a very crew-demanding fishery, and involves low travel and investment costs. This fishery is more accessible for migrants than others, as it does not require many skills. During the octopus season, firm owners depend on a high number of fishers; many come from rural inland areas or even from other states.

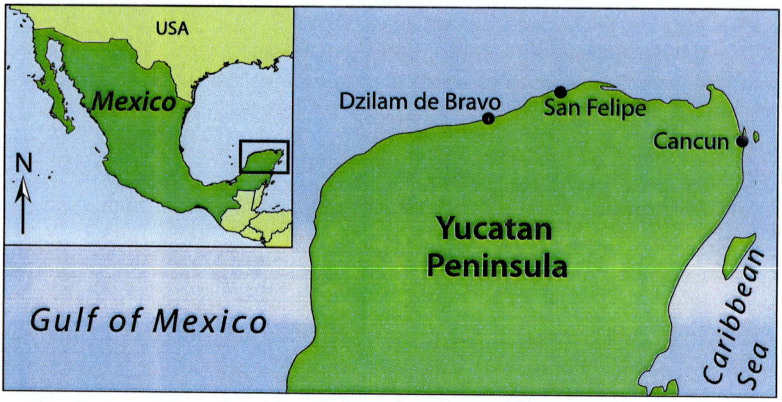

Fig. 10.1 Locations of the fishing communities San Felipe (SF) and Dzilam de Bravo (DB) in Yucatan, Mexico

Table 10.1 Characteristics of fisher groups interviewed in San Felipe and Dzilam de Bravo

Cooperative fishers	Independent fishers	Firm owners
Belong to an organization that holds the fishing permits and owns the boats	Work independently	They own boats and most fish-processing plants
They commit to comply with cooperative rules and get the benefits of being a member of the organization	Some of them can hold a fishing permit and own their own boats to work on their own; others get hired to work on other people's boats, which are mainly owned by firm owners	This allows them to participate in the extraction and commercialization process of the products at different levels
The organization sells their products directly or through the firm owners	Those who own a permit and a boat can sell their product to fishing cooperatives or to firm owners	

The three types of fishing participants organized under three different schemes in Yucatan are described in Table 10.1. While the fishers are understood as a homogenous group at the national level, fishing groups can differ in how they are organized, and also in the approach they follow to cope with stressors. Hence, we took as a basic premise the idea presented by Thorpe et al. (2007) – that fishers' households/communities are unequally vulnerable to stressors. We defined three groups (as presented in Table 10.1) as our target in this study.

According to the National Institute of Statistics, Geography and Informatics (INEGI 2005), the populations in DB and SF are 2,248 and 1,838 respectively, with a high percentage of people dedicated to fishing (75% in SF, and 65% in DB). In both communities, fishing is combined with other activities like tourism, ranching and agriculture, but the level of diversification among fishers is not homogeneous (Fraga et al. 2005).

We estimated a marginality index to obtain socio-economic indicators using the Stratification Technique that is generally used by INEGI (Dalenius and Hodges 1959).

Hence, this index was obtained by groups in contrast to a global value from the whole community as official records report.

To gather information in the field, we interviewed fishers in their homes, landing sites and processing plants, using semi-structured interviews and participatory observations (Bråten 2002; Ingles and Sepez 2007). A total of 101 fishers were interviewed in SF, and 159 in DB, having close to 10% of fishers from each community represented in the sample. We applied questionnaires to the three target groups defined in Table 10.1: cooperative fishers, independent fishers and firm owners.

We divided the survey questions into three main components related to: (a) fisher's perception on poverty and sense of well-being; (b) factors that generate a sense of vulnerability; and (c) coping strategies developed by fishers to face different shocks or hazards. For the first issue, we asked fishers if they considered poverty as a problem for them and their community; evaluated official indicators on marginalization; and explored perception of fishers regarding factors that provide a sense of well-being.

For the second issue, we evaluated vulnerability in terms of exposure to shocks that can affect fishers' livelihoods; what means they have to make a living; and what resources they depend on.

As to the third issue, we evaluated at an individual and a community level the strategies fishers have developed to face vulnerability. We applied content analysis to integrate information from the interviews for further evaluations. To compare information between communities, and among groups of fishers within the same communities, we employed a one-way randomization test of significance of pseudo F values through the statistical software R.

We also used information from scientific literature, technical reports and official statistics to determine the current status of resources and assess official marginalization indicators. In addition, a member of our research team spent about 6 weeks in each community making participatory observations.

10.4 Results

10.4.1 Poverty Conditions: The Official Picture and the Heterogeneous Reality

The indicators of marginalization reported by the Council of National Population (Consejo Nacional de Población, CONAPO) are frequently used by Mexican government agencies as a reference to evaluate the level of poverty in local communities, and to define poverty reduction interventions. We used the same technique as used by CONAPO to obtain the marginality indicators for the target groups we had selected. The purpose of doing this was to evaluate if the indicators applied to the whole community could be equivalent for all groups within the community.

As it can be observed in Table 10.2, according to official indicators, people in DB have a lower level of marginalization than those from SF. However, when we

Table 10.2 Socio-economic indicators, marginality index (MI) and marginality level (ML) for fishers in Dzilam de Bravo (DB) and San Felipe (SF) by target group: IF = independent fishers; COOP = fishers from cooperatives; FO = firm owners

			Interviewed fishers		IF[b]		COOP[c]		FO[d]	
	DB[a]	SF[a]	DB	SF	DB	SF	DB	SF	DB	SF
Population	2,248	1,838	159	101	119	41	24	55	16	5
Illiterate population above 15 years	6.9	6.7	9.7	6.4	12.9	4.8	0	7.3	0	0
Population above 15 years w/o elementary completed	33.0	38.9	41.0	38.2	53.6	46.5	46.6	32.7	8.16	0
People w/o sewage or sanitation in their houses	5.1	1.9	16.0	2.1	20.5	2.5	7.1	0	0	0
People w/o power in their houses	1.3	3.5	2.3	2.1	3.1	2.5	0	0	0	0
People in houses w/o drinking water	2.0	2.6	3.2	2.08	3.2	2.5	5.9	0	0	0
People in houses w/dirt floor	1.3	1.9	8.9	0	11.1	0	5.3	0	0	0
MI	−0.7	−0.6	−0.3	−0.7	−0.2	−0.4	−0.5	−0.8	−1.4	−1.5
ML	L	M	M	M	H	M	M	L	VL	VL

[a]Defined by CONAPO, other values estimated by the authors

[b]The total number of independent fishers is unknown but there are more independent fishers in DB than in SF

[c]In SF, there are two cooperatives; in DB there is one fishing cooperative

[d]16/22 firm owners in DB, 5/5 firm owners in SF. Range value for indicators: VL = very low (−2.366: −1.221); L = low (−1.220: −0.649); M = medium (−0.648: −0.077); H = high (−0.077:1.066) (INEGI 2005)

analyze the information by target group, we observe differences among them; a higher level of marginalization in DB for independent fishers is evident. For this group, the higher values of most socio-economic indicators show more disadvantageous conditions than in the case of other groups within the same community. This is even more prominent with regard to their level of education and living conditions – this group has a higher level of analphabetism (13%) than the general level in the community (6.9%).

In contrast, housing and education for the group of firm owners in both communities show a lower level of marginalization than the other groups; actually their index performs better than the population in general. Importantly, this shows that the fishers are a heterogeneous group – both within a community and between neighboring communities. The results in the "big picture," reported in official records, may be influenced by the weight of firm owners and ranchers in the community.

Despite official marginalization and socio-economic indicators demonstrating a certain level of poverty in the communities studied, close to 70% of the fishers that we interviewed do not define themselves as poor. However, many consider poverty as a problem in their communities. There were more people recognizing problems of this nature in DB (8 out of 10) than in SF (3 out of 10).

When fishers talk about poverty, they usually compare themselves with the inland communities nearby (people from rural areas, defined locally as "pueblos"): "*At least we can go out and fish for food, while in the pueblos, there is not money even to get food*!" Such remarks were typical in both communities. Official reports also confirm that in general terms, coastal communities have a better standard of living than rural inland communities in Yucatan (POETCY 2007).

These results make one wonder what it means to be poor for fishers in SF and DB. In both communities, poverty is perceived as an extreme situation, mainly defined by a lack of, or limited access to, food. Hence, working as a fisher is in itself a coping strategy to avoid poverty, since by fishing they have at least something to feed their families with. However, they recognize that their standard of living could be better if they could improve on other indicators such as income, health, education and access to other occupational opportunities.

Associated with income, the decline of fishing resources was an expressed concern by members from both communities. They see this situation aggravated by the immigration of people from rural and urban areas, looking for a source of income at the coast. Competition for limited resources has therefore increased lately. Lack of livelihood alternatives was reported as the most crucial issue. Having access to health services and education was defined by most respondents as the second concern in both communities. Interestingly, when fishers were asked if they considered participation in decision-making relating to resource management and community organization as relevant issues, most local fishers did not appear to value those issues highly. However, independent fishers (migrants) from DB expressed feelings of being rejected by local members of the community, and therefore felt unable to participate in local decisions, which for them generated a sense of vulnerability.

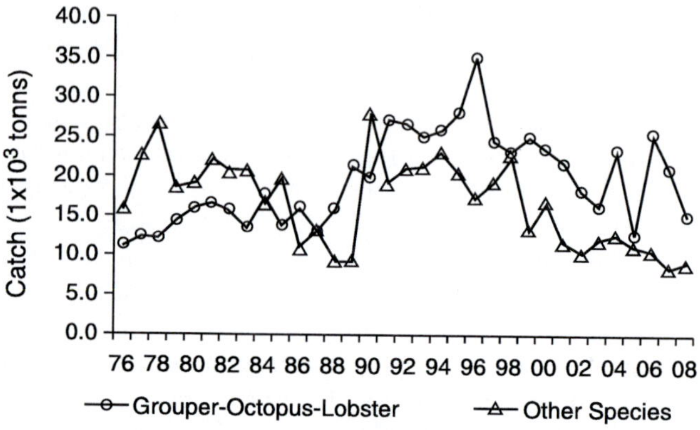

Fig. 10.2 Catch trends of main fishing resources (grouper, lobster and octopus) and other species caught in Yucatan from 1976 to 2008 (Source: 1976–1989 SAGARPA. Cuentas Mensuales de Oficina Regional. Subdelegación de Pesca. Delegación Federal en Yucatán. 1990–2007. Anuarios Estadísticos de Pesca)

10.4.2 Vulnerability

One of our aims here was to understand fishers' own perceptions about what factors increase their sense of vulnerability. Based on questionnaires and interviews, we were able to determine three areas of main concern: (a) resource availability given high variability in catches of main fishing resources, e.g. octopus, grouper and lobster, associated with governance issues (illegal fishing due to lack of compliance related to limited enforcement); (b) environmental factors (increased frequency and intensity of hurricanes, prevalence of red tides and strong winds); and (c) socio-economic issues (low fish price, increase in social problems).

The fishing resources in Yucatan include close to 60 species, but the economy of fishers depends largely on only a few of them (octopus, lobster, grouper and demersal fishes related to grouper). If fishers generally consider good and bad years and fluctuating catches as an intrinsic characteristic of the fishing activity, in the last decade they have noticed a decline in catch of the most important resources (Fig. 10.2). These conditions have also been reported by government institutions. Stock biomass reductions of grouper, lobster and other demersal resources have been reported by several scientists (Burgos and Defeo 2004; Salas et al. 2006; Solana et al. 2006; Ríos–Lara and Salas 2009). The critical condition of fishing resources was pointed out by all respondents in both communities as one of the most important sources of vulnerability they perceived.

The main reference fishers have in order to determine the status of their fishing resources is the volume of catches and number of species caught over time. However, the reasons people gave differ regarding why catches have decreased in the two communities. While in DB 56% of our informants associated lower catches with an increase in fishing pressure (due to an increase in newcomers), in SF only 36% of them agreed on this as a cause.

It is important to point out that a higher proportion of migrant people have arrived into DB than in SF. For instance, one of the interviewed fishers in DB stated: "*There are too many fishing boats chasing every day more limited resources.*" Although fishers in SF acknowledge that there has been an increase in coastal population and hence fishing pressure, they argue that environmental factors have also deteriorated the habitat of fishing resources, and that this could have an increased negative effect on fishing. In SF, more people (54%) placed a higher weight on environmental factors impacting their catches; they also complained about how those factors can limit the number of fishing days.

Beside the impact on fishing resources in both communities, hurricanes are perceived as a major threat to fishers' assets (boats, houses, community infrastructure and health).[1] The north winds and red tides were defined as the next most important environmental factors that affect fishers in both communities.[2] Red tide has an impact on the economy of fishers, not only because of the limitation on the fishing days, but also due to the customers' tendency to buy less seafood after such an incident.

At least one-third of our informants in both communities did not relate to social factors as a potential source of shock that could generate personal vulnerability. However, when we asked fishers about the impact on the community, their perception changed. For instance, alcoholism was pointed out as a common problem in both communities. Garbage disposal was mentioned by a third of the fishers interviewed in DB as one of the main problems (generating diseases), while in SF the equivalent proportion of people (30%) referred to drug consumption as a big community problem.

An increase in drug consumption, especially among the young, is becoming a major concern in SF. Few fishers from the SF cooperative mentioned their limited options for negotiating prices with the middlepersons as a factor that could generate vulnerability.[3]

Political conflicts were mentioned as a source of division between groups of fishers (between 6% and 7% in DB and SF communities, respectively); they inferred that in some cases different preferences for one political party could create conflicts even among family members. In Mexico, three political parties, PAN, PRI and PRD, have dominated.[4] It is not an everyday matter which party you vote on: according to

[1]Catch reduction was mentioned as one of the main concerns associated with those phenomena. This can happen while limiting them to operate during those days (cost of not fishing) and also because demersal fishing resources move away, as the sea bottom gets damaged by those impacts. Fishers also associate an increase in human diseases during the passage of hurricanes.

[2]The increase in frequency of hurricanes in the last decade in the region has been reported by several authors (Salas et al. 2007; Fraga et al. 2008; Mexicano-Cíntora et al. 2009), as well as an increase in frequency of red tides (Herrera-Silveira et al. 2004). This also coincides with the perception of fishers. In SF, fishers (57%) state that hurricanes have increased in intensity and frequency.

[3]This is remarkable because the social relationship between fishers and middlepersons is not based on supply and demand, but rather on debts fishers have with the buyer, so they have limited capacity to bargain. In short, fishers have a complex relationship with the middlepersons, where trust is an important dimension, while the middlepersons still gain more in terms of income.

[4]These are acronyms for Partido Accion Nacional (PAN); Partido Revolucionario Institucional (PRI); Partido de la Revolucion Democratica (PRD).

the informants, after the elections, the party in power tends to favour their followers with financial support through governmental programs. The conflicts generated by these conditions have become an increasing concern especially in SF.

In DB, on the other hand, rejection toward migrant people who arrive into the community is now considered to be a social problem. Migrants hardly get the chance to be integrated, mainly due to their different cultural backgrounds and because they are held responsible for most illegal activities that affect fishing resources.

To summarize, in both communities the main sense of vulnerability expressed was related to declining catches due to both natural hazards and fishing pressure. How severe the impact is from different sources is not totally clear, and it differs among fishing communities and groups of fishers. With regard to the social context, alcohol and drugs consumption was reported as a common problem generating worries at the community level. Such addictions are perceived to reduce peoples' capacity to save money and increase conflicts. Other sources of conflict had to do with political preferences, which can affect the construction of social capital in the communities. All these elements generate some propensity to reduce adaptive capacity and promote vulnerability.

10.4.3 Coping Strategies: How to Maintain Secure Livelihoods

According to official statistics (CONAPO 2005), people in Yucatan's coastal communities face a number of poverty problems such as inferior health services, fewer employment alternatives and limited access to education. Our study shows that vulnerability generated by factors like overfishing, natural hazards and social problems can be a concern of its own in addition to the indicators defined by CONAPO. As reported above, such stressors force people to develop strategies to keep vulnerability, and therefore poverty, at bay.

At an individual level, fishers from both communities use a range of strategies to secure their everyday lives. These strategies can be adapted through time when circumstances change. This adaptive capacity is what defines their vulnerability to shocks (Folke 2006; Smith and Wandel 2006). Taking as a base the concept of strategies (ex ante or proactive, and ex post or reactive strategies) defined by Fafchamps (2003), we observe this type of behavior in the studied target groups in both communities. As referred earlier, the ex ante strategies include a set of preventative actions in order to reduce vulnerability before a shock has occurred, whereas the ex post strategies include activities after a shock has occurred (Macfadyen and Corcoran 2002).

The data show that in comparison to other groups, firm owners have developed more ex ante strategies. All firm owners in SF, and close to 60% in DB, apply these types of strategies. These can include lending money to cooperative or independent fishers, which provides them with future bargaining power; in addition, they make their own savings (Table 10.3). It can also be said that cooperative fishers have an advantage over independent fishers, as belonging to an organization means that they will obtain some support from this organization in case of need. Thus, being a member of a cooperative is an ex ante strategy in itself.

Table 10.3 Individual coping strategies of fisher groups from San Felipe and Dzilam de Bravo

Strategies	Dzilam de Bravo (% fishers)			San Felipe (% fishers)		
Ex ante	IF	COOP	FO	IF	COOP	FO
Saving	11.3	9.5	15.8	5.6	19.3	80
Alternative activity [a]	0.0	0.0	15.8	0.0	3.5	0
Complementary activity [a]	15.4	9.5	21.1	5.6	26.3	0
Lending money to workers	0	0	5.3	0	0	20
Subtotal	26.8	19	57.9	11.1	49.1	100
Ex post						
Alternative activity [b]	28.9	28.6	21.1	38.9	15.8	0
Complementary activity [b]	11.3	14.3	0	16.7	8.8	0
Migration	2.1	4.8	0	5.6	5.3	0
Expenses reduction	7.2	23.8	5.1	16.5	10.5	0
Government support (food, temporal work programs)	7.3	00	0.0	8.3	1.8	0
Borrowing money (family, cooperative, middlepersons)	16.5	9.5	15.8	2.8	8.8	0
Subtotal	73.2	81	42.1	88.9	50.9	0

[a] Without external support
[b] With external support (government programs or borrowing money from middlepersons)

Despite the fact that they get less money for their catch, the cooperative provides members with important benefits which include medical and life insurance, a bonus at the end of the year, and funerary expenses in case of death. The end of the year bonus represents a form of saving for fishers. This helps them to obtain assets for their activity, to improve living conditions or to face times of crises, such as hurricanes or low catch season. Saving is also an important strategy developed by fishers in both communities for sending kids to school, as they do not want them to get into fishing given current conditions. In SF, 19% of fishers used this strategy. Interestingly, independent fishers from DB reported using saving as a strategy (11.3%) more than cooperative fishers (9.5%) in DB. This is because many of the independent fishers come from other regions, and either they want to save money to be able to get their own boats, get integrated into other activities or send money to relatives in other places.

Saving was a declared ex ante strategy reported in SF among cooperative fishers and independent fishers (Table 10.3). They indicated that they save money in order to be able to invest in complementary or alternative activities to maintain their livelihoods during periods of low catches. Some fishers also report saving as a strategy to send their kids to school out of town. Occupational plurality is an important ex ante strategy, as it allows for diversification of livelihood.

Another interesting issue came up in the interviews: Given that people are evacuated if there is a hurricane alert in their community, fishers with saved up money can afford to rent a house. This is an alternative to mass shelters, and their families can be more comfortable during the evacuation and recovery period.

Ex post strategies dominated among independent fishers in both communities (73.2% and 88.9% in DB and SF, respectively). Fishers who use these strategies generally have less means to support themselves and to respond to shocks.

They acknowledge high dependency on government or firm owners. Under those conditions, they are also unable to save money given the debt commitments they have with firm owners, which also limit their ability to negotiate fish prices. Pomeroy et al. (2006), Cinner et al. (2009), and Béné and Friend (2011) report examples of similar social traps in which people find limitations to mobilize the necessary resources to overcome either shocks or low-income situations, and consequently remain in conditions that weaken their status even more, including the possibility to fall into, or remain in, poverty.

As stated by Thorpe et al. (2007), fishing is just one component of the portfolio of activities to support fishers' livelihoods. Within this scheme, undertaking complementary or alternative activities (ranching, tourism, aquaculture or setting up a grocery store) was observed as a common strategy among fishers who develop ex ante and ex post coping strategies. The difference is in who provides the financial support to undertake such activities. For instance, 26.3% of fishers from cooperatives in San Felipe developed complementary activities without external support (government, relatives, firm owners), while 28.6% of fishers from cooperatives from DB undertook alternative activities encouraged and supported by government programs. Even if people in both cases see the advantages of having another source of income different from fishing, those from SF save money to invest in alternatives, while those from DB only do so if there is external support offered to them.

Interestingly, between 2% and 5% of fishers from cooperatives and independent fishers indicated that the best option to get out of crises is to migrate to another place and look for a different livelihood. Several fishers underlined that they do not want their children to become fishers, as they do not see a good future in this activity. This is becoming a common statement of fishers in several regions of the country (Fraga et al. 2008; Jiménez-Badillo 2008).

10.4.4 Networking and Cooperation

When looking into collective strategies, we used the dimensions defined by Bærenholdt and Aarsæther (1998) – namely networking, identity formation and innovation. The identity dimension in our case study can be understood in the sense of belonging, and this is related to the access to resources and dealing with newcomers. This is important according to our informants, since they relate the decline in fisheries, and hence in their income, to the increasing number of immigrants. Networking and cooperation appears to be an important strategy as well in the communities, especially in SF.

In both communities, fishers agree that belonging (identity) is important with regard to who should have access to fisheries, despite the official access regulations. Local fishers, especially in DB, argue that migrants do not have a sense of belonging, and therefore do not commit to protect the resources. The migrants are blamed for most of the illegal fishing activities. Firm owners in the same community, who dominate the fishing activity, do not totally agree with those statements. This could be explained by the fact that many of the immigrants have been brought to DB from other regions by the firm owners themselves. In San Felipe, there are two cooperatives and only two firm owners. Here, the cooperatives have more political power in the community than the firm owners.

As mentioned earlier, *a sense of belonging* can define a willingness to cooperate, especially under stressful conditions. Cooperation processes are understood here as part of the networking developed by people under such conditions, or when they work toward common goals. This strategy seems to be important for both communities. Several of our informants in SF indicated that those who belong to the category *from here* seem to help each other out in times of crises. Cooperative actions in DB have also been reported by Salas and Pitcher (1999), where fishers form teams to go fishing during the windy season.

The teams are comprised of two or three fishers, who go fishing using their own boats, but when they get back to port they share their catches regardless of who brought in more. In this way, all members of the team can be sure to maintain an average catch that can provide viable income when strong windy conditions limit fishers' operations. The authors reported this strategy in the 1990s, and when we asked fishers in DB if the strategy remains, they confirmed that it is still in place. The cooperative agreements between members of the team involve mainly relatives or close friends.

Other examples of cooperation are the relationships built up among both coastal communities and some inland communities. These are trust-based agreements for providing mutual support: fishers can seek shelter within the inland communities during the hurricane season, while the people from those communities can go fishing seasonally to the fishers' fishing grounds, especially during the octopus fishing season.

The cooperative actions among fishers do not seem to be random, but rather part of a complex system of coping strategies, where belonging, networking and innovation are important ingredients in order to be less vulnerable. This, of course, is making people with less networks and "different identity" – such as independent fishers – more vulnerable and with fewer options.

10.4.5 Generating New Options and Innovative Coping Strategies

According to Pomeroy et al. (2006), fishers mix a number of strategies to cope, and these vary according to season, skills, access to capital, education and risk preference. As noted above, the reduction of income due to low catches and low prices of fish has forced fishers in both communities to search for alternative or complementary activities; search for new fishing grounds by going further; spend more time at sea; or even undertake illegal fishing activities.

In the search for new alternative activities that generate an income for fishers, tourism is becoming one of the most popular strategies. With tourism, fishers can

Fig. 10.3 San Felipe fishers with their catch. Waiting for the octopus fishery to open, they are targeting other species such as sharks

take advantage of the natural beauty of the region. In both communities, cooperative fishers and often elderly fishers have started to guide tourists to mangrove areas, beaches, sink holes, as well as birdwatching areas. There is also one tourist cooperative in each community that is established by cooperative fishers and by other members of the community. Women from the fishing cooperative that targets crab in SF as bait for octopus also provide these services. In SF, the cooperatives work together to provide the tourists with the necessary services. In DB, on the other hand, cooperative actions are more common among relatives.

Competition can also take place under conditions of uncertainty. During times of catch reduction, people have to find ways to get more for their invested time (Fig. 10.3). Since many fishing grounds close to shore have been less productive, fishers have started to search for new fishing grounds by going farther and fishing in deeper areas. These actions can be risky for all fishers, but especially for divers who fish lobster. Diving for longer periods in deeper waters and without careful regulations can have an impact in the short term, with potential increase in health problems in the long term.

Officially, from 2004 to 2009, 250 decompression accidents and five deaths have been reported in Yucatan associated with lobster diving (Fig. 10.4). These records may be underestimated, as many fishers do not go to the hyperbaric chamber unless they feel really bad; some mild bends do not receive medical attention. More cases

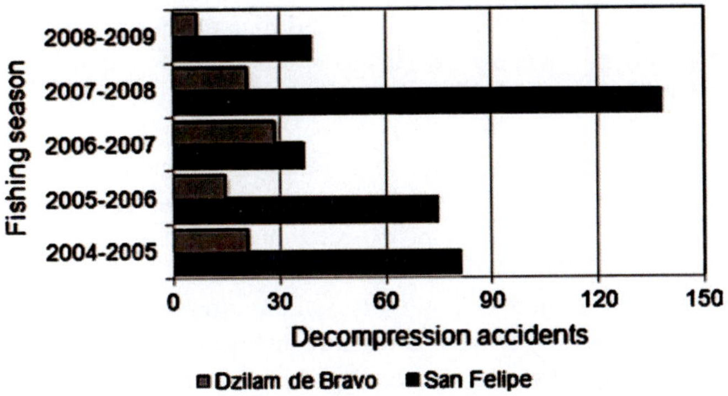

Fig. 10.4 Decompression accidents reported in San Felipe and Dzilam de Bravo between 2004 and 2009. Lobster fishery season (Source: IMSS: Unidad de Medicina Hiperbárica, Tizimín, Yucatan. Nov. 2009)

have been reported in SF than in DB, which can be related to a stronger diving tradition in this fishing community (Salas and Pitcher 1999).

The coping strategies employed by fishers also include illegal actions that can contribute to stock deterioration in the long term. In 2008, by the time we undertook the survey, the octopus (*Octopus maya*) fishing season was not generating good yield and fishers were claiming that their income was reduced at that time. Under such conditions, some fishers took the risk of fishing octopus by diving and employing a hook, which is a forbidden gear by regulations. Others employed chlorine to push the animals to leave their refuges. These actions have a tremendous impact on the habitat of fishing areas, especially on females during the incubation period. Females do not feed during this time, so they are generally not caught by the legal, traditional fishing method. The use of chlorine and hook facilitate the capture, which negatively impacts the female and the eggs she is guarding.

Another strategy with illegal implications is fishing for horseshoe crab (*Limulus polyphemus)* – considered a living fossil and protected by Mexican law. While the Longnose Spider Crab (*Libinia dubia*) is the preferred bait for octopus, some fishers turn to the horseshoe crab as an alternative when the former gets scarce. Interestingly, they use mobile phones and text messages as innovative techniques to contact potential buyers. This is risky, since fishing this crab is considered a federal crime and is penalized with high fines and jail. However, those who catch it are willing to take the risk to increase their incomes, or sometimes just to maintain it.

10.4.6 Governance Issues

The majority of fishers in both communities recognize that the sustainability of their fishing resources is threatened. However, they do not seem to perceive how they can contribute to a solution. Rather, it is expected that the Government will intervene in order to solve their problems. Some local management initiatives have been taken by cooperative fishers from SF in an attempt to reduce such vulnerability. For example, one initiative is linked to the self-control to avoid fishing small lobsters during the recruitment process, so they do not fish during February, despite losing 1 month of the fishing season (August to February). Introduction of artificial habitats to improve lobster habitat was another action implemented by those fishers to generate habitat for lobsters, with the expectation of increasing production over the long term (Salas et al. 2008).

In SF, a strong tradition for community participation linked to resource management has been reported by several authors (Chuenpagdee et al. 2002; Fraga et al. 2005; Bjørkan 2006). This has also been linked to a high level of social capital, which facilitates problem-solving actions and improves the capacity to face challenges in the community. In DB, low social capital appears to be a condition that limits cooperative actions between community members. Several of our informants in DB explained that they do not participate in issues that concern the whole community, because there is little cohesion among its members. This condition was also reported by Arceo (2005). Under such conditions, fishers in this community have to search for options at an individual level if they want to diversify their activities or contribute to resource protection.

10.5 Discussion

Cinner et al. (2009) state that there are two main explanations for the generation of poverty conditions in small-scale fishing communities: the lack of alternatives outside the fishery sector; and resource overexploitation. Both of these factors may apply under the context of the communities we studied. Competition for limited resources can be exacerbated over time if newcomers continue to enter the fisheries, with the consequential impact on local people. These people, therefore, need to develop different strategies under changing conditions. Poor understanding of how they deal with variability and which factors generate a sense of vulnerability can limit public policies that attempt to mitigate the impact of different kinds of stressors and reduce poverty.

Béné (2009) points out the relevance of considering exposure to risk as a condition that increases vulnerability, and contributes to reinforce poverty. A combination of factors can trigger vulnerability. Macfadyen and Corcoran (2002) present some of them which, in most cases, apply to the communities we studied. Table 10.4 summarizes an adapted list of such factors.

Uncertainty has been an increasing issue in the fisheries in Yucatan due to different factors which include reduction in resource availability, limited capacity to negotiate fish prices and hence to save money and lack of cohesion in the communities or groups which limits the cooperation processes. All these factors can be aggravated by triggers such as increase in coastal population; increase in sea food demand; and increase in social problems such as alcohol, drugs and political conflicts which can reduce the capacity of people to cope with vulnerability.

Table 10.4 Factors linked to vulnerability in fishing communities and factors that contribute to increasing vulnerability

Factors linked to vulnerability	Vulnerability triggers
• Fishers are prone to suffer accidents, and generally have insufficient health services • High fluctuations in natural resources and increase in risky and uncertain conditions for the activity • High fluctuations in fish price, and fishers have limited capacity to bargain • Increase in cost of fishing operations, reduction of fishing days, income reduction • Conflicts, lack of cooperation • Limited capacity to save money, increase in conflicts in communities	• Reduction in catches provides incentives to fish farther and deeper • Increase in fishing pressure as the number of fishers is increased • Changes in the relationship between middleperson and fishers • Entrance of newcomers • Extreme weather conditions affects fishing resources and the fishing activity • Increase in social problems given income reduction, changes in political context, immigration • Alcohol and drug consumption, women adopting some addictions

The perception of what triggers vulnerability is similar among people in both communities. However, differences were evident in the way each group of fishers faced disturbances. We argue that the groups showed different adaptive capacity – ability for renewal and reorganization of the system followed by disturbance (Folke 2006, p. 259) as a result of different levels of flexibility and self-organization, which are necessary to build such capacity in a continuous development while facing change (Gunderson 2002; Folke 2006).

The marginalization indicators and the strategies developed by fishers suggest that both the cooperative and the independent fishers in DB are more vulnerable groups than firm owners and cooperative fishers in SF. While the cooperative fishers have a higher level of organization, the firm owners have better means to face challenges. Hence, these groups tend to develop more proactive strategies to face different types of hazards, and they perform better in light of the marginalization indicators which allocate them as less vulnerable groups.

According to Folke (2006, p. 261), social capital (including trust and networking) and social memory (including experience for dealing with changes) are essential for socio-ecological systems to adapt to and shape change. Livelihood assets and capabilities of the population to adapt are also linked to resilience (Allison and Ellis 2001; Plummer and Armitage 2007; Cinner et al. 2009). Therefore, the low level of organization and the lack of social cohesion in DB, which are especially evident in the relations between locals and migrants, could contribute to increase vulnerability, and hence to reduce resilience within this community.

Most fishers in DB recognize that population growth in coastal areas has imposed more pressure on natural resources. These conditions make people's livelihoods more vulnerable; however, this is outside community control. In the same way, the illegal strategies reported earlier, which can provide an income during periods of low catches, are deteriorating the ecosystem in the long term. These issues can aggravate fishers' vulnerability in the long term. Still, they do not seem to find a way out of this vicious cycle, which reduces their capacity to face new challenges (Béné 2003; Venkatesh 2006; Béné et al. 2007; Cinner et al. 2009).

Identifying and characterizing the poor or vulnerable, their attitudes toward the conditions that generate such vulnerability, and how they face those challenges are crucial for designing and implementing actions to improve their situation, and to contribute to sustainability of natural resources (Adams et al. 2004; Fisher and Christopher 2007; Thorpe et al. 2007). While we found heterogeneity dominates small-scale fishers, government agencies that aim to reduce poverty tend to assume that poor, small-scale fishers are a homogeneous group (Pérez 2009). This means that they are ignoring the complexity of the heterogeneous group that the fisheries system comprises. Hence, it is essential to identify meaningful groups for policy and program actions (Thorpe et al. 2007; Olmos et al. 2008; Béné 2009; Béné and Friend 2011).

It is necessary to address sustainability of fisheries and deal with socio-economic problems to maintain sustainable communities. In this sense, temporary palliative solutions generated by government programs that encourage ex post strategies – predominantly developed by independent fishers in DB – cannot solve problems in the short term, but do not contribute to build resilient systems in the long term. In Yucatan, temporary jobs are one example of how to create incentives to generate ex post strategies. These jobs are central to fishers in times of crisis, as they can generate an income in the short term. Such times of crisis occur when hurricanes hit the coast, or during the closed season for grouper. However, this also creates a dependency, as now fishers do not cooperate unless there are economic incentives. These conditions reduce the possibility that fishing communities will build human and social capital, and furthermore increase resilience.

In order to break the cycles that generate dependence of fishers and instead promote adaptive capacity in the communities, it is necessary to acknowledge the link between social resilience and ecological resilience (Allison and Ellis 2001; Fafchamps 2003; Folke 2006; Janssen et al. 2007). We underline that the ability of ex ante strategies developed by several of the groups studied here had to do with a strong adaptive capacity. This capacity facilitates the development of strategies that allow them to increase, or at least to maintain, the quality of life while facing disturbances. They also appear to be searching for ways to deal with resource sustainability. It is natural that while facing conditions of stress, people respond reactively to solve immediate problems. However, there could be a risk getting trapped in a vicious cycle in which while trying to maintain an income, they respond mainly to immediate problems and hence they cannot build capacity to face new challenges, including resource deterioration, as suggested by Folke (2006) and Smith and Wandel (2006).

To guide efforts toward governance of the complex socio-ecological systems in which fisheries are embedded, we need to increase our understanding of the conditions that define fishers' strategies. This also applies to the interactions between the components of the socio-ecological system (Janssen et al. 2007). Based on the results of our study, we attempt to explain the interactions among different components of a conceptual model (Fig. 10.5). In this model, a combination of collective strategies (innovation, networking and identity formation) and individual strategies (ex ante and ex post) are understood as ways to deal with stress.

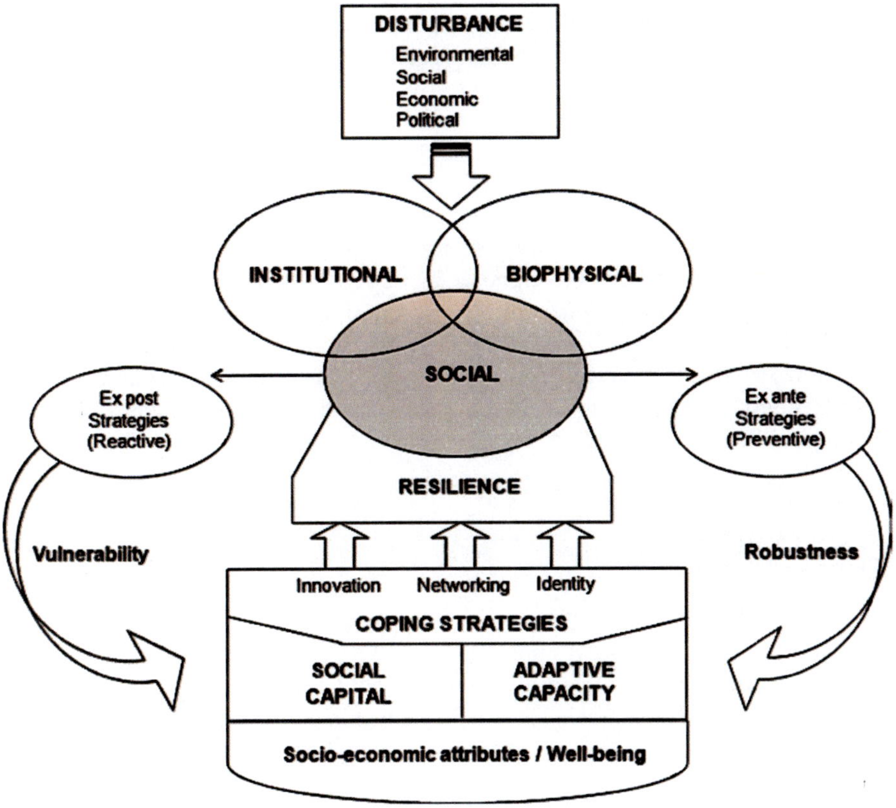

Fig. 10.5 Conceptual model of a social-ecological system that links social vulnerability and resilience to the coping strategies chosen by members of fishing communities

These strategies can be developed at the same time, or in alternative ways, as a dynamic process is in place.

The combination of strategies developed will define the level of adaptive capacity of the system. Socio-economic conditions within fishing communities, the level of organization of fisher groups, and the flexibility to adapt to changes can generate robust systems that can deal with the disturbances. We point out that ex ante strategies can contribute to building robust systems, while ex post strategies can expose the system to higher vulnerability in the long term. Social networks also represent an asset, as they can reduce transaction costs and increase trust among community members, thus building social capital (Schmid 2000).

We acknowledge that it is difficult to integrate a model that captures all potential interactions of a social ecological system. Here, we concentrate on the social system, not by dismissing the biophysical system, but because our study did not explore those issues directly. Despite this limitation, we consider that this can shed some light on the interactions within and between systems, and draw attention to how to promote a robust system that can favour higher resilience in coastal communities.

The model illustrates a path to increase adaptive capacity of communities (or groups). In this sense, resilience does not involve improving capacity only in the vulnerable groups; it also demands flexibility of the governance system.

Recognizing what kind of strategies people develop, and encouraging those that improve adaptive capacity of the groups and communities could lead to better forms of governance. In this sense, increasing social capital can improve people's abilities for self-organizing and for creating capacity; contribute to building resilience; and to develop more efficient ways to prevent people from falling into poverty conditions (Olmos et al. 2008; Béné and Friend 2011). It is important to define indicators that allow for understanding of the multiple socio-ecological interactions as well as the multiple sources of risk. These indicators could guide public policies and allow monitoring changes in the socio-ecological systems using an interdisciplinary approach considering social, political, economic and environmental issues (Ohl et al. 2007; Clay and Olson 2008).

10.6 Conclusions

Since fishers from SF and DB have access to food through their fishing activities, they do not consider themselves poor, despite acknowledging poverty issues in their community. They consider themselves in better condition than people from the rural sector. However, they acknowledge that they are vulnerable, due to different hazards and shocks they have been facing though time. Those hazards have recently increased both in frequency and intensity, negatively impacting their livelihoods. They report exogenous factors (i.e. hurricanes, red tides, entrance of newcomers, increasing competition) and endogenous factors (i.e. illegal fishing, limited opportunities to get an income outside fishing, different types of addictions), all of which represent potential sources of uncertainty, which can increase vulnerability.

At least 80% of fishers in DB and 30% of fishers in SF recognized poverty as a problem in their community, and acknowledged that better health services and education could help build up their sense of well-being at a community level.

Even though fishers from both communities, and different groups within the communities, share some problems, heterogeneity in terms of social and economic capacity affects significantly how these groups face the increasingly uncertain conditions related to their activities and lifestyle. In both communities, a combination of ex ante and ex post coping strategies has been developed among fishers to overcome exposure to vulnerable conditions. However, independent fishers (mainly migrant people) appear to be particularly vulnerable. This vulnerability can be worsened when community members in DB reject them. A key issue here is how to implement management programs to overcome vulnerability, and at the same time recognize the differences between groups to apply the programs accordingly. In light of the degradation of the resources fishers depend on and the heterogeneity of the groups, this becomes even more relevant.

In this context, the Mexican management framework needs to address these issues in order to move toward poverty reduction in coastal communities. While dealing with poverty, Mexican policies have been oriented toward changing some numbers of marginality indicators. However, they do not address crucial issues related to vulnerability, which can also expose people to poverty. It is necessary to recognize that minimal steps to reduce marginalization cannot necessarily improve population welfare. Rather, it is essential to increase fishers' adaptive capacity to deal with the increasingly risky conditions that the fishery sector is facing, especially within the small-scale fisheries sector. In this context, it is fundamental to encourage ex ante strategies from fishers if an improvement in local conditions of coastal resources is the goal. Public policies need to be oriented to strengthen the capacity of coastal communities and elsewhere, using a long-term perspective.

Temporary palliative solutions cannot build resilient systems. The solution, at least in part, is to generate opportunities – improve skills and assets of communities and those of the institutions in charge of fisheries management. Intervention programs should acknowledge risk and uncertainty conditions within the fishing sector and fishing communities. This includes diversification of livelihoods, and development of contingency programs to overcome the increasing challenges coming from outside (meteorological factors, market demands, urban development, migration, among others). Those conditions can lead to building resilient socio-ecological systems. A wide range of social, economic, environmental and institutional factors define the complexity of these dynamic socio-ecological systems. Given that uncertainty cannot be eliminated, knowledge improvement in several fields is necessary (Seijo et al. 2009). A careful analysis that accounts for the diversity of groups that integrate these systems, and the factors both inside and outside the sector that promote vulnerable conditions which can drive people into the poverty trap are required.

Acknowledgments We acknowledge the Norwegian Research Council through the Norwegian College of Fishery Science, University of Tromsø, for funding the PovFish project. The support from fishers and community members from San Felipe and Dzilam de Bravo fishing communities made this study possible. Eric May from SAGARPA and members of the fishing cooperatives kindly gave us access to statistical data. Special thanks to our field team, based largely on voluntary work from students and visiting scholars, who are: David Escalante, Luis Dzib, Miguel Duarte, Nayelli Torres (Universidad de Yucatan, School of Anthropology), Paula Sabido (Universidad de Quintana Roo, School of Resource Management), Reyes Emmanuel Méndez and Martín Marrufo (fishers from San Felipe and also students at the Instituto Tecnológico de Conkal), Oscar Pacheco (Universidad Nacional, Costa Rica) and Ximena Colombo (independent volunteer).

References

Adams WM, Aveling R, Brockington D, Dickson B, Elliott J, Hutton J, Roe D, Vira B, Wolmer W (2004) Biodiversity conservation and the eradication of poverty. Science 306(5699):1146–1149

Adger WN (2000) Social and ecological resilience: are they related? Prog Hum Geogr 24:347–364

Adger WN (2006) Vulnerability. Glob Environ Change 16:268–281

Allison EH, Ellis F (2001) The livelihoods approach and management of small-scale fisheries. Mar Policy 25:377–388

Allison EH, Horemans B, Béné, C (2006) Vulnerability reduction and social inclusion: strategies for reducing poverty among small-scale fisherfolk. Paper presented at the Wetlands, Water and Livelihoods Workshops, Wetland International, St. Lucia, 30 Jan–2 Feb

Arceo P (2005) El capital social en Dzilam de Bravo: marco conceptual y resultados de campo. In: Fraga J, Euán J, Salas S, Chuenpagdee R (eds.) Manejo comunitario de un área natural protegida en Yucatán, México. CINVESTAV-CIID, Canada/México

Bærenholdt JO, Aarsæther N (1998) Coping strategies in the North. In: Aarsæther N, Bærenholdt JO (eds.) Coping Strategies in the North: Local Practices in the Context of Global Restructuring. Nordic Council of Ministers, Copenhagen, pp 15–44

Béné C (2003) When fishery rhymes with poverty: a first step beyond the old paradigm on poverty in small-scale fisheries. World Dev 31:941–975

Béné C (2009) Are fishers poor or vulnerable? Assessing economic vulnerability in small-scale fishing communities. J Dev Stud 45(6):911–933

Béné C, Friend R (2011) Poverty in small-scale fisheries: old issue, new analysis. Prog Dev Stud 11(2):119–144

Béné C, Macfadyen G, Allison EH (2007) Increasing the contribution of small-scale fisheries to poverty alleviation and security. Fisheries Technical Paper No. 481. FAO, Rome, p 125

Bjørkan M (2005) Cooperatives and community failure in fisheries management: a Mexican case study on collective opportunism and social capital. MSc thesis, University of Tromsø, Norway, 122p

Bjørkan M (2006) Mexican case study on community empowerment. MSc thesis, University of Bergen, Norway, 127p

Bourdieu P, Wacquant LJD (1992) An invitation to reflexive sociology. University of Chicago Press, Chicago

Bråten E (2002) Studentfeltarbeidet i fare? En pragmatisk tilnærming. *Norsk Antropologisk Tisdsskrift,* 13

Burgos R, Defeo O (2004) Long-term population structure, mortality and modelling of a tropical multi-fleet fishery: the red grouper *Epinephelus morio* of the Campeche Bank, Gulf of Mexico. Fish Res 66:325–335

Chambers R (1989) Vulnerability, coping and policy. IDS Bull 20(2):1–7

Chuenpagdee R, Fraga J, Euán J (2002) Community's perspectives toward marine reserve: a case study of San Felipe, Yucatán, México. Coast Manage 30:183–191

Cinner JE, Pollnac RB (2004) Poverty, perceptions and planning: why socioeconomics matter in the management of Mexican Reefs. Ocean Coast Manage 47:479–493

Cinner JE, Daw T, McClanahan TR (2009) Socioeconomic factors that affect artisanal fishers' readiness to exit a declining fishery. Conserv Biol. doi:10.1111/j.1523-1739.2008.01041.x

Clay P, Olson J (2008) Defining "fishing communities": vulnerability and the Magnuson-Stevens Fishery Conservation and Management Act. Hum Ecol Rev 15(2):143–160

CONAPO (2005) Índice de marginación a nivel de localidad 2005. Consejo Nacional de Población. http://www.conapo.gob.mx/publicaciones/IndiceMargLoc2005.pdf

Dalenius T, Hodges JL Jr (1959) Minimum variance stratification. JASA 54:88–101

Fabricius C, Folke C, Cundill G, Schultz L (2007) Powerless spectators, coping actors, and adaptive co-managers: a synthesis of the role of communities in ecosystem management. Ecol Soc 12(1). http://www.ecologyandsociety.org/vol12/iss21/art29/

Fafchamps M (2003) Rural poverty, risk and development. Edward Elgar Publishing Limited, Northampton

Fernández JI, Álvarez-Torres P, Arreguín-Sánchez F, López-Lemus LG, Ponce G, Díaz-de-León JA, Arcos-Huitrón E, del Monte-Luna P (in press) Coastal fisheries of México. In: Salas S, Chuenpagdee R, Charles A, Seijo JC (eds.) Coastal fisheries of Latin America and the Caribbean. FAO Technical Report, Rome, Italy, 462 p (in press)

Fisher B, Christopher T (2007) Poverty and biodiversity: measuring the overlap of human poverty and the biodiversity hotspots. Ecol Econ 62:93–101

Folke C (2006) Resilience: the emergence of a perspective for social–ecological systems analyses. Glob Environ Change 16:253–267

Fraga J, Euán J, Salas S, Chuenpagdee R (eds.) (2005) Manejo comunitario de una área natural protegida en Yucatán, México. CINVESTAV-CIID, Canada/México

Fraga J, Salas S, Mexicano-Cíntora G (2008) La Pesca en Yucatán: de la abundancia a la escasez, a la fragilidad de las estructuras Institucionales. In: Fraga J, Villalobos GJ, Doyon S, García A (eds.) Gobernanza costera en México: descentralización y manejo ambiental en la península de Yucatán. CINVESTAV-IPN, Unidad Mérida, Universidad Autónoma de Campeche, IDRC, México, pp 179–201
Gallopín GC (2006) Linkages between vulnerability, resilience and adaptive capacity. Glob Environ Change 16:293–303
Gibbs MT (2009) Resilience: what is it and what does it mean for marine policymakers? Mar Policy 33:322–331
Gunderson LH (2000) Resilience in theory and practice. Annu Rev Ecol Syst 31:425–439
Gunderson LH (2002) Adaptive dancing: interactions between social resilience and ecological crisis. In: Berkes F (ed.) Navigating social-ecological systems: building resilience for complexity and change. Cambridge University Press, New York
Halpern D (2005) Social capital. Policy Press, London
Herrera-Silveira J, Aranda-Cirerol N, Troccoli-Ghinaglia L, Comín F, Madden C (2004) Eutrofización costera en la Península de Yucatán. In: Caso M, Pisanty I, Ezcurra E (eds.) Diagnóstico Ambiental del Golfo de México. Instituto Nacional de Ecología (INE-SEMARNAT), Mexico, pp 823–850
INEGI (2005) Conteo de Población y Vivienda 2005. http://www.inegi.gob.mx/est/default.asp
Ingles P, Sepez J (2007) Anthropology's contributions to fisheries management. NAPA Bull 28:1–12
Janssen MA, Anderies JM, Ostrom E (2007) Robustness of social-ecological systems to spatial and temporal variability. Soc Nat Resour 20:307–22
Jentoft S, Davis A (1993) Self and sacrifice: an investigation of small boat fisher individualism and its implication for producer cooperatives. Hum Organ 52(4):356–367
Jentoft S, Sandersen HT (1996) Cooperatives in fisheries management: the case of St. Vincent and the Grenadines. Soc Nat Resour 9(3):295–305
Jiménez-Badillo L (2008) Management challenges of small-scale fishing communities in a protected reef system of Veracruz, Gulf of Mexico. Fish Manage Ecol 15:19–26
Jóhannesson G, Skaptadóttir UD, Benediktsson K (2003) Coping with social capital? The cultural economy of tourism in the North. Sociol Ruralis 43(1):3–16
Macfadyen G, Corcoran E (2002) Literature review of studies on poverty in fishing communities and of lessons learned in using the sustainable livelihoods approach in poverty alleviation strategies and projects. FAO Fisheries Circular No. 979. FAO, Rome, p 93
Mexicano-Cíntora G, Liceaga MA, Salas S (2009) Uso de sistemas de información geográfica en pesquerías: La pesca en Yucatán, al sur del Golfo de México. Universidad y Ciencia 25:1–8
OECD (2001) Development action committee's guidelines on poverty reduction, vol May 2009. Organization for Economic Cooperation and Development, Paris
Ohl C, Krauze K, Grubeuhel C (2007) Towards an understanding of long-term ecosystem dynamics by merging socio-economic and environmental research: criteria for long-term socio-ecological research sites selection. Ecol Econ 63(2–3):383–391
Olmos ME, Salas S, Beltrán-Morales LF, Preceda-Solís A, Ortega-Rubio A (2008) Riqueza ecológica y pobreza económica en un área natural protegida en Baja California Sur. Región y Sociedad 42:133–164
Pérez UF (2009) Pobreza, políticas públicas y políticas sociales en Yucatán, 1995-2007. Rev de Economía 1:39–66
Plummer R, Armitage D (2007) A resilience-based framework for evaluating adaptive co-management: linking ecology, economics and society in a complex world. Ecol Econ 61:62–74
PNUD (2009) Informes sobre desarrollo humano. http://hdrstats.undp.org/es/countries/country_fact_sheets/cty_fs_MEX.html
POETCY (2007) Programa de Ordenamiento Ecológico del Territorio Costero del Estado de Yucatán (POETCY). Informe Final. CINVESTAV, ITC, ITM Y UADY, SECOL, SEMARNAT Y CBM

Pomeroy R, Ratner BD, Hall S, Pimoljinda J, Vivekanandan V (2006) Coping with disaster: rehabilitating coastal livelihoods and communities. Mar Policy 30:786–793

Råkjær Nielsen J, Degnbol P, Viswanathan K, Ahmedb M, Hara M, Abdullahd NR (2004) Fisheries co-management—an institutional innovation? Lessons from South East Asia and Southern Africa. Mar Policy 28:151–160

Ríos–Lara GV, Salas S (2009) Modelos estructurados para la evaluación de la población de langosta *P. argus* en la plataforma de Yucatán, México. Proc Gulf Caribb Fish Inst 61:451–460

Salas S, Pitcher T (1999) Fishing strategies of small-scale lobster fishermen in Yucatan, Mexico. Revista de Ciencias del Mar, UAS 16:30–38

Salas S, Mexicano-Cíntora G, Cabrera MA (2006) ¿Hacia dónde van las pesquerías en Yucatán?: retos y perspectivas. CINVESTAV Unidad Merida, México

Salas S, Chuenpagdee R, Seijo JC, Charles A (2007) Challenges in the assessment and management of small-scale fisheries in Latin America and the Caribbean. Fish Res 87:5–16

Salas S, Cabrera MA, Zapata-Araujo C, Euán J, Maldonado A (2008) Are artificial devices an option to improve lobster catches? The case of Yucatan fishery, Mexico. Proc Gulf Caribb Fish Inst 60:197–208

Schmid A (2000) Sources of social capital. Department of Agricultural Economics, Michigan State University, East Lansing. Staff Paper. 00–07, 19 p

Seijo JC, Cuevas A, Villanueva R, Duarte J, Villanueva J, Novelo G, Ancona P, Boa M, García I, Romero A, Aranda J, Mendoza E (2009) Estudio social de la pesca en México. Diagnóstico sobre la problemática social de los pescadores, su entorno y su visión sobre las posibles soluciones a los problemas que afectan la pesca en México. Proyecto FAO, UTF-079-MEX

Sethi G, Costello Ch, Fisher A, Hanemannc M, Karpc L (2005) Fishery management under multiple uncertainty. J Environ Econ Manage 50:300–318

Smith B, Wandel J (2006) Adaptation, adaptive capacity and vulnerability. Glob Environ Change 16:282–292

Solana SR, Pérez M, Wakida A, Burgos R, Santos J (2006) Evaluación de la población de pulpo (*Octopus maya*) en la Península de Yucatán 2005. SAGARPA, INP. CRIP-Yucalpetén, CRIP-Lerma

Thorpe A, Andrew NL, Allison EH (2007) Fisheries and poverty reduction. CAB Rev Perspect Agric Vet Sci Nutr Nat Resour 2, no.085, 12 pp

Venkatesh S (2006) Trends in poverty and livelihoods in coastal fishing communities of Orissa State, India. FAO Fisheries Technical Paper No. 490, 111

Chapter 11
Through Boom and Bust: Coping with Poverty in Sea Snail Fisheries on the Turkish Black Sea Coast

Ståle Knudsen and Hakan Koçak

Abstract Small-scale fisheries for the introduced sea snail (*Rapana venosa*) have seen booms followed by irreversible bust. This chapter focuses on the role of this fishery relative to poverty dynamics on the Turkish Black Sea coast, and explores how fishers cope with boom and bust, respectively. We consider poverty as a multi-faceted issue and analyze in some detail fishers' income, social security, health, education, housing, as well as people's own, culturally-informed perception of what constitutes poverty. Yet, our analysis aims beyond a descriptive account of poverty among fishers, and queries the epistemological status of the vicious circle model. Thus, we discuss how sea snail fishing has also constituted a way out of poverty; that it is uncertain whether overfishing can be blamed for the bust; that contextual factors, such as state welfare and agricultural policies, international organizations, and world economic dynamics, can have significant impact on poverty and wealth among the coastal population of the Black Sea coast of Turkey. The boom years of the sea snail fisheries clearly created a frontier situation inhibiting prospects of co-management between the state and communities of fishers. The observed lack of collective action among fishers and their concomitant incapacity to participate can be considered a dimension of poverty. Therefore, fishery development should not only go hand in hand with fishery management, but also with social policies aimed at reducing poverty and inequality.

S. Knudsen (✉)
Department of Social Anthropology, University of Bergen, 5007 Bergen, Norway
e-mail: stale.knudsen@sosantr.uib.no

H. Koçak
Department of Labor Economics and Industrial Relations, Kocaeli University, Umuttepe Campus, Kocaeli, Turkey
e-mail: hakan.kocak@kocaeli.edu.tr

S. Jentoft and A. Eide (eds.), *Poverty Mosaics: Realities and Prospects in Small-Scale Fisheries*, DOI 10.1007/978-94-007-1582-0_11,

11.1 Introduction

This chapter explores the role of small boat fishing of sea snail (*Rapana venosa*) relative to poverty dynamics on the Turkish Black Sea coast and asks to what extent, the thesis of the vicious circle can explain the developments in these fisheries. We focus particularly on how fishers cope with both boom and bust, and what contextual/structural factors impact upon their capacity to do so. In this context, we find it especially relevant to ask to what extent poverty in the fisheries is related to macro economic and policy developments in Turkey. This is related to our intention to explore the analytical utility of the vicious circle model. What is the epistemological status of the model? How much can we expect to be able to explain with it? Would examples of the vicious circle dynamic not taking place, even in the case of a lack of collective action or effective management, falsify the thesis?

To address these questions, we have explored the importance of sea snail fishing in two Black Sea communities. Terme, in the Province of Samsun, is still experiencing a boom situation; whereas Çarşıbaşı, in the Province of Trabzon, has had to cope with a new situation after the sea snail economy went bust a few years ago. How are fishers coping in boom and bust periods, respectively? What is the level of poverty, and how can it be measured? To what extent was poverty a driver for the growth in sea snail fisheries, and what are the effects on poverty of the bust? We also briefly survey sea snail management, and the role of communities and fishery cooperatives in managing sea snail fisheries.

In line with Amartya Sen and work inspired by his approach (such as UNDPs Human Development Reports), we consider measuring poverty only in terms of income to be insufficient; and at times this can be misleading if we are to understand the true character of poverty. Defining "capabilities" as "…the substantive freedoms [an individual] enjoys to lead the kind of life he or she has reason to value," Sen goes on to claim that "…poverty must be seen as the deprivation of basic capabilities rather than merely lowness of incomes…" (Sen 1999, p. 87). In this perspective, income is only one (among several) means to meet ends and the relation between low income and low capability is variable (Sen 1999, pp. 88–90).

Accordingly, we make an effort to assess and compare poverty by a range of indicators in addition to income – which often is difficult to measure anyway in the informal economies under study here. We particularly focus on education and social security, and to some extent on health. Additionally, while we acknowledge that there are some universal and "absolute" aspects to poverty (such as food security, health, education etc.), we also believe that the real import of poverty can only be grasped through people's own conceptualization and imagination of what poverty (and not being poor) means in everyday life.

Sen also stresses that in place of properties, we should conceptualize the right to resources as entitlements: "…the set of alternative commodity bundles that a person can command in a society using the totality of rights and opportunities that he or she faces" (Sen 1983, p. 754). With the limited character of marine resources and the common pool regime often characterizing access, entitlements related to fisheries tend to have particular qualities. Rights are established in very different ways from agricultural society; and the likelihood of a vicious circle dynamic ruining the resource base poses a particular management challenge.

Thus, economies and livelihoods based upon fisheries, particularly small-scale informal fisheries, have to manage a high level of risk and tend to be vulnerable (Acheson 1981; McGoodwin 1990). Yet, the dynamic character of such adaptations, as well as the fact that it is often an informal economy operating on the "fringes" of society, can also make it a sector of hope and opportunity, in effect a "frontier." "Kopytoff (1987) discussed the frontier as an undetermined space that is part of a regional system, on the margins of the state" (Mangahas 2000, p. 8). Tsing considers frontiers to be "...particular kinds of edges where the expansive nature of extraction comes into its own," "notoriously unstable," and "...deregulated because they arise in the interstitial spaces made by collaboration among legitimate and illegitimate partners..." (Tsing 2005, p. 27).

Such frontiers make for special challenges when it comes to analysis; we cannot suppose stability, transparency, and a fixed set of variables. The processes that drive or facilitate the opening up of new frontiers are often of a scale much larger than the frontier itself: The introduction of the sea snail to the Black Sea; the demand in Eastern markets; and the opening up of Turkey's economy during the 1980s. Yet, this also indicates that frontiers are insufficiently understood as the interaction between resources and resource users/extractors (e.g., fishers). This rhymes with Béné's assertion, in his review of studies of small-scale fisheries and poverty, where he draws on Sen's discussion of entitlement: "...to better understand poverty – as it relates to natural resources – it is necessary to redirect part of our attention and analysis effort away from the resources themselves...and to put greater emphasis on the role of politics of...access, control, and redistribution of these resources" (Béné 2003, p. 959). Additionally, we will later argue that even this expansion of perspective might be insufficient. It may also be necessary to account for politics and policies related to entitlements outside of those directly related to the natural resource.

In order to capture the way in which the various variables come together, and how poverty and wealth are produced, it is necessary to focus on how individuals and households combine, juxtapose, and switch between different entitlements as they make their livelihoods. Small-scale fishers' livelihoods often comprise more than fishing, and it will be necessary to view their coping strategies within the scopes of seasonal cycles and long-term individual careers (Allison and Ellis 2001). In this chapter, we illustrate this with a couple of cases; and also compare and summarize typical career paths and coping strategies over time in the two communities.

11.2 Methodology

This work leans on previous anthropological fieldwork in Çarşıbaşı since 1990 (see Knudsen 2009), and on multidisciplinary work (Knudsen et al. 2010) in the Province of Samsun during 2005–2006. The selection of the two study sites,

Çarşıbaşı and Terme, means that we have been able to monitor changes over a 20-year span (in Çarşıbaşı); and compare developments in two important, yet different fishing centers. Since we have already much insight into the situation in Çarşıbaşı, we have not applied the same methodology in both study sites. The fact that sea snail fishing was discontinued in Çarşıbaşı in 2004, further justifies this choice. What happened in Çarşıbaşı is a demonstration of what could – with some important differences – happen in Terme.

Overall, we have considered that the flexible and informal character of the small-scale fishing sector and its complex interweaving into regional, national, and global processes have necessitated a multi-faceted and flexible approach. Fieldwork was undertaken in May 2008 (2 weeks), September 2008 (1 week) and April 2009 (2 weeks, 4-person research team). These intervals of fieldwork contained a fair bit of participant observation: Much informal conversation in tea houses and in harbors; questionnaire surveys with fishers; and formal and informal interviews. Interviews were with managers at both the national and provincial levels, marine scientists, sea snail factory owners/representatives, an independent consultant, local and regional health service representatives, social service representatives, a local headman, sea snail buyers (middlemen), and fishery cooperative representatives at different levels of organization. While the questionnaires for male fishers ($N=36$), and semi-structured interviews ($N=26$) with women in fisher households in Terme were limited in scope, we believe that findings from these are still indicative of important tendencies; and of how different issues typically cluster together in individual lives and households. Finally, we have consulted public statistics (Turkish Statistical Institute, TURKSTAT); surveyed all relevant reports (which often rely upon TURKSTAT data); and read local internet sites and newspapers.

11.3 Poverty and Poverty Studies in Turkey

With a GNP/capita around 12,000 USD (2008), Turkey's economic development stands at the same level as the poorer countries of the EU, to which Turkey is an accession-candidate country. Turkey may thus not seem to be a poor country. Yet, there are very pronounced class inequalities and associated poverty in Turkey. Food poverty (2007) is estimated at 0.54%, and "food and non-food poverty" (income poverty) at 17.11% (TURKSTAT 2009). Turkey scores relatively higher on income than development indicators related to health, education, and gender.[1]

Poverty in Turkey is particularly prevalent in rural areas. While urban poverty fell significantly from 22% in 2002 to 9.39% in 2008, rural poverty has remained at a stable high level (34.62% in 2008) (TURKSTAT 2009). One study

[1] With an HDI value of 0.806, Turkey ranks lower (79) on this composite scale than on the GDP/capita ranking (63) (UNDP 2009).

notes that "…the rate of absolute poverty in rural areas is 3.5 times of urban areas" (Gülçubuk and Aluftekin 2006, p. 9). A decade ago, the majority of the nine million[2] people of Turkey in the lowest HDI category lived in rural areas (Akder 2000). There is a significant transformation taking place in the rural communities whereby the number of subsistence and property-owning households are decreasing. The poorest households (in terms of income) are subsistence households without land and day agricultural laborers (Köse and Bahçe 2009).

With the opening up of the economy in the 1980s, and the dominance of neo-liberal economic policies with de-regulation and privatization, poverty came to constitute a noticeable issue, both in public debate and in academic literature,[3] and the fight against poverty came high on the political and academic agenda. In 1986, the state established the *Social Assistance and Solidarity Fund* to supplement or replace "traditional welfare." The Fund's largest expenditures are related to health services. This has, since 1992, been administered through a so-called Green Card program. Green Card certificates give the holders right to basic health care free or at reduced price (Buğra and Keyder 2005, p. 29; Yoltar 2009).

11.4 Poverty and Sea Snail Fishing in the Black Sea Region

The Black Sea region presents a complex picture when it comes to poverty. On the one hand, only in Eastern and, especially, in South-eastern Turkey, is the proportion of the population that is poor higher than in the Black Sea region. On the other hand, unemployment rate is low and a large share of the population participates in the workforce. The region is also characterized by significant out-migration and a high degree of agricultural employment. Major cash crops are tobacco, hazelnuts, and tea. Due to WB-imposed structural adjustment schemes and concomitant retreat of state support, farmers relying upon these crops have been under particular stress during the last 15 years or so (Eruygur 2006).

In this context, fishing is one of the few options for income and career for the marginalized and poor coastal dwellers. Whereas there are important trawl fisheries

[2]Of Turkey's total population in 2000, 23 million were considered rural and 44 million urban. In 2009, this had changed to 17.8 million rural, and 54.8 million urban (TURKSTAT 2010).

[3]These developments have, especially during the last 10 years, led social scientists in Turkey to apply theoretical perspectives from the international poverty literature to studies in Turkey. These studies include the relations between globalization and poverty (Şenses 2001); social policies in the face of poverty (Buğra 2008); quantitative and qualitative studies of migration and urban poverty (e.g., Işık and Pınarcıoğlu 2001); and comparative qualitative studies of the "condition of poverty" in different parts of Turkey (Erdoğan 2007). Case studies of poverty and poverty alleviation initiatives in Turkey are collected in a recent anthology (Oktik 2008). Parallel to this, the Turkish Statistical Institute (TURKSTAT), and the State Planning Organization have intensified data collection directed at providing relevant information on the degree and character of poverty in Turkey. International organizations (esp. the World Bank and UNDP), as well as national research institutions, have often relied on data from TURKSTAT to explore and analyze poverty in Turkey.

Fig. 11.1 Typical medium-sized, multi-purpose boats. In front, sea sails in sacks. These boats have found shelter in Kozluk upstream from the mouth of the Yeşilırmak River. Note the net for bonito fishing at the deck of the closest boat. Fishers easily and regularly switch between different catch methods. Photo by Ståle Knudsen, September 2005

(mainly in Samsun) and purse seine fisheries (largely east of Samsun) along the coast (Knudsen 2009), sea snail fishing has been the most expansive sector during the last two decades (Fig. 11.1). For thousands of families along the Black Sea coast of Turkey, sea snail fishing has helped to improve the living standards.

The sea snail *Rapana* was accidentally introduced to the Black Sea in the late 1940s. Commercial fisheries started in Turkey in the early 1980s. Most catches of sea snails are taken by medium sized boats (8–16 m), which often switch between different kinds of gear, including trawling. All catches are delivered to one of the (currently) six processing plants in Turkey. The frozen meat is exported to the Far East, as there is no local demand (Knudsen 2006). In the eastern provinces, and especially around Trabzon (Fig. 11.2), this fishery experienced a boom during the 1990s. However, the average size of the sea snails gradually decreased to below marketable size to the extent that the processing plants stopped buying. Since 2004, there has been no fishery for sea snails in the Province of Trabzon and there are now no sea snail processing plants left east of Terme. In the Province of Samsun, sea snail fishing boomed after the fishery started to decrease further east, resulting in a 450% increase in small and medium sized boats' engine power in this province during 2000–2005 (Knudsen et al. 2010).

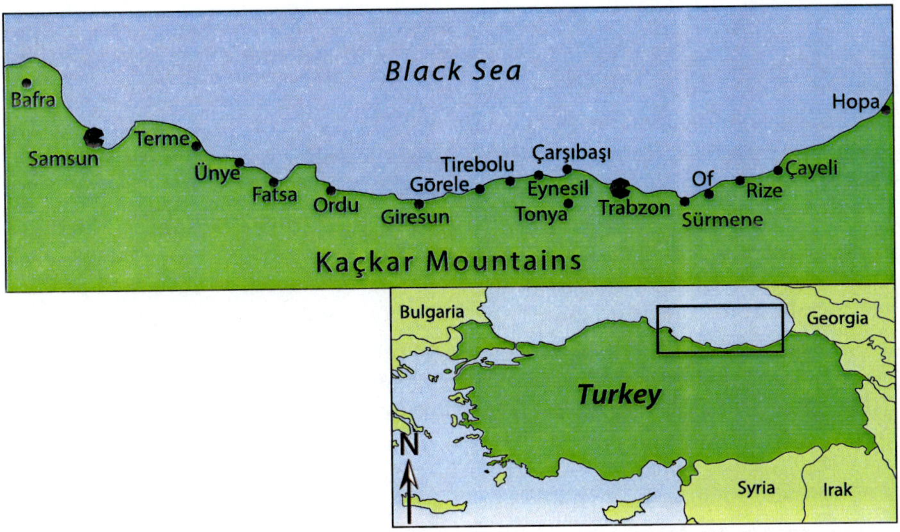

Fig. 11.2 Map of Eastern Black Sea region of Turkey. The two main field locations Terme and Çarşıbaşı are located relatively close to the major cities Samsun and Trabzon, respectively. The coastal highway passes through both field sites

11.4.1 Çarşıbaşı

The landscape in Çarşıbaşı is characterized by valleys and steep mountain sides, and the only cash crop that can be profitably grown here is hazelnuts (approximately 30% of the total area of the district). While hazelnut cultivation became the dominant cash crop in Çarşıbaşı from the 1960s onward, it was usually poor men, who could not make a living from hazelnuts, who became fishers. Some of these people developed spectacular careers and became owners of large purse seiners (Knudsen 2009). Çarşıbaşı developed into one of the major fishing centers along the Turkish Black Sea coast: Nine large family-owned and operated purse seiners, two fish meal and oil factories, five cold storage facilities, and a sea food and vegetable conservation plant. These factories together employ up to 100 seasonal workers. The district has long had two large fishing harbors.

Fishing and hazelnut cultivation, together with subsistence horticulture, has sustained a high population density (approx. 260 km^2); and the district population of 16,000 has been roughly stable over the last 20 years (TURKSTAT). Many indicators show that the economic situation of the district is relatively good.[4]

[4]Çarşıbaşı is the district in the province outside of the city that has the lowest unemployment rate (MEF 2008), and the highest proportion (13.12% in 2000 (SPO 2004a)) of industrial employment (MEF 2008, pp. XII-70). The service sector is also a major employer. The GNP/capita in Çarşıbaşı was 2,756 USD in 1996 (2,137 in the Province of Trabzon) (SAG 2005, p. 41); which was close to the national average (2,888 USD/capita), and made it the richest district in Trabzon after the city of Trabzon itself.

On the other hand, the number of individuals (3,542) holding green cards in 2009 amounted to 21% of the district population – well above the 11.6% average for the province (MHT 2009). This indicates that the positive economic figures disguise the existence of inequalities and poverty within the district. In the fisheries sector, there is a marked differentiation between owners of large boats and small boat fishers/crew.

In 1990–1991, small boat fishers in Çarşıbaşı typically employed 5–7 m boats with 9 Hp engines in diversified catches of fish and sea snails. The structure of the small boat fleet changed significantly with the establishment of a sea snail processing factory in the township toward the end of the 1980s, and the active support of the factory management toward construction of new boats. While the number of small boats in Çarşıbaşı during the years 1990 to 1997 increased from a little below 100 to approximately 130, most constructed new and larger boats (7.5–12 m) and installed more powerful engines (25–135 Hp) to more effectively dredge for sea snails (Knudsen 1997).

The fishery clearly experienced a kind of boom situation during the 1990s. It was not unusual to land 1/2–1 t of fairly large sea snails after one night's operation. Some also secured extra income from illegal bottom trawling. The rigging and engine power of larger sea snail boats made this trawling possible (at all times, illegal east of Terme). One to three men work on each boat (three during trawling). Sea snail dredging requires less training and knowledge than many other kinds of fishing, and is therefore relatively easily taken up by people with no previous experience in fishing. This became the major income for many households and made it possible for many men to stop working as crew on purse seiners. Most commonly, income from sea snail fishing was invested in house construction. Thus, this fishery helped to increase the standard of living and welfare, and decrease insecurity. However, these households remained relatively poor. Nobody came to possess a car, and nobody managed to climb a step and invest in other businesses. Being an informal sector of the economy, fishers had no health insurance, and were seldom able to afford to pay installments in one of the voluntary schemes (SSK, Bağ-Kur[5]).

Then, the unforeseen and unexpected thing happened. While fishers could still catch and land as many sea snails as before, the average size gradually decreased. From 2004, the factory in Çarşıbaşı stopped buying sea snails. Small individuals catch a much lower price on the international market. Not even exemption from the 40% tax on fuel from 2004 onward was enough to keep this fishery alive. With the close of the sea snail fishing, illegal trawling, which could only be ventured under the guise of dredging, discontinued. What happened to the sea snail boats and fishers? How did they cope with the new situation?

Contrary to our expectation, few have migrated. A handful have found jobs in other provinces, but these are men who already had strong family ties to those places

[5]SSK (Sosyal Sigortalar Kurumu) is the public social insurance foundation for all salaried employees outside of state civil servants. Bağ-Kur (Esnaf Ve SanatkâRlar Ve Diger Bagimsiz Çalisanlar Sosyal Sigortalar Kurumu) is a publicly administered social security foundation for the self-employed, including farmers and fishers.

Table 11.1 Number of multi-purpose/sea snail boats (8–12 m/35+ Hp) in two major quarters in the township of Çarşıbaşı. Note the significant decrease in boats suitable for sea snail fishing in both localities, from 34 to 15 in Keremköy; and from 9 to 3 in Burunbahçe. In Keremköy, many (14) have sold their multi-purpose boat and bought a smaller boat in its place

	1997/1998	2008
Keremköy	34	15 (+ 14 downsized)
Burunbahçe	9	3

and might have moved anyway. One major strategy has been to sell boats (many were sold to Samsun), sometimes replaced with smaller boats suitable for net fisheries but guzzling less fuel. Many have come to depend (again) more on work as a crew on large purse seiners, or in other seasonal maritime work (on yachts, in harbors). Other household strategies include not letting sons become fishers and reducing consumption.

In 1997/1998, we compiled lists of all sea snail fishing boats in two of the major fishing communities in the district, Keremköy and Burunbahçe. During our revisits in 2008 and 2009, we tracked the situation of each boat and its ownership, as well as updated the list with new boats/owners (Table 11.1).

In effect, the major choice facing poor people who want to continue being fishers in Çarşıbaşı concerns what they will do during winters: Work as crew on purse seiners (with potential good income, but being away from home for months on end); or staying at home fishing regularly for whiting on fishing grounds not very far from the harbor. This last option gives a meager income. Mehmet,[6] a middle-aged fisher from Çarşıbaşı, in an interview held on September 2008, described the following:

> With his brothers he bought, in 1995, a 12 m sea snail boat with the aid (credit) of the sea snail factory owner. They dived and dredged for sea snails. The fishery was good and in three and a half months they repaid the loan they made to buy the boat. "There are people here," he tells us, "who have made three, four, five storied buildings from sea snail catches. Fishers' standard of living increased somewhat."
>
> In 2005, after it was clear that the sea snail fishery definitely had busted, they sold the boat and bought a 7.5 m boat. For this, he borrowed 8,000 TL (approx. 5,000 USD). To show how much more profitable sea snail fishing was, he stresses that this far he has not been able to repay anything on this new loan. They simply do not earn enough.
>
> Small boat fishers usually keep no records of their economy, and it is therefore difficult to gain knowledge about their income. Mehmet helped us make a calculation of their income during one and a half month from early September to mid-October 2007. When all expenses are subtracted, each person made a profit of 411 TL during one and a half months, which equals 215 USD/month.[7] Mehmet concludes that it is now impossible to

[6]All personal names of informants referred to in this chapter are pseudonyms.

[7]During this period, he and his brother continuously set nets for whiting and delivered 31 landings to the auction in Trabzon. For the total of 412 k caught, they were paid 1,590 TL. Expenses: Fuel – 340, commission at auction – 92, transport to harbor and to auction –185, and wear and tear of nets – 150. One fishing trip lasts for approximately 3 h. In addition, comes travel time to the auction in Trabzon. This calculation has not taken into consideration investment and repair of the boat. It is worth noticing that, during this period, fish prices were relatively high since trawl fisheries had not yet started and because the Ramadan fell within this period (makes for higher demand of fish).

make a living from small boat fishing alone – you will have to have other incomes as well. All able fishers, those who used to depend on sea snails, now work as crew, even middle aged men. His youngest brother has started working as crew again. Mehmet himself planned, for the first time ever, to sign on a purse seiner. Like most other people in Çarşıbaşı, Mehmet's and his brothers' households own gardens and orchards which make them self-sufficient in vegetables, corn and potato, and some keep chickens and a cow or two.

11.4.2 Terme

11.4.2.1 Population and Economy of Samsun and Terme

The economy of the Province of Samsun is dominated by agriculture, processing of agricultural produce, and agricultural trade through the city's large port. In 2000, 63.4% of the population in the Province of Samsun (as against 34.5% in Turkey overall) had its main income from agriculture (SABEK 2005). However, relative incomes and employment in agriculture have decreased and industrial employment has declined (from 34,500 in 1990 to 32,500 in 2000 (GoS 2009). This has resulted in significant out-migration; and Samsun is actually among the provinces in Turkey with the highest rates of net negative migration (about 45‰ over the 5-year period since 1995 (TURKSTAT 2004)).

The District of Terme is, in many respects, representative of the province. It is situated on the coastal side of the large Yeşilırmak delta and consists mostly of high-quality,[8] flat agricultural land (86%). Most of the 80,000 inhabitants of the district population (80.29% in 2000) receive their main income from agriculture, making possible a relatively high population density (approximately 190/km^2). Like in Samsun overall, however, agricultural, husbandry, forestry, and fishery employment in the District of Terme has decreased, from 39,500 in 1990 to 38,000 in 2000 (TURKSTAT 2009). This has been accompanied by out-migration resulting in a decline in population (– 2.72‰ annually in the District of Terme, over the 1990–2000 period (SPO 2004a)). Although there are more than 20 factories in Terme (Termehaber 2010), these are small and only operate during a short season when they process agricultural produce (mainly hazelnuts and rice).

The main produce in Terme is hazelnuts, with hazelnut groves covering almost half (24,000 ha) of the total area of the district (Termehaber). The average agricultural net income of hazelnut farmers was 15,627 TL in 2005, of which hazelnuts contributed 85% (Alkan and Kiliç 2007). Income from hazelnuts is seldom enough to support a household. Yet, the average farmer cannot be considered very poor. Those lacking farmland are more prone to be poor. Like in Çarşıbaşı, many without sufficient land for making a living have become fishers.

[8] Category 1–2 of the total of six categories in Turkish authorities' categorization of agricultural lands, primarily based upon soil quality, degree of inclination, and exposure to erosion (SABEK 2005).

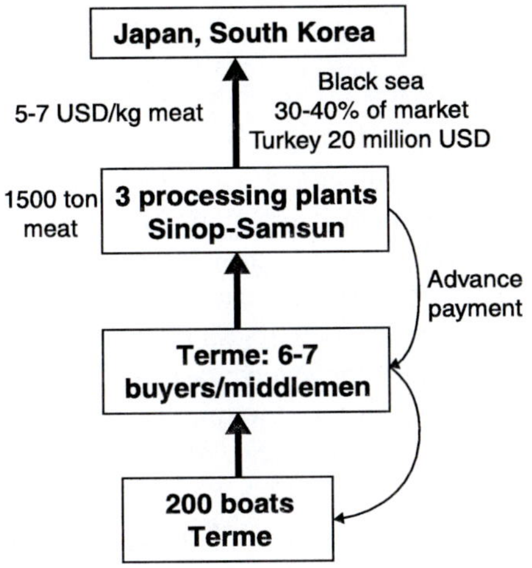

Fig. 11.3 Sea snail market chain, from Terme to East Asia. Note that the spatial scale expands from box 2 through 3 to 4 from bottom up. Factory owners usually obtain the sea snails through middlemen who do the actual collection and transport of the sea snails. From the factories, middlemen obtain advance payment which they forward to fishers. The three factories in Samsun produce probably more than half of the total Turkish production, almost all of which is exported to East Asia. Sea snail catches in the Black Sea stand for 30–40% of the world market in sea snails (*Rapana venosa*), and bring 20 million USD in annual export revenue to Turkey

11.4.2.2 Fisheries in Terme

Unlike Çarşıbaşı, fishing has no long history in Terme. Until the early 1990s, there were no large boats in Terme, but it was fairly common to work as crew on trawlers in Samsun or on purse seiners in Istanbul. There were also close to 150 small boats (up to 7–8 m) in the district in 1990 (Knudsen and Zengin 2006). During the early 1990s, the large fishing harbor in Terme was constructed, and it became a feasible strategy for fishers in Terme to invest in larger boats, since they were no longer dependent on landing their boats on the beaches or maneuvering them into shallow rivers. During the 1990s and early 2000s, numbers of boats increased, and grew to more than 400 in 2005, including 27 trawlers (typically 15–20 m).

Hundreds of boats in Terme dredge for sea snails. While sea snails are typically pursued by medium sized multi-purpose boats (8–12 m), also larger boats are active in this fishery during its main summer season, when trawling is illegal. The other way around, the smaller boats often trawl during winters (although illegal for boats under 12 m). All boats are family owned and operated, and few families own more than one boat. The population involved in fishing mostly lives in rural or semi-rural coastal areas of the district. Sea snails are delivered to factories by the aid of local middlemen who also take care of the economic transactions with the fishers (Fig. 11.3).

11.4.2.3 Poverty in Terme

We have studied four neighboring communities in Terme, all of which are relatively poor compared to the average of Terme, Samsun, and Turkey. One community is Kurdish, one Alewite, one Romani. We will here focus primarily on the more important of these: Yalı Mahallesi, a Sunni-Turk community within walking distance to the harbor, where most of the growth in number of boats has taken place. All trawlers and most small and middle-sized boats in Terme are based in this community. Yet, our questionnaire survey includes also some fishers in the neighboring communities. Our findings concerning poverty here conform roughly with a comprehensive study of social structure in the middle Black Sea region which found that: (1) the rural poor typically have little or no land, struggle to meet basic needs, receive social assistance, and live in poor houses; and (2) the most destitute members of the rural population typically posses no land, work as seasonal wage workers, and have problems meeting basic needs (SPO 2004b, pp. 2–124).

We surveyed poverty in this community through a discussion of income and capabilities such as education, social security, and health. The welfare system in Turkey is relatively poorly developed. Therefore, loss of land, health, or life usually leads directly to poverty. "Idiosyncratic shocks such as major illnesses, were the least prepared for and most difficult for the poor to handle, after the daily task of feeding the family was met" (World Bank 2003, p. i). This is well exemplified by the following life story which a woman in Terme told us during an informal interview in April 2009. This story will form a baseline for the following discussion.

> Hava Akçay was born to a poor family in Terme in 1968. At the age of 17, she married a man from Yalı Mahallesi who at the time of marriage was a baker. He started to work as crew on purse seiners, and later bought a small boat together with a partner. After a while, he was able to buy out his partner and, to earn more, ventured into the more risky work of dredging sea snails alone. The intention was to earn money to build a house. Then, one day in 1999, he had an accident - the boat sank and he drowned. Hava got no compensation – neither from the loss of her husband, nor for the boat. They had no insurance whatsoever.
>
> To get by and feed her three children (another child had died young), she worked seasonally in a hazelnut factory and picked hazelnuts. Now, she has had to stop working to look after her mother in-law who was paralyzed after her son died. She has a green card that entitles her to basic health care services, and she receives social assistance totaling 75 TL/month for children's schooling, for food, and for coal. This is now her only income. Her 16-year-old son, who is living with Hava's mother in law, is an itinerant *simit* (sesame "bagel," a kind of fast-food) seller in the Terme township. The house construction that her husband had started was completed by the help of neighbours. She complains that she is not able to travel outside of Terme.

Income

In 2009, a self-employed garbage collector in Samsun earned 25–30 TL/day (Samsun Halk Gazetesi (local newspaper), 30 December 2009), which is probably a good estimate for the lowest daily incomes in the informal economy in this region.

Thus, in terms of income, with 0.83 TL/day/person (if we count Hava and two of her children), Hava's household is not only far below the official absolute poverty line (2.15 USD/3.3 TL/day/person), but also below the official absolute starvation line (1 USD/1.53 TL/day/person). 35.4% of the individuals in Yalı Mahallesi, and 22.6% in the District of Terme, had green cards in 2009 (as against 12% in Turkey overall and 15.7% in the Province of Samsun); and 25% of households in Yalı Mahallesi received social assistance in 2009.

Beyond this, there exist no statistics of income or poverty in Terme, but the decreasing profitability and income from agriculture clearly indicates that Terme is relatively poor, and increasingly so. People have been, and still are, looking for alternatives to hazelnuts. For the coastal population, fishing has been one of the obvious alternatives for those with few entitlements (capital, land), and deprived of capabilities such as higher education. Hava's husband is one example of those who pursued sea snail fishing as a strategy to get out of poverty. With the accident and his death, however, his household was thrown into more abject poverty than before.

From our conversations and questionnaire survey, we can conclude that fishers (excluding trawl owners) typically earn between 4,000 and 10,000 TL annually, plus sometimes income from other sectors. This means that most are well below the official poverty line, but above the food poverty line. Fishers are thus poor, but overall perhaps not much poorer than the population in Terme that primarily rely on agricultural production and wage work for their income. It should also be taken into consideration that, like in Çarşıbaşı, most households own their own house and have some vegetable subsistence production.

Fishing is only one of many options, and people readily, yet often by necessity, switch between occupations and combine various income-generating activities.

> Ali[9], a man in his early forties, grew up in a remote village in Fatsa (in the Province of Ordu). His father worked in construction, and they moved to Yalı Mahallesi when Ali was young to make travel to seasonal work in Istanbul easier. Ali himself has pursued a range of different occupations, including: long haul truck/lorry-driver, construction work in Moscow, transport and sale of fish using own car, many years crew on purse seiner, three years settled on the Aegean coast where he worked as crew, sea-snail fishing from his own 8-9 m boat, crew organizer/middleman.[10] He owns only 2 decares of land, half of it now planted with poplar seedlings for sale, and four cows for producing milk for household consumption and sale. They produce vegetables from their own garden, and his 15-year-old daughter has worked seasonally in a hazelnut factory.

Not all have tried their hands at so many occupations as Ali, but it is evident that success as a fisher not only depends on success in the fishery itself. Boat economy and household economy are intimately intertwined. Moreover, economies of households of relatives are often overlapping (Ali's parents' household is separate from his). When first Ali's father and then his mother, who lives in a separate house, became

[9] Story extracted from conversations during participant observation in Terme, April 2009.

[10] "Kocareis," usually middle-aged men mobilising men from own village/region to work as crew on purse seiners in Istanbul. They are remunerated with agreed-upon shares from the boat's profit.

Table 11.2 Fishers' insurance – Province of Samsun (2005) and District of Terme (2009). The percentage of fishers insured with Bağ-Kur has increased from 22/24 in 2005 to 31 in 2009. There are significantly fewer fishers in Terme (3–6%) than in Samsun overall (17%) insured with SSK, yet many more fishers in Terme than in Samsun overall having green cards (26% in Samsun 2005, 49% and 58% in Terme in 2005 and 2009, respectively). Almost none are insured with the pensioners' chest, while the proportion without any social security has decreased significantly – in Terme from 24% in 2005, to 6% in 2009

Percentage of fishers insured	Province of Samsun 2005 (*N*=345)	Terme 2005 (*N*=59)	Terme 2009 (*N*=36)
Bağ-Kur	22	24	31
SSK	17	3	6
Green Card	26	49	58
Pensioners Chest	2	0	0
None	32	24	6

seriously ill and needed expensive treatment, he had to sell his boat and also take a loan, postponing the construction of a new house and taking up work as crew/*kocareis* again.

Fishers' income is typically variable and unpredictable. This makes it difficult to plan activities, and households often live through very stressful periods, months on end without any significant income. In such a situation, it is difficult to pay monthly installments on insurances or school fees. When the occasional high income occurs, it is typically spent on constructing part of a house (waiting for the next bumper income for the next phase) or upgrading boat and gear.

Social Security and Health

Hava's and Ali's stories bring out the striking importance of life, health, and the issue of social security for peoples' welfare. While approximately 90% of Turks have insurance (either directly or through close family members) with Bağ-Kur, SSK, or a private insurance company, only 27–37% of fishers in Terme have this kind of insurance (Table 11.2). Most fishers in Terme insured with Bağ-Kur are owners of trawlers. Moreover, the percentage of Terme fishers having a green card is above the overall average for Terme.[11]

In our conversations with fishers, social security clearly emerged as one of their major concerns, and one of the primary complaints they had about the fishery sector, especially the work as crew. One fisher succinctly summarized their thinking: "If we were able to secure a job that guarantees social security and comes with a monthly income of 800–900 TL (600 USD/month), we would not work as crew." People prefer the security of steady jobs, even though, for some, work as crew may at times bring higher monthly incomes.

[11]While the percentage of fishers having green cards does not seem to have decreased, the percentage of green card holders among the overall population went down from 2006 to 2009 in the Province of Samsun from 26% to 16%, and in the District of Terme from 36% to 23% (MHT 2009).

Table 11.3 Comparison of fishers' educational level, all figures in percentages. In Terme, the larger share of fishers only have primary education (or less) (94% in 2009, 86% in 2005) and only about 2% have completed high school. With 13% having completed secondary and high school respectively, fishers in Samsun overall (2005) are more highly educated than fishers in Terme

	5 years primary (or less)	Secondary (8 years)	High school (11 years)
Terme 2009 (*N*=36)	94	3	3
Terme 2005 (*N*=59)	86	12	2
Samsun 2005 (*N*=342)	74	13	13

We had problems getting sound official information about health issues. Overall, people find that health services have improved considerably over the last few decades. On the other hand, half the fishers report health problems stemming from the practice as fishers, and 6 out of 36 report having experienced work accidents. Approximately half of the fisher wives we interviewed report that their husbands have serious health problems, e.g., kidney and liver problems, rheumatism, back problems, and bronchial asthma. In addition to Hava Akçay's husband, we were also told about several other fishers who had lost their lives at sea.

Education

Educational level is certainly low in Terme.[12] Most fishers in Terme only have primary education (5 years), some even less (Table 11.3). The reason for the relatively high number of fishers with more than primary education in many Samsun communities is that men pensioned from salaried work (typically civil servants/teachers) supplement income by fishing. This class of people, often insured in the civil servant's SSK pension system, seems to be almost completely lacking in Terme.

The main point of the newspaper article about the garbage collector was that with his meager income he had, amazingly, been able to educate his four children – some of them had completed higher education. Household strategies are very much focused on this: Make sure that your offspring get a proper education, and give at least one of your children the opportunity to enter the university entrance exam. This will greatly improve the capabilities of the individual and, ideally, the rest of the family. Few fisher households succeed, however. With relatives already in the business, and seeing the occasional large shares earned as crew, many young males are tempted to drop out of school and start fishing, earning their own money already from the age of 13 or 14.

[12]In 2000, only 83.6% (SPO 2004a) were literate (national average in 2000 was 87%). A local newspaper article stressed that the high schools in Terme are particularly unsuccessful in getting their students into university (only 26 out of 1127 graduates over the period of 2005–2007 (Samsun Halk Gazetesi, 10 March 2008)).

Housing

Although housing generally holds a fairly good standard, there are variations within and between communities. Housing is especially poor among the 40–50 households in a small Romani community neighboring Yalı Mahallesi. Entitlements in this community are particularly restricted: No property (no title deeds even to their houses/land); no agricultural land, not even for subsistence production; a high degree of indebtedness; all have green cards and receive social assistance; and most women have no education. They rely primarily on irregular, often migratory, wage work for income: In agriculture; in factories in Istanbul; on purse seiners in Istanbul; or from picking blood leeches and frogs; but have recently also pursued some sea snail fishing from their own boats.

Until recently, they lived in tents. Most have constructed simple brick houses during the last few years. These small houses contain only one or two rooms, but no separate kitchen, no shower, and some do not have a toilet. Most use a few toilets set up on poles above the river. There is almost none of the furniture that you would find in most Turkish homes (such as carpets, chairs, tables, beds). By Turkish standards, these houses are clearly considered houses of the very poorest.

Summary of Poverty in Terme

In all four communities, but especially in Yalı Mahallesi, poverty has been an important driver for sea snail fishing. But poverty is not similarly distributed in the fisher communities in Terme. They display different poverty profiles. The Romani are, as a group, definitely the poorest. Entitlements are differently distributed between the communities, especially land. Involvement in fishing is, to a large extent, related to this, as is proximity to the harbor. The discussion above also shows that unpredictability in itself can hamper capabilities development. Thus, security and predictability – in the form of entitlements to health care or pensions, or regular income or ownership of the house you live in – often constitute capabilities that enable people to live the kind of life they desire.

Local Perceptions of Poverty

Above, we have tried to measure and compare poverty primarily on the basis of indicators that make sense relative to Turkish standards and Turkish society at large. Many of these issues are important also in people's own perceptions of poverty, but both in Terme and in Çarşıbaşı, some additional topics are added in their assessments of what poverty entails or what signifies poverty (or riches) (Table 11.4).

In our questionnaire survey, we asked people how they considered their own situation. Roughly, equal numbers of respondents considered themselves middle class and poor. Very few thought of themselves as being very poor. Not even all the

Table 11.4 Local perceptions of poverty and riches. This table summarizes some current and prevalent ideas about what it means to be, or what are indicators of being poor and rich, respectively. Information was gained during informal interviews and conversations

Poverty if	Not poverty if
Old and/or rented house	New, large own house
Being dependent upon usurers	Large hazelnut fields
"Green card"	Possess social security rights
Selling gold bracelets (for women)	(SSK, BağKur)
Early death	
Work as crew	Own trawler
Perform dirty work	Clean, new clothes
Female menial wage, work in fishing	
Being away from family for longer periods: Seasonal, migratory work, including work as crew	
Unable to educate children	Send children to private school, university
Little opportunity for social visiting	Car ownership
Not having enough food: "Our kids do not know what breakfast is"	
Being Romani	

Romani considered themselves poor. Those Romani thinking themselves not poor are definitely poor according to indicators of income, social security, education etc., but they may experience that they are better off than many others in their own community. Clearly, what people find to be a rich and fulfilling life, and what kinds of wants constitute poverty (capability deprivations), is to a large extent culturally constructed.

Turning the Corner

In Yalı Mahallesi, sea snail fishing has been important in increasing welfare and passing the threshold from poor fisher to owner of trawl vessel. What does it take to make this shift and the increase in wealth that comes with it?

> Yakup Korkmaz[13], in Yalı Mahallesi, has managed to improve the welfare of his household though fishing and is on the brink of passing into a situation where he is no longer dependent upon sea snail fishing. His father came from an inland mountain village in Ordu in the 1950 s, and most of his many brothers are fishers. Yakup, now in his late 30 s, says he has been a fisher all his life. They own only a small plot (1 da) of agricultural land on which they grow vegetables for their own use. They have no other property, no hazelnut groves.

[13] Yakup Korkmaz' story is based on informal conversation and response to questionnaire, Terme April 2009.

Working hard as a fisher during the 1990 s and relying especially on sea snail dredging, he earned well and was able to construct a two-story, six-room house for his family in 2000. He bought his first small boat in 1997, and his current 13 m boat in 2004. A few years ago, he upgraded the engine from 140 to 185 Hp. On his new boat, he not only has an echo-sounder, but also a GPS which is particularly useful during trawling. He took a 5000 TL loan from an acquaintance for this, but otherwise has no loan. His boat is typically multi-purpose, and he fishes with nets for a variety of species. But his main income is from dredging during summer, and bottom trawling during winter. Despite earning quite well from the fishing (15,000 TL in 2007, which is in the same range as the average hazelnut farmer), he and his wife sometimes engage in seasonal agricultural wage work.

Unlike many other fishers, he is able to send his two children to the expensive private tutorial *dershane*. His 16-year-old daughter is going to high school and preparing for the university entry exam. Another indication of their rise out of poverty is the fact that, while he in 2005 had a green card, he is now able to pay installments in the public Bağ-Kur health insurance scheme. His household receives no social assistance. He considers himself to be middle class (*orta halli*). He plans to buy a larger boat.

Yakup Korkmaz is typical of fishers owning boats larger than 12 m (which is the limit for having a trawl license). Of six fishers in our questionnaire survey who answered to owning boats larger than 12 m, only one had a green card (3 Bag-Kur, 2 SSK). Also, several of those owning larger trawlers (around 20 m) own a pickup car or small lorry which they primarily use for business purposes. Moreover, all fishers in this class are able to send their offspring to *dershane*, and all consider themselves to be middle class.

Yalı Mahallesi stands out as an interesting case of a community that sees population rising, but is still relatively poor. While the population of the district decreased by 6.2% between 2000 and 2008, the population of Yalı Mahallesi increased by 15.2% to 1,716 during the same period. The community is poor partly because people moving into the community tend to come from the lower echelons of agricultural society, especially from inland rural areas of the neighboring province of Ordu. People have moved there, and continue to come, because living in the community opens up some opportunities, potentially provides some capabilities, not available in remote hazelnut growing villages: (previously) Free or cheap land, proximity to major transport routes and the district centre; and the possibility of becoming a boat owner and being self-employed.

Poverty has clearly been a driver for sea snail fishing which has, overall, increased wealth in the community. As one elderly man in Yali Mahallesi commented to Yakup Korkmaz and his brothers:

> It was the sea snails that formed the start and basis for this. It was sea snails that "made them men" (*salyangoz adam etti*).

The relation of small-scale fishing to poverty is not static, and small-scale fishing does not necessarily "rhyme with poverty" (cf. Béné 2003). For some, success in sea snail fishing has also brought them out of a sea snail dependent adaptation; they have passed a threshold and are not so vulnerable to the potential vicious circle development in the sea snail sector. One of the attractions of owning (or working on) a trawler is that the men mostly stay at home. And, they are not being "bossed around." So, being able to invest in a trawler not only means higher income, but also brings other capabilities which are not easily measured by universal income indicators.

11.4.2.4 Recent Developments in Terme Sea Snail Fishing

Fishers in Terme report that, since approximately 2005, sea snail fishing has started to loose some of its attraction and profitability. Most importantly, there are indications of average sea snail size decreasing also in Samsun. Sea snails in Samsun seem to regenerate very quickly, and mean size of sea snails remained stable until 2005 despite increased catch effort (Knudsen et al. 2010). There exists no scientific data on recent developments of sea snail average size in Samsun, but there is some anecdotal information. Fishers in Terme complained about the gradual, although geographically uneven, decline of sea snail mean size. They increasingly find themselves shovelling undersized sea snails back to sea.

Catches are usually sorted into three categories for which fishers are paid differentiated prices: Small – 0.25 TL/kg; medium – 0.60 TL/kg; and large – 1.5 TL/kg. With more and more of the catch consisting of small sea snails, it becomes increasingly difficult to cover fuel costs, even though total catches in weight do not decrease.

Owners/managers of the three largest sea snail processing plants in Samsun complained during the autumn of 2008 about the increasing difficulty of finding buyers for their produce since average size had declined considerably during the past 2–3 years.[14] They also reported that they were experiencing the effect of the economic crisis with demand from South Korea falling.[15] During the 2007 and 2008 seasons, factories did not pay out advances any longer.

Sea snail fishing is also made less attractive by tightened inspection. This was a very common worry, and fishers find that whereas their livelihoods and the accumulated capital and skills that come with it were made by sea snail fishing, they are now made into lawbreakers when they pursue the same activity that earlier was tolerated, although formally illegal.

11.5 Coping with Bust

How are fishers in Terme coping with this new situation? We see some of the same coping strategies as in Çarşıbaşı, with decline in sea snail fishing (Fig. 11.4): sale of boat and a turn to work as crew on purse seiners. Although the exemption from fuel tax which has been in place since 2004 has eased the situation, the combined effect of sharpened inspection and reduced size of sea snails has made many reconsider

[14] In one processing plant, more than 50% of the processed sea snails were in the smallest of the seven size categories. Processed sea snails are sorted into the following size categories: 3L (Extra Large): 1–15 pcs/kg; LL (Extra Large): 15–20 pcs/kg; L (Large): 20–40 pcs/kg; M (Medium): 40–60 pcs/kg; S (Small): 60–80 pcs/kg; SS (Small): 80–120 pcs/kg; 3S (Small): 120 – up pcs/kg (Fora Food webpages: http://www.forafood.com/topshell.htm, accessed 08 Dec 2010).

[15] While the largest of the sea snail factories exported 988 t in 2007, export in 2008 had fallen to 422 t (Samsun Halk Gazetesi, 13 April 2009).

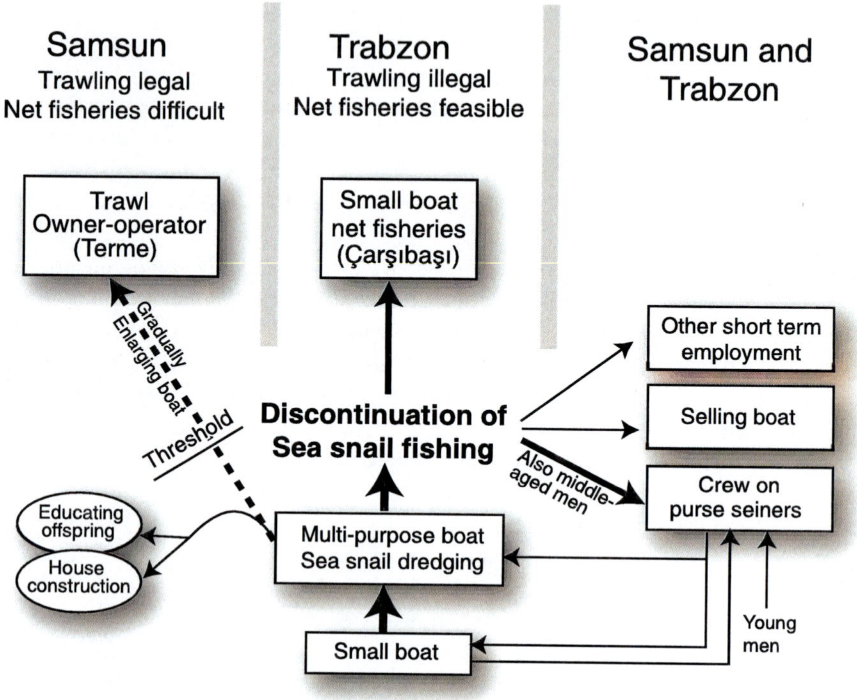

Fig. 11.4 Major career paths and coping strategies with bust and boom. This figure shows that ownership of multi-purpose boats has usually been gained through investing income from work as crew and/or on own small boat. In both Trabzon and Samsun, fishers would spend increased income on house construction and on the education of offspring. During a boom situation, some fishers in Samsun passed a threshold and became owners of larger trawl vessels. This has been a less realistic option in Trabzon where trawling is illegal. With discontinuation of sea snail fishing ("bust"), former sea snail fishers in Trabzon could switch to small boat net fisheries which is less feasible in Samsun where fishers are forced to seek other options

sea snail fishing. Many boats are for sale,[16] and fishers increasingly travel to Istanbul and the Sea of Marmara to dredge for sea snails there. Fishers in all of these communities now rely more on work as crew. Like fishers in Çarşıbaşı, fishers in Terme reduce consumption, including spending on social security. The trend toward a shift to wage labor, and reduction in consumption is likely a common tactic among self-employed households in Turkey, when faced with reduced income (see e.g., Özbudun and Başokçu 2004).

Whereas fishers in Çarşıbaşı could downsize and rely more on net fisheries, this option is less viable in Terme since the extensive trawling on the fishing grounds

[16]The Kurds in the Çangallar village owned 6 boats in this 12 m-p boat class, but sold them all in 2008. All Romani sea snail fishers were trying to sell out in 2009. The number of multi-purpose boats in Yalı Mahallesi decreased by 15, from 2005 to 2009.

makes use of set nets, especially bottom nets, difficult. Additionally, there can clearly also be cultural dimensions to coping strategies, household organization, and life prospects/chances which impact on poverty. All four fisher communities in Terme retain a strong sense of ethnic identity and rarely marry outside of it. Therefore, realistic coping strategies are, to a large extent, limited to what is possible to attain within the confines of the ethnically defined life-styles and social networks.

11.6 The Vicious Circle Revisited

Our discussion of coping strategies indicates that issues outside of those most immediately associated with the Vicious Circle model, and outside of fisheries management as well, should be considered when exploring interactions between poverty and small-scale fishing. Below, we discuss more issues at the "fringe" of the model, summarized here in Fig. 11.5.

11.6.1 Management and Collective Action

Management rules and regulations during the 1980s and 1990s have been informed by two scientific positions: (1) Those considering the sea snail as any other marine resource and worrying primarily about overfishing; dredging's harmful effects on habitat; the dredging facilitating illegal trawling; and (2) Those concerned about the sea snail being an introduced invasive top predator; worrying about its effect on the ecosystem; and wanting to control the stocks. Management has therefore been caught between different concerns.

Actual regulations have largely been a compromise between the two positions, perhaps with the "marine resource" perspective having an upper hand. In practice, lenience on the management side, insufficient resources for control and inspection (until approx. 2005[17]), and political influence of owners of sea snail processing plants, have resulted in the various restrictions put on sea snail fishing not being adhered to (Knudsen and Zengin 2006). In the Province of Samsun, 154 boats had licenses for sea snail fishing in 2005; but in practice, more than 400 boats were actually involved in this fishery in this province. Although illegal, most boats use two (or even three) dredges simultaneously, and often operate during nights (also illegal). Although dredging is illegal during summer, this fishery is most intense

[17] The tightening of inspection seems to be partly a result of EU-pressure in the process of adapting Turkish fishery policy to the Common Fishery Policy in the EU. In practice, policing authority was transferred from the Ministry of Agriculture and Rural Affairs' local Protection and Control branches to the Coast Guard under military command. They are better equipped and possibly less corrupt.

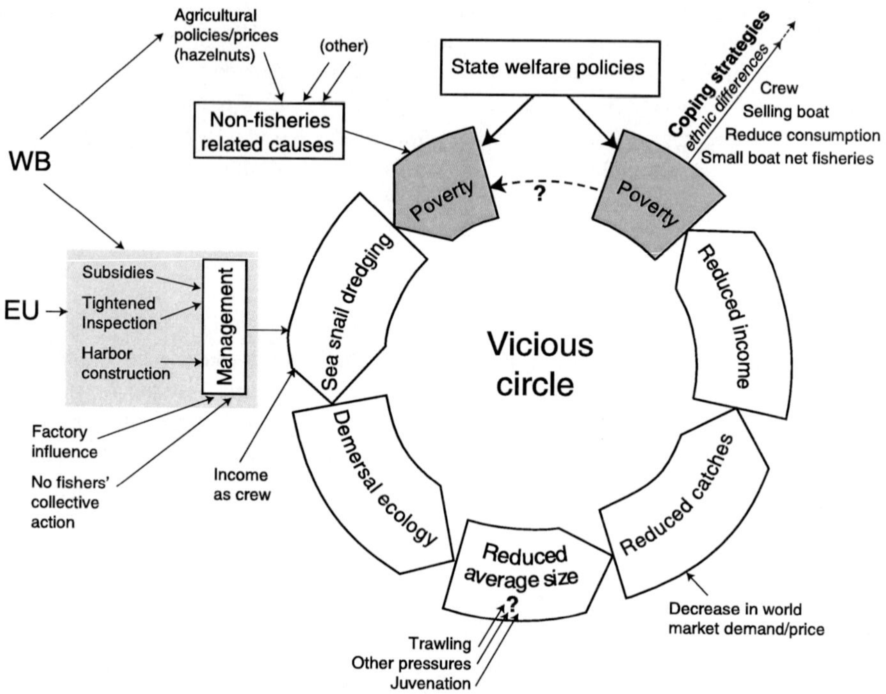

Fig. 11.5 Expanded vicious circle model, in a "bust" situation. The vicious circle thesis predicts that poverty will lead to increased fishing pressure, resulting in deterioration of the resource base and diminishing catches and incomes for fishers, thus resulting in more poverty. The main circle in this figure, starting from the *left* poverty box depicts this dynamic. However, the figure also demonstrates the many other factors that impacts on poverty, fishing activities and the resource base and which taken together distorts the logic of the vicious circle. Many of the processes are seen to be of a very large, sometimes global, scale. There are also many uncertainties involved: what causes the gradual decline in average size of sea snails? What is the effect of state welfare policies on poverty itself and on poverty as a driver for fishing effort? The review of coping strategies (*upper right*) shows that poverty does not necessarily lead back to increased fishing pressure

during the warm months when catches are best. In summary, most sea snail fishing has been illegal.

Yalı Mahallesi has been and is a land and coast of risk and opportunity, where people are willing to take chances. Many have "turned the corner." Yet, the community seems to be little incorporated, and many of the activities are informal, unregulated, and in the eyes of state representatives, often illegal. Terme fishers are known among Black Sea fishers and managers as being "Texas"; as being a region where regulation and law are constantly and routinely ignored. One fisher in Yalı Mahallesi typically summarized this as, "we have no relations with the state (*devletle ilişkimiz yok*)." Illegalities include below 12 m boats trawling, breaking regularly most regulations in sea snail fishing, and, most notoriously as seen from the outside, trawling in the closed area (border just east of Terme). From the perspective of state representatives, Terme fishers are criminalized. The same approach was seen in Trabzon during the 1990s, when fishers from Çarşıbaşı became infamous for illegal trawling.

Many of these illegalities are pursued in spaces/times outside of the reach of the state apparatus. However, some of the illegal activities, particularly summer time dredging and closed area trawling, are more conspicuous and therefore require some kind of protection. It is well known that some trawlers in Terme are protected by a group of powerful fish traders in Samsun with good political connections. They are regularly called "patrons" (*himaye*), but are sometimes spoken of as "mafia" or "government" (*hükümet*), i.e., those actually ruling the area. Illegal catches are channelled through them; they protect fishers against inspection or prosecution, and also provide fishers with goods and services such as cars, telephones etc.

While the sea snail sector has attracted many fishers, and it has at times had the character of a "frontier" situation where everybody rushes to exploit the resource as fast as possible, fishers have, to a certain extent, informally imposed various kinds of rules on the practice of sea snail fishing. These are basically of two kinds: (1) *Protection of "home territory"*: During the late 1990s, communities increasingly strove to restrict sea snail dredging in their "home" waters. (2) *Restrictions on the use of dredges*: Fishers in some small boat fishing communities have been less well equipped for sea snail dredging; and, partly in cooperation with State supervising bodies, opposed it because they believed it to be harmful (see Knudsen 2009). In some cases, such collective action was organized by the local fishery cooperative. Where this took place, the fishing communities were characterized by being homogenous small boat fisher communities with a strong presence of men with higher education and/or professional training, typically pensioned teachers and civil servants.

Like most fishery cooperatives in Turkey (Knudsen 1998; Knudsen 2009, Chap. 9), the cooperatives in Çarşıbaşı and Terme did not play any active role in implementing informal regulations. Fishers in these communities seem not to have tried to impose informal regulations of this kind. Neither do the four different communities in Terme each organize collective action based on their shared identity. Fishing grounds are shared between them (and fishers from other communities) in a wider region stretching from the mouth of the river Yeşilırmak to Ünye. Collective action that has taken place was organized outside of the cooperative, and was directed at protecting fishers against inspection. Fishers and the factory in Çarşıbaşı organized a patrol to warn fishers about approaching inspection vessels. The presence of owners of purse seiners or trawlers, the bonding to patrons, and the lack of resourceful educated men among them may have made the kind of collective action seen in some other communities more difficult in Terme and Çarşıbaşı.

11.6.2 State Policies

The Terme case study demonstrates that the character and quality of state welfare policies have significant impact on poverty and welfare. Until the 1980s, poverty was not considered a perceptible problem in Turkey, and "modern" poverty alleviation policies were not developed. Poverty was supposedly managed within a "traditional" welfare system (family, community, patron-client relations). With the accession to power of the AKP – a moderate Islamic party – after the serious economic crisis in 2001, poverty came onto the public agenda as never before. However, leaning on its conservative moral position, the AKP regime favors a family-centred solidarity and "Islamic style" philanthropy and charities (often organized by municipalities) to address issues of poverty and destitution which it regards to be "beyond the reach of [the] state's responsibilities" (Buğra and Keyder 2005, p. 31). With fundamental changes to health and social security policies, citizens' social rights (entitlements) were weakened. Thus, state welfare policies are one of the major causes for the character of poverty seen in Terme and in Çarşıbaşı. The same argument can be made for educational policies. On the other hand, "[c]oping mechanisms of the poor, especially relying on relatives and neighbours for in-kind and cash assistance, came under stress, and the poor reported a decrease in assistance from these traditional channels" (World Bank 2003, p. i). Almost none of the respondents to our questionnaire survey in Terme reported receiving any assistance from more wealthy relatives, although many had close relatives in Western Europe and in Istanbul.

11.6.3 Agricultural Policies

We have identified poverty as a major driver for the expansion in sea snail fishing. Underlying this poverty are changes in agricultural economy and policies. Turkey contributes 70% of the world production of hazelnuts, and the eastern Black Sea coast is the major production region. In 1999, Turkey adopted a World Bank supported structural adjustment and stabilization program, which from 2001 included the "Agricultural Reform Implementation Project" (Eruygur 2006, p. 18). One major objective and effect of the project was the phasing out of agricultural subsidies and the restructuring (privatization) of state-controlled sales cooperatives through which subsidies were directed.

Although the state has been unable to follow WB advice to totally discontinue price support, support is smaller and less predictable than before. This has resulted in hazelnut production in Çarşıbaşı and Terme, although still a security, being less attractive than before. Partly as a result of this, Terme fell dramatically on the socio-economic development ranking of districts in Turkey.[18] The relative decrease in income from the hazelnut sector is aggravated by the general diminishing value of and income from agricultural production in Turkey compared to other sectors.[19] In general, people engaged in agriculture earn much less than in other sectors, and have the lowest degree of social security coverage (9%).

[18]From place 240 in 1996 to 436 in 2004 (out of more than 872 districts in Turkey) (SABEK 2005).

[19]Share of agriculture in Turkish GDP fell from 17% in 1990 to 12% in 2005, and share of economically active population engaged in agriculture has decreased from 47% in 1990 to 27% in 2006 (Çakmak and Eruygur 2008, p. 174).

11.6.4 Why Ecological Change?

Did the boom in sea snail fisheries result in overfishing and the resulting bust? Many assume that extensive dredging is the reason for the decline in average size of sea snails, but there is no research to back up this assumption. There are actually very few studies undertaken and many unknowns. The nature of the sea snail stock, the role in the ecosystem, and the reason for the decrease in sea snail average size are not well known. It is unclear what effect sea snail predation has on mussels, what the impact of dredging is on habitat and other species, and what effect other pressures, such as trawling, have on the same habitat.

It appears that the same process toward smaller average size is taking place in most of the Black Sea region, but researchers find there is yet too little evidence to conclude whether it is juvenation,[20] or fishing pressure (or some other factor or combination) that is/are responsible for the problem. Whereas large specimens of sea snails are increasingly hard to come about, fishers and scientists alike report that there are plenty of sea snails on the sea bottom. They have no problems reproducing, and there seems to be no dramatic decline in biomass. This could indicate that, at least, fishing too many under-sized (i.e., not of reproductive age) sea snails is not the problem.

Thus, it is important not to rush to conclusions about the damaging effects of small boats' fishing pressure without evidence to substantiate it. It is often the case that causes other than small boat fishing pressure (e.g., big boat fishing) exerts more pressure on the stocks and habitat in question. This is not to say that dredging and small-scale trawling is not harmful, which fishers themselves readily accept.[21]

11.7 Conclusions

The findings of this study show that a range of variables not specified by the vicious circle thesis can have significant impact on resource use and poverty. Much of this is summarized in Fig. 11.5. Additionally, there are some issues of relevance for poverty in our case studies that are not accounted for in this figure.

It is wise not to too readily accept the implicit assumption of Malthusian overfishing implied in the vicious circle model (Pauly 1994). There has been a dramatic degradation of the resource (reduced average size), but it is not at all clear that this is caused by overfishing (and thus part of a vicious circle). It is, therefore, not possible to provide evidence for a vicious circle dynamic operating in this case.

[20] Internal dynamic toward younger age at reproduction, and smaller average size typical of introduced invasive species.

[21] After sea snail fisheries (and associated illegal trawling) discontinued in Trabzon, fishers report during conversations with us in 2008 and 2009 that catches of red mullet are again very good, after having been very meager during the 1990s and early 2000s.

Secondly, small-scale sea snail fishing does not always "rhyme" with poverty. Many fishers have moved out of destitution by engaging in sea snail fishing; and some have even built careers on sea snails that have made them into relatively wealthy, small capitalists (owners of trawlers).

State policies beyond those narrowly defined as fishery management, e.g., the construction of harbors, agricultural policies, and welfare policies have impact on sea snail fishing and poverty/wealth of fishers (Knudsen et al. 2010). Some policy developments, both within and outside of fisheries management, result from Turkey's engagement with supra-national bodies such as the WB and the EU.

The boom years of the sea snail fisheries clearly created a frontier situation, especially in Terme. In such a situation, with the race to extract the resource rent, the state constructing fishers' activities as criminal, and the fishers finding that they have "no relations to the state," co-management between state and communities of fishers is hardly a realistic option. Collective action among fishers also seems to be limited. The degree and character of fishers' collective action is shaped by the nature of the community (e.g., the frontier situation in Yalı Mahallesi), the educational level, and the politics of relations to external actors such as state representatives and business patrons.

There are cultural and political dimensions to fishers' poverty. That Romani fishers generally are poorer than fishers from the other communities in Terme is probably best explained by their cultural stigma and political marginalization which have been sustained throughout Turkey for decades. Also, poverty is sustained by intra-sectoral institutional mechanisms, such as the relations between crew and owners of purse seiners. This relation is characterized by power inequality, unpredictability, lack of social security, and in general a lack of formal regulation and control. The political marginalization of small-scale fishers and crew limits their participation and ability to influence the issues that impact on their capabilities. Thus, lack of capacity to participate can be considered a dimension of poverty (Baum 2001, p. 1843). While Sen shifts attention from means to ends, we can argue with Marshal Sahlins that, "[p]overty is not a certain small amount of goods, nor is it just a relation between means and ends; above all it is a relation between people" (Sahlins 1972, p. 37).

Political marginalization – political not only in the meaning of formal politics, but also including participation in informal networks of influence etc. – thus, bereaves people of capabilities, reduces their freedoms to be able to satisfy ends (cf. Sen 1999, p. 90). It can thus make sense to consider the ability to organize for collective action, and engage in co-management with the authorities; a capability which small-scale fishers in Terme and Çarşıbaşı largely seem to be deprived of, partly because of their limited education, and the lack of men with experience from public employment among them.

Based on this study, we argue that fishery development not only should go hand in hand with fishery management, but also with social policies aimed at reducing poverty and inequality. It is not developments in the fisheries that cause capabilities deprivation among small-scale fishers in Terme and in Çarşıbaşı. It is first and foremost exclusion from other capabilities, or entitlement breakdown

(Béné 2003, p. 968) – also from capabilities that we have come to expect modern states to provide – that make people poor. Yet, as was seen in the Çarşıbaşı case, changes in the resource base can also contribute to increased poverty. It makes sense to account for the degree to which the "vicious circle" operates, but analysis of the relation between small-scale fishing and poverty should not be restricted to variables included in the thesis.

In this respect, our assessment of the "vicious circle" thesis parallels Ottar Brox' assessment of the tragedy of the commons thesis, from which the "vicious circle" thesis can be said to be derived: "The common property theory should not be considered a statement about the world in the sense that it can be falsified by evidence, but rather as an analytical tool, a part of the language we use in describing and explaining the world" (Brox 1990, p. 227).

The thesis helps us ask good research questions, stimulates us to look for important causal relationships, and specify the conditions under which the vicious circle dynamics operate or not. Yet, to give good answers, we might find that our study should venture outside of the kind of variables assumed to be of importance in the model. There might be more important causes for poverty among fishers outside of the fisheries than within the fisheries. Especially the reasons why people are poor, to start with, need more comprehensive explanation. Why is the distribution of land unequal? Why are prices of agricultural produce falling? Why are the unemployment rates so high? Why are health and social services so poor, and access to quality education and access to universities so unequal? Poverty in fisheries is most meaningfully studied as an aspect of overall poverty in a region or country, yet with the added attention on the degree to which a "vicious circle" dynamic is in operation. After all, the entitlements established in small boat fishing – the more or less collective right to use finite, often unregulated natural resources – are of a quality different from most other entitlements.

Acknowledgments In addition to funding from the PovFish project, the research leading to these results has received funding from the European Community's Seventh Framework Programme [FP7/2007–2013]: Knowledge-based Sustainable Management for Europe's Seas (EC FP7 – grant 226675). The KnowSeas project is affiliated with LOICZ and LWEC. The authors want to thank research assistants Özlem Yeniay and Kayhan Ural for their very competent participation during fieldwork and data assessment. We are deeply indebted to all the informants who willingly shared time and information with us. This work would not have been possible without fishers, fisher wives, scientists, managers, factory owners, and others being so forthcoming. Thanks to them all!

References

Acheson JM (1981) Anthropology of fishing. Annu Rev Anthropol 10:275–316

Akder H (2000) Türkiye'de Bölgesel Eşitsizlikler ve Kırsal Yoksulluk. In: Akder H, Güvenç M (eds.) Yoksulluk. TESEV Yayını, İstanbul, pp 15–36

Alkan I, Kiliç O (2007) Samsun İli Terme İlçesinin ova ve yüksek kesimindeki fındık işletmecilerin karşılaştırmalı ekonomik analizi. J Fac Agric, Ondokuz Mayis Universitesi 22(2):171–178

Allison EH, Ellis F (2001) The livelihoods approach and management of small-scale fisheries. Mar Policy 25(5):377–388

Baum HS (2001) Citizen participation. In: Smelser NJ, Baltes PB (eds.) International encyclopedia of the social and behavioral sciences. Oxford, Pergamon, pp 1840–1846

Béné C (2003) When fishery rhymes with poverty: a first step beyond the old paradigm on poverty in small-scale fisheries. World Dev 31(6):949–975

Brox O (1990) Common property theory: epistemological status and analytical utility. Hum Organ 49(3):227–235

Buğra A (2008) Kapitalizm Yoksulluk ve Türkiye'de Sosyal Politika. İletişim Yayınları, İstanbul

Buğra A, Keyder Ç (2005) Poverty and social policy in contemporary Turkey. Unpublished manuscript, Istanbul

Çakmak EH, Eruygur HO (2008) Food, rural, agricultural, and fisheries policies in Turkey. Options méditerranéenns, Séries B 61:165–206

Erdoğan N (ed.) (2007) Yoksulluk Halleri. İletişim Yayınları, İstanbul

Eruygur HO (2006) Impacts of policy changes on Turkish agriculture: an optimization model with maximum entropy. Unpublished Doctor of Philosophy, The Middle East Technical University, Ankara

GoS (2009) Nüfus Durumu. Samsun: Governorship of Samsun homepages

Gülçubuk B, Aluftekin N (2006) Impact of international agricultural policies on rural poverty in Turkey. Int Bus Econ Res J 5(1):9–18

Işık O, Pınarcıoğlu MM (2001) Nöbetleşe Yoksulluk. İletişim Yayınları, İstanbul

Knudsen S (1997) A comparative study of fishing communities and public awareness in Turkey and Ukraine. Black Sea Environmental Programme, Istanbul

Knudsen S (1998) What role for the Turkish fishery co-operatives? Organisational preconditions for a new management regime in the Black Sea. In: Paper presented at the FISHECO'98: The Proceedings of the first international symposium on fisheries and ecology, Trabzon

Knudsen S (2006) Between life giver and leisure: identity negotiation through seafood in Turkey. Int J Middle East Studies 38(3):395–415

Knudsen S (2009) Fishers and scientists in modern Turkey. The management of resources, knowledge and identity on the Eastern Black Sea Coast. Berghahn Books, Oxford

Knudsen S, Zengin M (2006) Case study 2: Trawl and sea snail fisheries in Samsun, Black Sea coast of Turkey. Deliverable 5.2

Knudsen S, Zengin M, Koçak MH (2010) Identifying drivers for fishing pressure. A multidisciplinary study of trawl and sea snail fisheries in Samsun, Black Sea coast of Turkey. Ocean Coast Manage 53(5–6):252–269

Kopytoff I (1987) The African frontier. The reproduction of traditional African societies. Indiana University Press, Bloomington

Köse AH, Bahçe S (2009) Yoksulluk' Yazınının Yoksulluğu: Toplumsal Sınıflarla Düşünmek. Praksis 19(1):385–419

Mangahas MF (2000) Managing luck and negotiating change-ethnographies of fishing and sharing in the Philippines. University of Cambridge, Cambridge

McGoodwin JR (1990) Crisis in the World's fisheries. People, Problems, and Policies. Stanford: Stanford University Press

MEF (2008). Ordu-Trabzon-Rize-Giresun-Gümüşhane-Artvin Planlama Bölgesi 1/100.000 Ölçekli Çevre Düzeni Planı Araştırma Raporu. Ministry of Environment and Forest, Republic of Turkey, Ankara

MHT (2009) Yeşil kart kayıtlarının ilçelere göre dağılımı (İptal edilen yeşil kartlılar dahil değildir). The Ministry of Health of Turkey, Ankara

Oktik N (ed.) (2008) Türkiye'de Yoksulluk Çalışmalar. Yakın Kitabevi Yayınları, İzmir

Özbudun S, Başokçu EK (2004) Bizim Köyde Yoksulluk Yok Ki!' ya da Yoksunluktan Yoksulluğa Türkiye Kırsalı Üzerine Notlar. Toplum ve Hekim 19(1):39–47

Pauly D (1994) From growth to Malthusian overfishing: stages of fisheries resources misuse. Tradit Mar Resour Manage Knowledge Info Bull 3:7–14

SABEK (2005) Samsun İl gelişim stratejisi. Mevcut durum ve analizi raporu. Samsun Bölgesel Ekonomik Kalkınma, Samsun

SAG (2005) Trabzon yatırım ortamı ve yapabilecek yatırımlar araştırması. Sanayi Araştırma ve Geliştirme Genel Müdürlüğü, Republic of Turkey Ministry of Industry and Trade, Ankara
Sahlins M (1972) Stone age economics. Tavistock Publications, London
Sen A (1983) Development: which way now? Econ J 93(372):745–762
Sen A (1999) Development as freedom. Oxford University Press, Oxford
Şenses F (2001) Küreselleşmenin Öteki Yüzü Yoksulluk. İletişim Yayınları, İstanbul
SPO (2004a) İlçelerin sosyo-ekonomik gelişimlik sıralaması araştırması. Turkish Republic State Planning Organization, Ankara
SPO (2004b) Yeşilırmak havza gelişim projesi (Amasya, Çorum, Samsun, Tokat) Mevcut durum ve analizi. Turkish Republic State Planning Organization, Ankara
Termehaber(2010) Termenin Nüfus ve Sosyal Yapısı, 2010, from http://www.termehaber.com/haber-358-Termenin-Nüfus-ve-Sosyal-Yapısı.html
Tsing AL (2005) Friction. An ethnography of global connection. Princeton University Press, Princeton
TURKSTAT (2004) Turkey's statistical yearbook 2004. Ankara
TURKSTAT (2009) Results of 2008 poverty study. Turkish Statistical Institute, Ankara
TURKSTAT (2010) Adrese dayalı nüfus kayıt sistemi nüfus sayım sonuçları, 2009. TURKSTAT Haber Bülteni, 2010(15)
UNDP (2009) Human development report 2009, Turkey. The human development index – going beyond income, United Nations Development Programme
World Bank (2003) Turkey. Poverty and coping after crises (No. 24185-TR): World Bank, Human Development Unit, Europe and Central Asia Region, Washington, DC
Yoltar Ç (2009) When the poor need health care: ethnography of state and citizenship in Turkey. Middle Eastern Studies 45(5):767–782

Chapter 12
Community Response: Decline of the Chambo in Lake Malawi's Southeast Arm

Mafaniso Hara

Abstract In Malawi, the multi-gear, multi-species small-scale fishing sector lands more than 95% of the catch and employs over 95% of those participating in fishing, greatly contributing towards poverty alleviation and protein food security for the lakeshore communities and Malawians at large. Over the last two decades, catches of the chambo (*Oreochromis* spp.), the most valuable species in the Southeast Arm of Lake Malawi, have declined. This is a source of concern for the sustainability of the fishery as a whole, and the impact this could have on the dependent fishing communities, given that the devastated Lake Malombe fishery followed a similar trajectory. Fishers are ambivalent as to whether decline of the chambo should be a source of concern, especially if accepting this view would mean agreeing to new regulations aimed at reducing fishing effort. This study analyzes the strategies being used by fishers in response to the changing fishery dynamics as a result of the decline of the chambo. The responses include: investment in cheaper fishing gears; invention of new fishing techniques; introduction of new gear types; geographic and occupational mobility; business and livelihoods diversification; changes in relation to production within fishing units; and introduction of cage culture. Managers and development practitioners need to understand the changes taking place in the fishery in order to formulate appropriate and acceptable solutions, if the fishery is to continue to provide social-economic benefits for the fishing communities and Malawi.

M. Hara (✉)
Institute for Poverty, Land and Agrarian Studies (PLAAS), P/Bag X17,
Bellville 7535, South Africa
e-mail: mhara@uwc.ac.za

S. Jentoft and A. Eide (eds.), *Poverty Mosaics: Realities and Prospects in Small-Scale Fisheries*, DOI 10.1007/978-94-007-1582-0_12,

12.1 Introduction

Since the early 1990s, the estimated landed catch of chambo (*Oreochromis* spp.),[1] the most valuable species from the Southeast Arm of Lake Malawi, has declined. 1995 saw the lowest estimated catch of chambo ever recorded at 690 t compared to nearly 4,000 t at its height in the mid-1980s. The concern is that the demise of the Lake Malombe fishery started with the over fishing of the chambo. Thereafter, fishers switched to the less valuable but still abundant (at the time) *Lethrinops* (kambuzi) spp (Hara 2001). Within a period of 10 years, these other species had also been biologically and economically overfished. The whole fishery has never recovered to its former levels of productivity.

In view of the foregoing, the decline of the chambo fishery on the Southeast Arm over the last two decades has raised fears and concerns that the artisanal fishery of the area could follow the same trajectory as that of Lake Malombe (Hara 2001; Banda et al. 2005). This has great socio-economic implications for the dependent fishing communities in the area, and for other people in the ancillary sectors who greatly depend on the fishery for their livelihoods. The decline of the chambo on the Southeast Arm has increased government concerns about the sustainability of both, the chambo and the fishery as a whole; and most of all, the socio-economic impact this could have on the fishing communities.

The government concerns are such that it officially launched the *National Save the Chambo* campaign in January 2003 that resulted in the formulation of the Chambo Restoration Strategic Plan (Banda et al. 2005; Hara 2006a). To reiterate, the potentially disastrous socio-economic impact of a degraded artisanal fishery on fishers and the dependant fishing communities is epitomized by the devastation of the Lake Malombe fishery, which began with the overfishing of the chambo, the most valuable species, followed by the less valuable species. This could also happen on the Southeast Arm of Lake Malawi.

Are these government concerns necessary? If availability of the chambo has indeed declined, and the species composition of the fishery has changed as a result, how are fishers reacting in order to cope with these changing resource dynamics? How do they view the proposed government solutions?

The findings of this study are that in reaction to decline in profitability of inshore chambo beach seines, fishers have switched to a cheaper gear type (gill nets) and have invented a new method of catching the chambo (kauni), in order to continue targeting the chambo offshore. Other strategies include increased targeting of lower value species namely utaka (*Copadichromis* spp.), kambuzi (*Lethrinops* spp.),[2] and usipa (*Sardinella* spp.) and diversification of business interests and livelihood activities. In addition, the crew members have become more assertive in terms of business decisions, and also in terms of benefit sharing systems and formulas.

---

[1]Chambo is a general term for three (*lidole, squamipinnis*, and *karongae*) closely related species of tilapine cichlids of the genus *Orechromis* (FAO 1993).

[2]Utaka and kambuzi are haplochromine cichlids (FAO 1993).

As part of the chambo restoration plan (Banda et al. 2005), the government has introduced cage culturing on Lake Malawi as a way of trying to boost chambo production. The decline of the chambo, and attempts to restore the fishery to its former levels of production, present great management challenges as this has to be done while trying to improve or at least maintain the livelihoods of the fishers. Such a management approach calls for the incentivization of participation of fishers in finding solutions and sustainable exploitation practices.

The chapter is organized as follows: The preceding introduction outlines the problem and also summarizes the findings of the study; this is followed by a section on the theories of poverty in relation to human well-being (including Malawi's vision for development), and also small-scale fisheries. Next is a short overview of the role of fisheries in Malawi's economy, and presentation of the case study area and how it fits into Malawi fisheries. The methodology used for the study is then outlined. This is followed by the main section outlining the findings of the study. The last section, following the results, discusses the findings in relation to poverty and livelihoods; and whether the findings pertain to a virtuous or vicious circle in terms of poverty. The last section draws some conclusions from the study.

12.2 Theories of Poverty in Relation to Human Well-Being

Perceptions of poverty vary depending on culture, political ideologies, level of development, religion, and many other factors. In many cultures, "poor" is not just the opposite of "rich" (Rahnema 2007). The term could mean: falling from one's station in life, loss of one's status, loss of one's instruments of labor, lack of protection, exclusion from one's community, being abandoned, infirmity, public humiliation, etc. (Sen 1981; Rahnema 2007; Jentoft et al. 2010). It is with the monetization of economies and societies that poverty has increasingly been defined as lacking what those classified as rich have in terms of money and material possessions. This thinking had been strengthened by concepts that defined poverty on the basis of National Gross Domestic Product and per capita income (for example, the "dollar a day" threshold between being classified as "poor" or "not poor") by the multi-lateral development agencies such as the World Bank (WB) and the United Nations Development Programme (UNDP).

Conceptualization of poverty has, since the 1990s, changed to the use of more composite indices of human well-being such as the Human Development Index (HDI) by the UNDP (2009), as a result of recognition that poverty is multi-dimensional. While the basic necessities vary among cultures and societies, there is a minimum requirement, an irreducible core of absolute necessities, in the idea of poverty.

For the HDI concept, three dimensions are used to measure human development and human well-being: life expectancy, literacy, and standard of living.[3] Within the

[3] It is recognized though that the use of just these three dimensions is not comprehensive enough, since this excludes other important indicators such as gender, income inequality, human rights, political freedom, etc.

HDI concept is also the use of the Human Poverty Index (HPI). While the HDI measures the average progress of a country in human development, the HPI measures the proportion of people below a specific threshold in each of the dimensions of the HDI. The HPI thus represents a multi-dimensional alternative to the income (Purchasing Power Parity – PPP) poverty measure. For Malawi, a HPI value of 28.2% ranked the country 90th among 135 countries in 2007 (UNDP 2009). Rahnema (2007, p. 159) reiterates though that "among the various definitions and perceptions of poverty, the common denominator is the notion of *lack* or *deficiency*."

Malawi is signatory to the 2000 United Nations General Assembly Millennium Declaration. The country undertook to achieve the Millennium Development Goals (MDGs) by 2015 and lists the eradication of extreme poverty as the first and key of the eight goals (GoM 2008). The Government of Malawi defines extreme poverty as "the inability to meet the basic minimum food requirements" (GoM 2008, p. 2). Two targets were set by the government as key indicators for eradicating poverty by 2015: first, reducing by half the proportion of people whose income is less than one dollar per day; and second, reducing by half the proportion of people who suffer from hunger. One of the MDGs relates to ensuring environmental sustainability. The 2008 progress report (GoM 2008) states that Malawi continues to experience various forms of environmental degradation, caused by increasing population growth, poverty, impact of HIV and AIDS, and inadequate alterative livelihoods.

Following the end of colonization in Africa,[4] fisheries have always been linked to economic development and how the sector can contribute towards improving the lives of rural fishing communities (Ferguson et al. 1993; Hara 2001; Hara et al. 2009). As Béné (2003) points out though, there is an almost complete lack of references in the literature on poverty to cases on fisheries. The author attributes this to the nature of scholarship and analysis on fisheries and poverty and the way this attempts to explain the origins and causes of poverty in small-scale fisheries. According to Béné (2003), the relationship between poverty and small-scale fisheries has been explained in two contrasting ways. The first is to say that: "They are poor because they are fishermen;" while the other says that: "They are fishermen because they are poor" (Béné 2003, p. 949).

The first explanation has its origins from Gordon's (1954) classic paper that argued that fisheries are an open access resource, which was powerfully and famously re-interpreted by Hardin (1968) as the "tragedy of the commons." For both Gordon and Hardin, the open access nature of fisheries results in more and more people entering a fishery. Over time, excess effort results in overfishing, depletion of resource rents, and eventually the impoverishment of the fishers.

As Hersoug et al. (2004) point out, this argument has been very strong in fisheries, with both donors and scientists using it to explain poverty in small-scale fisheries. The second idea proposes that fisheries is an employer of last resort, which absorbs those falling out of other economic sectors (FAO 2000). Thus people enter into fisheries because they have no other option. In other words, it is a sector that absorbs those who are poor and have nowhere else to go (they are fishermen because they are poor). This

[4]The majority of African countries gained independence in the 1950s and 1960s.

argument proposes that small-scale fisheries act as a safety net, and closing participation might actually cause poverty (Hara 2001; Jul-Larsen et al. 2003; Jentoft et al. 2010). Thus, the widely accepted and conventional view in fisheries literature is embodied in these twin concepts: "Fishermen are the poorest of the poor;" and "fishing is an activity of last resort" (Béné 2003). Both these concepts convey the idea of small-scale fisheries being a sector characterized by structural chronic poverty (Béné 2009).

Jul-Larsen et al. (2003) demonstrate how these two approaches for explaining poverty might not be necessarily applicable in some of the inland fishing water bodies in southern Africa using Brox's (1990) concepts of *horizontal* and *vertical* fishing effort. Horizontal increase in effort pertains to increased effort as a result of more fishers using the same kind of technology entering a fishery; whereas, vertical increase of effort relates to increased fishing capacity as a result of new and improved technology. Because vertical increase in effort results in increased efficiency, it is regarded as being more harmful than a horizontal increase in effort, which is merely an increased effort of similar type and efficiency.

Jul-Larsen et al. (2003) argue that in most of southern Africa's small inland water bodies, horizontal increase in effort is more common since most fishing communities are not specialist fishers. Fishers enter and exit the fishery depending on opportunities outside fishing. Thus, fishing is usually one of a number of livelihood possibilities. In addition, fishers have flexibility in fishing strategies (e.g., targeting different species according to availability), and migrate geographically in pursuit of better fishing opportunities. Thus, mobility (occupational and geographic) is a key strategy that fishers use in order to overcome fluctuations in fishing fortunes. Jul-Larsen et al. (2003) argue thus that in most water bodies in southern Africa, increase in effort is usually horizontal; and that the increases have not usually been to the extent that this calls for limiting access. The authors further argue that limiting fishing effort in most of these communities might actually be harmful in terms of livelihoods based on adaptive mobility.

Hersoug et al. (2004) caution though that although Jul-Larsen et al.'s (2003) findings might be applicable to most small inland water bodies in southern Africa, Pauly's (1994) argument about "Malthusian over-fishing," and Panatoyou's (1982) argument about small-scale fisheries[5] being characterized by "easy entry and difficult to exit," points to the fact that even horizontal increase in effort can cause overfishing and therefore result in poverty.

12.3 Fisheries in Malawi's Economy

Fish contributes 40% to 50% of the animal protein in the diets of Malawians (Hara 2001; Njaya 2009). Despite its low contribution to GDP,[6] fishing is one of the main sources of livelihoods in the lakeshore areas of the major fish producing water bodies

[5] The two authors used the Asian small-scale fisheries for their analysis.

[6] Since the 1970s, the DoF has been putting out the figure of 4% as the contribution of fisheries to GDP (Hara 2001; Njaya 2009).

like Lake Malawi, Lake Chilwa, Lake Chiuta, and the Lower Shire River. According to Malawi's Department of Fisheries (DoF 2009), the artisanal fishing industry employed about 60,000 fishers (gear owners and crewmembers) in 2008. Over 95% of people were employed in the catching sector with landings of over 95% of the catch. Another 400,000 were estimated to have been working in the post-harvest sector as processors, traders, retailers, and also in the ancillary industries such as boat building and net-making. Given that the average household size in Malawi is five, this means that about 2.3 million people benefit from capture fisheries. Thus capture fisheries contribute substantially to the Malawi Growth and Development Strategy (MGDS) framework through protein food security and poverty reduction (GoM 2006).

12.3.1 *The Southeast Arm of Lake Malawi*

The Southeast Arm of Lake Malawi denotes the right arm (facing north) of the southern end of Lake Malawi (Fig. 12.1). Its total surface area is approximately 2,000 km^2. Although it represents only 8.4% of the total surface area of Lake Malawi, the contribution of the area to total production from Lake Malawi had usually been

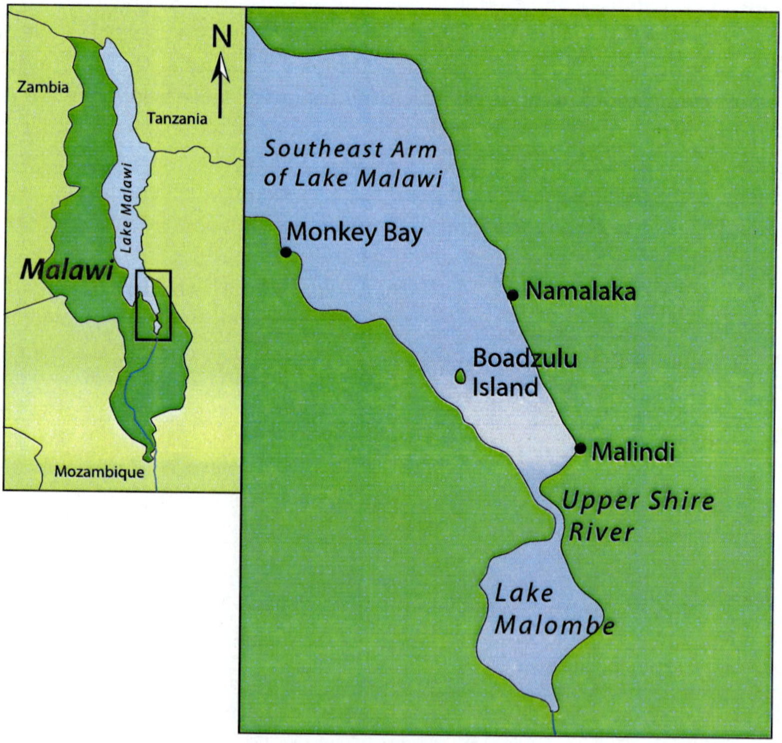

Fig. 12.1 Geographic location of the Southeast Arm of Lake Malawi

between 25% and 35%; while the area's contribution to total national landed catch had averaged over 20% since 1990. Within Mangochi District, the contribution of the area's artisanal sector had grown consistently; and by 1995, the sector accounted for 57% of the estimated landings. The high production of usipa (*Sardinella* spp.) from the area in recent years means that this contribution is even higher, reaching 82% in 2005; and 75% in 2009. It is thus the most productive area of the lake.

Mangochi District, where the Southeast Arm is located, had a population of 803,000 in 2008 (NSO 2008). The estimated density in the district was 191 people per square kilometer, with the density rising to over 500 people per square kilometer along the lakeshore area (NSO 2008).

Rain-fed agriculture is the most important sector in the local economy, with about 66% of the population being employed in the sector as family farm owners and unpaid family farm workers. Under Malawi's land tenure system, smallholder farmers normally hold land under customary tenure,[7] while commercial estates hold land under leasehold tenure.[8] In Mangochi District, customary land constituted 49.9% of arable land, while lease-held farms constituted the remaining 50.1% in 1995. In the same year (1995), there were 156,694 small holder farm households while lease-holders numbered only 650 (GoM/UNDP 1998). Because of this skewed distribution of land and the given population growth, there is an increasing shortage of customary land; so much so that most households in the small holder sector do not have enough land to grow adequate food for their annual needs.

12.4 Methodology

Fieldwork was undertaken between August 2008 and January 2010. Two main sources of data and information have been used for this study. Firstly, structured and unstructured interviews; and secondly, catch and effort data, and frame survey[9] data from the Department of Fisheries (DoF) for years 1977 up to 2007. For the unstructured interviews, 17 gear owners, 50 crew members (mostly as focus groups), 5 village headmen, and 8 traders (including three women traders) were interviewed. The gear owners were selected across all the important gears on the basis of the gear type owned and also their age/experience in the fishery. The crew members were also selected from across all gears on the basis of the main gear types that they were employed in. Also interviewed were six DoF officers in the following categories: Field staff working in the Southeast Arm area; the Fisheries Officer for Mangochi District; and the Director of Fisheries Research at Monkey Bay. Discourse analysis was used to analyze and interpret the views of fishers and DoF staff on what the state of the chambo means to them in terms of conservation of the fishery, and

[7] Customary land is owned by the state, but is left under the supervision of traditional leaders.

[8] Lease held land is under supervision of the government, but is leased out for a period of 99 years.

[9] A frame survey is a count of all gear owners, crew members, fishing gears, vessels, and engines deployed in the fishery. This is supposed to be done once annually.

its continued provision of socio-economic benefits for the fishing communities and other stakeholders. The District Agriculture Officer for Mangochi District, an official dealing with inheritance and wills in the District Commissioner's Office, and the Project Manager and Farm Manager for Maldeco Aquaculture Limited (cage culture production) were also interviewed. Four people (migrants) from the Southeast Arm area but currently working and living in Cape Town were also interviewed. All in all, over 90 people were interviewed during the three field trips undertaken between August 2008 and January 2010.

Malawi's Department of Fisheries' statistics on estimated catch, fishing effort, and beach prices were used in order to look at historical catch and effort trends; landed value and other variables of the fishery in order to look at the changing characteristics of the fishery.

12.5 Strategies for Adapting to the Changing Dynamics of the Southeast Arm Fishery

This section presents the study findings regarding some of the key adaptations and strategies being used by fishers on the Southeast Arm in response to the decline of the chambo (Fig. 12.2), catches, and variability in other main target species. These include: change in target species, adoption of new gears, invention of new fishing techniques, geographic and occupational mobility, broadening the portfolio of economic activities, reliance on remittances from relatives, change in production relations between gear owners and crew members within fishing units,[10] and introduction of lake cage culture.

12.5.1 Change in Target Species, Introduction of New Technologies, and Migration

For most fishers who can invest in the fishery (gear owners), the chambo remains the primary target species because it remains the most valuable species (Table 12.1). As the profitability of *chambo beach seines*[11] declined (the most profitable chambo fishing gear in the 1980s to early 1990s), gear owners disinvested from beach seining and switched to gill nets and chilimira nets (Fig. 12.3). Thus, the increase in the number of gill nets (by 517% between 1990 and 2005), and chilimira nets on the Southeast Arm from the mid-1990s into this millennium; while at the same time, the number of chambo seine nets declined (Table 12.2). Gill nets are cheaper in terms

[10] A fishing unit refers to a complete array of the equipment and persons with the skills necessary to undertake a fishing enterprise. Usually, a unit comprises of the gear owner (who owns the capital equipment – the vessel, net, and engine) and the crew members.

[11] The chambo beach seine is a rectangular net cast using a boat and then pulled to the shore from two ends by two groups of gang members (numbering between 10 and 30), one on each side (FAO 1993; Hara 2006b).

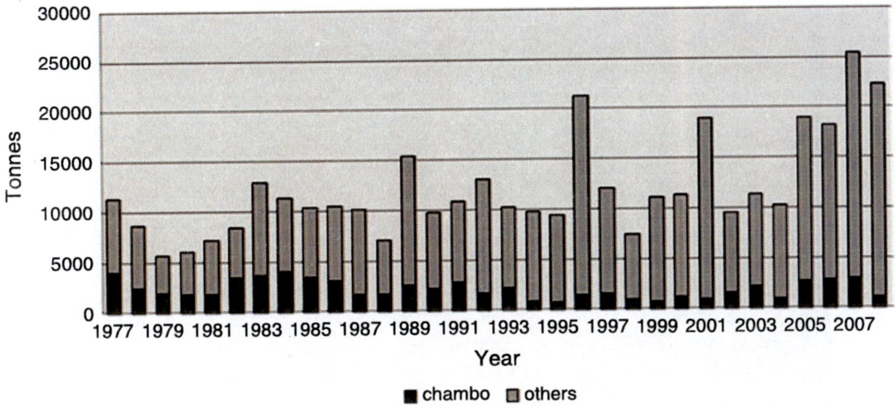

Fig. 12.2 Estimated catch for chambo and other species for the Southeast Arm (Source: Department of Fisheries, Lilongwe, Malawi)

Table 12.1 Estimated total annual catch (to the nearest ton) and landed value (in millions Malawi Kwacha (MWK))[a] for selected species and for selected years from Southeast Arm (Source: Department of Fisheries, Lilongwe, Malawi)

	Chambo		Kambuzi		Utaka		Usipa	
Year	Catch	Value	Catch	Value	Catch	Value	Catch	Value
2004	933	98.9	1,319	44.8	1,620	64.8	2,376	99.2
2005	2,646	365.1	1,672	76.9	4,336	186.4	3,203	176.2
2006	2,829	432.8	997	55.8	1,566	95.5	4,365	309.9
2007	2,882	631.6	1,572	108.5	2,215	166.1	8,237	593.1
2008	1,135	265.6	2,054	183.0	2,068	170.1	11,271	1,246.6

[a]1 USD = 150MWK (from http://www.xe.com accessed on 03/04/2010)

of required capital investment and also in terms of amount of labor requirements. The problems with regard to night set gill nets are first of all, destruction of the nets by trawlers; and second, increased theft thereby requiring crew members to stay out on the lake throughout the night to guard the nets.

Another critical development during the 1990s was the invention of *kauni fishing*.[12] This is a method of catching the chambo offshore by the use of the *chilimira*[13] using light attraction. This meant that as catches from beach seining declined, the chambo has mainly been targeted using either gill nets and/or kauni. Therefore, the increase in chilimira nets during the 1990s (Table 12.2) was partly

[12]In local (Chichewa) language, Kuwunika means lighting. The fishers on the Southeast Arm have termed the method of catching the chambo at night using the chilimira by attracting it to light as kauni.

[13]The chilimira is a conical shaped, open-water seine net first invented for catching utaka (FAO 1993; Hara 2001). By lining the bunt with a mosquito net, it can be used for catching usipa. The chilimira is operated from two boats by a total of nine crew members.

Fig. 12.3 A Chilimira seine being brought-in after a throw on Southeast Arm of Lake Malawi. Operated from two boats, the net is thrown in a wide semi-circle and then slowly pulled in a surround bagging motion from the two boats until the two boats come together. This photo shows one boat and one of the net ends (Source: Mafaniso Hara, 2009/11/02)

as a result of its use for catching the chambo. When the DoF discovered that fishers had invented kauni, the knee jerk reaction was to ban the method on the basis that the small mesh-sized[14] chilimira was catching undersized chambo. Fishers are adamant though that kauni does not catch undersized chambo, since most of the offshore chambo are adult.

A third major strategy by fishers has been to increasingly switch to less valuable, but still abundant, species such as utaka (*Copadichromis* spp.), usipa (*Sardinella* spp.), and kambuzi (*Lethrinops* spp.) as availability of the chambo declined (Table 12.1). Thus, the last 15 years or so has seen a dramatic increase in catches of usipa and to a lesser extent utaka (Fig. 12.2). The switch to these other species is also evident from the increase in the number of chilimira nets (the gear used to catch these species) by over 90% during the 1990s (Table 12.2). Thus, whereas the chambo used to be the main species caught from the Southeast Arm, other species have increased in importance in recent years. The increase in the number of chilimira nets has thus been as a result of the invention of kauni and also the increased targeting of usipa and utaka. In terms of versatility, this means that the chilimira can be used to target three species (usipa, utaka, and chambo) by simple technical adaptations to the same net.

[14]The minimum legal mesh size for gillnets and beach seines, the two gears meant to target the chambo, are 95 and 90 mm, respectively; while minimum mesh size of the chilimira is 25 mm (FAO 1993).

Table 12.2 Number of gear owners, assistants and gear units on the Southeast Arm counted during the frame survey (Source: Department of Fisheries, Lilongwe, Malawi)

	Number counted during frame survey							
Year	Gear owner	Assistants	Gill nets	Kambuzi seine	Chilimira seine	Chambo seine	Nkacha	Hand line
1990	979	6,655	1,685	157	378	52	0	1,138
1991	867	6,107	1,671	130	352	47	0	649
1992	1,090	7,025	2,383	183	381	43	1	622
1993	1,101	7,329	2,470	177	331	44	29	303
1994	1,226	8,062	2,416	119	475	32	43	1,413
1995	1,290	8,027	2,822	171	465	22	50	1,309
1996	1,153	8,268	2,566	89	512	25	17	365
1997	1,327	10,056	3,322	102	577	36	41	76
1998	1,337	9,069	4,554	100	689	24	12	201
1999	1,268	8,686	4,440	56	542	19	70	138
2000	No Frame Survey							
2001	1,716	10,167	9,432	57	572	7	83	18
2002	1,368	10,237	6,711	35	569	5	109	77
2003	1,693	10,796	9,612	60	539	6	89	46
2004	No Frame Survey							
2005	1,486	9,257	10,390	18	521	7	91	680

Utaka and usipa vary geographically in terms of abundance and availability. It is common therefore for chilimira fishers to migrate around the Southeast Arm, or even to other areas of Lake Malawi in pursuit of good usipa or utaka catches.

Notably, gear owners usually have more than one gear type, which can be deployed on the basis of availability and profitability of a specific fishery/species. This multiple gear ownership by individual gear owners can be discerned from the fact that while the total number of gear units in the fishery increased by 385% between 1990 and 2005, the number of gear owners only increased by 50% during the same period (Table 12.2).

The official figure of assistants (crew members) counted during the frame survey needs to be corrected by putting the number of crew members deployed in each type of gear into context. If we take the figures for the latest year available, which is 2005 (Table 12.2), the official number of assistants counted was 9,257. The following are the numbers of crew members that are deployed in each unit for the five main gear types: 2–4 (average 3) per unit in gill nets; 6–20 (average 13) per unit in kambuzi seines; 10–30 in beach seines (average 20); 9 in chilimira; and 8 in nkacha. Fishers further stated that usually, each chilimira and nkacha unit has at least two sets of crew members at a time. Using the average (or actual) number of crew members that are deployed in each unit for the five main gear types, Table 12.3 shows the total estimated number of crew members employed for each gear type, and therefore, the estimated total number of crew members employed for the five main gears.

Contrary to the official figure of 9,257 assistants therefore, 40,000–45,000 crew members is the more likely number of crew members that were actually employed in the five main gear types in 2005 (Table 12.3). Using the same formula, about

Table 12.3 Estimated total number of crew members employed on the Southeast Arm in 2005 (based on total number of units of each gear type, and average number of crew members per unit)

Gear type			Total crew members
Gill nets	Total units	10,390	31,170
	Average crew members/unit	3	
	Total crew members	10,390×3	
Kambuzi seine	Total units	18	234
	Average crew members/unit	13	
	Total crew members	18×13	
Chambo Beach Seine	Total units	7	140
	Average crew members/unit	20	
	Total crew members	7×20	
Chilimira	Total units	521	9,378
	Average crew members/unit	9×2	
	Total crew members	521×9×2	
Nkacha	Total units	91	1,456
	Average crew members/unit	8×2	
	Total crew members	91×8×2	
Total			42,468

15,000 crew members were employed in these gears in 1990 based on gear numbers in Table 12.2. This means that whereas the official figures show that the number of crew members increased from about 6,600 in 1990 to around 9,300 in 2005 (by about 39%), most likely the number increased by over 180% (from around 15,000 to between 40,000 and 45,000).

Recent years have also seen an increased number of *nkacha*[15] nets (Table 12.2). This gear targets kambuzi. An additional positive factor for the nkacha seines is the space left by semi-commercial pair trawlers, which mainly target kambuzi. Out of the seven semi-commercial pair trawlers that were licensed to operate on the Southeast Arm, only two have been in service in the last decade. Most of the nkacha nets currently operating on the Southeast Arm have migrated from Lake Malombe following the decline of the kambuzi fishery in Lake Malombe. Despite attempts to ban the use of the nkacha net on Lake Malawi by the DoF (on the argument that it had caused the destruction of the Lake Malombe fishery), the deployment of the net has increased. Nkacha fishers argue against the ban by pointing out that it is not possible to convert the nkacha to gears that have historically been used on the Southeast Arm such as the chilimira, since its configuration is totally different. They further pointed out that the nkacha is used to target the kambuzi rather than utaka or usipa and therefore is occupying a separate and different niche within the Southeast Arm fishery. The crew members also pointed out that their skills are around fishing using the nkacha, rather than other gears.

[15]The nkacha net is a rectangular net first invented in Lake Malombe to target the kambuzi in reaction to decline of the chambo in that lake. Its design is based on the kambuzi seine net. The special aspect of its operation is that one of the crew members has to dive to tie the two sides of the net together so that it forms a bag like a purse seine net before the net is pulled to the surface from two boats.

While fish traders also note with concern that the chambo has declined, they also see some positive aspects with regard to increased targeting of the other species. For the traders, the utaka, usipa, and kambuzi are easier to process, store, and transport (since these are usually dried) than fresh or smoked chambo. The low-income consumers in the urban areas where the fish is mostly sold also prefer the dried fish since this is cheaper and is also easier to store, given that most consumers do not have refrigerators.

12.5.2 Production Relations Within Fishing Units

In most instances, the gear owners do not go out fishing, rather they employ crew members. The way gear owners and their crew members relate to each other in a fishing unit is an important factor in gears operating on the Southeast Arm. This partnership has evolved from being based on wage employment in the 1970s and 1980s, to the sharing of proceeds from fishing based on agreed formulas in specific gear types (Hara and Jul-Larsen 2003; Hara 2006b).

In recent years, there have been further changes, especially in favor of crew members. What is clear is that crew members have increasingly asserted themselves within the fishing units when it comes to business and operational decisions. As examples: They insist on being involved (through their representative) in the auctioning of the catch to the highest bidder when the catch is landed. They demand that the gear owner provides them with breakfast or money in lieu of breakfast (they call this money *ya ndege*[16]), before they can go out fishing in the morning. They have a large say as to what species to target (whether chambo or utaka or usipa), when using the chilimira nets. Whereas with the chilimira, the gear owner still subtracts the daily operational expenses, in the nkacha net, the crew members refuse these deductions so that the sharing is based on the gross revenue. In the nkacha, therefore, the gear owner has to pay for all the operational costs for fishing. In choosing the buyer for the catch, the crew members stated that they started to insist on being involved in finding the highest bidder for the catch in order to counter practices of price collusion between the gear owner and the buyer that they claim had become common. Crew members also insist on being paid their portion of the share immediately after the sell of the catch rather than on a weekly or monthly basis as it used to be in former times (Hara and Jul-Larsen 2003).

In gears such as the chambo seine nets, gang members sometimes demand prepayment before they can operate the net (Hara 2006b). All in all, gear owners complained that crew members had become too powerful and are increasingly acting as if they are shareholders. Crew members argued that they were simply making sure that they got properly remunerated for their hard labor by demanding and insisting on a fair share of the catch proceeds. The crew members also pointed out that being

[16] Ndege means airplane. The context is that they need to eat before they can fly (go out fishing).

a crew member is not proper employment, since earnings vary from day to day on the basis of catch size, and the price of the catch (which depends on supply and demand and many other factors). They further argued that tenure as a crew member, within a given fishing unit, is very insecure and transient. For example, the gear owner might decide to suspend fishing, or the gear might get damaged and fall into disrepair due to weather and/or other factors that can instantly result in this source of income and livelihood being pulled away from under one's feet.

What has also become common is that most seine nets have more than one set of crew members per gear/fishing unit. This means that the number of days that a crew member goes out fishing is not for a full month. As far as crew members are concerned, therefore, it is important that they maximize earnings whenever they go out fishing.

12.5.3 Vulnerability

The issue of maximizing earnings by crew members also relates to how vulnerable they are to both short-term and major mishaps due to lack of easily convertible assets (those that can be sold in an emergency) and/or savings. Bad weather for a number of days, for example during the winter months when the *mwera* winds (southeast trade winds) are prevalent, can mean no source of income for those days. One of the major mishaps is when something happens to the fishing gear. This could be due to theft of the gear, wearing out of the gear, confiscation of the gear by the Department of Fisheries, or the death of the gear owner.

In case of the death of the gear owner, the relatives of the gear owner usually suspend fishing until the issues of inheritance have been sorted out. If there are disputes among the relatives, this means that the suspension of fishing can take even longer. This can mean that all of a sudden, a crew member and his family have no source of income. In most instances, the family members of the deceased gear owner sell the fishing gear and share the money, meaning that the gear changes hands. Crew members are not guaranteed employment by the new gear owner.

Wives are particularly vulnerable to the death of their gear-owning husbands. Thus, unless the gear owner had a will, the wife (or wives since in most instances gear owners practice polygamy) and relatives enter into dispute over inheritance and/or sharing of assets. In the past (even currently if the widow is not well informed), the usual practice had been that the male side relatives of the deceased gear owner grabbed all the assets including fishing implements, leaving the widow and her children destitute. In recent years, the government has developed legal provisions meant to protect spouses and their children in the case of the death of the husband/father. This means that even if the deceased husband did not have a will, the widow can take the matter to the Traditional Authority for settlement, if this cannot be settled within the family. If the widow is not satisfied with the outcome at the Traditional Authority level, then she can take the matter to the District Commissioner (DC). According to the provisions of the new Inheritance Act, the wife (or wives) of the deceased get 65% of the assets, and the man's male side relatives get 35%. Because of disagreements among those entitled to the estate of the deceased, the usual solution among the disputants is to sell the assets and share the money.

12.5.4 *Diversification of Livelihood Profiles*

Whereas most fishers used to rely almost solely on fishing and used to use income from fishing to buy food and other necessities, there has been an increased diversification in terms of businesses among gear owners, and sources of livelihoods among crew members. Most gear owners diversify into other businesses in order to spread their risks. The range of businesses they go into are cash-crop farming, building and running rest houses, public transport, keeping livestock (especially cattle), etc. (Hara and Jul-Larsen 2003; Hara 2006a). This way, investments are moved around a number of business portfolios, depending on the farming season and profitability of a specific business type at a given time.

12.5.4.1 Farming

For crew members, the most important alternative source of livelihoods is farming. Because of the high population density along the lakeshore area, there is very little land for farming along the lakeshore. Therefore, most households have increasingly acquired farming land further upland, between five and ten kilometers away from the lakeshore area. The provision of subsidies for fertilizer by the government in the last 5 years has helped improve productivity of small-scale farmers and thus improved household food security. In contrast, during the 1990s, the government was forced to strictly implement the Structural Adjustment Programme (SAP), and subsidies had been withdrawn resulting in exorbitant prices for fertilizer.

For those who have acquired or inherited customary wetlands next to the lakeshore or rivers, they usually establish dimba gardens[17] for vegetable and dry season maize production. Dimba gardens have thus become a very common sight on the shores of the Southeast Arm. Domestic animals (cattle, goats, chickens, ducks, rabbits, etc.) are another important source of household protein and income.

12.5.4.2 Other Livelihood Activities

Petty trading is another livelihood activity that has become increasingly common (Hara 2006a). Although the main centre for this type of trading is Mangochi Township, vending is increasing in rural areas also. Young men are also taking up

[17]Dimbas are small-scale cultivations for vegetables, bananas, sugar canes, maize, and other agricultural crops along the lake or rivers using water from the lake, rivers, or wells dug for irrigation.

transportation of people using bicycles as a livelihood activity, both in and around Mangochi Township and in rural areas.

12.5.5 Remittances from Relatives

Migrant relatives working in urban areas or outside the country usually send money back home, especially during times of need or stress. Apart from money, relatives in South Africa also send back goods that can be converted into cash such as vehicles, televisions, stereos, and bicycles, etc. Those interviewed said that such remittances were particularly important in times of stress or disaster such as when there is a death in the family (funeral expenses), for taking care of a chronically ill family member, or when households run out of food.

Regarding foreign migrant labor, especially to South Africa and former Rhodesia (present Zimbabwe), this used to be very common among most young people from Mangochi. In fact, most of those who became successful gear owners in the 1970s and 1980s had raised the income for investing into fishing from foreign migrant labor (Chirwa 1995; Hara and Jul-Larsen 2003). Following independence in Zimbabwe (1980), and the end of Apartheid in South Africa (1994) though, these countries have had to prioritize jobs for their own people; meaning that they put a stop on organized importation of unskilled labor from other countries. In addition, it has become very difficult and expensive to obtain a Malawi passport. Thus, although some people still go to South Africa for job hunting, they enter that country as tourists and simply overstay on their month-long visitor's visa, having chanced on a job. The recent 2008 xenophobic attacks on foreigners in South Africa (especially those from other African countries), have made potential migrants more circumspect about going to South Africa in search of jobs.

12.5.6 Chambo Cage Culture

One of the strategic actions suggested under the Chambo Restoration Strategic Plan (Banda et al. 2005) is production enhancement. This has included the introduction of Lake Cage Cultures[18] for chambo on the Southeast Arm, and other areas of the lake. In line with this, cage culturing has been introduced on the Southeast Arm as one of the technological innovations to enhance production and therefore supply of chambo. MALDECO, the only commercial fishing company in Malawi, which operates from a base on the Southeast Arm, are the fore-runners in the introduction of this technology. They plan to introduce over 150 cages on the lake, most of which will be located on the Southeast Arm. This would enable them to produce over

[18]This is a form of aquaculture practiced in the Lake. A cage is suspended in water, fingerings are introduced in the cage, fed artificially, and then harvested for sale when they reach the required size.

3,000 t of chambo annually at full production. As of 2009, MALDECO had 53 cages in the lake, and had harvested over 600 t of chambo in that year.

According to the Department of Fisheries, the cage culture technologies will eventually be made available to other private investors, and also communities. Regarding communities, the thinking is that villages along the lake could get involved in cage culturing, with the economic benefits being shared among the defined village communities. However, up to now, no communities are participating in cage culture activities. A number of concerns had been raised about cage cultures by fishers and other people interviewed, such as: the fact that fishing communities do not have the capital required to invest in such technologies; fishing communities do not have the technical knowhow for producing fingerings and then culturing them to market size in lake cages; communities lack the institutional and organizational set-up required to run such ventures as a business; and the environmental impacts of cage cultures have not been adequately assessed. For the artisanal fishers on the Southeast Arm, one of the concerns expressed was the issue of competition for fishing space out on the lake. Given that almost all the main fishing gears that are used by the artisanal fishers on the Southeast Arm are offshore, the increasing appropriation of space for fish cages could result in increasing contestation for space out on the lake between fishers and cage culturing.[19]

Therefore, an important consideration regarding the introduction of this technology is whether cage cultures would be acceptable to the majority of current and potential fishers to the extent that they would be willing to accept the trade-offs between loss of fishing grounds, and the benefits that will accrue from cage cultures instead. Also, it is not clear to village headmen how village communities will be organized for cage culturing. One possibility could be the formation of village fishing cooperatives, which could be used as the vehicle for funding and management of cages belonging to a village or area. The point is that currently and historically, investment in fishing has been by individual gear owners. Cooperatives have never been used.[20]

Communities would also have to learn to take a long-term view of benefits from such ventures, since it takes up to 3 years to produce a commercial-sized chambo through cage culturing. Therefore, apart from the technological challenges, there are issues of the institutional arrangements, the issue of funding for capital investments, and the possible environmental impact of cage cultures on capture fisheries over the long-term. All these issues would have to be resolved if community cage culturing is to become a reality, and one of the viable solutions to the decline of the chambo on the Southeast Arm.

---

[19] In addition to artisanal fishers, the semi-commercial and commercial trawlers also operate in the offshore areas of the lake.

[20] Although farmers' clubs (for provision of farming inputs and extension services by government and NGOs) are used in the agricultural sector, fishermen's cooperatives have never been popular. Fishers are very individualistic in terms of sourcing capital for investment into fishing and sharing formulas for benefits within fishing units. Besides, fishers launch from and land anywhere they want. They also need to follow good catches, and therefore migrate whenever there is need to. Thus, they do not limit themselves to specific areas or landing sites. This system of operation makes cooperatives unworkable in the fishing sector in Malawi.

12.6 Fishing for Profit and Livelihoods: Virtuous or Vicious?

The basis of concerns about the decline of chambo on the Southeast Arm fishery is that this spells out the decline of the fishery as a whole. This will have great socio-economic impact on the fishing communities in the area, just like the collapse of the fishery in Lake Malombe has had on the fishing communities in that area. The question to answer first is whether the chambo fishery has indeed declined to such a level to be a cause for concern. The last recorded stock assessment of the chambo fishery was in the early 1990s by the Chambo Fisheries Research Project (FAO 1993). The stock assessment suggested that the biomass for the chambo on the Southeast Arm was 9,883 t. Based on this, the Maximum Sustainable Yield (MSY) was computed at 3,510 t (FAO 1993, p. 40). If this is the only official and authoritative figure that we can use, then it is fair to say that based on the estimated catch figures, the landed catch for chambo from the Southeast Arm has been below this suggested MSY for most of the last 20 years (Fig. 12.2).

The high catches in the years 2005–2007 (Table 12.1) were the result of large catches in the northern part of the Southeast Arm, the Makanjira area. Fishers from that area point out that these good catches were mainly from the Mozambican waters just north of Makanjira, where Malawi fishers are fishing. The Mozambican area in question has very low population because most people moved away from rural areas during the civil war in the 1970s and 1980s. The waters had therefore been largely under-fished. Thus, the good catches of chambo that can still be obtained are from that area.

For the gear owners and crew members, the initial manifestation of this decline, was the decline in profitability of the main gear that was being used to catch the chambo in the 1970s and 1980s – the chambo beach seine. Fishers had experienced this change very quickly on the fishing grounds, and had to adapt to the changing biological dynamics of the resource in order to maintain and sustain their fish businesses and livelihoods. They have adapted to this by disinvesting from the more expensive chambo seine nets that had become unprofitable to operate, to use the cheaper gill nets. Also, they invented a new, more profitable method – kauni. In the context of both the increased switch to gill nets and invention of kauni, fishers realized that while there were less chambo available inshore, they could still catch chambo profitably offshore using gill nets and kauni. Because the chambo remains the most valuable species, fishers still use every possible means and method to catch whatever chambo can be caught, in order to cash in on its value.

The Department of Fisheries has blamed the decline of the chambo on growth and recruitment overfishing caused by use of illegal gears, fishing during the closed season, destruction of habitats, and increased fishing effort (FAO 1993; Hara and Banda 1997; Banda et al. 2005; Hara 2006a). The destruction of inshore habitats cannot solely be blamed on fishers though. Since the 1980s, there has been a boom in construction of hotels and private cottages on the Southeast Arm. In most instances, the hotels and cottage owners remove vegetation from the inshore areas in order to create clear and clean areas for recreation.

The contention that part of the blame for the decline of the chambo is due to increased fishing effort assumes that the argument put forward by Jul-Larsen et al. (2003) about the resilience of fisheries in most small water bodies in southern Africa to increased fishing effort[21] does not apply to the Southeast Arm. On the Southeast Arm, there has been an increase in both types of effort. In terms of horizontal effort, there had been an increased number of fishing gear units by over 300% between 1990 and 2005. Gill nets contributed the most (by over 500%) (Table 12.2). The increase in gear units has also meant a large increase in the number of crew members by at least 180% (Table 12.3).

In terms of horizontal fishing effort, therefore, there were increases in both the number of gears and also fishers in the fishery. The concept of horizontal increase in effort would argue that the increase in effort involved an increase in the type of gears (gill nets and chilimira) that were already in use in the fishery. Therefore, these could not have been harmful because it was an increase in the same type of effort. There is also a possible argument about "Malthusian" overfishing. The entry of so many crew members, and deployment of so many gears (even though of similar type) intensively fishing day and night in such a small area could be harmful. There has also been the case of vertical increase in effort on the chambo through the deployment of a new and more efficient type of gear that targets the species offshore – the kauni.

Even if increased effort (both horizontal and vertical) has contributed to the decline of the chambo fishery, other factors such as destruction of habitats argue against putting the blame on increased effort alone. Until more and thorough investigations are undertaken, it cannot be said conclusively what factors have contributed to the decline of the chambo the most, and by how much. Just like the chambo, there is little scientific data and information of the biology, stock size, species interactions,[22] and other important factors of the other major and important exploited species (utaka, usipa, and kambuzi) for informed management decisions. The concerns and management actions that have been, or are being currently proposed by the DoF, such as the banning of kauni and nkacha, are thus mainly on the basis of the precautionary principle (FAO 1996).

From a livelihoods perspective, the decline of chambo beach seines and kambuzi beach seines has meant more crew members being employed in the fishery as a result of increased number of gill nets and chilimira nets. If we assume that each of the crew members' earnings benefit households of an average of five people, then the total number of people benefiting directly from the fishery had increased almost threefold from about 75,000 (14,940 × 5) in 1990, to over 210,000 (42,468 × 5) in 2005.

[21] Because most of the time the increase in effort is horizontal rather than vertical (Brox 1990).

[22] All the four species (chambo, kambuzi, usipa, and utaka) are planktivorous. While both the chambo and kambuzi are bottom feeders, the chambo mainly feeds on bigger particulates (compared to the kambuzi). The usipa and utaka are pelagics. The usipa is more off shore in open waters, while the utaka live more near shore. Although there are likely to be interactions among the various species, their feeding habits differentiate them into separate niches, thereby lessening competition (M. Banda, personal communication, 10 January 2010).

It is from this perspective that for crew members, the demise of the chambo might not be viewed so negatively. For gear owners, the increased use of chilimira nets has enabled them to spread fishing business risk across three main target species – chambo, utaka, and usipa. Therefore, gear owners are also ambivalent about the decline of the chambo – i.e., whether it is a bad thing or not. For gear owners and crew members, the chilimira is a versatile gear that can be deployed for catching the most profitable species at any given time, allowing flexibility in terms of business and livelihood opportunities.

As Mr. Yezayeza Nkhwazi (a chilimira gear owner based at Kela beach, Mwawa village on the Southeast Arm) put it:

> Mmene wa perekera Mulungu ndi momwemo

The nearest literal English translation of this statement is:

> God decides what and how much to give. We just have to accept what he gives.

This philosophical and enigmatic statement sums up the often expressed attitudes of most fishers when asked whether the chambo fishery has declined and if this is a problem. For them, it is not for anyone to determine how much fish they catch. It is God's will. Sometimes God gives and sometimes God keeps it for them for another day. This philosophical attitude has many meanings. It is a refusal to accept that there are less fish out in the lake than in the past. It is demonstrating denial to take responsibility for any possible overfishing. It brushes aside government calls for limiting fishing. It can be viewed as a resignation to the situation. It can also be real belief that there is no problem with the fishery.

Thus, whether the existing fishing practices and level of effort on the Southeast Arm is virtuous or vicious in terms of providing sustainable socio-economic benefits, depends in the end, on which side of the coin you choose to see. For the Department of Fisheries, the current fishing practices on the Southeast Arm are unsustainable, and therefore will eventually come back to haunt the fishers. For fishers, the fishery continues to provide fishing profits, livelihoods, and socio-economic benefits, thereby alleviating poverty. What the future holds is in God's hands.

12.7 Conclusions

The Department of Fisheries continues to use technical (input) regulations (mesh size restrictions, closed seasons, closed areas, and minimum sizes of chambo) rather than output regulations for management. Even if fishers were using the correct and legal types of gears and adhered to regulations, it is unlikely that the use of technical and input regulations alone can result in sustainable utilization of a fishery (FAO 1984). Some levels of limits on the output from the fishery have to be used in combination with input regulations. This is partly historical, since the fishery was seen as one of the key sectors for rural development (Hara 2001). It was, therefore, left open access in order to encourage those who could invest in fishing. This has resulted in engrained attitudes about the fishery being open access.[23] It has also resulted in chaotic organization of catching and landing activities, in that fishers launch and land wherever they want. This makes control of fishing activities and enforcement of regulations difficult and expensive.

The way the fishery is currently organized calls for a strong form of a cooperative management system based on use of participatory research, participatory data collection, and peer enforcement. Such a management system could probably provide the most workable solution to the problem of a poorly resourced Department of Fisheries that can never be present everywhere at all times to undertake the duties under its current mandate. The fishers also need to understand and accept that it is in their own best interests to ensure the biological and economic viability of the fishery into the future (God helps those who help themselves). Continued viability would entail accepting and putting into place measures that could control fishing effort, rather than thinking that the fishery has boundless capacity to absorb any levels of fishing effort. In addition, this would entail fishers taking up and taking on tasks and responsibilities that have hitherto been viewed as being the sole responsibility of the Department of Fisheries.

An idea that has been broached is the introduction of some kind of rights-based system, a move away from the present open access system. This aims at developing a sense of responsibility for, ownership of, and stewardship for the fishery within fishing communities. In this context, the Department of Fisheries is seriously thinking of dividing the lake into zones, and putting these under the responsibility of the traditional authorities directly adjacent to each area (A. Bulirani, personal communication, 15 April 2009). While such a system presents serious practical and political challenges in a fishery largely based on offshore fishing, and the way the sector is currently organized, the debate and consultations between government and communities that this has initiated can only be a good and positive development.

The role of the Southeast Arm fishery and its ability to provide profit and livelihoods for the majority of people in the area and beyond is of paramount importance. This brings into sharp focus the need to conserve the resource into the future so that it continues to provide socio-economic benefits for fishing communities and the nation as a whole. This is the dilemma that the fisheries managers, fishing communities, and development planners have to grapple with.

Acknowledgments Research for this chapter was made possible through a research grant from the Norwegian Research Council for the PovFish project under the coordination, leadership, and partnership of Professors Svein Jentoft and Arne Eide at MaReMa, Norwegian College of Fishery Science, University of Tromsø. I am also greatly indebted to all the fisher folk, Malawi Department of Fisheries staff, and many others for granting me interviews and use of secondary data and information. Without the generosity of these people and organizations, this chapter would not have come to fruition.

[23] Although local level controls exits, in principle anyone can buy a net and start fishing. The annual license fees are part of revenue collection, rather than as a management tool.

References

Banda M, Jamu D, Njaya F, Makuwila M, Maluwa A (eds) (2005) The chambo restoration strategic plan. In: World fish center conference proceedings 71, Mangochi, 13–16 May 2003. World Fish Center, Penang

Béné C (2003) When fishery rhymes with poverty: a first step beyond the old paradigm on poverty in small-scale fisheries. World Dev 31:949–975

Béné C (2009) Are fishers poor or vulnerable? Assessing economic vulnerability in small-scale fishing communities. J Dev Stud 45(6):911–933

Brox O (1990) The common property theory: epistemological status and analytical utility. Hum Organisation 49(3):227–235

Chirwa WC (1995) Fishing rights, ecology and conservation along Southern Lake Malawi; 1920–1964. Afr Aff 95:351–377

Department of Fisheries (2009) Census of fisheries, fishing gears and fishing crafts in lakes and rivers of Malawi, 2008, unpublished report. Department of fisheries, Lilongwe

FAO (1984) Expert consultation on the regulation of fishing effort (fishing mortality). Rome, 17–26 Jan 1983. Food and Agriculture Organisation of the United Nations (FAO) Fish Report No. 289, FAO, Rome

FAO (1993) Fisheries management South-East Arm of Lake Malawi, Upper Shire River and Lake Malombe. Prepared by GOM/FAO/UNDP Chambo Fisheries Research Project. CIFA technical paper 21. FAO, Rome

FAO (1996) Precautionary approach to fisheries. FAO fisheries technical paper 350. FAO, Rome

FAO (2000) Poverty in coastal fishing communities. In: Advisory committee on fishery research 3rd session, 5–8 Dec 2000. Food and Agriculture Organisation of the United Nations (FAO), Rome

Ferguson AE, Derman B, Mkandawire RM (1993) The new development rhetoric and lake Malawi. J Int Afr Inst 3:1–18

GoM (2006) Malawi growth and development strategy: from poverty to prosperity 2006–2911. Government of Malawi, Lilongwe

GoM/UNDP (1998) Management for development programme: revised Mangochi socio-economic profile. Mangochi. United Nations Development Programme, Lilongwe

Gordon HS (1954) The economic theory of common property resource. The fishery. J Polit Econ 62:124–142

Government of Malawi (GoM) (2008) Malawi millenium development goals report. Ministry of Economic Planning and Development, GoM, Lilongwe

Hara MM (2001) Could co-management provide a solution to the problems of artisanal fisheries management on the Southeast Arm of Lake Malawi?' PhD dissertation, University of the Western Cape

Hara MM (2006a) Restoring the chambo in Southern Malawi: learning from the past or re-inventing the wheel? Aquat Ecosyst Health Manage 9(4):419–432

Hara MM (2006b) Production relations and dynamics among user-groups in the artisanal fisheries of Malawi: implications for representation in co-management arrangements. MAST 2006 4(2):53–71

Hara MM, Banda M (1997) Habitat degradation caused by seines on the fishery of Lake Malombe and Upper Shire River and its effects. In: Remane K (ed) African inland fisheries aquaculture and the environment. Fishing News Books (for FAO), London, pp 305–310

Hara M, Jul-Larsen E (2003) The "Lords" of Malombe; an analysis of fishery development and changes in fishing effort on Lake Malombe, Malawi. In: Jul-Larsen E, Kolding J, Overa R, Raakjaer Nielsen J, van Zwieten PAM (eds) Management, co-management or no management? Major dilemmas in Southern African freshwater fisheries. FAO fisheries technical paper 462/2, FAO, Rome, pp 179–200

Hara M, Turner S, Haller T, Matose F (2009) Governance of the commons in southern Africa: knowledge, political economy and power. Dev South Afr 26(4):521–537

Hardin G (1968) The tragedy of the commons. Science 162:1243–1248

Hersoug B, Jentoft S, Degbol P (2004) Introduction: the fisheries development challenge. In: Hersoug B, Jentoft S, Degbol P (eds) Fisheries development: the institutional challenge. Eburon, Delft, pp 11–17

Jentoft S, Onyango P, Islam MM (2010) Freedom and poverty in the fishery commons. Int J Commons 4(1):345–366

Jul-Larsen E, Kolding J, Overa R, Raakjaer Nielsen J, van Zwieten PAM (2003) Management, co-management or no management? Major dilemmas in Southern African freshwater fisheries. FAO fisheries technical paper 462/1, FAO, Rome

National Statistical Office (2008) Malawi population and housing census. National Statistical Office, Zomba

Njaya FJ (2009) The lake Chilwa fishing household strategies in response to water level changes: migration, conflicts and co-management. PhD dissertation, University of the Western Cape

Panatoyou T (1982) Management concepts for small-scale fisheries: economic and social aspects. FAO Fisheries Technical Paper 228, FAO, Rome

Pauly D (1994) On malthusian overfishing. In: Pauly D (ed) On the sex of fish and the gender of scientists: essays in fisheries in science. Chapmann in Hall, London

Rahnema M (2007) Poverty. In: Sach W (ed) The development dictionary. Witwatersrand University Press, Johannesburg, pp 158–176

Sen A (1981) Poverty and famines. An essay on entitlements and deprivation. Claredon Press, Oxford

United Nations Development Programme (UNDP) (2009) Human Development Index Report 2008. UNDP, New York

Chapter 13
To Make a Fishing Life: Community Empowerment in Small-Scale Fisheries in the Pearl Lagoon, Nicaragua

Miguel González

Abstract This chapter explores management from the perspective of a fishing community located in the Pearl Lagoon basin of the Caribbean Coast of Nicaragua. The chapter seeks to address the following questions: How do fishing households in the Pearl Lagoon area respond to management plans designed by regional agencies and national authorities? How is poverty understood and experienced by fishing families and individuals? How is access to land – meaning securing land and aquatic rights – affecting the livelihoods of the people living in fishing communities of the area? Which coping strategies have people undertaken to reduce the vulnerability of their livelihoods?

13.1 Introduction

We depend on the fish, even if the price went down. We cannot buy a pound of rice with a pound of fish, but we have to keep fishing, because there is no other source to have an income, to make a life. And the agricultural production you cannot sell. So you have to just do it. There is no other alternative. Even if this fish is 8 Córdobas (50 cents US) per pound, you have to sell it. What can you do?

Leroy Bennett, fisherman

M. González (✉)
International Development Studies, Department of Social Science, York University, 309 Founders College, 4700 Keele Street, Toronto, ON M3J 1P3, Canada
e-mail: migon@yorku.ca

S. Jentoft and A. Eide (eds.), *Poverty Mosaics: Realities and Prospects in Small-Scale Fisheries*, DOI 10.1007/978-94-007-1582-0_13,

> *Tasba (paunie) gave us authorization to grant fishing permissions over our territorial waters. Our community has laws, but there is no enforcement to support us. We need the government to help us enforce this law.*
>
> Asolin Chang, fisherman and president of the fishing cooperative, Marshall Point

The study focused the research in Marshall Point – a coastal fishing community of indigenous/Afro-descendant origins – located on the west margins of the Pearl Lagoon basin. The research found that Marshall Point seems to be confronting critical dilemmas related to the sustainability of its vital natural resources – land and water. Historical accounts indicate that the community has struggled to secure land and aquatic rights from its surrounding communities (Tasbapaunie and Pearl Lagoon communities), as well as from the Nicaraguan state. These efforts have been made in the context of ever-increasing exploitation of the resources of the Pearl Lagoon basin, particularly fish. At the community level, the research revealed an incipient process of socioeconomic differentiation among individuals and families, which originates from competing dynamics among community members in the light of declining resources. In addition, conflicts over resource use (for instance, cattle-ranching versus small-scale agricultural activities and the use of unsustainable fishing gears) have begun to challenge deep-seated concepts, practices, and norms concerning the sustainable use of natural resources.

Faced with these predicaments such as undefined property rights, internal conflicts over resource use, the exhaustion of the fishing stocks, and the continuous marginalization from spaces in which relevant policies originate (i.e., the design of management plans, and the land titling process), fishermen from Marshall Point have developed a variety of mechanisms to cope with vulnerability and mounting poverty. These mechanisms include, though are not limited to: (1) strategizing toward securing land and aquatic rights; (2) shifting labor from fishing to agricultural production with the aim of securing food supplies and basic dietary needs; (3) organizing a fishing cooperative aimed at timely access of national funding for fisheries development; and, (4) implementing informal community-based actions inspired by sustainable principles to manage the resources of the Lagoon.

However, as the chapter demonstrates, in order for these mechanisms to be effective in the long run, they require sound governance in the area – which includes (though is not limited to) a proactive central state; as well as purposeful local (communal and municipal) and regional authorities. Internally, strategies to overcome poverty, upon which some families have embarked (e.g., overfishing, cattle-raising) run the risk of further marginalizing and impoverishing vulnerable groups (elders and women) in the community.

This study examines the rationale through which community members have responded to policy initiatives aimed at implementing management systems in the Pearl Lagoon basin. In doing so, the research associates policies toward management systems of the resources of the Lagoon to ongoing processes of land surveys intended to grant communal land to indigenous and Afro-descendant peoples. Community members' attitudes and rationale toward management are placed in a broader scope, with the purpose of comprehending livelihood strategies and the community efforts in confronting historical socioeconomic changes, vulnerability, and marginalization. The chapter delves into the tenets that underlined management designs during the 1990s as promoted by competing policy actors in the area; and seeks to comprehend whether such schemes were able to incorporate the needs and aspirations of the people from Marshall Point. In understanding the perspectives of the community, we draw from local perceptions of poverty, on the assessment that community members formulate with regard to their natural environment and resource base, and their strategies toward empowerment.

At the theoretical level, we hypothesize that management is better understood within the concepts of social power and the livelihood approach (Mukherjee Reed 2008, pp. 26–27). As we hope to demonstrate throughout this chapter, management has limited possibilities for success when the agency for sustainable development is affected solely by outside actors (both governmental and non-governmental), within a precarious governance framework; and through partial participation by the owners of the resources. The chapter argues that the potential of management in promoting sustainable fisheries in the Pearl Lagoon has been further restricted due to the narrow understanding of poverty that has transpired, affecting policy initiatives in the area over the last two decades.

A case in point has been the lack of integration of land and aquatic rights within management schemes. Both land claims and aquatic rights have had an historical significance in Marshall Point's struggle against political marginalization and poverty. Having continuous access to land and aquatic resources might explain the community resilience in the light of socioeconomic and ecological transformations. It can also be argued that this question is also significant for other communities in the area (Nietschmann 1973; Hale 1994). Hence, we suggest that the need to preserve a sustainable resource base has forced Marshall Point people to develop a multi-faceted approach to cope with the further impoverishment of their livelihoods.

The chapter identifies four decisive actions in which individuals, families, and community institutions have engaged in their efforts to overcome poverty, vulnerability, and marginalization: (1) strategizing toward securing land and aquatic property rights; (2) shifting labor from fishing to agricultural production with the aim of securing food supplies in times of economic vulnerability and fluctuating market prices for fish and shrimps; (3) organizing a fishing cooperative aimed at accessing national funding for fisheries development; also, a process toward reinvigorating community governing bodies is also noticeable. Finally, (4) implementing informal community-based actions to manage the resources of the Lagoon. This last effort has emerged as a creative innovation between management plans designed by outside actors, and local/communal imperatives for protecting the resources of the Lagoon.

At a more general level, the chapter attempts to scrutinize the relationship between environmental insecurity and poverty. We contend that the effect that environmental insecurity and degradation have on poverty (and vice versa) should be understood within a rights-based approach. For instance, in some particular contexts – such as the case study we are discussing here – securing individual access and collective rights over land and aquatic resources might be seen as necessary conditions for fisher communities to cope with, and eventually overcome, poverty. This question calls upon the design of management systems sensitive enough to take into account the voices of the owners of the resources, the role of their collective governing institutions in decision-making processes, and the livelihood strategies households have unfolded to confront poverty and marginalization.

Management systems unable to capture the multi-dimensional features through which a fisher community understands, copes with, and experiences poverty and marginalization would be certainly limited, based on their lack of capacity, to address the needs of providing a sustainable resource base (Berkes et al. 2000). By the same token, empowerment should be analyzed as the capacity of the community to mobilize the resources at their disposal toward social and economic transformation. Thus, empowerment is for fisher communities *an enabling process* through which poverty can be tackled. The notion of social power captures the agency "from below," through which we try to explain Marshall Point's strategizing in coping with poverty.

The chapter is organized as follows. Section 2 is dedicated to the theoretical framework in which we highlight the theoretical literature as the basis for discussing artisanal fisheries, poverty, and empowerment. The livelihood approach, we propose, better captures the local dynamics in which individuals, families, and the community confront environmental insecurity, resource degradation, and marginalization. We suggest that the livelihood approach should be supplemented with a rights-based understanding of how local community members cope with social and institutional constraints. Therefore, opportunities for local organizing and agency are conceived as devices of social power that explains community-based mechanisms and actions to fight poverty. In Sect. 3 we describe the research methods. Section 4 provides background information on the Pearl Lagoon basin, its natural and sociocultural environment. Overexploitation of natural resources, population increase, and the precarious governance setting are highlighted in this section. Section 5 discusses property rights and commercial fishing in the study area. Section 6 describes and assesses the management initiatives in the Lagoon, as proposed by development actors. This section also examines how residents of Marshall Point cope with poverty and disempowerment. Finally, Sect. 7 offers the research findings.

13.2 Theoretical Framework

The literature on poverty is extensive. That being the case, we found it important to discern a working definition able to guide our research. Income-based and "basic-needs" approaches to poverty have been considered insufficient to capture the dynamics of artisan fisheries. For instance, although a "basic needs" approach might help us to determine absolute and relative poverty, it might not help us to assess the way in which it is conceived, experienced, and acted upon by local people. Such approaches may also be limited in grasping historical transformations a social group has experienced with regard to its understanding of poverty, or to assess the rationale of its adaptive strategies. Therefore, we decided to utilize a notion of poverty receptive to material needs (such as income), but supplemented with a multi-dimensional focus (for instance, social exclusion and discrimination). This conceptual integration allowed comprehending a community's adaptive strategies for coping with the instability of food supply, for example, due to the volatile market prices for fish and shrimps; or as a result of the overexploitation of the Pearl Lagoon resources. Stressing the multi-dimensional character of poverty also provides a better exploration of social manifestations of poverty and power relationships, which play a critical role in understanding marginalization. Furthermore, social exclusion and ethnic discrimination may explain the ambivalence through which Marshall Point's communal authorities have handled claims toward land and aquatic rights, historically.

13.2.1 Livelihoods and Small-Scale Fisheries

We studied Marshall Point's socioeconomic transformation and current challenges in confronting poverty from the perspective of the sustainable livelihoods approach (Allison and Ellis 2001). As pointed out in the literature, fisher households confront crucial imperatives in their pursuit for an increased well-being. The capacity to do so is dependent on their access to various sorts of capital assets, which may be natural, physical, human, financial, or social. Access in turn, is conditioned upon social and institutional factors, which may constrain or provide opportunities for livelihoods strategies to unfold.

In coming to terms with the literature on sustainable livelihoods, we reflected upon the need to relate access (to capital assets) and capabilities (for mobilizing adaptive strategies and/or coping mechanisms) with empowerment. In doing so, access and capabilities were conceptualized within a social power perspective (see diagram below). Based on Friedmann (1992), we grasp the interplay of various types of social power, as they are integrated within artisanal fisheries. Friedmann identifies eight bases of social power: (1) defensible life space; (2) surplus time; (3) knowledge and skills; (4) appropriate information; (5) social organization; (6) social networks; (7) instruments of work and livelihood; and (8) financial resources (Friedmann 1992, p. 69). Empowerment, in Friedmann's view, may occur as the result of an increased access to single elements of these bases. Rather than assuming that these elements are "structurally" interconnected in the life of social groups, we propose to test their actual interactions as they materialize in livelihood strategies as processes of agency and empowerment (Jentoft 2005). This is, to be attentive to "the processes through which social and political powers are acquired," and to conceptualize human development as "the processes of mobilizing social and/or political power to affect social relationships of structural inequality" (Mukherjee Reed 2008, p. 28).

Mukherjee-Reed also proposes to focus attention on four dimensions through which social relationships might be altered: (1) "the division of labor and the processes of material reproduction/production; (2) decision-making processes;

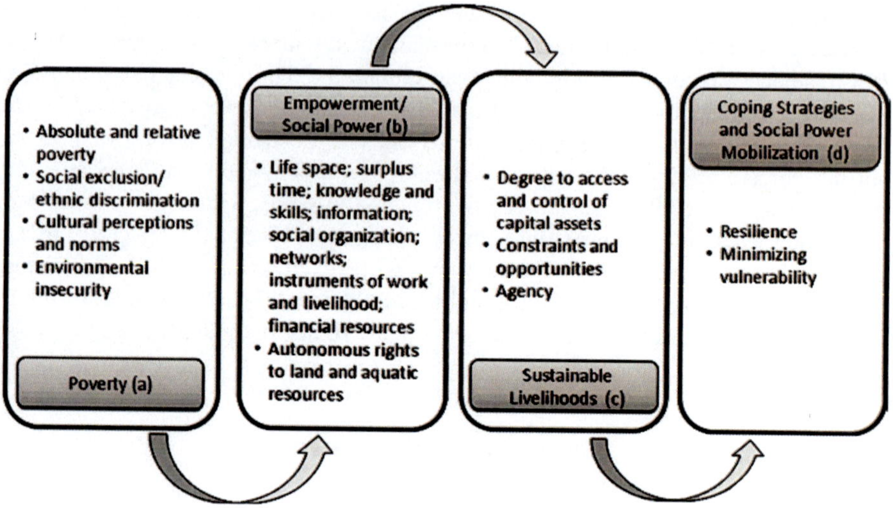

Fig. 13.1 Empowerment from below and human development. Poverty is multidimensional (**a**), communities have access to material and immaterial capital assets, which create conditions and possibilities for empowerment (**b**, **c**). In turn, they often result in multifaceted strategies which can be analyzed from a social power approach (**d**)

(3) the realm of norms/culture/values; and finally, (4) the ownership of knowledge production" (2008, p. 28). For the purpose of this study, we find it relevant to explore the extent to which these dimensions of social relationships have changed over time, through the history of the community. We would argue that various degrees of changes in these realms inform the variety of mechanisms that Marshall Point has turned to in order to cope with poverty and vulnerability.[1]

In synthesis, we propose studying agency and empowerment "from below," without disregarding the effects development initiatives "from above" may have in generating sociopolitical changes conducive to empowerment and minimizing vulnerability (Fig. 13.1).

13.3 Research Methods

With regard to methods, the study included two phases. First, a literature review on the topic of small-scale fisheries and poverty was conducted, looking at current debates in the field. In addition, we delved into studies developed in the Pearl

[1]Vulnerability is understood here as "the degree to which a system or unit, such as a human group or a place, is likely to experience harm due to exposure to perturbations or stresses." Kasperson and collaborators highlight three dimensions of vulnerability: exposure (to perturbations and shocks); sensitivity (of people and places, and the capacity to anticipate and cope with the stress); and resilience (ability to recover and adapt) (Kasperson et al. 2010, p. 236).

Lagoon basin. We focused our pursuit on research related to fisheries, recent development initiatives, and available reports on population increase in the study area. While up-to-date official (government-based) sources were limited, we found a significant amount of information (published and unpublished) generated by two major development/research initiatives in the Pearl Lagoon region, implemented during the 1990s: The *Proyecto para el Desarrollo Integral de la Pesca Artesanal de la Región Autónoma Atlántico Sur* (Project for the Integral Development of Artisanal Fisheries in the South Atlantic Autonomous Region, (DIPAL, for its Spanish acronym)) and the Coastal Area Monitoring Project and Laboratory (CAMPLab), both discussed later in the chapter. Substantive secondary information both from DIPAL and CAMPLab have been incorporated into the present chapter, particularly with regard to the design of management systems.

The second phase of the study consisted of field research. This was conducted in two parts – a 2-week period during the dry season (February 2009); and a 1-month period in the rainy season (July/August 2009). Field research during both periods consisted of ethnographic research, mostly through observations, participant observations, interviews, focus groups, and field trips to fishing and agricultural areas in Marshall Point and its surroundings.[2] Two open-ended questionnaires for interviews were designed. The first one, utilized during the first research period, aimed at understanding historical transformations on the use of natural resources, socioeconomic activities, and sociocultural aspects of the community.

This questionnaire was used in interviews with elders – male and female – and adult fishermen. The second questionnaire, used during the second visit, was designed to capture locally held notions of land and aquatic rights and poverty; as well as to inquire into the livelihood strategies families and individuals have adopted in order to cope with poverty and marginalization. This second questionnaire was mostly used in interviews with adults, young adults and community members who held positions of authority in local governing bodies and community institutions.

While ethnographic techniques provided valuable qualitative information, our study also included quantitative data, gathered through both a community census (conducted twice, for the purpose of reliability), and a socioeconomic survey. The survey, aimed at obtaining data from fishing and agriculture as well as income-generation activities, was conducted with 95% of the community households. The gathered data (both qualitative and quantitative) also benefited from the collaborative research approach our study framed from the onset.

Through a series of community assemblies, as well as meetings with fishermen collectives, our research project's objectives received significant insights.[3] At the same time, the study envisioned research data that can potentially be used for the purpose of contributing to the organizing process of the community's fishermen

[2] All interviews were conducted in Creole English.

[3] This was particularly evident on issues related to community land rights. Our research team was invited to attend various community meetings in which a strategy for dealing with land claims was intensely debated among community members.

who at that time were seeking financial support from the Nicaraguan Institute for Fisheries (Instituto Nicaragüense de la Pesca (INPESCA)).[4]

13.4 Nicaragua and Its Caribbean Coast – A Brief Historical Note

Nicaragua is the second poorest country in Latin America. According to the World Bank, the incidence of poverty is 46.2%, while the population living in conditions of extreme poverty is 14.9%.[5] Economic recovery since the end of the *Contra War* during the 1980s has been troublesome.[6] Although poverty levels have been reduced in the last decade, and the country has made significant progress toward improving its overall economic performance, human development gaps, social inequality, and unemployment remain high.

Development gaps are particularly acute in the Nicaraguan Caribbean Coast regions (from here on referred to as "the Coast"), which comprise approximately half of the Nicaraguan territory, and where 12% of the national population lives. UNDP data from 2005 estimated the human development index (HDI) to be 0.466 and 0.454 for the Región Autónoma del Atlántico Norte (North Atlantic Autonomous Region, (RAAN)); and the Región Autónoma del Atlántico Sur (South Atlantic Autonomous Region (RAAS)), respectively (PNUD 2005, p. xxi; Fig. 13.2). These indexes are comparable to current data reported on Gambia (0.471) and Zambia (0.453), in West Africa. Nicaragua as a country ranks position 124 (0.699) on the HDI scale for 2009 out of 182 countries with data.[7] However, the Pearl Lagoon basin has a relatively higher human development index (0.622) in comparison to other municipalities in the RAAS (PNUD 2005, p. 68). This might be explained by improvements in access to social services (particularly health and education) in the area over the last decade.

Historical relationships between the Coast society and the Nicaraguan state have been characterized by contention and mutual distrust. Pacific Nicaragua was originally colonized by the Spanish in the sixteenth century, while the Coast had the early arrival of English colonizers almost a century after (around the 1600s). During the nineteenth century, the British established a Protectorate, promoted trade with local indigenous peoples, and later in 1860, instituted an indigenous/Creole self-governing entity (the *Moskito* Reserve). The Reserve lasted until 1894 when it was dismantled by the Nicaraguan state, which got hold of formal sovereignty in the Coast region.

[4]For instance, with the support of the research project, a community workshop was conducted at the end of September 2009. This workshop was designed with the purpose of increasing local capabilities of fishermen and fisherwomen on cooperative management (legislation, operation, etc.). This topic seemed relevant for local fishers in the face of a promised loan on fisheries development promoted by government officers from the Nicaraguan fishing authority.

[5]The World Bank. Website: http://data.worldbank.org/country/nicaragua. Accessed 30 September 2009.

[6]The Contra War was sponsored by the US against the Sandinista Revolution that ousted the Somoza dictatorship in 1979.

[7]Human Development Index for 2009 can be consulted online at: http://hdrstats.undp.org/en/countries/country_fact_sheets/cty_fs_NIC.html.

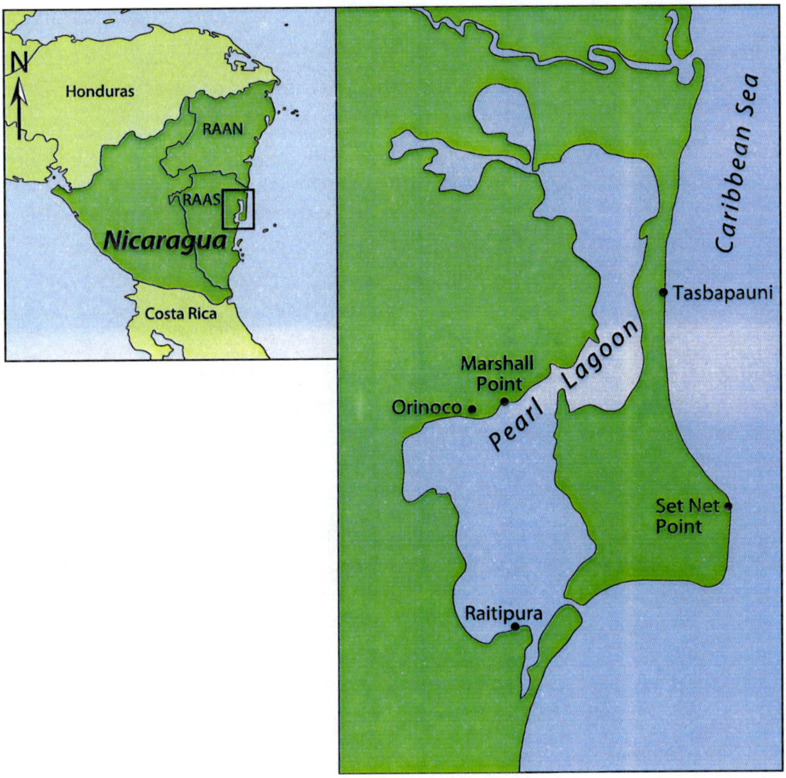

Fig. 13.2 Location map. The Pearl Lagoon Basin. Also shown are the boundaries for RAAN (Región Autónoma del Atlántico Norte (North Atlantic Autonomous Region)), and RAAS (Región Autónoma del Atlántico Sur (South Atlantic Autonomous Region))

Disputes for sovereignty revolved around the Coast's natural resources. The British presence was substituted by the United States, who instituted an enclave economy in connivance with Nicaraguan governments. The region's forests and mines – gold and silver – were intensively exploited. During the twentieth century, the Somoza regime (1934–1979) continued this trend of seeing and acting upon the Caribbean Coast as a reservoir of natural resources, one that must be exploited for the benefit of outside economic agents. The Nicaraguan state has therefore instigated historical animosity and mistrust on the part of the Coast society.

In 1979, an armed revolutionary movement, the Sandinista Front of National Liberation (FSLN), overthrew the Somoza regime. One of the FSLN government's first policy initiatives was to nationalize the country's natural resources. However, this and other policy decisions prompted frictions with the Coast population who had perceived Nicaraguan governments with suspicion, in particular on issues related to the use of the region's natural resources. Land claims, access and control over natural resources, and political participation were crucial demands brought to the fore by culturally diverse indigenous and Afro-descendant peoples and their organizations in the context of a national democratic opening.

In the context of external hostilities against the Sandinista government by the US administration, tensions along the Coast with indigenous and Afro-descendent organizations evolved from political conflict toward an open-armed confrontation. The conflict, which lasted for about 6 years (from 1981 till 1987), was extremely divisive, caused displacement of large segments of the regional population, and many people died. In 1987, and after various rounds of negotiations and consultations, the Sandinista government granted regional autonomy to the *Costeño* people. Specific cultural, economic, and political rights were recognized for the inhabitants of the Coast, and self-governing regulations were instituted on matters related to the protection of communal lands (which cannot be sold or levied), the use and control of natural resources, and political representation.

Autonomy, for the Coast society, is considered a political platform of self-governing rights. In 1990, under the egis of the autonomous regime, two regional multi-ethnic governments were inaugurated, and indigenous, Afro-descendants, as well as Mestizo peoples achieved political representation. However, after almost 20 years, autonomous rights still remain to be materialized. The performance of regional governing institutions has been feeble while their actual influence over regional decision-making is limited. Recognition of communal land has advanced at a low pace; eastward migration of Mestizo colonizers has altered the ethnic composition of the autonomous regions – with consequences in terms of political representation. Furthermore, effective control over natural resources is still an unresolved issue for indigenous and Afro-descendant peoples (Gonzalez 2008).

In synthesis, autonomy represents a decisive legal framework for the Coast population. It has aimed at healing centuries of mutual distrust between two societies, the Caribbean and the Pacific. Nonetheless, levels of poverty and development gaps along the Coast remain significant, and there is little that can be achieved by regional authorities with regard to increasing the well-being of the population in the absence of a collaborative state. In this context, Pearl Lagoon communities have endeavored to assert their collective rights in managing and protecting the resources of the Lagoon in the light of critical internal (e.g., securing food supply and basic sustenance), and external imperatives (e.g., market forces, population increase). Indeed, it is crucial to consider this backdrop from communities' current strategies in coping with and overcoming poverty and marginalization.

13.4.1 Pearl Lagoon Basin – Population, History, and Natural Environment

The Pearl Lagoon Basin constitutes an area of great natural and sociocultural diversity. Various studies have characterized the basin's natural environments as formed by wetlands, pine savannas, mangrove, and tropical rainforests (Christie et al. 2000, pp. 26–32). The Pearl Lagoon itself covers an area of approximately 625 km^2, and is formed by an interconnected system of brackish lagoons and rivers, with at least two passages to the open Caribbean ocean to the North and the South (Sánchez 2001, p. 7).

Studies have pointed out that this system allows a complex influx of salted and freshwaters, ideal for the reproduction of various species of fish, shrimps, and crabs (Sánchez 2001). According to Christie, 46 species of fish have been identified in the Lagoon (Christie et al. 2000, p. 32).[8] The relative abundance and distribution of these species varies according to season, salinity, and the levels of oxygen dissolved in the waters (Pérez and van Eijs 2002, p. 21).[9] In addition, the rich biodiversity that characterizes the natural environment surrounding the Lagoon is the source of animal and vegetal species, highly used by the local population (Fig. 13.2).

The inhabitants of the Pearl Lagoon basin are also culturally and ethnically diverse. English-speaking Creoles and Garifuna peoples constitute two Afro-descendant communities who reside in communities of different sizes around the Lagoon. In addition, Miskitu and Ulwa, both indigenous peoples, have also occupied for time immemorial various scattered communities in the surroundings of the Pearl Lagoon and the Rio Grande delta. Creole and Miskitu communities comprise around 50% of the total population in the basin.[10] In recent decades, Spanish-speaking Mestizo peasants have migrated from the Pacific part of Nicaragua and new settlements have been founded, mostly up rivers, not in the margins of the Lagoon.[11] Some studies attribute the main cause of the population increase in the area as resulting from the eastward migration from interior Nicaragua. The inter-census data reports a 40% population increase in a decade (1995–2005) in the Pearl Lagoon basin. However, the natural growth of local indigenous and Afro-descendant peoples should not be underestimated.[12] Increase in population has also resulted in pressure over the natural resources of the Lagoon, as reported by various studies (Table 13.1).

[8]Fish of commercial value include: snook (*Centropomus* spp.), catfish (*Bagre marinus*), snapper (*Lutjanus* spp.), stripped mojarra (*Eugerres plumeri, Gerres cinereus*), whitemouth croaker (*Micropogonius furnieri*), mackerel (*Scomberomorus brasiliensis*), crevalle jack (*Caranx hippos*), and coppermouth (*Cynoscion* spp.). In addition, five species of shrimps are found in the Lagoon: brown shrimp (*Penaeus aztecus*), white shrimp (*Penaeus schmitti*), pink shrimp (*Penaeus duorarum*), and Atlantic seabob (*Xiphopenaeus kroyeri*) (Christie 2000, p. 32).

[9]For instance, Pérez and van Eijs note that the *B. marinus*, *C. hippos*, the bull shark (*Carcharhinus leucas*), and sardines (*Opisthonema oglinum*) are predominant in the dry season (from November through April/May); while snooks, whitemouth croaker, black mojarra (*Lobotes surinamensis*), tarpon (*Tarpon atlanticus*), and mackerel can be found over the whole year. Though, they are also predominant during the rainy season (from May until October) (Pérez and van Eijs 2002, p. 21).

[10]The communities of Awas, Raitipura, Kakabila, and Tasbapaunie are mostly inhabited by Miskitu people; while Brown Bank, Haulover, the town of Pearl Lagoon, Set Net and Marshall Point have historically been considered Creole-inhabited communities.

[11]However, during the second period of field research, a small group in Mestizo families was given temporary permission by Tasbapaunie for lodgings and to cultivate a plot of land in the western part of the Lagoon. The nature of this informal agreement was not entirely clear to us.

[12]For instance, unofficial estimates from 1992 reports 4,749 Afro-descendants and indigenous inhabitants (Christie et al. 2000, p. 22). In the 2005 census, this population is 6,394 (Williamson and Fonseca 2007, p. 59). This is about 26% of the population growth in a 12-year period. In addition, local population, particularly youth, have engaged more intensely in migration abroad as temporary workers on shipping cruisers in the US. These data might be unreported in official national or regional censuses.

Table 13.1 Population increase in Pearl Lagoon, 1995–2005. This data shows the population increase of Pearl Lagoon at a rate of 40% every 5 years since 1995. Data for 2010 was not yet available

Year	1995	2000	2005
Population	6,253	8,936	10,676

Source: INEC (1995), and Williamson and Fonseca (2007, pp. 57–58)

13.4.2 Why Study Marshall Point?

There were important considerations in choosing Marshall Point as a research area. First, the population size: 361 inhabitants; 54 houses concentrated in a relatively small coastal settlement allowed us to pursue more engaging ethnographic research techniques with the various segments of the local population (Fig. 13.3).[13] Second, the scant research that is currently available for the community, and the reasons that might explain this situation, highlighted the need for scholarship on the area. While significant research efforts – both historic and contemporary – have been conducted in Tasbapaunie and Orinoco, the two neighboring communities, Marshall Point has received just marginal attention.[14] In part, this might be explained by the intermediate sociocultural (also spatial/geographical) position of Marshall Point, which found itself between two poles of assertive and contrasting ethnicities in the Pearl Lagoon, the Miskitu (Tasbapaunie) and the Garifuna (Orinoco).

It is noticeable that, in both communities, a great deal of scholarly attention has focused on processes of socioeconomic transformation due to market forces, ethnic relationships, and cultural change. Although Marshall Point has been described as a Garifuna community, current inhabitants prefer to emphasize the mixed character of their socio-ethnic origins: Garifuna, Miskitu, Kukra, and Creole.[15] In the past, labeling the community by outside agents as belonging to one of the regionally identified ethnic groups has often meant political and social discrimination.[16] Our research also sought to investigate the specific role identity has played within community strategies in the direction of securing land and aquatic rights.

Marshall Point's community authorities also demonstrated early interest in the study, at the first round of consultations. The research coincided with the organizing

[13]Two censuses were conducted: the first in February, and the second in July 2009. Our data registered 30% population growth in Marshall Point between 1992 (Christie et al. 2000, p. 22) and 2009. Multi-family households characterize Marshall Point's social structure. Our survey revealed an average of 6.5 persons residing per household.

[14]For Tasbapaunie, see Nietschmann (1972, 1973); Kindblad (2001). For Orinoco, see Davidson (1980).

[15]Kukras indigenous peoples inhabited the area of what is now called Kukra Hill, south of the Pearl Lagoon basin. Ethnographic studies suggest that Kukras are now extinct. However, families in Marshall Point still track their ancestors to the "kukras" from the Kukra Hill area.

[16]Nicolas Gutierrez Bennett, locally known as Uncle Pi, personal communication, Marshall Point, 26 February, 2009.

Fig. 13.3 Fishers from Marshal Point, July 2009

process – certainly not the first one – to form a community-based fishing cooperative. Therefore, the study presented potential collaboration of mutual interest with the fishermen. Overall, studying Marshall Point's fishery gave us an opportunity to advance knowledge about the community; but also to engage in a collaborative research methodology suitable for the scope of the present study.

13.5 Early Ambivalence on Property Rights and the Impact of Commercial Fishing

Marshall Point was populated in the second half of the nineteenth century, probably around the 1870s.[17] In this sense, it is one of the newest settlements to be populated in the Pearl Lagoon basin. Original inhabitants of Marshall Point arrived from Brown Bank and Kukra Hill. They were from mixed ethnic

[17] This data is based on historical accounts provided by the community's elders. Local narratives made references to Mr. Henry Patterson from Pearl Lagoon (referred to as "Mr. Patisson"), who served as Vice President of the General Council of the Moskito Reserve (see Von Oertzen et al. 1990, p. 322).

origins, blacks (or Creoles), Miskitu, and possibly remnants of Kukra Indians, now extinct.[18] Later on, Garifunas – from La Fe and San Vicente, two surrounding communities, joined them.[19]

Marshall Point had been a small agricultural plantation previously occupied by a US citizen ("Mr. Marshall, a white man"), and it was abandoned when the new settlers arrived into the area.[20] A common practice of the Moskito Reserve's authorities was to grant land plots to foreigners (to individual entrepreneurs and firms) with the purpose of forest extraction and commercial agriculture.[21] The new inhabitants of Marshall Point did not have to request "permission" from the Reserve's authorities to settle in the former plantation since they "were people from the Atlantic Coast."[22]

At an unspecified date during the 1930s (probably 1932), Marshall Point inhabitants were invited by public authorities to the town of Pearl Lagoon – which once served as the headquarters of the *Reserve* – to join a newly emerging administrative municipal jurisdiction. However, representatives of Marshall Point were rapidly dismissed by Pearl Lagoon political authorities arguing that residents of Marshall Point were "*karob* people," and therefore they were no longer welcomed "in our community."[23] Historical narratives reported that Orinoco – which has been a mostly Garifuna settlement founded in the early to mid-ninetieth century – have also made references to this momentous episode (Figueroa 1999, p. 39).

The incident has been incorporated in local narratives to demonstrate an open discrimination experienced by the community due to an externally imposed ethnic identity. It also provides historical justification for explaining why the community turned to Tasbapaunie – a larger and older Miskitu-inhabited neighboring community – seeking guarantees for their continuous access to land, natural resources, as well as for securing political inclusion within the new governing framework which was in the making. As stated by Uncle Pi, an 86-year-old, respected elder in the community:

> We join to get power from them, to control the land, the beach, all the timber, all the log, because you understand, Kurinwas is a big river and the Kurinwas River they say gone to Matagalpa.[24]

---

[18] "Wild Indians" was the term used by Uncle Pi to emphasize the indigenous/black ancestry of the founding families of Marshall Point (personal communication, Marshall Point, 26 February 2009).

[19] Men from Marshall Point took wives from Garifuna communities in the area. According to Davidson, Garifuna communities were founded between 1880 and 1912. Square Point was the first to be inhabited around 1880/1881, and Orinoco the last one, in 1912 (Davidson 1980).

[20] Three families were the first to set foot in Marshall Point: The Goff, the Bennett and the Peralta. They "divided" the community into three sections, uptown, middle town and downtown. It is interesting to note that these areas are said to be "private property" of the founder families who also hold original title deeds on their land.

[21] Several land grants were later annulated by the Nicaraguan government once it took over the Moskito Reserve in 1894.

[22] Nicolas Gutierrez (Uncle Pi), personal communication, Marshall Point, February 26, 2009.

[23] Ibidem. "Karob" in Creole refers to "Carib" which was a term commonly used to refer to the Garifuna people from the Caribbean Coast of Nicaragua. The term might have been used during the colonial period, when Spanish authorities used the term to label all the non-colonized (non Christianized) natives of the lowlands of Central America.

[24] Uncle Pi (personal communication, Marshall Point, February 2009). Kurinwas refers to the river that signals the west border of the Tasbapaunie's land claim.

The act of inclusion on the part of Tasbapounie's authorities has been also conceptualized in the community's social memory as recognition of the Indian ancestry and social identity of Marshall Point's inhabitants.[25] Marshall Point's bonds with Tasbapounie have further been fortified by a sense of common struggle for land rights. Indeed, as a result of these struggles, several land title deeds were secured early in the twentieth century.[26] Up to the present, Marshall Point has remained as a "sister community," included in the Tasbapaunie's communal land claim (CCARC 1998, p. 336).

13.5.1 The Impact of Commercial Fishing

Commercial fishing along the Caribbean Coast began around the 1950s, and became intensified during the late 1960s and 1970s (Kindblad 2001). Nietschmann (1973) provides a comprehensive account of the impact of commercial fishing of lobster and turtle on Tasbapaunie's moral economy. Christie and collaborators report that: "Until about the 1960s, Pearl Lagoon fishers usually used hooks-and-lines and harpoons to strike fish in the shallows; a practice that was only effective when fish stocks were abundant" (Christie et al. 2000, p. 39).

The abundance of fish and shrimps before the introduction of commercial fishing was also reported to us by adults and elder fishers of Marshall Point. They also commented upon the earlier capacity for economic self-sufficiency the community had on some of the basic products that form the daily dietary needs. "We never used to buy rice because we had the land and the farmers to cultivate it."[27]

This was also the case for other agricultural products such as cassava, bananas, pineapple, and dasheen.[28] It was also reported that the community had once the capacity for producing modest surplus of agricultural products, particularly rice, which could be traded with the surrounding communities. This capacity is now gone, rice is rarely cultivated, and it has to be purchased in Bluefields or the town of Pearl Lagoon.

It also seems that commercial fishing had also triggered a shift in the sexual division of labor. As Ms. Adleen puts it:

> In first time men hardly fishing, they do more farming, cut the ground and plant it, and then the women do more fishing, they catch fish and shrimps.[29]

---

[25] "We joined with them (Tasbapaunie) because they recognize us as Indians. Yes, Indians for Indians" (Uncle Pi, personal communication, Marshall Point, February 2009).

[26] Local accounts also attest the common efforts in which both communities have been engaged with in order to protect their land and natural resources (Christie et al. 2000, p. 26) citing data from IPADE reports that between 1917 and 1957 Tasbapaunie was able to secure several title deeds over plots of land in their territory, as well as in the Pearl Cays.

[27] Adleen Bennett (personal communication, Marshall Point, February 2009).

[28] Dry-salted fish was also produced locally and sold in Bluefields or Managua (Joice Cayasso, personal communication, Marshall Point, February 2009).

[29] Ibid.

New fishing gears and techniques were introduced by commercial fishing – such as gill nets – and men were dragged out from agriculture toward the cash economy encouraged by the newly established private processing centers. This shift has also been reported by studies conducted in other communities of the Pearl Lagoon basin (Christie et al. 2000).

The introduction of new fishing techniques, in particularly multi-filament gill nets, in turn impacted severely in the fishing stock, which were now placed under a greater catching effort.[30] Christie notes that, "although gill nets seem to have been introduced in the 1960s, they remained relatively scarce until the late 1980s when government and free-funded development programs supplied them to fishers" (Christie et al. 2000, p. 39). Most community members consider the introduction of gill nets as a critical turning point in the capacity of the Lagoon in providing a stable resource base for food, and a defining moment that brought about both the loss of self-sustainable agriculture and internal socioeconomic differentiation. Local traditional norms have prescribed that hook-lines are to be used in the dry season, while gill nets should be used during the wet/rainy season, when more fish are available. Nonetheless, a common complaint in Marshall Point today is that "people are now fishing with gill net the whole year around,"[31] which negatively impacts on the sustainability of the lagoon's resources.

Commercial fishing generated several incentives over local subsistence economies toward cash-based market relationships. Food supplies were now made available in local markets, and therefore a significant effort – in terms of labor – was directed toward fishing, so as to get "cash for the day." Uncle Pi noted:

> When fishing was good, we could buy lots of food. Now that the fishing has gone down low is hard, is better if we did plant rice for the whole year to eat, so you don't have to buy. But now the rice price is higher, because when rice comes from the US it costs more.[32]

13.5.2 *The Expansion of the Agricultural Frontier*

Several authors have noticed a trend along the Coast, described in terms of the expansion of the "agricultural" frontier. Although this is not a new process, it might have been intensified over the last two decades. At the end of the war, development "poles" along the Coast were promoted by national governments, and the expansion of cattle-raising – due to an increased demand for beef – generated an eastward migration of mostly poor peasant colonizers. Indeed, the Coast population grew

---

[30] Gill nets are used to catch fish in the rainy season. However, community members complain that gillnets have now also been used in the dry season. Shrimps are captured with cast nets mostly during the dry season. Trawling for catching shrimps is not allowed within the Lagoon (municipal ordinance). Lobster fishing is done by diving or by setting pots (or traps) out at sea – around the Cays, which are located between 1 and 10 miles away from the coastline.

[31] Leroy Bennett, personal communication, Marshall Point, 20 July, 2009.

[32] Uncle Pi, personal communication, Marshall Point, February 2009.

4.9% between the 1995 and the 2005 censuses. This rate of growth constitutes a record for all the municipalities of the country. The Pearl Lagoon basin has experienced the impact of this population increase (Table 13.1). New settlements have been established "up rivers" in Wawashang and Patchi Rivers, and land use experienced important transformations (e.g., from forest use to agricultural and cattle-raising production).

This expansion has also instigated tensions between indigenous and Afro-descendant communities, and Mestizo peasants with regard to land tenure. Two concepts of land property have often conflicted: on the one hand, community-held land; and on the other, privately held land as upheld by Mestizo peasants. This tension is even more problematic in a context in which communal lands are yet to be surveyed and demarcated, and therefore competing claims emerge over land property rights between *campesinos* and indigenous/Afro-descendant peoples.[33]

Until recently, Marshall Point had not faced this tension in a critical manner, which seemed to be the exception in the Pearl Lagoon basin. Though, this has changed. Due to the fact that its land claim has been made in conjunction with Tasbapaunie's land claim, negotiations with illegal Mestizo occupants is now a matter of common concern of both communal authorities. Indeed, they agreed to form a unified territorial government to oversee their land and to advocate demarcation.[34]

13.5.3 Socioeconomic Changes and Outward Migration

At the end of the war (around 1986), Marshall Point began to be once more re-inhabited. The armed conflict had provoked displacement with a number of families fleeing the country, and others moving to Bluefields, or "up river" (to Wawashang) seeking refuge. We have estimated that approximately 60% of the population left the community around 1983; and the following year when the armed conflict intensified in the area. Several of these families did not return to the community but others did. Returning to the community meant the possibility for reconciliation but also for economic recovery. However, in 1988, hurricane Joan devastated the area, which made things difficult for returning families to cope with the post-war period

[33] State recognition of communal property rights on land has been a contentious issue in the history of the Coast. Lack of land surveying and titling sparked conflicts between coastal communities and the state since the Moskito Reserve was dismantled and placed under Nicaragua's full sovereignty. Although some efforts were conducted to survey communal land early in the twentieth century, undefined property rights had been the norm until 2003 when a communal land bill was passed by the National Assembly.

[34] It is important to note that Marshall Point has identified its specific land claim within the overall territory that is claimed by Tasbapounie. Marshall Point's authorities had initially decided to join the "Pearl Lagoon block" which is formed by 10 communities to pursue its land title deed. Afterwards, the community changed its position on the matter and reunited with Tasbapaunie. A common sentiment of collective struggle for land rights between both communities has made Marshall Point's claim a relatively consensual process.

Table 13.2 Shrimp, fish, and lobster production for the Caribbean Coast and Pacific Nicaragua, 1994–2007 (in millions of pounds). This table compares seafood production – scale, shrimp and lobster – for the two coasts of Nicaragua. It is noticeable that the Caribbean Coast makes a substantial contribution to national volumes, in particular, shrimps and lobster

	Shrimp (*Penaeus* spp.)	Scale fish	Lobster (*Panulirus argus*)
Caribbean Sea	54.2/80%	51.6/41%	38.5/99.7%
Pacific Ocean	13.6/20%	74.3/58.9%	0.1/.3%
Total	67.8	126	38.6

Source: INPESCA 2007, pp. 49–53

(Envío 1989). Fishing represented a potential resource – and perhaps the only one – which communities could rely on. Christie reports that "fish harvest decreased in the 1980s allowing fish stocks to grow, but now have risen again to prewar levels" (Christie et al. 2000, p. 38).

In 1990, the FSLN lost the general elections to a right-wing coalition, which made matters more problematic for the actual implementation of the autonomous regime. The new national government sought to develop a market-based, export-oriented extractive economy. The imperatives and stimuli this shift in policy brought about, induced significant changes along the Coast, and particularly for the Pearl Lagoon economy. New seafood processing plants entered into the area, and a new cycle of market-based, cash economy began. From 1994 until 2007, market for fish, lobster, and shrimps did provide a relatively stable source of cash income for fishermen in the Pearl Lagoon basin. It also provided a source of food supply for their families. This came at the expense of the sustainability of the resource base, and continued the long-term trend of undermining the subsistence agriculture.

Historically, the Coast's contribution to national seafood supply has been significant. It contributed 35% of national fish production between 1994 and 2007. More concretely, between 1994 and 2007, artisanal fisheries along the Coast contributed 12% of shrimps and 51% of lobster (*Panulirus argus*) to national volumes (Table 13.2). Seafood products originating in Pearl Lagoon accounted for 6.9% and 4.5% of national volumes in 1998 and 1999, respectively (Hostetler 2005). These levels of extraction might be arriving at their limits.

Over the last 2 years, the effects of overfishing over communities' well-being have been felt more intensely. This has raised concerns *between* communities in the Pearl Lagoon areas as well as *within* communities. In addition, over the last 2 years, international market prices for shrimp, fish, and lobster have declined as the result of contractions in demand, particularly in the US market, where most of Nicaraguan seafood exports are directed.[35] Processing plants that entered the area early in the 1990s have either shut down or reduced operations, or have moved to Bluefields to minimize operational costs. Itinerant buyers are now sporadic and fish prices are significantly low as compared to previous years. Seeking economic

[35] In addition to contraction in demand, for instance, for lobster, the state of Florida has also banned the imports of snook. For the Coast, this resulted in falling prices for snook about 100% (Ibarra 2009).

alternatives, individuals and families have started to search for stable jobs outside the community.

Although outward migration is not new, a new wave can be traced over the last 5 years. Our survey estimated that around 10–15% of Marshall Point's active labor force – youth and adults between 18 and 35 – are now working abroad on international cruise ships or have found seasonal employment in the English-speaking Caribbean. The survey also registered an increase in outward migration the last 5–6 years. Access to remittances from abroad has supplemented the income of a few families in Marshall Point, and to some extent this is contributing to increasing levels of social differentiation internally.

13.5.4 *Increase in Social Differentiation and Changes in Social Organization*

Traditional, self-governing authority in the community has relied on two local entities: the communal board and the sorority group (also called "the society"). While the "society" provides material and moral support to community members in times of illness or death, the communal board oversees decision-making with regard to the community life, including matters related to the use, exploitation, and access to natural resources.[36] The regular functioning of both entities was interrupted during the war, but nowadays they perform an essential role in community life. Churches also exert influence in local matters, though this is done through the direct participation of religious leaders in communal governing bodies, such as the community board.[37]

Internal socioeconomic differentiation does not seem to have been a factor of internal disputes or tensions two decades ago, during the pre-war years. A relatively equitable sexual distribution of labor within households, self-reliance in subsistence agriculture (including some levels of surplus exchange with neighboring communities), and stable access to fish and shrimps from the Lagoon, did not propel internal differentiation in the degree of individual or family's access to capital assets. Apart for commercial fishing, there was no other market mechanism that produced substantive change in local economic dynamics.[38]

This has now changed drastically. Cattle-raising, which has historically been marginal in Marshall Point is now being promoted by a handful of relatively

---

[36] Since the approval of Law 445, communal authorities have "formalized" in written form their communal regulations. These regulations establish the norms and procedures through which local governing bodies are formed and operate. They also include some regulations regarding the use and exploitation of natural resources (Community of Marshall Point 2003).

[37] For instance, the local priest of the Evangelical church is also the head of the "society" directive board. The actual community coordinator – a woman – is a founding member of the Pentecostal church in town.

[38] Private banking did explore a few ventures for commercial agriculture, though with limited results.

better-off families. The expansion of pasture fields (*potreros*) is competing with areas traditionally used for agricultural production. In addition, several fences have been erected which contradict the norms set forth on communally held land traditionally upheld by Marshall Point's inhabitants. This emerging process of competition over communal land is more critical in context in which land claims are yet to be adjudicated (or demarcated), and when the community is also confronting encroachment by external occupants (also competing with agricultural areas), including from its neighboring community, Orinoco.

In synthesis, over the last decade, the Caribbean Coast has experienced important socioeconomic and political changes. Some of these may be explained by internal factors and long-term trends and dynamics – such as the contrasting development gaps between Pacific Nicaragua and the Coast. Others are related to external processes – such as the fluctuating market prices for sea products – which are better comprehended by looking at the changing patterns in the global political economy. Both processes – internal and external socioeconomic and political factors – have represented constraints and opportunities for the subsistence economy and social organization of the Pearl Lagoon basin.

The long-term trend of overexploiting the resources of the Lagoon is probably reaching its limits in terms of guaranteeing a secure resource base for future generations. For Marshall Point, as well as for other communities in the area, this represents a critical challenge for coping and eventually overcoming poverty and marginalization; considering the slow pace through which communal land recognition has advanced. Though, livelihood adaptive strategies (for instance, outward migration or shifting toward agriculture in times of crisis) are exhibiting a certain capacity for community resilience in light of external constraints.

13.6 Management – What Needs to Be Done and for Whom? Conflicting Visions

During the 1990s, two opposing development visions brought about major management initiatives to deal with the conservation of the natural resources of the Pearl Lagoon basin. DIPAL, a bilateral funding initiative (between The Netherlands and Nicaragua), began its activities in 1994 and ended in 2001. The project was aimed at, "creating the conditions for improving fishermen's wellbeing as well as the living conditions of their communities." The project sought to achieve this by providing, "a sustainable use of the fisheries' resources [based on] equal opportunities, and local participation" (Pérez and van Eijs 2002, p. 3).[39]

Since its inception, DIPAL promoted a vision of management that emphasized the commercial use of the fisheries of the Lagoon. During its first phase (1994–1997), DIPAL devoted its activities to generate baseline information on the hydro-biological resources and the economy of the Pearl Lagoon basin. During its second phase

[39] My translation.

(1998–2001), efforts were made in order to influence policy making, in particular with regard to the implementation of a management plan in the Lagoon.

CAMPLab, an IDRC-funded program inaugurated in 1993 through a research alliance between CIDCA (The Center for Research and Documentation of the Atlantic Coast, and affiliated with the Jesuit-led Central American University, UCA), national/local and international researchers, and various local communities, had the aim of developing a management plan for the Pearl Lagoon basin. The goal of CAMPLab was to, "develop a knowledge base to inform a formal management regime for the area's coastal resources." This regime was to be designed through a participatory methodology (Vernooy 2000, p. 9).[40]

Both management initiatives somehow ran parallel to each other, with little instances of collaboration between them. During its second phase (in 1997) DIPAL proposed an "integrated" management plan for the fisheries of the Pearl Lagoon, which was later adopted as a government official regulation (Pérez and van Eijs 2002, p. 3).[41] However, the proposal was not widely consulted with the resource's users, which reflected DIPAL's top-down development strategy and intervention principles. The consequence was open ambiguity on the part of the fishermen in adopting the regulations established in the management plan and therefore its potential for managing the resources of the Lagoon has remained with marginal effect to the present day. In addition, the governmental agency that oversees the enforcement of fishing norms, INPESCA, lacks the human and financial resources to exert supervision and control in the fishing areas.[42]

In 1998, CAMPLab also produced a management plan. The scope of the plan went beyond the aquatic ecosystems, to also include terrestrial resources, in correspondence with community consultations (Christie et al. 2000). The plan was "adopted" by the municipal government through a local ordinance. At the regional level – within the South Atlantic Regional Autonomous Council – representatives recommended that both DIPAL and CAMPLab management regimes be integrated and should result in a "unified" plan (Christie et al. 2000). This collaboration did not render positive outcomes, and relations between these organizations became "confrontational" (Christie et al. 2000).

By and large, the two management schemes still remain today as an illustration of opposing development views with regard to *what* needs to be done, *with whom*, and *for what purpose* in managing the fisheries of Pearl Lagoon (Table 13.3). In constructing management designs, the perspectives and participation of local communities are critical factors in determining feasibility.

[40] Since its inception, the project design adopted a participatory methodology which included, "a 6-month period of exploratory research resulting in the identification by local people of the need for a management plan for the basin's natural resources" (Vernooy et al. 2000, p. 8).

[41] Government decree No. 043-98 issued by the Ministerio de Fomento, Industria y Comercio (MIFIC). The plan, according to DIPAL's directors, "establishes the designated fishing zones, as well as the basic designs and manufacturing guidelines for the fishing gears permitted" in the Lagoon. See Pérez and van Eijs (2002, p. 3).

[42] Karen Joseph, regional delegate of INPESCA, personal communication, July 2009.

Table 13.3 DIPAL, CAMPLab, and Marshall Point Management Plan comparison (fisheries)

Issue	DIPAL	CAMPLab	Marshall Point
Gill net mesh size	4 in. – exception of mackerel and mulit nets used in times of high productivity specified by the government	4 in.	3–3.5 in. (no consensus)
Shrimp nets	2.25 – exception of sea bob nets used in times of high productivity	1.5 in.	1.5 in.
Shrimp size	71–80 per pound	60–70 per pound	Varies 55–60 per pound
Fishing in canals	Only with line and cast net	None specify no sinking of nets in canals	Allowed for local fishermen
Refugee/Breeding places ("the hole")			Only hook-lines permitted (dry season)
Outsiders	No fishing by outsiders	No fishing by outsiders	Allowed under certain conditions (taxation), mesh-size to be inspected
Closed season (Vedas)	N/A	N/A	Gill nets should not be allowed in the dry season
Trawling	N/A	N/A	Not allowed in the Lagoon
Registration	Register fisherman and boats		For outsiders
Sardines	2.25 in. nets	No massive fishing for species	
Motors	70 hp max		
Limits to the sea		No industrial boats within 5 miles	No industrial boats within 5 miles
Area	3 miles to sea 5 miles around cays	5 miles to sea (no industrial boats) 15–20 around cays	5 miles to sea
Coral		No removal of dead coral	Corals are removed for ornament purposes
Lobster diving		No lobster diving	Lobster diving allowed
Lobster season		Closed February–May	February–April
Waste water		No disposal of waste fish or contaminated water	
Limits in the lagoon			Community exclusive area for fishing should be regulated

Source: DIPAL and CAMPLab data based on Hostetler (2005, p. 331). Data from Marshall Point, based on field research

13.6.1 *Perspectives from Marshall Point*

"Protecting the Lagoon? Indeed, but we cannot do it alone." The research inquired about the responses of the people of Marshall Point to the management plans as designed and proposed by DIPAL and CAMPLab. The first noticeable finding was that, overall, community members were not able to distinguish between the two management proposals. Among fishers, references were made to the "DIPAL project" when referring to commercial fishing initiatives – which were widely promoted by DIPAL.[43] For CAMPLab, comments were made with regard to "workshops" on natural resources of the basin being held in the town of Pearl Lagoon or Haulover. Overall, few adults – mostly men – had participated in some of the meetings organized by DIPAL or CAMPLab. Technical details – about norms and regulations being proposed in the management plans – were largely ignored. Nonetheless, rather than stressing opposing interests between the two schemes, deep concerns were expressed with regard to the lack of implementation on the part of the government (municipal, regional, and national). The following comment from Manuela, a founding member of the Orinoco-based women's cooperative, is telling:

> I remember that was when DIPAL was working in the Pearl Lagoon area. Yes, I still have my management plan book from that time. No government came out strictly about it, they slack, and don't care. You know what I got to say? Them eat bread with butter, so them have no interest in we the poor people. They are getting thousand and thousand (of dollars) and we no get nothing, and when the fish go away from we, they getting a thousand there in the office and we don't going have fish, we don't going have nothing.

It is also noticeable that Marshall Point's inhabitants were not consistently consulted about management designs as proposed by the two development initiatives. This does not mean that residents opposed the idea of regulations and control over the resources of the Lagoon. On the contrary, during the research activities community members explored a variety of ideas for securing the resource base. These ideas ranged from completely banning resource extraction (through *vedas,* or closed seasons); using a number of selected techniques (gill nets) on certain seasons; limiting access to outside fishers (from Bluefields, or Rama); protecting distinctive reproductive areas; to locally managed regulations (Table 13.3).

However, management and ecological protection cannot be done by Marshall Point in isolation from the surrounding communities, and especially it cannot be

[43]For instance, in 1999, DIPAL placed an ice box (cooler) in the community for storing fish and shrimps. It also provided a boat and fuel. Lack of proper maintenance and management skills resulted in the shutting down of the local acopio. The person responsible for the ice box told us in an interview that "local community needs" were as important as storing fish. For instance, he provided fuel to families in need, and also advanced small cash amounts to fishermen. He saw no contradiction between community needs and the role that DIPAL was aiming to play in improving the well-being of the fishermen.

done in the absence of a collaborative government. This view is clearly expressed by Herbert Bennett, a local fisherman:

> I feel like, if the government and the people get together, with the people of the community, you have power. *The community decides and the government put the force*, because that's the government. The community alone I don't feel like we can get no way, we need the government. The government has to be in it.[44]

Marshall Point has not had to wait for the government to take action. A multi-faceted, innovative approach for protecting the resource base of the Lagoon is underway. However, the rationale that informs the community's strategies for coping with poverty are deeply embedded in peoples' perceptions of poverty as well as on their assessment of the relative resource depletion.

13.6.2 Not So Poor, But Not So Rich Either

Having a secure, even if minimal, resource base for food supply, particularly fish and shrimps, is the key determinant for the community's perception of someone being in a situation of extreme poverty. One is said to be "poor" if there is "nothing in the plate to eat." This is a situation that, according to local perceptions, no family in the community has yet ever experienced. This is what places Marshall Point at a relative advantage with regard to other settings in Nicaragua – references are made to urban areas – which suffer from extreme poverty, and where there "is nothing to eat."

In the community's view, the relative availability of resources capable of providing some "food in the plate" is being threatened by internal as well as external practices and dynamics. For instance, local fishermen use detrimental fishing practices. Overfishing with gill nets by the few affluent families in town encroaches upon the rights of the local poor, and menace their livelihoods. As cogently stated by Uncle Pi:

> You might have 20 (gill) nets and you put out the 20, and I might not even have a net, none at all, so when you put out 20 nets, you are harming me, because when the net go and set out maybe you catch 400 pound of fish and you have and I don't have no fish. So my family starve, is that truth? You starving me and my family, when I go with my hook line there is no fish in the lagoon.

In local perceptions, poverty is more than just lack of access (and availability) to a (traditionally) stable resource base. It is also explained in terms of marginalization from relevant policy decision-making and politics at the municipal regional and/or national levels. For instance, over the last 5 years, there has been an increase in basic public services available in the community; in particular, electricity (established in December 2007), schools (elementary), a health clinic, and communication (through privately owned satellite telephone services). These services have not come as the result of the "good will" of the government or private providers. Community members

[44]Personal communication, Marshall Point, February 2009; emphasis added.

explain that the majority of these are the result of active community's advocacy before the government and non-governmental organizations. Other matters, equally relevant for the community's well-being, are not easily subject to influence. For instance, achieving political representation in the municipal or regional governments is complicated due to exclusionary mechanisms that, according to local residents, reproduce historical marginalization.[45]

13.6.3 Marshall Point – Coping with Poverty and Disempowerment – A Multi-faceted Strategy

As it has been explained, Marshall Point confronts a vulnerable situation characterized by undefined property rights, and internal conflicts over resource use that are the source of growing tensions, the exhaustion of the fishing stocks, and the continuous marginalization from decision-making of relevant policies. Facing these challenges, inhabitants have developed a variety of mechanisms with the purpose of coping with vulnerability and poverty. Livelihood strategies, as we would like to argue, make sense within a comprehensive scenario of agency and social power displayed by individuals, families, and community institutions. We have tried to illustrate this multi-faceted approach in Fig. 13.4:

Six strategies for this multi-faceted approach displayed by Marshall Point merit a closer examination: (1) strategies toward securing land *and* aquatic rights; (2) efforts in the direction of reorganizing a fishing cooperative; (3) shifting labor from fishing to agricultural production; (4) outward migration and educational opportunities; (5) accruing social and political power; and (6) implementing informal community-based actions to locally manage the resources of the Lagoon.

13.6.3.1 Securing Aquatic and Terrestrial Rights

For Marshall Point, securing unambiguous terrestrial and aquatic rights constitutes a leading livelihood strategy toward empowerment. Christie reports that the "concept of sea tenure does not exist in the strict sense" in Pearl Lagoon. He observes that, "each community has its preferred fishing grounds, but many of the most popular sites are used by a number of communities" (Christie 2000, p. 37). We did not observe this apparent lack of explicit definition of tenure over aquatic sources in

[45]For instance, the electoral law includes Marshall Point within circumscription 7, which comprises the "Garifuna" district. The law establishes that the first candidate heading the three-member list should be a "Garifuna." Political parties and organizations have traditionally chosen residents from Orinoco, which is the largest Garifuna settlement in the Pearl Lagoon basin. This practice has meant that Marshall Point's residents are continuously marginalized from the possibility of being elected into the Autonomous Regional Council. Since 1990, only two persons from Marshall Point have served as members of the Council.

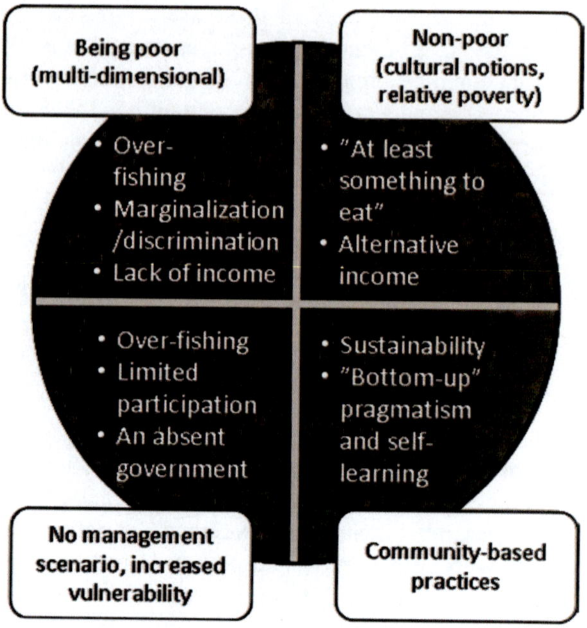

Fig. 13.4 Marshall Point, a multifaceted approach to poverty

Marshall Point. Though, the community's sense of rights to the Lagoon's waters and resources have undergone transformation due to reassessment of current perceived pressures over the resource base.[46]

Today, Marshall Point has articulated a clear sense of rights over its terrestrial and aquatic area, and has eagerly pursued negotiations with Tasbapaunie to have these rights recognized "on paper."[47] This is, once Tasbapaunie secures the title deed over its historical territory, Marshall will request its "own land."[48] Such development will certainly increase the capacity of Marshall Point to oversee and protect its resource base. Currently, Marshall Point claims approximately 7.5% of Tasbapaunie's territory (CCARC 1998, p. 335). It seems that Marshall Point's communal authorities have expressed these concerns to Tasbapaunie's territorial board and have found a positive response. As suggested by Uncle Pi:

> Well the other day we had a little trouble, and some of the people got the other way and we ask for an amount of land (from Tasbapaunie), and to "give we the documents" and they say "ok, no trouble, we will give you'll because you deserve it," and we deserve it. They give us some paper that say you can take from *Key Suta* Point, from there come right to *Justo Point*, that is our *mojon* (recognized physical limit).

[46]For instance, it is noticeable that the community's regulations establish the notion of fishing "exclusive areas" that are said to be located "in the jurisdiction of the community." See Community of Marshall Point (2003, p. 7, Chapter II, Article 30).

[47]Indeed, Tasbapaunie has "authorized" Marshall Point communal authorities to levy taxes over incoming fishing boats (fishers and buyers), from Pearl Lagoon and/or from Bluefields.

[48]Adleen Peralta, personal communication, Marshall Point, August 2009.

Over the last 2 years, Tasbapauni has experienced internal political conflicts that resulted in two competing territorial boards being elected. This has delayed progress toward land demarcation and has made communication with Marshall Point's authorities difficult.

13.6.3.2 From "Grupo Solidario" to "Cooperative," and Back to Just "a Group of Fishermen"

In 1999, DIPAL promoted the creation of *grupos solidarios* (or solidarity groups) among fishermen in Pearl Lagoon intended as organizational units supposedly suitable for commercial fisheries. These "groups" were basically a top-down imposed organizational model, with the purpose of channeling limited funding to supposedly "efficient" and "true fishers."[49]

In Marshall Point, the *grupo solidario* did not develop independently, and rapidly disbanded in 2001, once DIPAL concluded its activities. Outside-promoted forms of economic organization have been part of Marshall Point's history. The formation of fishing "cooperatives" by fishermen had also been made conditional for receiving governmental loans during the first Sandinista administration in the 1980s. Once more, the new FSLN administration had promised to provide loans to fishers on the condition that they form a cooperative (Galeano and Silva 2009). Ignacio Casildo, a member of the fishing cooperative, sees this move as *déjà vu* with a twist:

> They (the government) come to the fishermen and say "if you people need help all you have to do is to form a cooperative and make a project," that how they say. So ok, we remember like in the 1980s with the government that was in power, is something similar: they didn't ask for fifty cents. With this project, they say the bank will lend the money, but you have to get the group form so we gone through all that.[50]

The cooperative is largely being formed out of the previously disbanded members of DIPAL's *grupo solidario*. It does make sense for fishermen (and some fisherwomen) to get the group organized in light of imminent funding.[51] For community members, it is clear though that they are, above all, just "a group of fishermen," with the aim of accessing funding and seeking local economic development. After submitting the proposal to government officers at INPESCA, the "project" was returned

---

[49] "True fishers" for DIPAL, were distinguished from "occasional fishermen" who did not devote to full-time fishing. In 1997, DIPAL estimated that 100 fishermen were working on a full-time basis (Bouwsma et al. 1997). Our data from Marshall Point indicates that income from fishing is regularly supplemented with other household economic activities. Even in cases of well-known "traditional" fishermen, other sources of income are regularly pursued over the year. The most common economic activities in the community besides fishing are selling gasoline, petty trade (small stores), and occasional employment at housing construction. Waged-labor remains minimal and remittances from relatives abroad are becoming important. In the past, as suggested in various interviews, "people could live off their (hook) lines, fish and shrimps" (Herbert Bennett, personal communication, Marshall Point, February 2009).

[50] Ignacio Casildo, intervention in local community assembly, February 2009.

[51] As for August 2009, at least three cooperatives have already received the loan.

to the fishermen for amendments.[52] According to Asolin Chang, President of the fishing cooperative:

> The project went to Managua. From Managua they send it back, because it was not complete, one of the questions from them was: where is the constitution of the cooperative? And, we don't have that because we just forming it, this is just a group of fishermen, this is not a cooperative. The project what we send, is good but is not complete.

In synthesis, the fishing cooperative makes sense within the community efforts in securing capital assets in context of renovated opportunities for fisheries development. However, the government decision has not gone without criticism by community members who are not part of the "group of fishermen" – due to their lack of start-up financial resources which were set as a condition for membership.[53]

13.6.3.3 Shifting from Fishing to Agriculture and Cattle-Raising – A Decision Not Absent of Conflicts

Within the livelihood strategies Marshall Point inhabitants have put forward, shifting labor from fishing to agriculture and cattle-raising constitutes a coping mechanism that is also infringing upon traditional norms and cultural arrangements with regard to resource use. Data generated from the survey indicates that more than 65% of households are now cultivating small plots of adjacent agricultural lands in the community. Interviewees suggested that, over the last 2 years, in response to rising food prices, households have returned to farming activities. Another reason given is the decreasing levels of fish and shrimps in the Lagoon in the dry season, when hook lines are commonly used by people who do not own gill nets.

At the same time, cattle-raising is on the rise and has involved some better-off families in town. Subsistence livestock has traditionally provided savings in time of illness or hardship. Besides, wandering cattle in the community "keep the grass at shape."[54] Cows are rarely milked or used for local consumption, unless there is a necessity. However, pasture fields (*potreros*) have been established in areas commonly used for farming. Herbert Bennett complains that:

> Now people make *potrero* and you can't even open the man's fence and get to go in. He say "this is my business" and like that then, and it shouldn't be like that. That is what I crying about, but you hear, this thing is a big problem.[55]

---

[52]The "project" was basically a form to be filled out by the "fishing cooperative" members (38 in total). Fishermen were asked to request a list of fishing gears and other related materials for the reactivation of the cooperative. The project requested 78,000 US dollars. See: Cooperativa de Pesca Artesanal. United Brothers and Sisters of Marshall Point (2009, p. 6).

[53]For instance, some community members voiced concerns with regard to the impact of this new fishing effort over the sustainability of the Lagoon. In addition, they argued that the loan might have a negative effect in the community, by increasing the existing socio-economic differences among fisher and non-fisher members of the community.

[54]This contradicts formal community regulation, which prohibits cattle from wandering within the community-inhabited areas.

[55]Herbert Bennett, personal communication, Marshall Point, February 2009.

Cattle-raising is now developing in opposition to community norms with respect to the use of certain areas for agricultural activities. This competition over resource use also involves encroaching cattle-owners from the neighboring Orinoco.

13.6.3.4 Outward Migration and Educational Opportunities

As mentioned earlier, outward migration – in particular to temporary jobs on international cruise ships – has been a traditional practice that is used to supplement families' income. This is a practice in which most communities in the Pearl Lagoon basin are involved. Indeed, for some families in Marshall Point, remittances now constitute the main source of income. In addition to outward temporary migration, opportunities for attaining higher levels of education have also been sought as a strategy for social mobility, and therefore to cope with poverty. To this effect, regional universities have also expanded outreach programs in the Pearl Lagoon basin, and scholarships have been made available to families.

13.6.3.5 Accruing Social and Political Power

Marshall Point's coping strategies are not reduced to socioeconomic elements. Facing a new round of regional elections (to be held in March 2010), community's authorities are also seeking to increase political power on supra-community levels. Political opportunities have opened up for Marshall Point in context with political parties trying to broaden their political base.[56] The community board has brought this question to open assemblies in order to strategize about taking advantage of these opportunities.

13.6.3.6 Community-Based Management Actions

The Pearl Lagoon basin constitutes a semi-open resource system in which customary norms and occasional state-enforced regulations exert some level of control over access and exploitation by outside fishers. Generally, communities do not oppose access to fishing grounds by the surrounding indigenous and Afro-descendant communities of the basin. However, communities believe that outside fishers should be subjected to taxation; or that eventually, access to outsiders must be openly denied. Indeed, some communities have pursued an active enforcement of these rules, as reported by some studies (Henriksen 2008).

In Marshall Point, access to aquatic resources by non-community members is subjected to constant evaluation and learning. It can be argued that, to some extent, the community approaches the matter with pragmatism rather than with permanent

[56]For instance, for the first time in history, since the inauguration of the regional councils, the FSLN has invited Marshall Point to propose pre-candidates for its first selection round to be held in Orinoco.

guidelines. For instance, some customary norms and regulations are in place – such as taxation if fishers or buyers from Bluefields or Pearl Lagoon enter the community's waters. However, if the presence of outside fishers or buyers represents an opportunity for income generation by local fishermen, taxation would be omitted.[57]

Community regulations indicate that fines should be issued over recurrent breaches of fishing norms – for instance, persistent omission of the required authorization on the part of external buyers or fishers. However, no fines have been issued since the community bylaws were approved.[58] Community members are more eager to oversee that certain fishing areas within Marshall Point's territorial waters are not subjected to resource extraction. This is the case with a place called "the hole," which is said to be a "cave" where fish reproduce and seek refugee from overexploitation. The "hole" is located approximately 4 km north of the community, close to the center of the Lagoon. Fishing at the "hole" with gill nets – even by surrounding fishers from neighboring communities – can cause intense disputes between local fishermen and would certainly involve other community members as well. Fishing with hook-lines at the "hole" is more often allowed during the dry season.

Fine-mesh gill nets (3 in. and smaller) are frequently used by some fishermen in Marshall Point. This practice contravenes norms stipulated in proposed management regimes both from DIPAL and CAMPLab. However, these fishermen confront increasing criticism from other members of the community who see such practices as a violation of their rights to access a sustainable resource base. Seeking to avert starvation, multi-family households that have limited access to fishing gears usually share them during the rainy season. The diagram below (Fig. 13.5) offers a composite of the six strategies displayed by Marshall Point. Rather than linearity, the idea is to represent the mutual reinforcement and cyclicality of the various strategies.

13.7 Conclusions

Marshall Point shares with other indigenous and Afro-descendant communities of the Pearl Lagoon basin similar concerns and challenges with regard to preserving a sustainable resource base for the long-run. Increasing over-exploitation due to commercial fishing has placed local fisheries on a critical path in their capacity for meeting this end. Livelihood strategies displayed by Marshall Point exhibit certain commonalities with adaptive strategies and trends observed in other communities in the area (for instance, finding jobs abroad or supplementing households' income with seasonal work available in other parts of the region).

However, at the same time, the community seems to be engaged in pursuing strategies that are somehow qualitatively different from its counterparts. For instance,

[57] Adlene Peralta, community coordinator, personal communication, Marshall Point, August 2009.

[58] Adlene Peralta, community coordinator, personal communication, Marshall Point, August 2009.

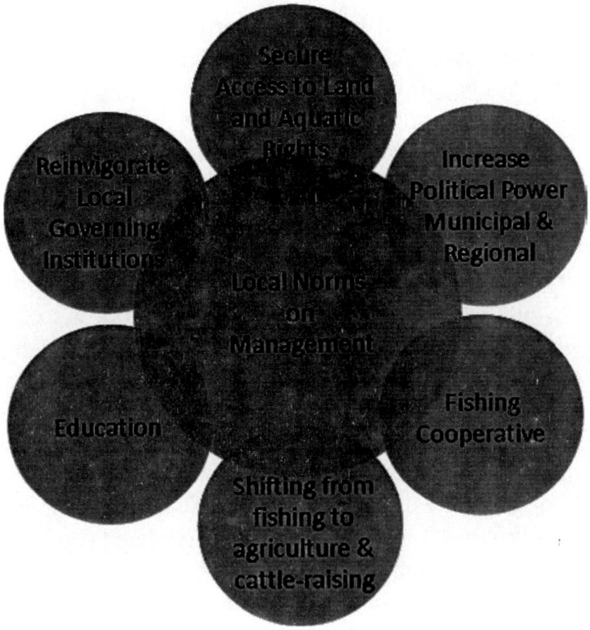

Fig. 13.5 Marshall Point: reinforcing strategies. This diagram offers a composite of the six strategies displayed by Marshall Point. Rather than linearity, the idea is to represent the mutual reinforcement and cyclicality of the various strategies. Enforcing local norms on management depends on the positive interaction of internal and external factors over which the community has relative influence (for instance, securing access to land)

with regard to collective action toward securing terrestrial and aquatic resources as a means to increase political and social status *vis-à-vis* Tasbapaunie and Orinoco. This strategy also seems judicious in relation to increasing the political community leverage with reference to other sources of power in the area/region, such as seeking public office in municipal and regional governments.

Moreover, Marshall Point shows a creative integration of management regimes and locally managed norms in relation to fisheries. Enforcing these norms, however, seems to be shaped by a pragmatic and self-learning approach on the part of the community's authorities. Specifically, community-based management, even if strongly supported by local fishermen and other resource users, might still be at a great disadvantage in context of increasing pressure and contestation over the use of natural resources in the Pearl Lagoon basin. A collaborative national state capable of reversing what is now a precarious governance environment is constantly invoked as a condition to the successful integration of management regimes in the area.

We suggest that Marshall Point's capacity for coping with poverty should be understood within a social power and livelihood approach. Locally held, cultural understanding of poverty as a social relationship is keenly integrated by community members within an assessment of their capacity for satisfying tangible material needs. From this emerges a multi-dimensional concept of poverty, which we utilized in grasping Marshall Point's strategies to cope and overcome poverty and minimize vulnerability. For instance, management systems that emphasize commercial fishing run the risk of not capturing the comprehensiveness within which a community creates and develops livelihood strategies. Within these strategies, fisheries constitute a crucial one, though not the only component for securing communities' well-being. Diversifying sources for food production, targeting other fisheries, supplementing household income (by traveling abroad), and assigning priority on educational achievement (human capital) are all discerning strategies within the spectrum of possibilities that Marshall Point is exploring to overcome poverty.

Finally, we suggest the need to theorize livelihoods strategies to overcome poverty in connection with empowerment. As we have examined, by looking at Marshall Point livelihood strategies, to be empowered at the individual and community levels is to have the resources (material, political), the capacity (the leadership, consensus, and self-determination), and the opportunity (autonomous rights) to mobilize for social transformation. In this sense, we propose to conceptualize empowerment for fisher communities as an enabling *process*, through which higher levels of social power can be accrued.

From this theoretical insight, we can conclude that in coping with poverty, Marshall Point has been engaged in pursuing a strategy of empowerment "from below;" and in doing so, has drawn from a variety of available resources and opportunities. Some of these resources have been advanced by outside actors, others are locally-generated. This strategy has rendered significant gains for the community as a sociopolitical entity. Nevertheless, crucial challenges lie ahead in the pursuit for a more equitable distribution of resources *within* the community. Our study suggests that strategies to overcome poverty should then be thought of as having two levels: one that is external and another that is internal, that are intimately connected. Without taking both dimensions into consideration in tackling poverty, just marginal outcomes can be achieved.

Acknowledgments This chapter would not be possible without the invaluable assistance of Angie Martinez, my former sociology student at URACCAN, who generously introduced me to the people of Marshall Point. Along with Floyd Martin, they helped me conduct the interviews, provided valuable insights throughout the research process, and also offered comments to early versions of this chapter. Angela Fletes and Brenda Lopez provided support in data gathering through a household census. The community board as well as the fisherfolks' cooperative of Marshall Point were supportive in all the phases of the research. Dennis Mairena (senior researcher, Centre for the Development of Indigenous Peoples, CADPI) offered advice on the socioeconomic and organizational dynamics of the Pearl Lagoon area. The initial research design benefitted greatly from conversations and intellectual exchanges with Svein Jentoft (Norwegian College of Fishery Science), and Camila Andreassen (PhD Candidate at the University of Tromsø). Georges Midre and Hector Andrade also provided substantial comments to early drafts of this chapter, and for that I am very grateful.

References

Allison E, Ellis F (2001) The livelihoods approach and management of small-scale fisheries. Mar Policy 25:377–388

Berkes F, Mahon R, McConney P, Pollnac R, Pomeroy R (2000) Managing small-scale fisheries. IDRC, Ottawa

Bouwsma H, Sanchez R, van der Hoeven JJ, Rosales D (1997) Plan de Manejo Integral de los Recursos Hidrobiologicos de la Cuenca de Laguna de Perlas y la Desembocadura de Río Grande. Proyecto para el Desarrollo Integral de la Pesca Artesanal en Laguna de Perlas (DIPAL) and Centro de Investigacion de los Recursos Hidrobiologicos (CIRH), DIPAL/CIRH, Laguna de Perlas

Caribbean Central American Research Council (1998) Diagnóstico General sobre la Tenencia de la Tierra en las Comunidades Indígenas de la Costa Atlántica. Informe Final. CCARC, Austin

Christie P (2000) The people and natural resources of Pearl Lagoon. In: Christie P et al (eds) Taking care of what we have. Participatory natural resource management on the Caribbean Coast of Nicaragua. CIDCA-IDRC, Ottawa, pp 17–46

Christie P, Simmons B, White N (2000) CAMPLab: the coastal area monitoring project and laboratory. In: Christie P et al (eds) Taking care of what we have. Participatory natural resource management on the Caribbean Coast of Nicaragua. CIDCA-IDRC, Ottawa, pp 99–126

Community of Marshall Point (2003) Internal regulation. Ford-BICU, Marshall Point

Cooperativa de Pesca Artesanal. United Brothers and Sisters of Marshall Point (2009) Proyecto: Adquisición de Materiales para Elaboración de Nasas y Fondos para Material de Trabajo. Marshall Point. Manuscript

Davidson WV (1980) The Garifuna of pearl lagoon: ethnohistory of an Afro-American enclave in Nicaragua. Ethnohistory 27(1):31–47

Envío (1989) Atlantic Coast pearl lagoon: back from wars and winds. Revista Envío, 92, March, Managua

Figueroa D (1999) "Historia del Pueblo Garífuna en Nicaragua". In: Obando V (ed) Orinoco: Revitalización Cultural del Pueblo Garífuna de la Costa Caribe Nicaragüense. URACCAN, Managua

Friedmann J (1992) Empowerment: the politics of alternative development. Blackwell, Cambridge

Galeano L, Silva JA (2009) Hambre Cero Desvirtuado. El Nuevo Diario, 28 Sept 2009. http://www.elnuevodiario.com.ni/especiales/58070

Gonzalez M (2008) Governing multiethnic societies in Latin America: regional autonomy, democracy, and the state in Nicaragua, 1987–2007. PhD Dissertation. York University, Toronto, Canada

Hale Ch (1994) Resistance and contradiction. Mískitu Indians and the Nicaraguan State, 1894–1987. Stanford University Press, Stanford

Henriksen K (2008) Ethnic self-regulation and democratic instability on Nicaragua's Atlantic Coast: the case of Ratisuna. Eur Rev Latin AmCaribbean Stud 85:23–39

Hostetler M (2005) Enhancing local livelihood options: capacity development and participatory project monitoring in Caribbean Nicaragua, PhD Dissertation Thesis. York University, Toronto

Ibarra E (2009) Procuraduría ayuda a Fiscalía de EU a perseguir empresas pesqueras en Nicaragua. El Nuevo Diario, 9 Jan 2009. http://impreso.elnuevodiario.com.ni/2009/01/09/nacionales/93052

Instituto Nacional de Estadísticas y Censos, INEC (1995) Censos Nacionales. INEC, Managua

INPESCA (2007) Anuario Estadístico. Instituto Nacional de Pesca, Managua

Jentoft S (2005) Fisheries co-management as empowerment. Mar Policy 29:1–7

Kasperson JX, Kasperson RE, Turner BL II (2010) Vulnerability of coupled human-ecological systems to global environmental change. In: Eugene AR, Diekmann A, Dietz T, Jaeger C (eds) Human footprints on the global environment. Threats to sustainability. MIT Press, Cambridge, pp 231–294

Kindblad Ch (2001) Gift and exchange in the reciprocal regime of the Miskito on the Atlantic Coast of Nicaragua 20th century. Department of Sociology, Lund

Mukherjee Reed A (2008) Human development and social power: perspectives from South Asia. Routledge, London

Nietschmann B (1972) Hunting and fishing focus among Miskito Indians, eastern Nicaragua. Hum Ecol 1:41–67

Nietschmann B (1973) Between land and water: the subsistence ecology of the Miskito Indians, eastern Nicaragua. Seminar Press, New York

Pérez MM, van Eijs S (2002) Plan de Manejo para los Recursos Pesqueros de la Laguna de Perlas y la Desembocadura del Rio Grande de Matagalpa. DIPAL, Laguna de Perlas

PNUD (2005) Informe de Desarrollo Humano 2005. Las Regiones Autónomas de la Costa Caribe. ¿Nicaragua asume su Diversidad? PNUD, Managua

Sánchez BR (2001) Biología Pesquera del Camarón y Camaroncillo en las Lagunas Costeras de la RAAS. DIPAL, Laguna de Perlas

Vernooy R (2000) The setting, issues, and research methods. In: Christie P (ed) Taking care of what we have. Participatory natural resource management on the Caribbean Coast of Nicaragua. CIDCA-IDRC, Ottawa, pp 1–15

Von Oertzen E, Rossbach L, Wünderich V (eds) (1990) The Nicaraguan Mosquitia in Historical Documents 1844-1927. Dietrich Reimer Verlag, Berlin

Williamson D, Fonseca G (2007) Compendio Estadístico de las Regiones Autónomas de la Costa Caribe de Nicaragua. CIDCA, Managua

Chapter 14
Learning from the Experts: Attaining Sufficiency in Small-Scale Fishing Communities in Thailand

Ratana Chuenpagdee and Kungwan Juntarashote

Abstract Small-scale fishing communities in Thailand cannot be readily classified as poor when compared to other non-fishing sectors in rural areas, or the "urban poor." Rather, fishers have often referred to the concept of "sufficiency" as a measure of life satisfaction, which often means making ends meet and having a supportive network in case of emergency. While all were faced with changes brought about by industrialized fishing, coastal development, and globalization, some fishing communities seem to possess higher levels of capability to stay afloat, thus maintaining a satisfying level of sufficiency. From a governance perspective, learning about why some communities are better at coping and averting poverty, is useful to help those who are less able, as well as to prevent others from falling into a poverty trap. This chapter reports the findings from a study conducted in small-scale fishing villages in four provinces in Thailand that differ in geography and context, with the aim to understand their coping strategies and the poverty-averting potentials.

14.1 Introduction

As in many countries around the world, fisheries resources of Thailand are heavily exploited by both large-scale and small-scale fishing activities. Modernization of fishing fleets, gears, and technology took place early in the 1960s, and resulted in

R. Chuenpagdee (✉)
International Coastal Network, Department of Geography, Memorial University of Newfoundland, St. John's, NL A1B 3X9, Canada
e-mail: ratanac@mun.ca

K. Juntarashote
Coastal Development Centre, Faculty of Fisheries, Kasetsart University, 500 Phaholyothin Road, Jatuchak, Bangkok 10900, Thailand
e-mail: kungwan.j@ku.ac.th

S. Jentoft and A. Eide (eds.), *Poverty Mosaics: Realities and Prospects in Small-Scale Fisheries*, DOI 10.1007/978-94-007-1582-0_14,

Table 14.1 Number of fishing households and level of employment in 2000, categorized by the use of boat and boat power. Almost 90% of these (all categories except the last) are considered part of the small-scale fishing sector, which relies more heavily on household members as crew members than the large-scale fishing sector (Source: DoF 2002)

Type of fishing households	Number of households	% of households with hired labor
Without boat	3,550	0
With non-powered boat	2,559	7
With outboard motored boat	41,225	17
With in-board motored boat, less than 5 GRT	3,249	31
With larger boat (large-scale fishing category)	7,218	86
Total	57,801	

the rapid and unplanned expansion of industrialized fishing fleets, especially trawlers (Pauly and Chuenpagdee 2003). Also, similar to the majority of the world's nations, the small-scale fishing sector in Thailand dominates the fisheries – constituting about 90% of the total fishing households. Small-scale fisheries take place predominantly in coastal waters within 3–5 km from shore, using several gears such as gill nets, traps, and hand-lines. Small-scale fishing can be done without a boat or with a non-powered boat, but mostly it involves uses of outboard powered boats and, to a lesser extent, inboard powered boats with hold capacity of less than 5 gross registered tonnage (GRT).

According to the most recent census conducted by the Department of Fisheries in 2000, there were 57,801 fishing households distributed around the Gulf of Thailand and the Andaman Sea. About 87% of them were considered small-scale. This included a small minority (12%) of fishers who fished without boats or with non-powered boats; a majority with outboard powered boat (82%); and the other 6% were fishers using inboard powered boats with hold capacity of less than 5 GRT (Table 14.1). In addition to GRT, the amount of hired labor is another criteria differentiating small-scale from large-scale fisheries. For example, only a small percentage of fishers using non-powered boats, and those using outboard powered boats (7% and 17%, respectively) hired labor. Together with those not using boats, they are the typical small-scale fishers. Of the fishing households using inboard powered boats with the hold capacity less than 5 GRT, about 31% hired crew members and were thus considered to resemble the large-scale fishing sector, while the rest may still be considered small-scale.

In 2007–2008, the total fisheries production was estimated at almost 4 million tons, contributing about 1.2% of the national gross domestic product (GDP), with a total worth of about 3,121 million USD (FAO 2009). About 58% came from marine capture fisheries, of which 80% were small-scale. The importance of the small-scale fishing sector to the Thai economy is undoubtedly high, especially in terms of jobs and livelihood contributions, but it also means that the currently poor state of the marine ecosystem may be attributed to both small- and large-scale sectors. While highly destructive gears such as bottom trawls are usually large-scale, other non-selective gears, including set bag nets and push nets which scrape and destroy bottom habitats, are still considered small-scale.

Until recently, push nets were allowed to operate within the 3 km zone reserved for small-scale fisheries because of the small engine size. The present prohibition of trawlers and push-netters in near-shore areas helps relieve some pressure on fisheries resources. Yet, small-scale fishing communities are still vulnerable to changes in the marine ecosystem brought about by activities which have taken place in the area, including industrialized and destructive fishing, unsustainable aquaculture practices, unplanned coastal development, as well as factors such as market globalization and global warming.

With few exceptions in remote areas where subsistent fishing may be found, the majority of small-scale fishers in Thailand engage in commercial fishing. Most fishing communities are properly connected with facilities and infrastructure, such as roads, electricity, telephone service, schools, and a health clinic. This implies that the Thai small-scale fishers are not geographically, economically, and politically detached from the attention of the state, market, and society; and thus are not as marginalized as those in other countries. These fishers are better off when compared to Thai farmers and other rural people in upland areas who continue to be drawn to the coasts in search of better livelihoods.

As previously indicated, this does not mean they are not susceptible to changes that threaten their wellbeing. How the small-scale fishers cope with the existing pressure and competition from industrial fisheries and new entrants is the first research question linking to poverty. What begs for further investigation is whether they could stay ahead of poverty, and under what conditions. In this chapter, coping strategies and poverty averting potentials among small-scale fishers are examined in the pre-harvest, harvest, and post-harvest part of the fisheries production chain.

In the following, we first describe key aspects of the interactive governance framework which we use to guide our analysis of poverty. Next, we present the study sites and the data collection methods. The results of the study are then reported, followed by the discussion of the findings and policy implications. Drawing from the lessons learned in this study, we conclude with some recommendations for improving poverty averting potentials among small-scale fishers in other areas.

14.2 Theoretical Framework

Studies about poverty in small-scale fisheries often begin with one of two premises: Either people fish because they are poor; or fishers are poor because they fish. The former reflects a strongly held perception about fishing as an "occupation of the last resort," which may still be true in some areas, but is now much less applicable given the generally poor state of the fisheries. With the rapid development in coastal areas, shellfish gathering and fishing without a boat or with a non-powered boat is becoming less subsistent. Certain investment is often required to obtain boats and gears that would enable fishing to take place in this increasingly competitive environment. In the case of Thailand, poor people from rural areas migrating to the coasts in search of fishing livelihoods will also need to compete with legal and illegal foreign migrant workers from the neighboring countries who are willing to work harder for lesser wages. As will be shown in this study, new entrants to the fisheries find success when supported by existing social networks.

Related to the latter premise, Béné (2003) referred to it as the old paradigm centered on the "poverty-environment" nexus that links overexploitation and resource depletion with poverty. While this may be true, he suggested that the investigation about poverty in fisheries should be broadened to consider the "socio-institutional" dimensions, which may underlie the real causes of poverty. This call for the understanding of rules, norms, principles, institutional arrangements, social organizations, and other aspects of social and institutional relationships aligns well with the interactive governance framework (Kooiman et al. 2005) that we draw upon in our examination of the Thai small-scale fisheries.

This framework posits that poverty needs to be understood in the context of the inherent and constructed qualities of the system-to-be-governed (e.g., the fisheries resources and the fishing communities), the governing system, and their interactions. The concept of "fish chain" (i.e., pre-harvest, harvest, and post-harvest) (Kooiman et al. 2005) is also useful in the examination of where in the system poverty arises; and of how fishers cope with the challenge. As earlier discussed, ecosystem degradation is an issue in Thailand that contributes to poverty in small-scale fisheries. The causes of such degradation, as well as the socio-political conditions of the fishing communities are, however, multi-faceted and could be linked to poverty in several ways, including poverty-averting potentials.

As we began this project to research poverty, it became apparent that our earlier observation that small-scale fishers were generally poor and in need of assistance from the government (Juntarashote and Chuenpagdee 1991) may no longer be valid. Thailand is among the majority of countries in the world that fall under the "medium" category of the United Nations "Human Development Index," which takes into consideration life expectancy, education attainment, and adjusted real income (UNDP 2000). According to the statistics from the Ministry of Interior (MOI 2008), the 2006 poverty line for the country was 41 USD/month/person,[1] with a higher average of 50 USD/month/person for urban areas, and a lower average of 38 USD/month/person for rural areas.

Based on unofficial records and casual conversations with several fishers along the coasts during the exploratory phase of the research, we learned that the average household income of small-scale fishers is about 298 USD per month (gross income), or a net income of about 179 USD per month. For a family of four, this level of income (about 45 USD/month/person) puts them slightly above the 2006 national poverty line, and also above the average net income for the country (about 43 USD/month/person). Additionally, the data suggests that small-scale fishers do much better than people in non-fishing occupations in other rural areas whose average net income is estimated at about 26 USD/month/person.

[1]One USD is about 33.55 Baht (according to the 2009 average exchange rate).

This observation brings up a question about how poverty is defined, and the suitability of the existing poverty indicators. In the case of the Thai small-scale fishers, they do not seem to conform to the assumed image that fisheries rhyme with poverty (Béné 2003). Is it because their socio-economic condition has indeed improved; or is it because there is no real benefit associated with the poverty label? In other words, playing the "poor" card may not attract the same kind of support from the government as it seemed to have done in the past. Instead, it may be the "sufficiency" card that is worth playing in the current climate.

We argue here that it is worth examining how fishers perceive poverty, and what image they have of themselves and their present situation. Such inquiry about "images" is another research dimension in the interactive governance framework. The image and its formation are crucial elements that need to be clearly understood, discussed, and communicated in the same fashion that other governing components such as instruments and actions are (Kooiman et al. 2005). Jentoft et al. (2010a) further argue that an exploration of images helps define problems and present solutions. It is under this premise that we propose to learn from the "experts" about the role that certain images play in preventing fishers from falling into a poverty trap.

As Narayan et al. (2000) put it, discourse about poverty and poverty alleviation is mostly dominated by people who are generally not poor. The poor people, or the so-called poverty experts, are usually not directly involved in the examination of the problems and in finding the solutions since this task seems to belong to the other group of experts (i.e., university, government, and non-government scientists and researchers). We consider Thai small-scale fishers as experts, particularly because of their knowledge contribution to possible mechanisms to avert poverty. Learning from some of these experts about how they cope with poverty, and importantly how to stay ahead of it, may help inspire others about what they may be able to do to resolve their unfavorable conditions.

14.3 Site Selection and Data Collection

In the first phase of the research (April to October 2008), we conducted literature review, compiled information, made preliminary field visits and had informal conversations with key informants and local fishers in order to select study sites. The main information sources were the National Statistics Office (Rural Community Division), Department of Fisheries, and provincial fisheries offices. National censuses and household surveys were also consulted, although they generally do not contain specific information about fishing households. Based on this information, we selected four provinces that are similar in terms of the importance of small-scale fisheries, but differ in the geographical and biophysical settings and in lessons about poverty (Fig. 14.1) – Chanthaburi, Prachuap Khiri Khan, Ranong, and Krabi.

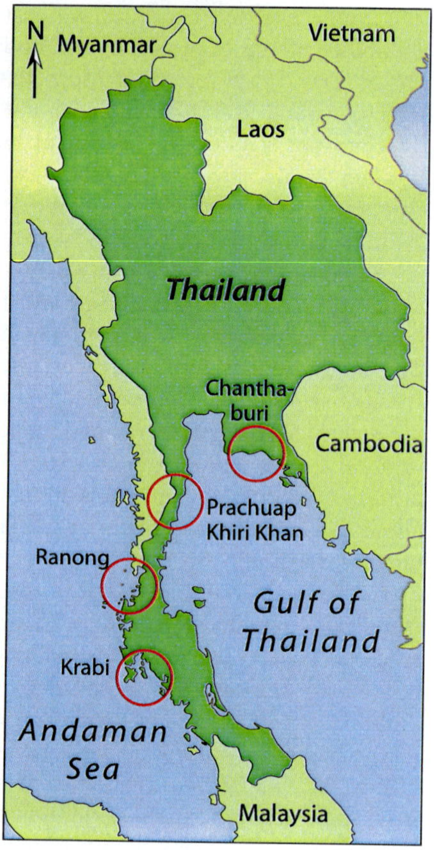

Fig. 14.1 Map of Thailand showing the four study provinces, Chanthaburi on the east coast and Prachuap Khiri Khan in the south, both are part of the Gulf of Thailand, and Ranong and Krabi on the southern coast, part of the Andaman Sea

Chanthaburi, on the east, is closest to Bangkok (about 250 km) and is well-developed in terms of roads, infrastructure, public services, and other amenities. Coastal water of Chanthaburi is shallow with gradual slope and mixed-substrate bottom. Parts of the coastal areas are lined with mangrove forests, which are relatively abundant, due partly to the reforestation program promoted by the government. Also on the Gulf of Thailand but in the southern part is Prachuap Khiri Khan which is about 350 km from Bangkok. This area is famous for its long stretch of sandy and shallow bay area with abundant fisheries.

Strong tidal wave currents are experienced along this coast, but nothing compared to the Indian Ocean Tsunami of 2004[2] that hit Ranong and Krabi, the other two provinces selected for the study. Ranong, being more exposed than Krabi, suffered greater levels of damages to beaches and coral reefs. In both provinces, the Tsunami resulted in major changes, not only to the land and seascape,

[2]Hereinafter referred to as the Tsunami.

but also to the social fabric and people's quality of life, which will be focused on in this study. Ranong is about 200 km south of Bangkok. It is well known for fruit production, particularly pineapple and durian. While not as popular as a tourist destination, Ranong has a number of excellent dive sites and nice sandy beaches. Krabi is further south from Ranong. Its proximity to the famous tourist town, Phuket, brings good infrastructure to the area. Tourists generally visit the islands within Krabi (for example, Phi Phi Islands). The small-scale fishing communities are less frequented by visitors. Krabi was included in the EU-Government of Thailand collaborative project referred to as CHARM (Coastal Habitat and Resource Management[3]). Initiated in 2003, the project covered two major coastal areas, Ban Don Bay on the Gulf of Thailand and Phangnga Bay on the Andaman Sea, where Krabi province is located.

Visits to the four provincial fisheries offices were made during the preliminary phase (February 3–9, 2009) of the project. General information about the fisheries were gathered and used to identify two to three small-scale fishing villages considered to be relatively poor according to the local officers and the local key informants for the detailed study. The question "How is village 'X' compared to village 'Y' and village 'Z' in terms of poverty?" was used to systematically compare the candidate villages. Once the villages were selected, the official local leaders (normally the village head or the chair of the local administrative authority) were approached for an overview of the area, and for recommendation about who the "poor" fishers were.

In the following trips, unstructured interviews were conducted with these fishers, as well as with other fishers who were available and willing to be interviewed. The informal interview was helpful in encouraging fishers to speak freely about various issues concerning their livelihoods, while the opportunistic interview allowed for an inclusion of other fishers who may not be currently poor but may be vulnerable to poverty. In some cases, group discussions with fishers, government officials, and researchers were conducted instead of individual interviews. The conversations normally began with questions about the current state of the fisheries resources, any recent changes observed, and the effects on the livelihoods. These were followed by specific questions about poverty, including what the term meant, what caused poverty, who among them were most vulnerable, and what they did to avert poverty.

Data reported in this chapter is based on individual interviews with 25 small-scale fishers, and two focus group discussions in the four provinces. Field observations were also made to compare the level of infrastructure and facilities (e.g., housing types, road conditions, availability of amenities such as water, electricity, telephone, etc.) within the selected villages, and with others villages that were not identified as poor. The bulk of the field data collection took place from February 2009 to June 2009, and went smoothly with generally good reception and cooperation from the local people.

[3] www.charmproject.org.

14.4 Key Findings

14.4.1 Poverty Versus Sufficiency

The growth and development that Thailand experienced in the past decades have contributed significantly to improving the standard of living of its people. Small-scale fishers, while they may still be economically marginalized compared to industrialized fishers, enjoy benefits from the social and economic development that comes in the form of infrastructure such as roads, telecommunication, and landing facilities. In all the villages we visited, even those considered poor by the government officials and the local leaders, there are reasonable roads (mostly paved with concrete) and electricity, and the majority of the households have proper roofing (mostly made of tin or tin and zinc alloy), basic amenities, including toilets, and in many cases, refrigerators, televisions, and telephones. Some villages rely on water from wells and rain, while some have potable water provided through pipelines by the government. The poorer houses are noticeable in the cheaper materials used for roofs and walls, the shabby way the houses were built, the smaller house size, the location (far from the village center), and the general untidiness of the household area.

Prior to the introduction of trawls in the early 1960s, catches from small-scale fisheries constituted about 23% of the total landings. Since then, their contribution has been steady at about 8–10%. Accompanying this decline was the shift from catching fish using raking devices or grasping by hands which dominated the catches until 1970; to using gears such as bag nets and seines. Despite the decline in the catch share, small-scale fishers are better off than in the past, as they target shrimp and crab which fetch high prices. In Prachaup Khiri Khan Province, for example, catches of small anchovy-like fish bring good revenue to small-scale fishers because of the export market of the fish. The majority of small-scale fishers in this province are able to create a market niche for their products that make it possible for them to maintain a relatively high income level.

Rather than applying certain indicators to gauge the level of poverty in the studied villages, we forward the argument by Narayan et al. (2000) by asking fishers to define what poverty meant. Common references emerged in the interviews when fishers talked about the meaning of the word "poverty" and the context surrounding it. According to many fishers interviewed in the study, poverty was related to "lack of opportunity for education." While local ecological knowledge was widely recognized as an important asset, formal education was considered a crucial step required to stay away from poverty. Poverty was also linked to the "lack of capital" for investment in fisheries, or for sending children to school.

As mentioned by one fisher, "*poor people have brain but lack capital.*"[4] Some fishers noted that fishing was hard work and it was not a job for lazy persons. Further, poverty was attributed to the lack of land ownership, the lack of family planning (too many children) and inability to put aside money. While inference was made to

[4]All quotes were translated to English from Thai.

the connection between hard work and poverty, they acknowledged that poverty was also linked to the inability of some fishers to save the money that they earn. In the words of one fisher:

> An opportunity to get rich might be slim, but through hard work, and being thrifty and modest, we will not be poor but instead we will attain sufficiency.

In the context of the small-scale fishers in the south of Thailand where the majority are Muslim, they regard the sea as "God's fish pond" to be shared by all, including the commercial fishing sector. Finally, many fishers referred to fishing as an occupation of choice, something they preferred and enjoyed, even though income may be variable. Many of them also explicitly indicated how they chose not to expand their fishing operation (such as buying a more powerful engine or bigger boat), even when they had an opportunity to do so. This freedom to determine their own lifestyles resonates what Sen refers to in his paper "Development as Freedom" (Sen 2000), reiterated in the context of fisheries by Jentoft et al. (2010b).

Although there are certainly poor fishers in Thailand, most small-scale fishers we interviewed did not think that the concept of poverty applied to their present situation. Quite the contrary, fishers generally considered themselves "rich" and "blessed" with the abundance of resources. Even when some of them admitted to not having enough to feed their families on certain days, they still did not think of themselves as poor. As noted by one fisher:

> …there is always food from the sea. Some days we catch more fish, some days we catch less, some days we cannot go fishing [because of the weather], but we don't need to buy food, so we're always okay.

Fishers referred to this way of thinking as living a "sufficient life." According to several fishers, the word "sufficiency" suggested a condition of being neither rich nor poor, but of "having enough to get by." Another fisher explained that:

> When we have less money, we spend less. If we have more money, we may buy a pick-up truck.

The sufficiency concept was often mentioned in the same context as sustainability, as in "living sufficiently for sustainability." One (Muslim) fisher further explained that, "the sea is the God's given gift. If we don't use it properly, we will run out of places that feed us. Thus we need to use the sea in a sustainable way." The sustainability principle has been promoted through the outreach efforts by the Department of Marine and Coastal Resources, a new government unit responsible for conservation of natural resources. Many fishers talked about conservation measures that they voluntarily practiced, such as releasing the female berried crab; and also their effort to protect mangrove areas (by refraining from cutting them).

The idea of "sufficiency" originates from the revered King Bhumibol of Thailand, and is promoted through various government programs to improve the quality of life of the rural people. The concept of sufficiency is explained by His Majesty the King as a philosophy to guide a modest but reasonable living which also offers security and provides protection against future threats (TRF 2006). It is often referred to as the "middle way" or the moderate and reasonable path to pursuing economic development, while keeping with the globalized world. It should be noted that sufficiency does not imply self-reliance. Rather, it refers to a production system (including capture fisheries and fish farming) that produces enough food to satisfy the dietary needs of the households. It provides a sufficient amount of excess supply that can be sold locally to avoid high transportation costs, and minimizes dependency on outside buyers. The "Philosophy of Sufficiency Economy," as it is commonly known, has been used as the guiding principle in drafting the current 9th National Economic and Social Development Plan (Krongkaew 2003).

The widespread acceptance of the sufficiency concept among fishers could be due to the popularity of the King, a reflection of the emphasis in the government policy, or the desire by fishers to present a better image. Yet, some expressions by the small-scale fishers interviewed in this study hinted at their approval of the concept. One village leader submitted, for instance, that:

> Villagers want to have principles and practical knowledge that help improve their livelihoods; this is more important than receiving money from the government.

He presented three examples of practices transferred through local knowledge and government extension service; i.e., how to make use of all parts of sugar palm trees, how to make home-made fertilizer, and how to locally process fish sauce. This has linked them with the sufficiency principle, and has credited them for their contribution to lowering household expenditures.

Another fisher indicated the preference in the fishing lifestyle; even though he could do other work that may earn more income, noting savings and additional employment opportunities as key enabling factors for following the sufficiency principle.

> Being a small-scale fisher is about freedom and independence. I can go fishing and come home every day as I like. Income from fishing is usually enough to support the family. We can save money and cut some spending, like drinking rain water [instead of buying bottled water]. But if there was any shortage, it is still possible to get additional money from working in the pineapple plantation.

14.4.2 System Effects on Poverty

As indicated above, the sufficiency principles may be one of the fundamental coping strategies that small-scale fishers in Thailand use to avoid falling into poverty. They are nonetheless susceptible to other stressors operating throughout the fish chain. The following captures three main challenges that fishers face related to change in the natural, social, and governance systems, which weaken their ability to avert poverty and maintain their livelihoods.

14.4.2.1 Ecosystem Alteration

The prohibition of trawlers and push-netters in the near shore area was considered by small-scale fishers to be one of the most significant contributions to their well-being.

Fig. 14.2 A fisher preparing the nets

Fishers indicated that they recognized the linkages between healthy ecosystems and good fisheries, and they have taken an active role in monitoring that these destructive gears are not used in their nearby waters and by members of their communities. They also expressed concern about other destructive practices such as the use of small mesh size (less than 2.5 cm) in gill nets and the use of light luring devices. Fishers in Bang Saphan village, Prachuab Khiri Khan, in particular, have been supportive of a community-based fisheries management project initiated in 1999 as part of the government program to mainly address gear conflicts, and have agreed to the demarcation of about 240 km^2 for protection.

This, in essence, extends the government-designated no trawling zone from 3 km to about 12 km from shore. They have also been involved in the artificial reef construction program promoted by the government to help improve the productivity of the marine ecosystem. In addition to halting the destruction of the marine environment, the absence of trawlers helps reduce direct conflicts between the use of mobile gears (mostly large-scale) and small, stationary gears in the same area. Indirectly, fishers viewed the big boats as owned by rich people who are able to use equipment like sonar and echo sounder to catch more fish. The prohibition of large-scale trawlers, in this case, is thus also seen as a justice issue.

Small-scale fisheries in the four studied provinces, as in the rest of the country, are multi-species and multi-gears, with fishers depending on the seasonal availability of the fish species (Fig. 14.2). This versatility, also reflected in the fisheries regulations, is an important characteristic contributing to their ability to gain sufficient income from fisheries throughout the year. It also helps distribute pressure on fisheries resources, allowing the protected stocks to recover. The rotating seasonal and area closures for Indian mackerel fisheries, for example, are considered an effective measure since the prohibition applies only to this species. Fishers are still able to fish for other species during the closure and earn reasonable income. Another example that fishers gave was the switch of target species between fish and shrimp due to weather conditions. With the rich traditional and local knowledge that fishers possess, they are able to decide whether to catch fish or shrimp on certain days, and deploy the gears accordingly.

In one fishing village in Krabi, we observed an interesting characteristic about fishers, not seen in other places. There were some fisher-women who were well-regarded for their very good fishing skills. One woman worked alone on a small boat using fish gill nets and always returned with reasonable catches of big and small fish. Another fisher-woman was an old lady of 70 years who was renowned for her ability to catch shellfish using a small net. Other fishers in this village used shrimp gill nets, which also caught fish, crab, and squid. This species diversification of catches and the specialization of some fishers may be one of the reasons that helps maintain fisheries abundance.

Catches from this small-scale fishery were also quite small. Each fisher generally gets less than 1 kg of shrimp and another 1 kg of other species each day. Here, their practice of the sufficiency concept was again revealed. They all indicated that this amount was sufficient for them because they were able to sell their catches at reasonable prices, leaving them with no desire to fish more. The fisheries in this village exhibited small-scale fishing practices that seemed highly sustainable.

Not all small-scale fisheries are sustainable, however. The most contested fishing gears employed by small-scale fishers, especially in Chanthaburi province, are set bag nets for shrimp fisheries, which have been used for at least 50 years. Shrimp bag net is a stationary gear which could cause minor harm to life at the seafloor, as the bottom of the net needs to be closed and dragged when retrieved. The gear is also highly unselective, especially when used with very small mesh size. It is usually positioned in the channel to capture shrimps that drift with the high tides. Lots of juvenile shrimps and immature fish often end up in the net and are used to feed groupers and snappers reared in cages. The Department of Fisheries recognizes the destructive nature of this gear, but it is not able to impose a total ban on the gear, as fishers claim traditional use of it. The only thing that the government can do at this time is to control the number of bags from increasing. Similar concern applies to the use of wooden stake traps in Ranong province, also because this practice often involves the cutting of mangroves to use as stakes and to build traps.

14.4.2.2 Climate Variation

Fishing is subject to climatic variability which can cause heavy wind and high waves that inhibit fishers from going to sea. Fishers who we interviewed noted that several days of bad whether and no fishing would put some serious strain on their livelihoods. At no other time has this situation rendered more hardship to the fishers of Ranong and Krabi provinces, than after the Tsunami. They referred to this as the "incident that robbed them of their livelihoods." Some of the observations made by fishers about changes in the fisheries and the marine ecosystem since the Tsunami were the decline of shrimp fisheries due to water turbidity, destruction of seagrass beds leading to low production of shellfish, and alteration of water channels, making it impossible to apply local knowledge for navigation and net setting.

14.4.2.3 Social Change

As part of the National Economic and Social Development Plan, the Rural Development Division developed a quality of life index to assess the living conditions of people in rural areas (MoI 2008). A total of 28 indices and 31 areas of concern are included in the assessment comprising six main categories: i.e., basic infrastructure (such as roads, water and electricity); occupation and employment (e.g., jobs, productivity, and home-based business); health and safety; education; social capital; and environment and natural resources. According to the 2007 assessment (MoI 2008), about 85% of the villages in the four provinces included in the study were considered to have high quality level (i.e., having been identified with less than 5 out of 31 areas of concern). The areas of primary concern were local capacity building, availability of sport facilities, accessibility to financial capital, soil quality, and education.

Secondary issues were employment rate, water quality, land use, and agricultural production. While these factors underlined some of the problems that fishing communities face, there were also certain livelihood conditions that were advantageous to fishers. For example, fishers in the Krabi and Prachuab Khiri Khan provinces were able to diversify their portfolio by working in rubber, palm sugar, and coconut plantations. Those in Chanthaburi province were engaged in small-scale fish cage culture or shellfish culture (blood cockles), which brought additional income to the families. Some of the fishers in Chanthaburi were involved in tourism businesses by selling their catches directly to tourists, or by providing "homestay" services. Opportunities for livelihood diversification were less available for fishers in Ranong province, but at least they were able to use multiple gears to target different fish species according to their seasonal availability.

14.4.2.4 Role of Women

In all cases, women played an important role in obtaining income supplement to support the fishing families. In Chanthaburi province, for example, fishers targeted small shrimp for in-house processing into dried shrimp (i.e., cooked and sun dried for a couple of days). Ten kilograms of fresh shrimp priced at 1.50 USD per kilogram on average when processed into 1 kg of dried shrimp could be sold at 6–30 USD per kilogram, depending on the size. Demand for dried shrimp has been high because of the rise in both domestic and export markets. Women also helped with gear making and maintenance, especially of gill nets, which need to be fixed or reassembled every 3–4 months. In Ranong and Krabi provinces, women received training to participate in non-fisheries-related activities such as handicraft making, batik painting, and sewing, earning them supplementary income for the households. This skill development was part of the post-Tsunami recovery program provided by the governments and non-governmental organizations. Many women in the fishing households in Prachuab Khiri Khan province, on the other hand, worked on the farm and the plantation.

14.4.2.5 Relationship with Fishmongers

Our study revealed that one of the unique qualities of the small-scale fisheries playing a significant role in poverty aversion was the long-term relationship that fishers had with local fishmongers. In each studied village, there were often two or three fishmongers who lived there and had good rapport with fishers and the villagers. There was a sense of loyalty that fishers had toward their fishmongers, who often provided them with credits to buy nets, ice, and fuel. All interviewed fishers indicated that they got fair prices from the local fishmongers, and were generally happy with the arrangement. Because there was usually more than one fishmonger in the village, as well as those in the nearby areas, fishmongers did not have market monopoly and thus tended to offer comparable prices. This situation greatly benefited fishers, making it possible for them to obtain reasonable income for their catches.

14.4.2.6 Market Cooperatives

Another post-harvest strategy that fishers in Krabi used was to form locally-based market cooperatives. These cooperatives were established to enable fishers to set their own prices for their catches and to take control of the sale. Such mechanisms helped reduce fishers' dependency on the fishmongers and the middlepersons, as well as allowed them to gain bargaining power and increase income. Support from the governments has been provided in terms of training to fishers on business and financial management, thus building their capacity for organization. In addition to the market cooperatives, other forms of local organization and initiatives have been implemented in many fishing villages. For example, several local financial saving groups were organized to offer low interest loans to fishers and other villagers.

14.4.2.7 Migrant Fishers

Thai small-scale fishing communities are generally small (about 100–150 households) and they mostly convey strong social ties. While many households are related, they do extend their care to non-related village elders and those who require assistance, provided that the persons are considered good members of the communities. The demographic change in the communities is normally small, with few new fishers entering the villages either through marriage or recommendation by the earlier settlers. Newcomers are mostly welcome to the area, as long as they follow the fishing regulations and the informal local rules. There seems to be limits, however, to what the locals will tolerate, and with the declining resource status, competition from outsiders is a concern. For example, conflicts with the new entrants in the fisheries were mentioned in some of the villages in all four provinces, due largely to the use of destructive and more powerful fishing gears by these newcomers.

In Ranong and Krabi provinces, the problem was further complicated with migrant workers coming from Burma, many of whom worked as crew members for the industrial fishing companies. Additionally, some of them engaged in small-scale fishing activities, using illegal fishing gears in some instances, thus competing directly with the local fishers. The lack of familiarity and connectivity to the sea and to the local communities of these foreign fishers raised doubts in the minds of the local people about their interests in the protection of fisheries ecosystems and resource sustainability. The added dilemma was the change in the marine environment after the Tsunami, as previously mentioned, which minimized the advantages of local fishers in their local knowledge about fishing grounds and the sea condition. Local Thai small-scale fishers post-Tsunami have been put at par with the Burmese migrant fishers and other new entrants when it comes to knowing where and when to fish with what gears. Several fishers also commented that the migrant workers seemed to work a lot harder than the locals, adding to their fear of being taken over in the fishing livelihoods by these newcomers.

14.4.2.8 Tsunami Aftermath

To many, the true aftermath of the Tsunami was not about the change in the amount of available fish in the sea, but the number of boats being put on the water and their ownership. Boat building projects were the most popular form of assistance to the Tsunami victims provided by international organizations and individuals who were concerned that direct cash contributions might end up being misused. Unfortunately, the lack of coordination and transparency in aid distribution coupled with the poor official records of small-scale fishers meant that the number of boats post-Tsunami was higher than before, and several boats were given to non-fishers. While the case of the *bona fide* fishers having to rent boats from non-fishers is not as common as in other countries facing similar problems (like Indonesia and Bangladesh), rental and sales of the new fishing boats were major factors enabling newcomers, legal and illegal migrants to enter the fisheries. It should be noted, however, that many of these boats ended up not being used, not only because of the lack of engine which few people donated, but because of their poor quality. Wooden boats, for example, were built in haste using woods that were not sufficiently dried. Fishers also indicated that they did not like fiberglass boats that were often donated because they were heavy and too hot and slippery to operate in.

14.4.2.9 Education

The final point about the change in the social system in the Thai small-scale fishing communities is related to the next generations of fishers. The importance of education expressed by many fishers led them to encourage their children to go to schools. While high schools are available locally, youths need to leave home to attend technical colleges and universities, and most of them do not return to their communities to continue with the fishing livelihoods. Yet, in many villages we often found young people who decided to stay at home after completing secondary schools to work under the apprenticeship of their fathers, male-relatives, or neighbors until they are able to buy their own boats. Fishing tradition and local knowledge, as well as the work ethics of saving and sharing, are thus passed on.

14.4.3 Governance Reform

Administratively, Thailand comprises 75 provinces with governors as administrative heads. Each province is further sub-divided into districts, sub-districts, and villages, according to the area and population size. Similar to other countries in the region such as the Philippines, Thailand recently went through a process of decentralization in 1997 with the establishment of a sub-district administration office (known locally as "Au-Bor-Tor"). Au-Bor-Tor is a governance arrangement at the sub-district level with representatives elected by the communities to take responsibility over the day-to-day management of the sub-district. They are also tasked with the development and implementation of the management plan for the local natural resources, including fisheries. With funding obtained from local taxes and provincial government and through a grassroots operation, Au-Bor-Tor is expected to respond more timely and efficiently to people's needs than the central government (Nasuchon and Charles 2009). While the success of Au-Bor-Tor as the decentralization scheme has never been properly assessed, it is noted that some Au-Bor-Tor leaders are well-regarded for their genuine interests in the well-being of the local communities, and for holding good governance principles such as transparency, inclusivity, and accountability.

One notable example of a success story was found in Bang Saphan Bay, Prachuab Khiri Khan. Here, fishers worked collaboratively with the Au-Bor-Tor and the Department of Fisheries, along with the Food and Agriculture Organization of the United Nations, and the Southeast Asian Fisheries Development Center, on a pilot project to create shared rights over fishing areas adjacent to the nine local fishing communities (a total of 392 fishing households). The entire project ran from 1999 to 2007, with the initial phase taking place in the first 3 years. Indicators of success included an increase in the designated fishing areas for small-scale fishers, a high level of community participation in management and collaboration with government officials, increased knowledge about fisheries management and fisheries law, and increased awareness about marine conservation.

Several meetings between fishers, community members, and government officials were held to discuss fisheries management issues, and through inputs from the local communities, they resulted in participation of fishers in monitoring the restricted inshore area from violating fishing activities (such as trawlers and push nets).

During the project period, 100 artificial reef sites were constructed and installed to serve as habitats for marine organisms, resulting in a noticeable increase in catches, and thus fishers' income. From the amount of distinguished national awards they received, their effort was highly recognized as a role model for others; for example, in terms of the use of local knowledge and the collaboration in monitoring of illegal fisheries (Anuchiracheeva et al. 2003).

The focus group discussion with the local leaders in Bang Saphan revealed some of the difficulties fishers faced in sustaining the project. After the official project completion in 2007, all fishing groups involved in the project agreed to continue with the existing conservation measures. Specifically, they prohibit large-scale fisheries and destructive gears, such as trawls and light luring purse-seine, from operating in the 240 km^2 area that they have been able to protect. This effort is remarkable, given that in most cases, sustaining community-based resource management programs after the donors withdraw and the funding ends has proven to be a major challenge (Christie 2005). The success of the Bang Saphan project is attributable partly to having good leadership with vision for the well-being of the communities. This resonates well with other fishers we interviewed who regarded good leaders as those who provided people with sound guiding principles. It was in the similar context of leadership quality that the King's sufficiency philosophy was endorsed.

Like the project in Bang Saphan, but of larger scale, the CHARM project was a research and capacity building initiative aiming to halt environmental degradation and to promote marine conservation. The project began (in 2003) at the time when the institutional reform for fisheries took place in Thailand with the establishment of the new Ministry of Natural Resources and Environment. In addition to the Department of Fisheries, the CHARM project involved the Department of Coastal and Natural Resources from the new ministry, extending thus the traditional focus on the harvesting and natural aspects of the fisheries to social sustainability and conservation. According to the fishers in Krabi, the relationship between the government officials and the local communities had greatly improved with the CHARM project, and they were very eager to see it continued. Like the fishers in Bang Saphan, they have been trying to keep the spirit of the program going despite its official ending in 2007; for example, through their own continuation of the monitoring, surveillance, and control activities. The high cost of implementing these activities is, however, the main deterrent that they will need to overcome.

14.5 Discussion

From the interactive governance perspective, the meta-order governance, including values, images, and principles, plays an important role in determining governing direction and influencing governance outcomes (Kooiman and Jentoft 2009). For many small-scale fishers in Thailand, the alignment of the fishing lifestyle with the sufficiency principle may well reflect the positive image that they would like to portray along with their conservation conviction. Based on the interviews, poverty is not only a condition that fishers try to avoid for their own well-being and that of their family members, but also an image that they do not wish to be associated with, especially because of the connection to laziness. There is an exemption to this generalization, however. In the four provinces we studied, the "poorer" fishing households are usually those with elderly couples without children or grandchildren to care for them. While their living conditions may be poor, they generally receive food and other support from others in the villages. Caring for the elders is a social characteristic typical of the Thai people in closely knitted communities, which contributes significantly to alleviate impoverished conditions of the vulnerable members of the population. Similar social attributes have been reported in other settings, not only in the south, or only among the indigenous communities as often documented.

Some may argue that the shift from poverty to sufficiency is part of the "image makeover" that small-scale fishers were willing to undergo in order to benefit from the government's sufficiency economy program. Yet, following the King's advice to live sufficiently may also be seen as a tribute to help move the country along the path to sustainability. While there is limited literature on the concept of sufficiency, Princen (2005) refers to it as "simple, intuitive, and rational," as well as "an idea, a principle, and ethics for sustainability." He further argues along the same line with what is being promoted in Thailand that, "sufficiency is a class of principles sensitive to critical environmental risks, to the needs of management and self-management, when it is otherwise all too easy to evade responsibility for such risks" (Princen 2005, p. 19). Self-imposition of limits, expressed through sufficiency, forms a basis for discussion about society's reflection about its own well-being and a reduced emphasis on economic expansion (Sachs 1999). Questions that follow from this debate, as suggested by Sachs (1999), for instance, whether the appreciation of limits leads to a more flourishing society; and whether such self-limitation is part of self-liberation, are directly related to the discourse in poverty and poverty-coping strategies.

If the sufficiency philosophy offers an alternative image of those being good at what they do and being satisfied with their lifestyle, it can then be seen as an implicit strategy for fishers to stay ahead of poverty since it signifies modesty and simplicity of the conduct of their profession. It may, for example, encourage diversification of livelihoods to farming in unfavorable fishing conditions, rather than investing in expensive gears (such as trawls and push nets) that may bring them high but unsustainable income. The principle also supports the implementation of long-term solutions such as protection of seagrass and mangrove forests and release of egg-bearing females, which enable fishers to cope with changes in the natural systems at the pre-harvest part of the fish production chain. Livelihood diversification and conservation strategies are tactics to preserve the natural resources and protect the asset base that are considered critical to reducing and averting household poverty (Scherr 2000).

Part of the recovery effort after the Tsunami has had unintentional consequences through an increased number of boats, and the change in ownership to non-fishers and migrant fishers; thus creating new challenges to the governing system. However, it has heightened the attention to conserve mangrove forests. Thailand has a long history of mangrove conversion for charcoal production and housing construction in the 1960s, followed by clear-cutting for tiger prawn farming in the 1990s (Dierberg and Kiattisimkul 1996). Such massive destruction has, however, discontinued with the increased knowledge and the understanding about the importance of mangroves. Several reforestation programs have been in effect over the last 10 years, and received renewed interests from the public and funding agents after the Tsunami (Chang et al. 2006).

It is common these days to find community-based mangrove conservation projects in schools and in communities along the coast, reinforcing the sense that the mangroves belong to the people. From the natural, pre-harvest system perspective, the protection of mangrove forests along with the prohibition of destructive gears and the artificial reef program, have all contributed to the restoration of the marine ecosystem that small-scale fishers depend on. These and other examples, such as new knowledge about fisheries like spawning of crab in the deep sea, and a recent interest among fishers in the use of sails instead of engines, all indicate the potential for marine conservation as a poverty aversion strategy.

In the harvest system, prohibition of destructive gears is a coping strategy shown by some of the studied fishing communities, which also aligns with the sufficiency principle. Studies have shown that mobile bottom-tending gears such as trawls and dredges cause devastating impacts to the seafloor and its inhabitants (see, for example, Watling and Norse 1998; NRC 2002). While the impacts of these gear uses have not been studied in Thailand, there is no reason to think that they will not result in similar consequences, given how they are deployed.

Comparatively speaking, gears that operate at the bottom of the seafloor are likely to cause more habitat damages, and in some cases bycatches, than gears operating at mid-water level or at the surface (Morgan and Chuenpagdee 2003; Fuller et al. 2009). The efforts of the fishing communities in Prachuab Kiri Khan Province in maintaining the extended no-trawling zone are examples of initiatives that correspond well with conservation and sustainability principles that should be promoted in other areas.

This is, however, easier said than done. In Chanthaburi Province, prohibition of trawls and push nets in the normal 3 km nearshore area is not as fruitful with the on-going use of set bag nets. Scientific research needs to be conducted to properly examine the ecosystem impacts of the use of this traditional gear, in order to provide a basis for discussion with small-scale fishers. Although a study by Chuenpagdee et al. (2001), conducted in the south of Thailand, shows that fishers recognize the importance of marine ecosystems such as coral reefs and seagrass beds, and were concerned about the impacts of different activities, it is uncertain that they would agree to abandon the use of set bag nets, had the impacts been known. The regulation of set bag nets is further complicated by the fact that the control in the number of bags set has resulted in high market value for the "real estate" where the bag nets are set. This *de facto* property right basically sets precedent for unlawful ownership by private citizens of state-owned areas. The implication of this, as in other property markets, is the discrimination against some fishers, usually the poorer and the less powerful.

Some of the factors affecting the harvest part of the fish chain, and thus fishers' ability to avert poverty, fall outside of the control of fishers, such as the weather or the fluctuation in the global oil prices. On a larger scale, concerns due to the changing demographics with foreign migrant workers involved in the fisheries, while seemingly manageable, may not be easily handled. The case of the Burmese refugees (the Rohingya) taken up by international human rights communities well illustrates the difficulty that the Thai government may face in its attempt to restrict the access of the Burmese fishers to the fisheries.

The most unique feature of the Thai small-scale fisheries in averting poverty appears in the post-harvest system, where the relationship between fishers and fishmongers is generally positive. In other parts of the world, we often hear about the high dependency of small-scale fishers on middlepersons, and how such dependency results in a disadvantageous situation; with fishers having no bargaining power for the prices they get for their catches, or with the high interests on the loans. Several conditions enable Thai fishers to avoid this vicious relationship, and thus poverty.

Among them are the high demands for fisheries products, created partly through tourism development and the lack of market monopoly. The fact that many of the fishmongers are considered members of the communities, not outside buyers, helps create trusting and mutually beneficial relationships. Good physical infrastructure (e.g., roads and transportation systems), improved communication technologies (e.g., cell phones) and the institutional support that fishers have in the organization of the local market cooperatives are conditions favoring fishers in obtaining reasonable income.

Fishing communities have largely benefited from the infrastructure development and other amenities geared to accommodate foreign and Thai tourists who like to visit beaches and islands. Good transportation systems and facilities in these areas mean better access to markets for the fishing communities, as well as opportunities for supplemental income. Similar to what Barrett (2008) observed in the farming system in eastern and southern Africa, participation of small-holders in marketing activities often make it possible for them to break out of the poverty trap and become more subsistent. It should be noted that while fishers' well-being is possible to attain under the local market cooperative, they are still vulnerable to the price fluctuations driven by export and global markets. The local market cooperative helps mitigate such effects, but only to a certain extent. An exploration of new domestic markets and the development of other value-added products would also be required.

In terms of the governing system, the establishment of the decentralized Au-Bor-Tor, and the new government agency responsible for natural resource conservation are two major institutional changes with direct impacts on the livelihoods of small-scale fishers. The effectiveness of these reforms depend largely on factors such as good leadership; collaboration between communities and governments; and at least at the initial stage, external funding. In accord with the interactive governance theory, our study shows that participation of fishers in resource governance is an important factor contributing to poverty-aversion potentials. The attempt of fishers in Krabi Province to maintain monitoring, surveillance, and control activities, and the continuation of the community-based fisheries management project in Prachuab

Khiri Khan Province are examples of initiatives that help build local capacity and strengthen social organizations. The changing attitude between communities and government officials observed throughout the study is an encouraging sign to support other community-based programs that focus on building capacity for non-fishing alternative livelihoods.

The general interest expressed by some fishers in nature tourism businesses is an example of portfolio diversification that may be critical to help alleviate and avert poverty. Research on tourism carrying capacity and careful planning would be required to minimize impact on the natural environment, to promote sustainable uses of marine resources and to ensure that local communities benefit from such development.

14.6 Conclusions

Like many others elsewhere, Thai small-scale fishers face challenges in maintaining their livelihoods and staying out of poverty. We learned from the field observations, focus group discussions, and individual interviews that many of them possessed a certain philosophy and several traits that enabled their poverty-coping and averting potentials. The sufficiency principle played an important role in guiding fishers toward a satisfying lifestyle that also corresponded well with resource conservation and sustainability principles. Sufficiency was generally referred to as a condition of being neither rich nor poor, but of having enough to get by. It suggested modest living, believed to offer security and protection against future threats, while allowing for reasonable enjoyment of the economic development. While their proclaimed acceptance of the King's philosophy may suggest the stigma about poverty, many of the fishers interviewed explicitly stated how they chose not to expand their fishing operations. Many fishers refrained from using illegal fishing gears and instead participated in the protection and restoration of coastal and marine ecosystems. These practices contributed to keeping the natural environment relatively healthy and thus sustainability of the fisheries.

The majority of the Thai small-scale fishers diversified their livelihoods both in fishing- and non-fishing-related activities. Most of them engaged in multi-species and multi-gear fisheries, which allowed them to switch their target species according to the fisheries regulations. In unfavorable weather conditions, they were able to work in farming and as labor on plantations. Further, the growth in coastal tourism has not only increased the local demand for fisheries, but also resulted in infrastructure development that has benefited the fishing communities. Some fishers have even included homestay businesses as part of their livelihood portfolios. Involvement of women in net-mending and income-generating activities including post-harvest processing and marketing, as well as in craft making, contributed to poverty-averting potentials among fisheries households. Yet, the most advantageous condition for small-scale fishers was the good relationship with fishmongers who, often as members of the communities, offered fair prices. In some cases, fishers also organized local market cooperatives to reduce their dependency on the middlepersons and to increase their bargaining power and market control. Through this initiative, fishers also obtained organization and business skills useful for other purposes.

From the interactive governance perspective, fishers' active involvement in the management of their resources, as shown in the case studies, is one of the most important factors contributing to poverty alleviation. Accelerated by the decentralization scheme in the form of Au-Bor-Tor; and the external funding support for the community-based management in Bang Sapan; and the habitat restoration program (through CHARM); fishers in these communities were empowered to continue with the initiatives when they came to an end. The working collaboration and partnerships established between the communities and government officials during these projects offers a solid platform for small-scale fishers to protect their resources and sustain their livelihoods.

Some of the conditions that can help fishers cope and avert poverty such as livelihood opportunities, improved infrastructure and government support can be created in other areas. Training and capacity-building programs to help diversify livelihood and income-generating activities, for example, are already part of the standard development package. Markets and marketing strategies, as well as linkages to the tourism industry can be developed to benefit small-scale fisheries. In many places, awareness building for resource conservation and conditions for sustainable fishing practices are being promoted. Yet, it may be the generation of images for the sufficiency philosophy, based on the understanding of the underlying values and principles that fishers and fishing communities have, that may be required to avoid the poverty trap.

Acknowledgments We are grateful to the Norwegian Research Council for funding the PovFish project, and the PovFish team members, especially Svein Jentoft and Maria-Victoria Gunnarsdottir, for their inputs. We thank all the key informants who provided us with information about their fisheries and their lives. We also appreciate the help from Daracha Thiammueang in identifying key informants and in data collection. Ratana Chuenpagdee acknowledges support from the Social Sciences and Humanities Research Council of Canada and Canada Research Chairs program.

References

Anuchiracheeva S, Ganesh HD, Shivakoti P, Ruddle K (2003) Systematizing local knowledge using GIS: fisheries management in Bang Saphan Bay, Thailand. Ocean Coast Manage 46:1049–1068

Barrett CB (2008) Smallholders market participation: concepts and evidence from eastern and southern Africa. Food Policy 33:299–317

Béné C (2003) When fishery rhymes with poverty: a first step beyond the old paradigm on poverty in small-scale fisheries. World Dev 31:949–975

Chang SE, Adams BJ, Alder J, Berke PR, Chuenpagdee R, Ghosh S, Wabnitz C (2006) Coastal ecosystems and tsunami protection after the December 2004 Indian Ocean Tsunami. Earthquake Spectra 22(S3):S863–S887

Christie P (2005) Is integrated coastal management sustainable? Ocean Coast Manage 48:208–232

Chuenpagdee R, Knetsch JL, Brown TC (2001) Environmental damage schedules: community judgments of importance and assessment of losses. Land Econ 77(1):1–11

Dierberg FE, Kiattisimkul W (1996) Issues, impacts, and implications of shrimp aquaculture in Thailand. Environ Manage 20(5):649–666

DoF (2002) Census of fishing households 2000. Statistical Unit, Department of Fisheries, Ministry of Agriculture and Cooperatives, Bangkok

FAO (2009) FAO country profile – Thailand, national fishery sector overview 2009. Food and Agriculture Organization of the United Nations, Rome

Fuller SD, Picco C, Ford J, Tsao CF, Morgan L, Hangaard D, Chuenpagdee R (2009) How we fish matters: addressing the ecological impacts of Canadian fishing gear. Vancouver

Jentoft S, Chuenpagdee R, Bundy A, Mahon R (2010a) Pyramids and roses: alternative images for the governance of fisheries systems. Mar Policy 34:1315–1321

Jentoft S, Onyango P, Islam MM (2010b) Freedom and poverty in the fishery commons. Int J Commons 4(1):345–366

Juntarashote K, Chuenpagdee R (1991) Small-scale fisheries of Thailand. In: Durand JD, Lemoalle J, Weber J (eds) Research and small-scale fisheries. IFREMER/ORSTOM, Paris

Kooiman J, Jentoft S (2009) Meta-Governance: values, norms and principles, and the making of hard choices. Pub Adm 87(4):818–836

Kooiman J, Bavinck M, Jentoft S, Pullin R (2005) Fish for life: interactive governance for fisheries. Amsterdam University Press, Amsterdam

Krongkaew M (2003) The philosophy of sufficiency economy. Kyoto review of Southeast Asia, Issue 4, Regional Economic Integration, Oct 2003

MoI (2008) National report on quality of life in rural areas. Information Centre for Rural Development, Ministry of Interior (in Thai)

Morgan L, Chuenpagdee R (2003) Shifting gears: addressing the collateral impacts of fishing methods in US waters, Pew science series. Island Press, Washington DC

Narayan D, Chambers R, Shah MK, Petesch P (2000) Voices of the poor: crying out for change. Oxford University Press, New York

Nasuchon N, Charles A (2009) Community involvement in fisheries management: experiences in the gulf of Thailand countries. Mar Policy 34:163–169

NRC (2002) Effects of trawling and dredging on seafloor habitat. National Research Council, National Academy Press, Washington, DC

Pauly D, Chuenpagdee R (2003) Development of fisheries in the gulf of Thailand large marine ecosystems: analysis of an unplanned experiment. In: Hempel G, Sherman K (eds) Large marine ecosystems of the world: trends in exploitation, protection, and research. Elsevier Science, Amsterdam, pp 337–354

Princen T (2005) The logic of sufficiency. MIT Press, Cambridge

Sachs W (1999) Planet dialectics: explorations in environment and development. Zed Books, London

Scherr SJ (2000) A downward spiral? Research evidence on the relationship between poverty and natural resource degradation. Food Policy 25:479–498

Sen A (2000) Development as freedom. Anchor Books, New York

TRF (2006) Economic sufficiency. The Office of Thailand Research Fund, Government of Thailand, Bangkok

UNDP (2000) Human development report 2000. United Nations Development Programme, Oxford University Press, New York

Watling L, Norse EA (1998) Disturbance of the seabed by mobile fishing gear: a comparison with forest clear-cutting. Conserv Biol 12:1189–1197

Part IV Changing

For small-scale fishers around the world, a lot of things need to be changed. For that to occur, we also need new ways of thinking about them. This is regardless of their poverty and vulnerability status. In some instances, small-scale fisheries represent a kind of wealth that makes fishers well off. In other instances, their livelihoods are far from secure, and they have no guarantee that they will be able to fend off poverty.

The sector as such may be in a precarious state that may threaten long-term viability. Even if people do not perceive themselves as poor, they may still be vulnerable, even if they do not see any immediate problems concerning their livelihoods. For various reasons, such as overfishing and illegal fishing practices, as mentioned in the Vietnam, South Africa, and Guatemala chapters in this section of the book, the sector may be volatile; their ecosystems are fragile and may not be able to sustain the pressures they are under. Climate change is hurting coastal people. The coast is booming with new industries and increasing populations that may affect small-scale fishing people negatively; even if, in many instances, this also provides opportunities for livelihood diversification that will give people more options and more secure economies.

If people find other ways of sustaining themselves, it may also relieve the pressure on the resources, provided that other users do not replace them. How to bring about this change, which would not only bring greater prosperity, but also social justice for the poor, and what local people can do are issues discussed in this section.

Sri Lanka was victim of the tsunami that devastated communities and made recovery an urgent issue. But it soon brought new pressures on the resources to an extent not witnessed before the disaster, which calls for more effective resource management. This is a role that cooperatives play due to the social capital they represent, but which do not do so to the extent possible. Change is therefore also organizational and about leadership, which is also well-illustrated in the case of Mozambique, where poverty among small-scale fishers is prominent. Here, several political regime shifts have not always helped the poor, but poverty alleviation for small-scale fishing communities has gained political standing, preparing them thus for the emerging user conflicts.

Chapter 15
Facilitating Change: A Mekong Vietnamese Small-Scale Fishing Community

Kim Anh Thi Nguyen and Ola Flaaten

Abstract As one of the poor provinces in the Mekong Delta, Ben Tre has the three poorest coastal districts, namely Binh Dai, Ba Tri, and Thanh Phu. Thanh Phong is one of the 18 communes of Thanh Phu. The objectives of this chapter are: (1) to understand poverty and discover its causes among the poor fishers in Thanh Phong; (2) to evaluate the insufficiency of poverty-alleviating policies for small-scale fisheries in Thanh Phong; (3) to suggest tailored solutions and poverty-alleviating policies for Thanh Phong. Currently, Thanh Phong has a community of 56 fishing households making a living from small-scale fisheries. Illegal fishing gear is still being used, like "day-mung" nets with a very small (1–1.5 mm) mesh size. With small-scale boats and simple fishing tools, the productivity as well as the commercial value of the fish caught are very low. There is awareness among the households that inshore fisheries' resources are limited, and that their illegal practices are exhausting the resources. However, the dependence on fisheries' resources and the absence of alternative incomes force people to use destructive kinds of fishing gears. As a consequence, the vast majority of the poor fishers in Thanh Phong have to face livelihood uncertainty and the threat of increasing poverty. Compared with the poverty line adopted by the government of Vietnam, the poverty rate of 12.7% in 2008 is considered high, and the rate of 13.3% in 2009 is a signal of poverty persistence. Government support is needed to combat poverty, including upgrading the infrastructure and assisting in creation of alternative jobs and better education.

K.A.T. Nguyen (✉)
Nha Trang University, Nguyen Dinh Chieu St, Nha Trang, Vietnam
e-mail: sonanhcc@gmail.com

O. Flaaten
Norwegian College of Fishery Science, University of Tromsø, 9037 Tromsø, Norway
e-mail: olaf@nfh.uit.no

S. Jentoft and A. Eide (eds.), *Poverty Mosaics: Realities and Prospects in Small-Scale Fisheries*, DOI 10.1007/978-94-007-1582-0_15,

15.1 Introduction

This research project is about a small fishing community in Ben Tre, one of the poor provinces in Vietnam.[1] Fishing households in this community seem to be caught in the transitional loss traps where individual actions taken for long-run gains imply suffering and loss in the short run. The province of Ben Tre is located in the Mekong Delta and has a complex system of internal rivers with a total length of over 382 km. These branches of internal rivers originate from the Mekong River system and meet the East Sea, also called the South China Sea, at four big estuaries. Accompanying this system are fertile lands, bordering 65 km of the coastline and a privileged sea area, which is part of the exclusive economic zone of approximately 20,000 km^2.

On the map, Ben Tre has the shape of a pied fantail; the upstream source divides into dispersed smaller branches and follows the eastern direction to meet the sea. The province has a natural area of about 2,300 km^2. It borders Tien Giang Province in the north by the Tien River, Vinh Long Province and Tra Vinh Province in the west, and in the south the Co Chien River. Four big rivers, Tien Giang, Ba Lai, Ham Luong, and Co Chien, surround and divide Ben Tre into three separate land areas (Fig. 15.1).

In the past years, Vietnam gained significant achievements in reducing poverty. Its determination was highly appreciated by international organizations and other nations. According to the Vietnamese General Office of Statistics, the general poverty rate decreased sharply from 58% in 1993 to 37% in 1998, 29% in 2002, 20% in 2004, 16% in 2006 and only 14.8% in 2008. During the last 15 years, the number of poor people has decreased by three-quarters (MFA 2008); and the per capita income in 2008 was 960 USD (GSO 2008). However, the poverty rate change slowed down from 2002 to 2008, and the income of the majority of the population was close to the poverty line. Hence, a small increase in the defined poverty line also makes more people fall below it. Another poverty issue is that the gap between the rich and the poor, between coastal and urban areas, is becoming worse (MARD 2004).

According to the Vietnamese poverty census in 2008, the poverty rate in the Mekong Delta was 11.1%, Ben Tre was 13.0%, and Thanh Phu district was 17.5%, compared with the national average of 14.8%. Thanh Phu and the two other coastal districts (Binh Dai and Ba Tri) are the poorest parts of the province. Compared with the poverty line adopted by the government of Vietnam, which is under VND 300,000/person/month,[2] the poverty rate of 12.7% in 2008 (TPCPC 2008) is considered high; and the rate of 13.3% in 2009 (TPCPC 2009) is a signal of poverty increase.

---

[1]Several Vietnamese official statistical sources have been used for the facts presented both in this, and in other sections of this chapter. A complete list may be obtained from the corresponding author. Some general information presented, in addition to the data presented in the tables, has been collected by the authors through interviews with officials and people from the fishing industry in Ben Tre.

[2]VND 300,000 was equivalent to USD 17.65 (2008) and USD 16.67 (2009).

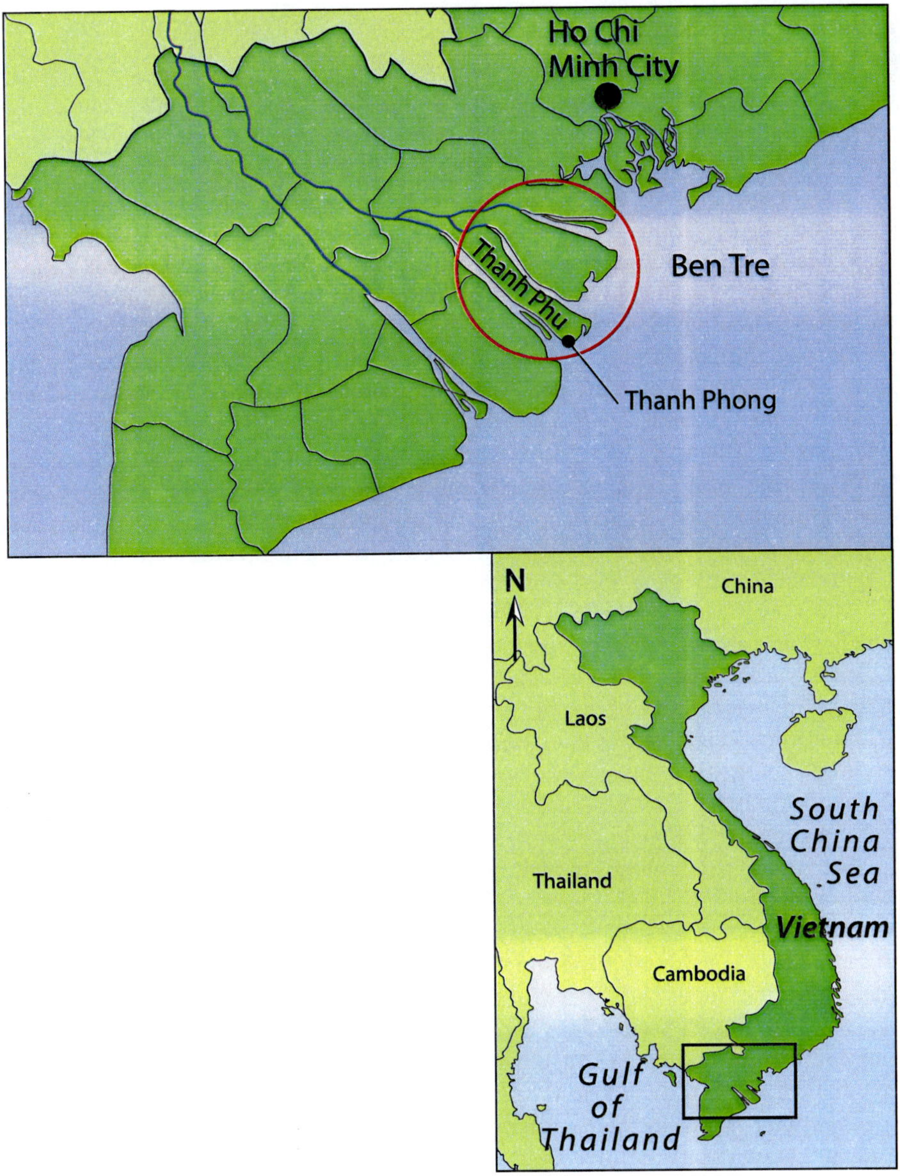

Fig. 15.1 Thanh Phong Commune, Thanh Phu District, Ben Tre Province, Mekong Delta, Vietnam. Thanh Phong Commune is located in the poorer outskirts of the Mekong Delta

15.2 Background

Thanh Phu district has 18 communes, of which Thanh Phong, the study site of this chapter, consists of 5 hamlets. This commune shares borders with Thanh Hai commune in the north, the Co Chien River in the south, the sea in the east, and Giao Thanh commune in the west. The centre of the commune is located 25 km east of the centre of the Thanh Phu district. There are about 6,490 ha of land of which 2,860 ha is for agriculture, and 920 ha for forestry. There are 17 km of coastline in Thanh Phong. The alluvium area is deposited by the two rivers, Co Chien and Ham Luong, and hence widens over time.

Overall, the landform of the area is low – about 1.8–2 m above sea level – and most of the land is wet in the rainy or high-tide season, creating problems for agriculture, communication, and other economic activities. The population of Thanh Phong is 9,250 – with almost 50% females – living in a total of 2,380 households, with a density of 6 people per km^2. The district is dominated by Kinh people, who came from the nearby provinces, first to work and then to settle down. Fishing and aquaculture (80%) and agriculture (15%) are the main income activities in this commune, in addition to 5% labor in services and commerce.

Currently, Thanh Phong has a community of 56 households making a living from small-scale fishing activities, some of them illegal. The size of the fishing households is, on average, not much bigger than that of farming households. The main inshore fisheries' resources of the area are Perciformes, Siluriformes, and Clupeiformes; clam (*Meretrix lyrata*); blood cockle (*Anadata granosa*); marine shrimp such as black tiger (*Penaeus monodon*), cat tiger shrimp (*Parapenaeopsis hardwickii*), and greasyback shrimp (*Metapenaeus ensis*); slipper lobster (*Ibacus ciliatus*); mud crab (*Scylla serrata*), and swimming crab (*Portunus pelagicus*).

The vessels have engines of mainly 20–24 HP. Approximately 71% of the households use bottom nets with a very small mesh size (about 1 mm), called "day-mung" to catch very small shrimp. These nets are illegal when the mesh size is smaller than 10 mm. Shrimp are dried in the sun, sold mainly to middlemen, and used as feed in the aquaculture industry. The rest of the households use river bottom nets, gill nets, and traps. To finance their fishing activities, many households depend on loans from the middlemen, and as part of the deal, they have to sell what they catch to the middlemen, often at a much lower price than otherwise.

Regarding aquaculture, the main products are now tiger shrimp (*P. monodon*) and crab (*S. serrata*). Before 2009, the culture of many crops per year resulted in widespread failure. From 2009 on, Thanh Phong people started to concentrate on the aquaculture of just one crop per year; the rest of the year is for cultivating rice (which is mostly used for everyday meals instead of being sold because of its low quality). In addition, there are alluvial sediment areas where people have established co-operatives to culture clams. In recent years, the increase in the use of destructive fishing gear, such as raking the sea bottom with nets of very small (1–1.5 mm) mesh size; the increase of water area usage for aquaculture at the expense of fishing grounds; and the transmission of diseases from aquaculture to wild fish have all contributed to causing an unstable income for the people here (Ben Tre DARD 2009). As a consequence, the vast majority of the poor people in Thanh Phong have to face livelihood uncertainty, and the risk of an increasing poverty level.

The infrastructure is very poor in Thanh Phong. The transport of goods encounters various difficulties, especially in the rainy season, because of the wet pathways. Clean water is yet to be available for the Thanh Phong fishing community. Regarding fisheries, aquaculture, and agricultural goods, fishers who are able to travel to the central Thanh Phu district may sell their products at a much better price than those who sell to middlemen, who often force fishers to lower their price due to the difficulty of transportation. However, most products from fisheries, aquaculture, and agriculture are sold locally to middlemen.

Fishing families seem to be conservative in the sense that the knowledge about their resources and occupation is transferred from generation to generation, and few leave the fishing activities to change their lives. Most of them have very limited formal education, which is more demanded in other trades. There is one primary school in every hamlet of the Thanh Phong commune, but only one junior high school for the whole commune. In recent years, people in some fishing households have moved into aquaculture and agriculture as part-time occupations. This earns them some extra income. With small-scale boats and simple harvesting technology, the productivity, as well as the commercial value of the fish caught is very low, but this is still the main income source for fishers in Thanh Phong.

Fishers seem to be aware of the illegality of the fishing gear they use, and the fact that this practice is exhausting the fish resources. However, the dependence on fisheries' resources and the absence of alternative livelihoods forces people to use those kinds of fishing gears. They are caught in the transitional loss trap: to gain economically in the long run would require loss in the short run. Even if each fishing household was to recognize the need to change the total fishing effort in the area, the technical characteristics of the gear or the age and species composition of the catches, such changes would require collective action. For example, some fishers have suggested that they should be allowed to continue to use "luoi mung," but only in some months, combined with a ban on fishing in other months to restore the fisheries' resources (Fig. 15.2). However, this has not been implemented.

Is poverty alleviation in this small-scale fishery possible, or will it come as a result of the general economic development of the country? The poverty alleviation policies for Vietnam and the Mekong Delta have not been adapted to the special small-scale fishing in Thanh Phong. Even though poverty-alleviating policies for the small-scale fisheries may have been sufficiently formulated, the implementation seems to be weak. In this fishery, people depend critically on the inshore resources, which are quite unstable due to the variation in the environment and the fisheries' resources. Some fishers possess neither a piece of land to cultivate, nor alternative livelihoods, nor a high education level. Also, as mentioned before, Thanh Phong's natural conditions are not favorable. The inshore fisheries' resources are decreasing; and the landform is low and often wet in the rainy season. Moreover, the brackish

Fig. 15.2 A fisher prepares to set his trawl. With the small mesh size, he hopes to catch both fish and shrimp

water with a mixture of salt and fresh water varies in salinity, due to the weather and other stochastic processes. Therefore, the agricultural and fisheries' outputs are difficult to predict. The transport infrastructure in Thanh Phong places restrictions on the trade of fish and fish products, thereby limiting the market access and value creation of the fisheries. In addition, economic and social norms may prevent the efficiency in implementing poverty-alleviating policies for the fishing population.

Against this background, several questions regarding poverty in a Mekong Delta community may be raised. First, what are the facts about poverty in Thanh Phong? Second, what are the causes of possible poverty in Thanh Phong? Third, are the current poverty-alleviating approaches suitable in this case? If not, how may they be changed? Fourth, are the current policies appropriate for the specific needs of the poor fishing community of Thanh Phong, and if not, why not? In addition, with the unique characteristics of the small-scale fishing community, should the government provide more suitable assistance and specific policies to support this area in order to raise the local income level?

The main objectives of this chapter are to understand poverty among poor fishers in Thanh Phong and to discuss the causes. Further, it aims to evaluate the poverty-alleviating policies for small-scale fisheries in Thanh Phong; and, based on this, to indicate directions for poverty-alleviating policies for this and similar communities. The chapter is organized as follows: The next section presents the research methods used and the results of the investigations. Then the limitations of pro-poor policies as well as strategies are discussed. Subsequently, several sustainable pro-poor proposals are dealt with; and finally some conclusions are reached.

15.3 Research Methods

15.3.1 Methods

To answer the research questions posed above, this chapter uses a combination of methods and approaches. First, it studies theories and applied work on poverty, including a description of small-scale fisheries and poverty, as well as fisheries and poverty alleviation in Vietnam. The literature used includes documents from the Ministry of Agriculture and Rural Development (MARD), the Prime Minister's office, and other central and provincial governments and public offices.[3] Several important international reports on economic and social development in Vietnam were examined (see e.g., World Bank 2007, 2008, 2009). Second, meetings were arranged with the Women's Association, Farmers' Association, and Fishermen's Association of the Thanh Phu District. Third, study-site investigations were made by the use of questionnaires designed to conduct in-depth interviews with fishers.

The field work was done in December 2008, and the sample consisted of all 56 households in the Thanh Phong fishing community. Fourth, the authors met local, regional, and national experts for discussions and workshops to ascertain their ideas. The participating partners included fishers, local associations, the local government (commune, district, and provincial authorities), the local Departments of Agriculture (district and province levels) and scientists. Based on the scientific discussions, policy analyses, and problem-decision analyses, the chapter also includes the authors' indication of policy directions for poverty alleviation in the small-scale fisheries in Thanh Phong and similar communities.

15.3.2 Poverty Indicators for Thanh Phong

The economic growth rate of Ben Tre is relatively low, so the provincial government has an insufficient budget to conduct poverty-alleviation programmes related to infrastructure, welfare, and education. Hence, support from the central government's budget for poor provinces is in urgent need in order to help eliminate the poverty trap. The distribution of economic growth and income among the inhabitant groups widens the poverty gap between the provincial capital and the coastal districts (Table 15.1). Ben Tre city has few poor households, according to the national definitions by the Ministry of Labour, Invalids and Social Affairs (MOLISA). Rural households with an income of VND 300,000/month/capita (approx $15/month/capita) or lower; and urban households with VND 390,000/month/capita (about $20/month/capita) or lower are classified as poor. These poverty lines have been in place in 2009 and 2010.

[3] A complete list is available from the corresponding author.

Table 15.1 The rate of poor[a] households in the town and districts in Ben Tre from 2006 to 2009

		2006		2007		2008		2009	
No.	Town/district	No. poor house holds	%	No. poor house holds	%	No. poor house holds	%	No. poor house holds	%
1	Ben Tre (town)	1,065	4.18	854	3.09	758	2.75	576	2.04
2	Chau Thanh	6,794	17.02	5,553	13.83	3,993	9.5	3,685	8.90
3	Binh Dai (coastal district)	7,026	19.94	7,693	22.91	4,615	13.43	4,206	12.13
4	Ba Tri (coastal district)	11,017	25.01	9,717	21.70	7,587	16.26	6,918	14.68
5	Giong Trom	9,201	20.09	8,158	17.57	5,554	11.48	4,959	10.29
6	Mo Cay	14,392	23.01	13,236	20.46	10,700	16.14	9,092	13.61
7	Cho Lach	5,545	18.03	4,788	14.09	3,776	10.88	3,842	10.96
8	Thanh Phu (coastal district)	7,749	26.03	7,258	23.03	5,646	17.46	5,400	16.56
Total		62,789	20.02	57,257	17.74	42,629	13.01	38,678	11.58

Sources: Ben Tre Department of Labour, Invalids and Social Affairs (DLISA) 2006, 2007, 2008, 2009

[a]According to national definitions

However, in the coastal district of Thanh Phu, more than a quarter of the households were poor in 2006. This decreased to one in six in 2009. Table 15.1 indicates that in the years 2006–2009, the poorest households in Ben Tre, and also in the coastal districts, had positive economic development.

Even though Table 15.1 demonstrates a reduction in the number of poor households in the districts of Ben Tre according to the national definitions, many other households would be classified as poor according to international definitions.

15.4 Results and Causes

There are three indications/manifestations of the fishing community in Thanh Phong: low income, low basic living standard, and poor social services availability. Some households do not even have enough food in days with bad weather like storms and floods (14.3% of the households interviewed). Sheltering conditions are quite temporary in the way that the roofs are too weak to deal with violent weather (64.3%). Clean water is not currently provided, so local people have to use water from wells and rivers. Only 1 of the 56 houses has a filter system for well water, and 37.5% of the houses have toilets with standard hygiene. The rest depend on their neighbors or nearby nature. In the rainy season, the roads are in a very bad condition, and it is very difficult for local elementary school children to walk to school – a distance of 3–4 km. Many children of secondary school age have to drop out of school, and very few of them finish high school. Of the interviewed families, 64.3% of the children quit school while at the age of high-school attendance. Illiteracy and a lack of reading materials are the main reasons that many people do not read newspapers or rarely have the chance to read one (57.2% of the cases). The free health-care service cards provided to the poor fishers are not often used, as the traveling cost is much higher than the cost of the medical treatment received.

The 56 fishing households in Thanh Phong have some common characteristics[4] related to poverty, including, in a somewhat arbitrary order:

- The per capita income per day is less than 1 USD.
- The food quantity and quality available are insufficient.
- Shelters with bamboo roofs are in poor condition.
- The transportation means are limited.
- Clean water is not available yet.

Poverty is a complex, multi-faceted concept reflecting a low level of well-being (World Bank 2000). To put it another way:

> Poverty is a complex problem, and it is now widely accepted amongst policy-makers, development practitioners, and academics that a good understanding of the extent, nature, and determinants of poverty is a precondition for effective action to reduce deprivation in fishing communities (Neiland and Béné 2004, p. 2).

[4]The questionnaire and the detailed results are available upon request from the corresponding author.

Therefore, in order to provide proper solutions that suit the special conditions of the community, it is necessary to locate the exact causes that trigger poverty. Only then can poverty alleviation be solved in a sustainable way. The following are the causes considered to make fishers poor in Thanh Phong, according to the investigations undertaken, including the survey.

15.4.1 Indication of the Causes of Poverty in Thanh Phong

15.4.1.1 Geographical Location and Natural Conditions

It is a challenge to achieve economic development and poverty reduction in regions with unfavorable natural conditions and geographical location. Thanh Phong is far from the centre, so it is difficult to travel to obtain information. Also, there are few jobs there (100% agreed, according to our interviews). The inshore fishery resources are decreasing (78.6% agreed), so the income from fishing activities is decreasing accordingly (100% agreed). The fresh water is mixed with sea water. Most of the surface water here is for aquaculture. However, a lack of capital and a large amount of damage from diseases make the aquaculture practice ineffective – all 20 households engaged in aquaculture activities agreed with this in the questionnaire. The other 36 households do not perform aquaculture and they marked "No idea" in the questionnaire. In bad weather, fishing activities cannot be undertaken; in addition, damage due to the bad weather makes fishers face even worse conditions afterward (100% agreed).

15.4.1.2 Fishing Methods

The main types of fishing practices in Thanh Phong are described in Table 15.2. In general, destructive gear means gear types that catch fish that are too young (growth overfishing); do not allow enough fish to spawn (recruitment overfishing); or have unwanted effects on non-targeted fish (by-catch). Fish refers to all the marine species available for fishers. In the case of Vietnam, with hundreds of marine species within small coastal areas, scientific "proof" in the form of stock assessment, predictions of stock development, effects of gear selectivity, etc. are very limited (Do and Nguyen 2005). This is also the case for Thanh Phong. The distinction between destructive and non-destructive gear is not precise, but indicative. A trap, for example, could be turned into destructive gear if it was to be fitted with very small mesh.

Fishing with a bottom net with a very small mesh size (1–1.5 mm), called "day-mung" and river bottom net, make up the highest rate (83.9%), and these are considered destructive fishing gears (Fig. 15.3). Despite the popularity of these types of fishing gears, the economic values from it are the lowest (the subjects of stow draft nets are too-small shrimp, which are of very low economic value). Even if these households wanted to change their fishing gear, so that they could catch other

Table 15.2 Fishing gear in Thanh Phong[a]

No.	Fishing gear	Number of households	Rate (%)	Notes
1	Bottom net with a very thin mesh size, called "day-mung"	40	71.4	Destructive fishing gear
2	River bottom net	7	12.5	Destructive fishing gear
3	Trap	7	12.5	Non-destructive fishing gear
4	Gill net	2	3.6	Non-destructive fishing gear
Total		56	100	

Source: Own survey

[a]The Ministry of Agriculture, in charge of fisheries, promulgated the documents concerning destructive gear and mesh-size requirements for each particular kind of fishing gear. In each province, based on the Ministry's documents, depending on the specific situation of the applied fishing gear, the provincial Agriculture Department has also published documents for its management. Therefore, Table 15.2 has been classified according to Ben Tre Fisheries Department's regulation documents. Bottom nets with a very thin mesh size (1–1.5 mm), called "day-mung," and river bottom nets are both categorized as destructive gear, due to the violation of the required mesh size

Fig. 15.3 Photo showing the destructive 1–1.5 mm mesh size of nets

aquatic products that are of higher economic value to alleviate poverty and improve their living conditions, they are caught in the transitional loss trap. Sustainable improvement requires both collective action (to avoid overfishing) and private investment (in non-destructive gear). Fishing gear other than "day-mung" nets and river bottom nets can better protect the inshore marine resources because they will reduce the likelihood of catching and killing the very young fish and shrimp (Ben Tre DoF 2006).

15.4.1.3 Risk and Access to Capital

In the survey, 100% of the fishers agreed that the lack of access to production capital is the fundamental cause of low income since this prevents them from boosting their income. Without capital, the people cannot invest in boats and fishing gear for offshore fishing or other industries. Only 19 of the 56 households (33.9%) had access to credit; whereas 33 (58.9%) of poor households agreed that it is difficult for them to approach official loan institutions as they do not have property to mortgage. There are, however, in the community, unsecured loans for fishers through associations, but the limit of five million Vietnam dong (VND), about 250 USD, is too low to make it possible for them to replace their existing fishing gear. Therefore, changing their financial situation through this small loan is impossible. Let us take one example of households who fish with "day-mung" nets: they need from 22.5 to 30 million VND to convert to trap fishing, as it costs 150,000 VND for one trap and there need to be an average of 150–200 traps for each household to achieve effective catching.

Fishing and aquacultural activities heavily depend on climatic and natural conditions, and therefore are prone to risks like disasters and epidemic diseases. The people, as a result, lose their investment capital, fall into bad debts, and lose the boat to invest in reproduction. Since they lack capital, the number of fishers using inexpensive but destructive fishing gear, the "day-mung" and river bottom nets, together makes up about 84% of the households in Thanh Phong. Consequently, the inshore fish source at the surveyed sites is exhausted (Ben Tre DARD 2009). The results of our investigations prove that the households using destructive fishing gear are subject to being poor.

The reason why many households are using destructive and illegal fishing gear is the low cost of this tool. The administration of Thanh Phong commune acknowledges this fact, and encourages people to use other fishing methods, such as fish traps (TPCPC 2009). However, the biggest problem the people face is a lack of capital to change their fishing gear. Consequently, the households in Thanh Phong cannot eliminate poverty; as destructive fishing tools are still being used, and the aquatic products obtained from these fishing methods are of very low economic value. Continuing to use fishing nets with an illegal mesh size/destructive gear that violate fishing regulations will contribute to the depletion of coastal fisheries' resources, and will therefore result in keeping the household income inherently low.

15.4.1.4 Difficult Access to Fish Markets

Poor households in Thanh Phong have difficulties in approaching fish markets and receiving market information. The market prices are unstable, and living without saving buffers is a constant worry. As the middlemen are the ones who mainly decide the input and output prices of fishing and aquacultural products, these households are the targets of negative effects from the market. The fishers are forced to sell fish at a lower price and buy foodstuff and equipment at a higher price. As expressed in the Lake Victoria, Tanzania case study on poverty:

> If we can break our dependency on fish agents, it would probably be possible to get better prices for our catches (Jentoft et al. 2010; Onyango, Chap. 6).

15.4.1.5 Lack of and Poor Quality of Land

Among the 56 families, only 12 (21.4%) have farming land, from which most of the products are for their own consumption. In general, fishing communes on the islands and outskirts of Vietnam possess little agricultural land, contrary to those in more favorable areas. Thus, across the country, the available agricultural land in and nearby fishing communities varies, but good statistics for comparing Thanh Phong with other communities are lacking. However, an example can be provided, based on other research activities of the authors. At the other end of the scale is, for example, Nha Phu Lagoon in Khanh Hoa province (with the fast-developing coastal city of Nha Trang as the province capital), whose agricultural land occupies about 95% of the natural land.

A lack of land is a repeated cause of meetings between fishers and local authorities. Those who have the land reason that the land quality, which is poor in nutrition, wet in the rainy season and dry in the dry season, and high tides sometimes make agricultural activities difficult. There are only one or two seasons in which people can grow crops in a year, thus the productivity is low. In addition, the production cost is high and epidemic human diseases occur quite often, making people remain in poverty (TPCPC 2008). There are three main reasons for the relatively high production costs: (1) harsh weather conditions and impoverished soil; (2) high costs for inputs such as seeds, fertilizers, insecticides, and electricity, which are a general problem in Vietnam; and (3) more occasional animal and plant diseases affecting productivity. Until now, some households have switched to raising aquatic products, yet the lack of capital and the disease problems make that ineffective. Others, who do not have many culturing opportunities, work part-time to increase their income, but they are still poor.

15.4.1.6 Low Education Level and Limited Knowledge of Technology

As the survey results show the educational level of the householders is low: 14.3% of them are illiterate (compared with the national average of 2%); and 64.3% have completed only primary school. In total, the two groups account for 78.6% of the total number of households. This high percentage of fishers with a low educational level can be explained by the influences of their jobs; i.e., to perform inshore fishing well, fishers mainly use their experience, not their education level. Therefore, when the source of the inshore fishery is exhausted, the fishers here have difficulties in changing their job. They lack effective working experience, methods and knowledge of technology. Furthermore, their awareness of other matters in life, especially of protecting the environment and the marine ecosystem, is low (Ben Tre DARD 2009).

15.4.1.7 Lack of Alternative Jobs

The lack of alternative, stable, and local jobs is one fundamental cause of low income among poor households. Fishing activities can be carried out only in the morning due to the tidal changes and fish availability, and for the rest of the day the fishers have little else to do. There are several reasons for the lack of on-site jobs: the long distance from the centre; limited land area; poor infrastructure; low education level and technological skills. In addition, the local commune has an ineffective economy and limited social policies. Therefore, in the short run, people here have few possibilities of leaving this place and heading for more central areas. However, in the long run, it is likely that at least some fishers or their children will swap to other jobs or income-generating activities. It remains to be seen whether this will be enough to help reduce labor in the open-access fisheries (further discussion below).

15.4.1.8 Lack of Infrastructure

The poor infrastructure actually prevents Thanh Phong from having external investors, which therefore diminishes the job opportunities for local people. In Thanh Phong, transportation is bad and many muddy roads exist. The only backbone road that crosses the commune is Highway 57. The bad condition of transportation negatively impacts on the lives of the people here. In rainy seasons, they have great difficulties in transporting the products harvested, and have to sell their catch to middlemen at a much lower price. The secondary schools are far from their residence and there are no container depots, grounds, or markets for people here to preserve and trade their products.

15.4.2 Further Discussion of Some Typical Causes of Poverty

From the above-listed causes, some fundamental reasons behind these causes of poverty in the small-scale fishing community in Thanh Phong are analysed below.

15.4.2.1 Natural Resources and Capital

Due to the lack of capital assets (natural, physical, human, financial, and social capital), poor fishers in Thanh Phong have great difficulties in developing offshore fishing, which promises better harvesting (Ben Tre DARD 2008). They normally choose inshore fishing, and only spend half the day fishing due to the tidal changes and fish availability. This has negative short-run effects on their income. Fishers use destructive fishing gear and traditional modes of production, which contribute to exhausting the resources and their low income in the short run.

Poor fishers are not able to invest in their human capital; and in turn, low human capital prevents them from changing their job. However, in reality, because of their low education level, the deficiency in cultivating land, and the difficulty in obtaining credit, not all the fishers can change their "mode of production." Consequently, the fishers who have been involved in fishing practices for a long period have limited choices in changing their job. As discussed above, 12 out of 56 households possess land and they have had few opportunities to obtain agricultural extension services; that is, cultivation technique training, guidance on disease prevention and treatment, etc. The increase in production costs of inputs like electricity, water, seeds, breeders, and fertilizers thus boost the total cost, and leave the farming people with less net revenue from their small cash crops.

As discussed above, the obstacles that hinder the fishers in Thanh Phong from investing in offshore fishing, renovating fishing gear, and applying science and technology include the limited capital and the rare opportunity to gain access to credit sources. One reason for this is that because they do not have valuable property to mortgage to the bank for a large amount of loan money, the fisher's households have to apply for small unsecured loans that are too small to invest efficiently. On the other hand, most fishers have no feasible project or production plans to present to the government in order to borrow money. The poor fishers in Thanh Phong also lack information about law, policies, and the market, which makes their lives even more difficult.

15.4.2.2 Education

The income of poor fishers is hardly enough for basic nutritional needs and leaves little for investment in their human capital to improve their knowledge to alleviate poverty in the future. According to survey figures from 2008, the per capita income of 43 out of the 56 fishing households in Thanh Phong is less than 1 USD per day. With such a low income, they can hardly pay for their children's education. Therefore, people's low education level not only affects their income but also negatively influences their decisions related to bearing and rearing their children, as well as later generations. Malnutrition of children and newborn babies is also a factor affecting the chance for the children of poor households going to school (Ben Tre DLISA 2006). These challenge the strategy of reducing poverty through education and training. Low education is the cause on one hand; and the effect on the other of poverty and vice versa.

No member of the fishing households here receives higher education, and it can be said that difficulties in approaching opportunities for learning are a big barrier to the poor. Their low educational level prevents them from finding jobs in other areas, in non-agricultural sectors, or jobs that are stable and well paid. New industrial zones, export processing zones, and new urban areas are mushrooming in Ben Tre, especially in the suburbs of Ho Chi Minh city. However, due to their lack of skills, the people in Thanh Phong are not able to seek skill-requiring jobs (TPCPC 2009). This is a widespread reality, not only in Thanh Phong, a coastal community, but also nationwide.

15.4.2.3 Demography

The household size is a crucial factor which affects the average income of the household members. Having many children is the cause of poverty (per capita income) on the one hand; and the consequence on the other. The proportion of the poor fishing households who have large families is high, and this is one of the specific characteristics of the poor. According to the survey results, there are 5 members in a poor household on average; but only 3.5 members in a non-poverty household. What explains the high birth rate of the households in the coastal communities, especially the poor households in Thanh Phong, is restricted access to information on reproductive health and contraceptive methods. Poor couples have a limited understanding of hygiene, safe sex, and the connection between poverty, reproductive health, and the negative effects of having many children. On the other hand, having many children may be a rational strategy for poor parents, considering the very limited public pension and social security schemes. The children are their insurance and pension.

The rapid population growth and large-sized families in Thanh Phong puts pressure on job opportunities and poverty reduction (TPCPC 2008). In addition, the high rate of dependents (25.4%) in poor households makes their economic situation worse. Recently, many socio-economic investigations have shown that in addition to the impacts of socio-economic factors, the rapid population growth rate and large number of children in each household (though lower now than in the past), continue to be impediments to poverty alleviation at the national and local levels.

15.4.2.4 Economic Structure and Natural Risk

In Thanh Phong community, due to its natural characteristics, people mainly live on inshore fishing. The manufacturing industry, commerce, and services here are underdeveloped, and the economic structure is very slow. In recent years, the minimal infrastructure in Thanh Phong has hampered the economic development and reduced the effectiveness of implementing poverty-reducing policies. Indeed, areas with poor communication conditions normally have a low socio-economic structure. On the contrary, good communication conditions will enable aquatic and agricultural products to reach the market more easily, which will then improve the income of fishers, in general, and especially help to reduce poverty.

Additionally, the poor fishers in this area have difficulties gaining access to other services, information, markets, etc. Therefore, their chance of escaping poverty becomes slimmer. Also, because of the unfavorable natural conditions, it is difficult to train and improve human resources, or take advantage of the improved human capital to reduce poverty. In addition, the high costs of production that result from the lack of public transport also make their products less competitive in urban markets, due to quality deterioration and the cost of transport. The households carrying out small-scale fishing are vulnerable to daily problems and sudden environmental changes. Most of the poor in this area earn their living by fishing, so they are badly affected by unfavorable weather conditions (TPCPC 2008).

Apart from the above-mentioned causes, the poor fishers have a low ability to cope with mishaps, because of their low and uncertain income and limited saving ability. As a result, they can hardly handle sudden risks in life (e.g., natural disasters, failure of crops, loss of health and human resources). With their lean economic ability, the life of the poor households in the coastal areas remains vulnerable to these mishaps, and people may become even poorer as they grow older. That is one of the reasons for this conclusion:

> Households along the coast are critically dependent on fishing for their living, food, employment and income. These fishing communities have very often been characterized as, "among the poorest of the poor" (Neiland and Béné 2004, p. 1).

15.4.2.5 Public Policy

The investment rate in the fisheries sector of Ben Tre province is low. In 2008, only 3.6% of the total investment in the province was set for the fisheries sector (Ben Tre PPC 2009). There has not been a reasonable investment plan for developing the fisheries; especially for poor fishers to improve their production capacity and fight poverty; either through changing the fishing gear for offshore fisheries, or replacing inshore fisheries with non-destructive fishing gear. This is a finding that is clearly expressed by the FAO Advisory Committee on Fisheries Research Joint Working Party on Poverty in Small-Scale Fisheries: "The better integration of the fisheries sector within national poverty reduction strategies is desirable" (FAO 2002, p. 4). In addition, to make it worse, as confirmed by all the interviewed households, the administration has not planned any project to change the livelihoods of the fishing community residents of Thanh Phong.

15.5 Discussion of the Pro-Poor Policies and Strategies

15.5.1 Limitations

As indicated in the national report in 2005 (MPI 2005), Vietnam has achieved two of the eight Millennium Development Goals before the deadline, namely poverty reduction and universal primary education. As evaluated by the United Nations,

Vietnam met the goal of fighting poverty by 2015, 10 years ahead of schedule (UN 2005). The government has embraced the necessary policies to reduce the poverty level. These bring positive effects to poverty alleviation.

However, coastal and remote districts and communities with high rates of poor households still exist – for some examples see Table 15.1. Besides the unfavorable natural and socio-economic conditions, the policies and strategies that are inadequate and unsuitable for the coastal communities dismiss the effectiveness of rapid and firm poverty reduction in the small-scale fishery communities like Thanh Phong.

The main limitations of the present poverty-reducing policies applied in Thanh Phong, a small-scale fishery community, are discussed in the following section.

15.5.1.1 Poverty Reduction Policies and Small-Scale Fishing

The current governmental top-down approach to poverty reduction has not proved suitable for Thanh Phong, a poor fishing community far from the district centre. The state's present pro-poor programmes, such as Program 135 (GoV 1998, 2006), are being run in a "top-down" mode. Poor fishers are not genuinely the focus of this present approach. Hence, it is necessary to change the mode of the pro-poor programmes in the future. For example: (1) the policies do not encourage the administration of Thanh Phong to assist the poor fishers to reduce poverty by themselves. Quite a large number of the poor fishers still have the tendency to rely on the government's support, and are not active in fighting poverty. (2) The poor fishers in Thanh Phong have little voice in making poverty-reducing policies and strategies. The role of the fishers in planning and conducting strategies to alleviate poverty is not encouraged or developed to a high degree. Therefore, it is obvious that the government needs to decentralize decision-making, and involve the poor fishers – co-management may be a key word.

Most of the fishers in Thanh Phong, as well as the local government, are fully aware of the exhaustion of the coastal fisheries' resources, as well as the fact that most fishing nets are illegal, depleting the resources. However, because of the lack of alternatives to the current low income from fishing, they have to ignore resource protection. As a consequence, there can be neither a cease of illegal catching for a long time nor any punishment or fine reinforced by the government, and thus the poverty trap still remains unchanged.

15.5.1.2 Small-Scale Fishing Households and Policy Means

As the survey results and meeting outputs with the householders indicate, the assistance rates of a series of policies are not sufficient to be effective. There are at least three reasons for this. (1) The credit to assist production is low. The highest loan amount that a household can apply for is about five million VND with a 0% interest rate. This sum is too small for the fishers to change their fishing gear or their occupation. Some households can only receive a loan of 2–3 million VND, and not all poor fishers' households are provided with loans because of the very limited capital source (only 23% of the poor households in Thanh Phong can obtain this loan). (2) The financial aid is not coordinated with a strategy for rebuilding the resources. Thus, support to improve the productivity of fisheries may have a negative effect on the long-run efficiency and resource sustainability. (3) The norm of the financial aid for vocational training, 1.2 million VND/person/year, is too low for skilled workers or trained workers to obtain a job. Thus, the level of financial aid needs to be improved to help the poor fishing households in Thanh Phong, in particular with education, training, and the development of resource-friendly fishing methods and practices.

Some characteristics of the policy system evidently do not remove the obstacles facing the poor fishers.

1. Most of the fishers have not yet received any formal training at vocational schools to create alternative sources of earning.
2. Some of the fishers who participate in aquaculture are trained and provided with credit to invest in production. However, they are not able to sell their products or they lose money due to the relatively high costs of breeding fish and shrimps, purchasing foodstuff and chemicals, and the relatively low price received for their products (both partly due to the high transportation cost and the role of the middlemen).
3. It is hard for children in the commune to go to high school since Thanh Phong is an area with many muddy roads, and the high schools are far from their residence.
4. Although the tuition fees are exempted or reduced, many poor fishers' households cannot afford their children's learning expenses. This is one of the causes of the high dropout rate among local children.
5. Because of the weak soil composition and interlacing system of rivers and streams, the investment expenses for building schools and roads in Thanh Phong are 1.5 times to twice as high as those in other areas. However, the investment amount allocated by the government to Thanh Phong is only the same as that for other areas.

15.5.1.3 Province Policy and Community Needs

The socio-economic reform and development policies of Ben Tre province do not effectively support the target of reducing poverty in Thanh Phong community. The ecological conditions and aquatic resources have been weakened by destructive fishing gears which are not strictly forbidden (Ben Tre DARD 2008). The not-yet-improved infrastructure, and the market structure that is not accompanied by risk dispensation solutions, such as strategies for stable livelihood alternatives to help poor coastal fishers, a master plan for non-disease culturing areas, and strong enough social security policies, may make the poor even poorer (TPCPC 2009).

In reality, the central authority and Ben Tre administration have gathered investment resources for Ben Tre province and coastal districts in order to bring into play the strong points of each region, foster economic growth, and lessen the difficulties of the inhabitants. However, the limited resources hinder uniform investment in the province. As a result, Thanh Phong has not been properly invested in, which has resulted in poor infrastructure, lack of on-site jobs, deprived living conditions and risky fisheries and aquaculture activities.

15.5.2 Development and Poverty-Reduction Policy in Small-Scale Fisheries

In recent years (2000–2007), the socio-economic development of Vietnam has made great achievements, and the GDP has increased between 6.5% and 8.5% annually.[5] Thanks to the economic growth, the fight against poverty has succeeded to a certain extent at the national level. However, poverty persists in some areas, including small-scale inshore fishing communities. This fact raises a question: Are the present poverty-reducing approaches suitable for coastal communities living on small-scale fisheries?

In the development process, Vietnam has applied different approaches that are suitable for each development period to reduce poverty, including top-down and bottom-up approaches. The present approach, mainly of the top-down type, is gradually showing some shortcomings and inefficiencies. To attain the goal of stable poverty reduction in the area of fisheries, it is necessary to follow a new approach that includes both governmental institutions, and the local organizations of fishers in a cooperative manner.

It is essential to promote the roles of community organizations like co-operative groups, as cooperatives play an important role in helping poverty reduction. The community needs to be considered an indispensable factor in the partnership relations (the administration, the community, and the fishers) and in the process of forming, implementing, and supervising poverty-reducing policies. Poor fishers should not passively receive support, but should be active partners who themselves have to find the causes of their poverty and suggest solutions. Then, they must define what they have, and to what extent and how they need support from the poverty-reducing programmes.

There is a distinction between poverty reduction and poverty prevention, as the following quotation indicates:

> While major effort has been made recently to better understand the nature and cause(s) of poverty in fishing communities, a more recent focus includes a parallel effort to look at the "other side of the coin" and attempts to understand how small scale fisheries can also contribute to poverty alleviation (Béné et al. 2007).

In the case of Thanh Phong, it is mainly a question of poverty reduction. The State may concentrate on supporting the poor in small-scale inshore fisheries during

[5]http://www.gso.gov.vn/default.aspx?tabid=388&idmid=3&ItemID=9897.

a transitional period to improve the conditions of the ecological environment, the inshore fisheries resources, knowledge, and infrastructure. The support that works against these long-term goals should be abolished. Giving money to poor fishers just as a means of support without offering them professional advice or guiding them on how to use the money and to manage the resources will not be a sustainable solution. Reviews and supervision need to be conducted on a regular basis to avoid support becoming just a transitional gain for the recipients.

When designing a poverty-reduction policy for poor Vietnamese fishing communities, including Thanh Phong, there are two main groups of issues to be discussed[6]: First, within the area of fish harvesting and resource management; second, within the alternative livelihood area (jobs and income), education, public health, housing and transport, and other forms of communication. Policies related to the second area are very important for creating alternatives to the still open-access fisheries for the current and the might-be fishers and their families.

15.6 Conclusions

To sum up, the root causes of poverty in Thanh Phong are found to be the combined free-for-all inshore fisheries, and the lack of alternative employment opportunities. The open-access small-scale fisheries operated in this community seem to suffer from overfishing and the use of destructive fishing gear with illegal mesh sizes, though the scientific basis for this conclusion is weak. The very small mesh size used by the majority of fishers makes the nets retain the smallest juvenile shrimp and fish. For one fisher to increase the mesh size of his gear, to allow the juveniles to grow, would imply increased costs and reduced income in the short run. Thus, he is caught in the transitional loss trap that to a great extent hinders his economic improvement through his own actions.

The alternative to invest in other types of gear to fish other available species and sizes of fish (inshore or offshore) may not be possible; partly due to the lack of acceptable collateral in the fishing households, and partly due to the lack of banking facilities in the poor villages.

Alternative sources of income for fishers in Thanh Phong are meager since this village is far from the district centre, and has unfavorable conditions for agricultural production due to natural and economic factors such as alkaline soil, poor infrastructure, a bad traffic system, and a lack of clean water for daily activities. People in Thanh Phong will not be able to free themselves from the poverty trap without initial support from the government, and this support has to be sufficient to upgrade the infrastructure and assist in creating alternative jobs. Especially for migration to more central areas, better education is needed. Together with the general poverty-reducing policies applicable nationwide, it is also necessary to supplement or adjust

[6]A discussion paper in Vietnamese on these issues is available from the corresponding author.

some policies to make them more effective in this fishing community. These policies must be appropriate for alleviating the particular causes of poverty in this area, related to ecological and economic overfishing.

The poverty-reduction programmes to be developed, whether run by the government or by non-governmental organizations, should encourage the involvement of local people and institutions in cooperation with governmental institutions. The involvement of the community and fishers – in co-management – needs to be institutionalized in the government's development programmes to improve the effectiveness and ensure the stability of their activities. In particular, in the fisheries, such collective action is crucial to overcome the transitional loss trap on the road to ecological, economic, and social sustainability.

Identifying the specific causes of the high poverty rate in some small-scale fishing communities plays a key role in helping the central government, as well as local administration and non-governmental organizations formulate and conduct suitable policies and solutions for the sustainable reduction of poverty in these communities. In addition, basic knowledge of the functioning of local ecosystems, economic behavior, and the interaction between fishers and the resources is necessary.

Acknowledgments First, we would very much like to offer our sincere thanks to the staff of Ben Tre Provincial People's Committee, Thanh Phong Communal People's Committee, and the Department of Agriculture and Rural Development. Additionally, we extend our thanks to members of the Women's Association, the Farmers' Association, the Fishermen's Association of Thanh Phu district; and the 56 fishing households in the Thanh Phong community who enthusiastically supported us, exchanged ideas, and participated in local meetings and workshops. The chapter has benefited from the insights and comments of all of them. Next, we must acknowledge the Norwegian Research Council for the financial support of the PovFish project and the kind support and encouragement from the PovFish team. We would also like to thank our students from Nha Trang University who have completed their theses funded by the PovFish project, and have contributed three research cases on poverty measurement in the Thanh Phong fishing community, Bich Dam Island, and Nha Phu Bay, to help us gain a comprehensive view of poverty-related issues in small-scale fisheries in Vietnam.

References

Ben Tre DARD (2008) Report on performance of the fisheries sector. Ben Tre Department of Agriculture and Rural Development

Ben Tre DARD (2009) Report on performance of the fisheries sector. Ben Tre Department of Agriculture and Rural Development

Ben Tre DLISA (2006) Report on the rate of poor households in the town and districts in Ben Tre based on the Vietnamese poverty line in 2006. Ben Tre Department of Labour – Invalids and Social Affairs

Ben Tre DLISA (2007) Report on the rate of poor households in the town and districts in Ben Tre based on the Vietnamese poverty line in 2007. Ben Tre Department of Labour – Invalids and Social Affairs

Ben Tre DLISA (2008) Report on the rate of poor households in the town and districts in Ben Tre based on the Vietnamese poverty line in 2008. Ben Tre Department of Labour – Invalids and Social Affairs

Ben Tre DLISA (2009) Report on the rate of poor households in the town and districts in Ben Tre based on the Vietnamese poverty line in 2009. Ben Tre Department of Labour – Invalids and Social Affairs

Ben Tre DoF (2006) Report on five year performance of Ben Tre's fisheries sector 2000–2005. Ben Tre Department of Fisheries

Ben Tre PPC (2009) Report on performance of Ben Tre's 2008 social-economic development plan. Ben Tre Provincial People's Committee

Béné C, Macfadyen H, Allison EH (2007) Increasing the contribution of small-scale fisheries to poverty alleviation and food security. FAO Fisheries Technical Paper. No. 481. FAO, Rome

Do VK, Nguyen CH (2005) Protection of environment and fisheries resources: achievements, challenges, orientations and main solutions. In: Proceedings of the national workshop on protection of environment and fisheries resources, Hai Phong, Jan 2005. Agriculture Publishing House, Ha Noi

FAO (2002) Report of the sustainable fisheries livelihoods program (GCP/INT/735/UK) and FAO advisory committee on fisheries research joint working party on poverty in small-scale fisheries – promoting the contribution of the sustainable livelihoods approach and the code of conduct for responsible fisheries in poverty alleviation, 10–12 April 2002. FAO Fisheries Report, No. 678, FAO, Rome

GoV (1998) Prime Minister's Decisions on Poverty Reduction Program 135 - Phase 1 (1997–2006). Government of Vietnam's Gazette. Gazette Publisher, Hanoi

GoV (2006). Prime Minister's decisions on poverty reduction program 135 – phase 2 (2006–2010). Government of Vietnam's Gazette. Gazette Publisher, Hanoi

GSO (2008) Vietnam living standards surveys 2006 – statistical abstract. General Statistics Office. Statistical Publishing House, Hanoi. www.gso.gov.vn/default.aspx?tabid=388&idmid=3&ItemID=9897

Jentoft S, Onyango P, Islam MM (2010) Freedom and poverty in the fishery commons. Int J Commons 4(1):345–366

MARD (2004) Report on poverty reduction strategy implementation in rural areas. Ministry of Agriculture and Rural Development

MFA (2008) Vietnam continues to achieve the millennium development goals. Ministry of Foreign Affairs. http://www.mofa.gov.vn/vi/vd_quantam/economics/ns080116153539#2oDM0Fw1nTk1

MPI (2005) Report on Vietnam's social-economic situation of 2000–2005. Ministry of Planning and Investment, Hanoi

Neiland AE, Béné C (2004) Poverty and small-scale fisheries in West Africa. Kluwer Academic, Norwell

TPCPC (2008) Report on the socio-economic status and the rate of poor households in Thanh Phong in 2008. Thanh Phong Communal People's Committee

TPCPC (2009) Report on the socio-economic status and the rate of poor households in Thanh Phong in 2009. Thanh Phong Communal People's Committee

UN (2005) Millennium development goals in the Vietnam social-economic development plan for the 2006–2010 period. United Nations in Vietnam

World Bank (2000) World development report 2000/1. Oxford University Press

World Bank (2007) Vietnam development report 2007. World Bank, Hanoi

World Bank (2008) Vietnam development report 2008. World Bank, Hanoi

World Bank (2009) Vietnam development report 2009. World Bank, Hanoi

Chapter 16
Creating Action Space: Small-Scale Fisheries Policy Reform in South Africa

Moenieba Isaacs

Abstract The main argument of this chapter centers on whether formalizing governance processes and drafting a small-scale fisheries policy will decrease vulnerability and improve the livelihoods of small-scale fishers. Findings suggest that with no one organization representing fishers in the communities of Struisbaai and Arniston along the southern coast of South Africa, the space is wide open for the elite (rights holders) to capture the benefits. The inability to access rights through formal channels has forced a situation where many fishers resort to poaching, even within the marine protected areas (MPAs). The analytical framework draws on concepts related to the institutional dimensions of fisheries governance; the formal and informal action space for developing a new small-scale fisheries policy for South Africa; and the vulnerability of fishers with weak agency. Data were collected mainly through qualitative methods from key informants, focus group interviews, household interviews, and participatory observations at the local community level.

16.1 Introduction

The main argument of this chapter centers on whether the governance processes involved in drafting a small-scale fisheries policy and the declaring of marine protected areas (MPAs) will reduce vulnerability and improve fishers' livelihoods.

M. Isaacs (✉)
Institute for Poverty, Land and Agrarian Studies (PLAAS), University of the Western Cape, Private Bag X17, Bellville, Cape Town, South Africa
e-mail: misaacs@uwc.ac.za

S. Jentoft and A. Eide (eds.), *Poverty Mosaics: Realities and Prospects in Small-Scale Fisheries*, DOI 10.1007/978-94-007-1582-0_16,

16.1.1 Conceptual Framework

The conceptual framework draws on the concepts of fisheries governance and the formal and informal nature of "action space" in the context of developing a new small-scale fisheries policy for South Africa. Barberton and Kotze (1998) place their understanding of action space in the broader political context, and use the notion as a metaphor for the reform process in South Africa that symbolizes new possibilities and new opportunities for the poor through the democracy. These alternatives for the poor are created through the new governance structures and organizations to shape the social, economic and political reform processes.

Formal provisions of the Marine Living Resources Act (MLRA) (Act No.18 of 1998) created opportunities more for the elite than the poor to access fishing rights in the post-apartheid era from 1994 to 2006. This left the *bona fide* fishers outside the formal allocation process (Isaacs 2003, 2006; Isaacs and Hara 2007; Isaacs et al. 2007). The formal provisions also include the impact of marine protected areas on adjacent fishing communities; e.g. Arniston and Struisbaai.

Formal action space will be explained in the context of broader governance processes, which dictate the formal action space for drafting a small-scale fisheries policy, this time with the participation of civil society and community representatives who previously formed part of the informal action space in voicing their discontent with the formal rights allocation process. Informal action space refers to the marginalized, poor and vulnerable fishers' with organizational structures, and their abilities to implement small-scale policy to alleviate poverty.

This chapter situates the concept of action space within the institutional dimension of governance process and structures, to address poverty and reduce vulnerability of the poor and marginalized fishers. The works of Scott (1985, 1990), Ostrom (1990), Agrawal (2002), Wilson et al. (2003), Jentoft (2006), and Paavola (2006) are all useful in understanding and situating these concepts. The qualitative methodology focused on research through fieldwork, and the findings are presented by case studies followed by analysing the key themes. The discussion brings together the findings and conceptual framework by highlighting the politics involved in formalizing the informal, and how institutions are created and recreated to suit the needs of the elite.

More specifically:

- To what extent has the new action space created by the drafting of the small-scale policy made fishers less vulnerable?
- What formal and informal institutions were created to manage the resource?
- To what extent have marine protected areas and the rights allocation system impacted on the livelihoods of fishers in local communities such as in Struisbaai and Arniston?

The conclusions look into the implications of elite capturing; poaching in marine protected areas; the impact from the lack of institutions representing fishers; and what effects these have on the small-scale policy. Finally, recommendations are offered for areas of further research.

16.1.2 Formal Action Space

South Africa has a long history of commitment to marine protected areas, dating back to 1964 with the proclamation of the first marine protected area in Tsitsikamma. MPAs in South Africa are guided by a number of legislations, but their mandate is in the Marine Living Resources Act (MLRA) of 1998 (Act No. 18 of 1998; section 43) – namely, to conserve species and ecosystems, to rebuild fisheries stocks and to facilitate proper management of an MPA through reducing potential conflict between competing uses.

Parts of the National Environmental Management Protected Areas Act of 2003 (Act No. 57 of 2003) also apply to MPAs; e.g. protected areas should have integrated consultative management and cooperative governance, promote ecotourism and yield benefits to local communities, while still prioritizing the conservation of biodiversity.

The National Environmental Management Biodiversity Act of 2004 (Act No. 10 of 2004), provides for regular assessment of the nation's biodiversity, protection of threatened and endangered species and ecosystems, and where applicable, assists local communities to benefit from biodiversity though activities including sustainable use and bioprospecting.

Finally, the Integrated Coastal Management Act of 2009 sets out frameworks for consultation, particularly where part or most of an area is not contiguous to a terrestrial protected area.

In South Africa, there are currently 20 declared MPAs, with 3 more having been formally proposed. One of these, Cape Agulhas, is located next to one of the case study sites, Struisbaai. These current MPAs offer a degree of protection to 20% of the coastline, and a high level of protection to approximately 9% of the habitat. The statuses of certain coastal marine stocks, in particular line fish, are very poor in South Africa, with a number of stocks regarded as endangered. Fishing access, to some degree, occurs within 12% (of the overall 21%) of South Africa's coastline that falls within MPAs.

Recognition of the need to secure that traditional communities obtain an equitable share of the benefits of protected areas is a well-entrenched principle in the legislative framework. The same is true for recognizing traditional rights; and the need to ensure that the communities participate in, and benefit from, these areas. This is particularly so for the terrestrial areas covered by the Protected Areas Act but less so for the MPAs, as the MLRA is not specific on these aspects (Sunde and Isaacs 2008). The MLRA allocates rights on an individual basis to commercial, recreational and subsistence fishers. Fishing communities are not recognized as rights holders, and neither are the artisanal fishers.

The emphasis on social and economic justice was important in bringing about a paradigm shift in conservation in South Africa leading to attempts to balance conservation and development needs. A more inclusive approach of community-based conservation gained ground as the solution to exclusive state control. However, the marginalization of the small-scale sector within fisheries policy and management in this country, in general, spills over into the management of marine protected areas. This contributes to the further exclusion of these fishers and undermines their traditional livelihoods.

With the advent of the new democracy in 1994, the Reconstruction and Development Plan (RDP) supported equitable redistribution of resources, especially to those communities who derive a livelihood from the sea. Governance structures must be in place to assist people with access and management of resources. However, South Africa had a well-established fishing industry; and to maintain its competitive nature at the same time, the state had to redistribute access rights equitably to existing and potential new entrants. Reallocation of fishing rights was soon construed as a means of creating a space for all previously disenfranchised individuals and groups, but not necessarily in a way that the impoverished fishers and coastal communities would benefit. The goal of reallocating fishing rights was to transform the sector based on gender and race; not to alleviate coastal poverty (Isaacs et al. 2007).

Those fishers who did not fit into the formal rights allocation process formed part of the informal category of fishers, and they were defined by the Marine Living Resources Act 18 of 1998 as subsistence fishers. Hence, the Department of Environmental Affairs and Tourism (DEAT), through their line agency Marine and Coastal Management (MCM), allocated Individual Transferable Quotas (ITQs) to commercial right holders, and subsistence and interim rights to individuals mainly as a social relief programme to poor fishers. The informal fishers and communities on the west coast of South Africa were discontented with the ITQ rights allocation system and decided to challenge this system legally.

In 2004, the Artisanal Fishers Association, Masifundise, and the Legal Resources Center, with the support from academics, launched a class action suit against the Minister of the Department of Environmental Affairs and Tourism (DEAT). This case, Kenneth George and Others vs. the Minister, used the Constitution and the Equality Act (2004) to litigate on the social and economic impacts of the reform process (allocation of fishing rights). This case was to be heard in the Equality Court. In April 2007, the claimants of Kenneth George and Others agreed to put the case on hold, provided that the small-scale fisheries and subsistence policy would be reassessed with broader participation and input from various stakeholders.

16.1.3 Informal Action Space

Before 1994, there was no formal space for poor fishers to access rights, and they used the recreational permits to fish and harvest species in the open access area. The fisheries sector has been industrialized since the early 1900s, and the resources are fully utilized, which means that the reallocation of access rights is much more limited.

In the coastal communities of Struisbaai and Arniston (the case study areas of this chapter), those individuals with the necessary social, political and human capitals (i.e. the elite), were able to succeed in the formal action space in accessing fishing rights.

The vulnerable poor groups (i.e. the subordinates), without the necessary capitals failed to do so and are without rights. They form part of the crew on the boats of rights holders.

The works of Scott (1985, 1990) help to understand the complexities of the informal action space – through his notion of hidden transcripts; and the onstage and offstage performances of actors. Scott (1990, p. 4) argues that members of the subordinate group construct hidden transcripts as they congregate *offstage,* to speak and act outside the purview of the elite. Because these private conversations are unknown by the elite (in terms of their existence and content), subordinates can speak more freely without fearing the consequences of being overheard. This freedom results in a hidden transcript, somewhat conspiratorial in nature. The hidden transcript provides a means for subordinates to covertly express their antagonism toward the dominant, and to strategize their actions, while overtly complying with the rules of the dominant outside of their hidden transcript.

Scott's (1985, 1990) concepts of "onstage" and "offstage performances" can help understand how the poor exercise and structure their acts of defiance (e.g. poaching for abalone in marine protected areas) against the formal fishing rights application process. His work on hidden transcripts helps to relate the acts of the marginalized poor fishers as agency from below, according to fishers in Arniston:

> We poach during the night and day. Spring tide, during full moon is the best time to go for abalone as we are able to see. We take whatever we see. We do not have any other options, no fishing rights, and no restricted areas to fish. We know it is illegal, but we are struggling to survive.

In South Africa, the poor and marginalized fishers have weak agency, they have few assets and are capital poor, with low numeracy and literacy skills. These fishers also fit into Hogan and Marandola's (2005) conceptualization of vulnerability, by describing them in the context of the social disadvantages they experience, which makes them products of poverty. Here, the concept of vulnerability is situated in the social, political and economic processes, creating a useful tool to analyze the socio-economic vulnerability of SA small-scale fishers.

In Hogan and Marandola (2005, p. 458), the Economic Commission for Latin America researchers identify that: "If left to the mercy of the market of opportunities – the citizen is without rights and is forced to negotiate assets and empowerments, according to his or her management capacity." Furthermore, vulnerability is therefore expressed as restriction of rights, whether economic, political or social. It is here where poverty and exclusion come in – the restriction of the right to dignity, to health, decent housing, to be respected, to be allowed political participation, to be represented, to speak and to be heard.

According to the livelihoods framework, as described by Leach et al. (1999), success depends on the resilience of the households, their access to resources and their ability to shape and reshape institutions. This could imply that the new action space created by the drafting of the small-scale policy will generate benefits to those with the necessary agency, assets and skills. Hence, the nature, design and dynamics of locally based institutions to manage the implementation of the new small-scale policy are critical to the marginalized and vulnerable fishers, and the ability to effectively deal with the escalating poaching activities.

16.1.4 *Formalizing the Informal – Drafting of the New Small-Scale Fisheries Policy in South Africa*

In November 2007, DEAT committed to address access rights to poor fishers through developing a small-scale fisheries policy, which, as of the date of this chapter, is still under development. The court challenge of drafting a new small-scale policy resulted in formalizing the informal. The governance processes involved in declaring marine protected areas is another form of formalizing the action space, which has impacts on the livelihoods of the "informal" actors. The new draft for small-scale policy made no mention of marine protected areas and the impacts on coastal livelihoods.

The creation of a new small-scale fisheries policy is an instrument to bring the hidden transcripts to formal space or, in other words, formalizing the informal. The formal processes include forming alliances (same as above) to put pressure on the government (court application) to recognize the poor fishers who were left outside the formal allocation system; and to show their discontent with the formal ITQ rights allocation process. The governance processes involved in getting the MCM to recognize small-scale fishers, and to develop a small-scale fisheries policy for South Africa resulted in the informal action space becoming formal.[1]

16.2 Institutional Dimensions of Fisheries Governance

Agrawal (2001, p.1653) states: "The local community efforts to manage and govern resources depends on four sets of variables: (a) characteristics of resources; (b) nature of groups that depend on the resources; (c) particulars of institutional regimes through which resources are managed; and (d) the nature of the relationship between a group, and external forces and authorities such as markets, states and technology." According to the works of Ostrom (1990) and Agrawal (2002), governance functions relate to the rules that provide for exclusion, create entitlements, regulate users, provide for monitoring of the resource and structure participation and decision-making.

Jentoft (2006) argues that diversity, complexity, dynamics and vulnerability are key structural properties in governance systems. Wilson (2003) states that since formal and informal institutions are always changing – even when such changes are not dramatic shifts in written laws and basic organizational structures, marginal changes are happening through evolving interpretations and shifting degrees of compliance.

---

[1] See Sect. 1.2 for what happened with the out-of-court settlement and small-scale policy processes.

These constant changes are formed by competitive processes in which different groups seek to push institutions in the directions they desire (Wilson 2003). This statement is also supported by Leach et al. (1999, p. 12): "Social actors alter their behavior to new social, political and ecological circumstances"… hence, "institutional flexibility and dynamism are essential." They also warn us that some institutions claim to promote collective good, and then reproduce exclusion and marginalization of certain actors. Institutions are thus shaped by politics and power.

Paavola (2006) argues where there are conflicts over environmental resources, the choice of an institution is a matter of social justice more than efficiency. Here, "social justice includes the values and motivators of agents' influence, and what is considered just in a particular institution" (Paavola 2006, p. 96). Hence, the legitimacy of environmental decisions or governance outcomes involve both distributive and procedural justice, which include recognition, participation and distribution of power. "Distributive justice matters in a broad sense to whose interests and values will be realised by the institution; whilst procedural justice plays a role in justifying decisions to those whose interests and values are sacrificed to realise some other interests and values" (Paavola 2006, p. 97).

Paavola's arguments on social justice are useful in situating action space created by the informal actors through the litigation process. Social justice is key in legitimizing the allocation of rights to those who were excluded from the ITQ process; and could possibly lead to reducing poaching activities, provided access in marine protected areas is considered.

16.3 Methodology

Struisbaai and Arniston, along the southern cape coast of South Africa, were selected for this study. The selection criteria were based on their small-scale fishing activities, the low technology gear, the species they target, their dependency on marine resources, their poverty and unemployment levels, and that they reside next to or within a marine protected area. The methods used were qualitative in nature, and included the use of participatory observations, the use of key informants, four focus group interviews and 15 selected household interviews. Fieldwork was conducted over a period of 2 weeks in November 2008; 2 weeks in May 2009; and 1 week in March 2010. Interviews were with fishers, traditional line-fish rights holders, boat owners and entrepreneurs.

In each community, a focus group meeting was held to introduce the study and to discuss the key issues around governance and management of the resource, such as conflicts. Subsequently, 15 in-depth household interviews were conducted. The study also made use of key informants to assist with arranging interviews, and to serve as contacts when not conducting fieldwork. The study was guided by an interview schedule around the themes of governance, local institutions, small-scale policy awareness and process, and the impacts of MPAs on livelihoods. The interviews addressed a number of specific questions such as: (1) What do people perceive to be the main issues that small-scale policy should consider? (2) What are the resource constraints and conflicts? (3) How were they able to resolve conflicts over scarce resources? (4) What is their relationship to (if any), or problems with large-scale operators? (5) What explains their successes and failures?

16.4 Case Studies

South Africa has many settlements along the West Coast, dependent on harvesting marine resources, either for sale or directly for human consumption. Although most of these settlements were established around the processing industry, people in the two communities presented in this chapter, Struisbaai and Arniston in the Southern Cape region of the Western Cape Province (Fig. 16.1), have traditionally fished for sale and subsistence. From the time the waters adjacent to these communities were declared marine protected areas, their livelihoods have been negatively impacted,

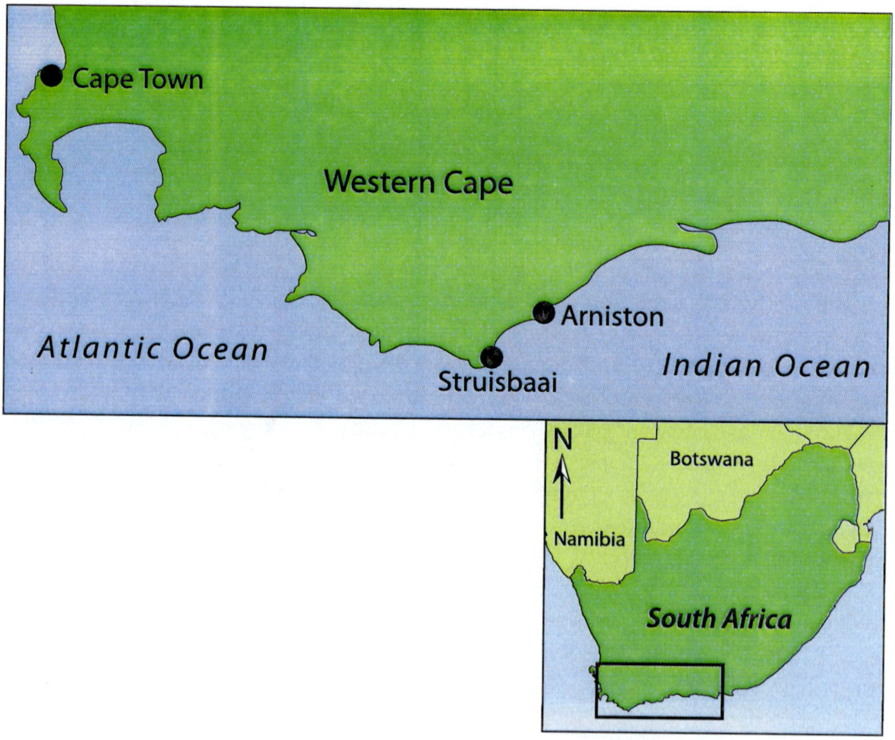

Fig. 16.1 Field sites along the southern cape of South Africa. The two communities where research was conducted – Arniston and Struisbaai – are in close proximity to the southernmost point in Africa, Cape Agulhas of the Western Cape of South Africa

leading to the increase of poaching. Sunde and Isaacs (2008) argue that in MPAs with adjacent small-scale coastal communities, it is imperative for management to consider local and indigenous knowledge and management practices in the zoning process. Currently, fishers do not acknowledge the findings of scientists, as they were not consulted. They also enter the MPA area regularly to harvest in the no-take zones. A multiple uses zoning approach can provide high levels of protection for specific areas, while allowing sustainable harvesting activities to occur in other zones – thereby separating out conflicting uses.

Isaacs (2003) argues that the formal action space created by the MLRA failed to respond to the fundamentally heterogeneous social, political and economic nature of fishing communities in South Africa, particularly neglecting the importance of creating institutional structures to interface with poor communities. With the informal action space becoming formal with the drafting of the small-scale policy, the need for locally based institutions to implement policy is imperative. This is no less true for the governance of MPAs.

The two communities where research was conducted are in close proximity to the southernmost point in Africa, Cape Agulhas of the Western Cape of South Africa. In Arniston, the population is 1,373 with 458 households; whereas Struisbaai has a population of 2,052 with 1,588 households. The majority of the fisher households fall in the low-skilled category with basic literacy and numeracy skills. The unemployment rates in both areas are staggeringly high, at 61.5%. Twenty percent of the households in the two communities are on the government welfare system; and 85% of the households earn less than 426 US dollars per month (Cape Agulhas Municipality's Draft Integrated Development Report 2009/2010).

Income generating activities are mainly seasonal with fishing as the most important source of livelihood followed by farm work and domestic cleaning services. Other livelihood activities include harvesting of sour figs to make preserves, and berries to make cleaning wax. These species are now in protected areas and 100 permits were allocated to women in Struisbaai who either sell to buyers from Cape Town or makes fig preserves for the local market. In Struisbaai and Arniston, poor people have to cope with both land and sea protected areas. In both areas, the women form part of the government public works initiative *Working for the Coast.*[2]

Both Struisbaai and Arniston are important tourist destinations for international and domestic tourists. Yet, tourism opportunities rarely filter through to fisher households. Fishing trips are organized by entrepreneurs from outside the area who recently moved to Struisbaai. At the harbor, fish is processed and sold, but no colored rights holders are involved. In Arniston, some fisher households are involved in lodgings, crafts, and restaurants to cater to incoming tourists. These opportunities are individually operated by households.

[2]*Working for the Coast* is a Department of Environment and Water Affairs initiative to address coastal poverty. Workers are paid 8 US dollars a day to clean the coast from debris, rubble and waste.

16.4.1 Struisbaai

Fishing in Struisbaai (and Arniston) can be traced to the first nations *KhoiSan* through the use of *vyfers* (fish traps) in the intertidal zone. Species such as elf (*Pomatomus saltatrix*), harder (*Liza richardsonii*), kolstert (*Diplodus sargus capensis*), strepie (*Sarpa salpa*) and galjoen (*Dichistius capensis*) were caught in these traps. The traps were maintained by clans and families, and women would harvest, gut, cut and cook. As part of the MPA community development program, there is a proposal to rehabilitate the fish traps and allocate to fishers to diversify their livelihoods from harvesting in restricted areas. The fish harvested in the traps will only be for subsistence. However, fishers in Struisbaai do not agree with this initiative, as they need to sell their harvest and engage in commercial activities to sustain their families. Currently, they are interested in using faster boats to target high-value species.

Many fisher families who reside in Struisbaai were evicted from an area called Skipskop, for the use of Armscor (now Denel) as a weapons testing ground. Currently, the area forms part of the de Hoop marine protected area declared in 1989.[3] Fishers recollect thc abundance of marine resources they harvested in the open access inshore zone to sustain their livelihoods. Since 1998, a new fisheries policy was enacted, which contains new regulations on all inshore resources. Fishers now have to have permits to harvest species, which before were accessed openly for subsistence. Furthermore, they have to comply with regulations in terms of quantity, size and area restrictions. Although marine and coastal scientists make it clear that these restrictions are based on scientific research, fishers feel they were not consulted and local and indigenous knowledge was not considered to be relevant.

In the past, line-fish species used sail boats in the inshore zone to target multiple species. Currently, most of the fishing activities in Struisbaai during summer target yellow tail and cape salmon. In winter, beach seine was used to catch harders (*L. richardsonii*) – the main fishing activity and source of livelihoods. Before, the stock was sold to local businesses in Bredasdorp and carted on horses. Processing methods such as drying and salting were common to provide protein during winter. Since 1994, rights allocation was informed by an ITQ approach and the supply of raw fish began to be sold at the Cape Town market for higher prices. Traditional methods of drying and salting were therefore lost to the more lucrative demand for fresh fish in the greater Cape Town market.

16.4.2 Arniston

Arniston is in close proximity to the Armscor, now the Denel missile testing grounds; and when testing happens (four times a year), the boat onshore is compensated

[3]Both the MPA and weapons testing site were declared at the same time in 1989, and fishers used to fish in these zones.

R4800, if at sea R7500, which gives the fisher R200 for the day. Fishers still criticize the rationale for a missile testing site in a marine protected area. They claim that: "The sounds disrupt the spawning, and influence the growth of juveniles." They also say that: "Testing not only impacts on fishing areas, but also livelihoods." Fishers call this fishing area "dead sea."

Currently, individual rights were allocated in west coast rock lobster, hake, abalone, line fish and wild oysters. However, these rights do not provide economic opportunities for women or fishers, as most of the species are harvested outside the area, and it is expensive to transport crew from the area to the fishing ground.

In the past, fishers used beach seine to target harders (*L. richardsonii*), which they salted, dried and sold to local farmers. The harders are locally known as *bokkoms,* and provide protein to farm workers in the Western Cape Province. Processing methods of salt dried harders and other species were common during winter. In 2006, no beach seine permits were allocated to Struisbaai and Arniston. With no winter species allocated in the area, many fishers and local authorities experienced a negative impact on the income, and an increase of poverty levels.

Angling, beach seine and fish traps were traditional sources of subsistence during winter. In both communities, fishers and rights holders echoed the need to reinstate the beach seine rights. They also raised concerns about the price increases of the recreational permits from the 2010/2011 season of 500%. Fishers use the recreational permits for angling. They state that the exorbitant increase in permit tariffs will result in fishers resorting to illegal fishing.

In Arniston, *Kassiesbaai* refers to the colored fishers who reside in the traditional stone and thatch roof houses. In 1932, a *visserunie* (fisher committee) was established to manage the land allocated to the community. The committee's main function was to deal with basic services to households and to pay property tax to the community. Each title deed is privately owned, and the owners pay their services and property tax to the committee, who in turn pays to the municipality. The committee has been working for several generations and often recruits new and younger members. The old fisher stone houses with thatch roofs were proclaimed as a South African Heritage site. This proclamation does not, however, mean much to the community, as they do not get any income from tourism.

The committee does not discuss any issues relating to rights allocation, permits, poaching, etc. Further, the committee has not capitalized on the tourism opportunities. However, fishers are providing accommodations in fisher households, craft centers, family restaurants and two mobile "fish and chips" outlets at the harbor. The committee feels they need to ask visitors for a fee for viewing their houses and for taking pictures of the homes. There has also been some talk about establishment of a tour guide for visitors; but none of these initiatives have been implemented. A challenge for the committee to diversify tourism opportunities is that each household is privately owned, and this adds to the difficulty in developing proposals and funding as a community. One fisher alludes to, "the running of the fisher union to that of a church and not as an economic interest."

16.4.3 *Mismatch Between Policy and Practice*

In Struisbaai, there are 6 ski boats and 16 chukkies (motorized wooden boats) with a crew of 8 fishers on each boat. The allocation of permits is based on Total Allowable Effort (TAE). Regulations are also based on the number and size of the boats. Marine and Coastal Management (MCM) would like to decrease the effort on the line-fish species, and to decrease the number of boats to 450 due to the resource constraints.

Chukkie boat owners argue that they need to convert their motorized boats to ski boats due to the distance to the fishing grounds, and the time it takes to get there. According to them, they need to cover 15 nautical miles offshore from Struisbaai harbor, and with the chukkies it can take up to 5 h. Fishers indicate that the fish are no longer within the 4 mile inshore zone, but have moved to deeper waters due to ocean temperature changes. They also blame the commercial pelagic boats targeting the food (anchovy, sardine and mackerel) of yellow tail and cape salmon. The pelagic boats are moving to their areas to harvest anchovy, sardine and mackerel, which is what the yellow tail and cape salmon feed on.

Rights' holders and fishers in Struisbaai agree that the Territorial User Rights Fisheries (TURF) system promoted by the small-scale fisheries policy will not work for them. They explained: "We are hunters and the fish we target are migratory so we should be too." They therefore want access to other areas, to target line fish. With ski boats, they can fish from the 5–8 mile zone, and even reach the 20 mile zone and also other areas. In Arniston, the rights holders and fishers are content for now to continue fishing with the chukkies in the 3–8 mile zone, provided MCM allows them to harvest inside the MPA.

In Struisbaai, the biggest issue for rights holders and boat owners is the urgency to convert their chukkie permits to ski boats. The main motivation for doing so is economics. According to the chukkie rights holders, the white rights holders were allowed to harvest with ski boats, and they yield around R200,000 per week in the season. The crew on the boats earn to the maximum of R3,000 per week for a period of 8 weeks, as yellow tail and cape salmon are sold raw for R20-30 per kg domestically. The chukkie rights holders are aggrieved by the amount of money the mainly white boat owners earn. For example, 500 tonnes of yellow tail, which equates to R10 million, were harvested in 2008, and mainly by ski boats.

In 2006, as part of the new permit regulations (given to rights holders in the traditional fish sector), boat owners were required to place a Vehicle Monitoring System (VMS) onboard. The VMS costs between R3,700 and R8,000, and R300 per month for air time. This was too expensive for the rights holders, and in Struisbaai they decided to protest against the installation of these units. In 2006 and 2007, they applied for an exemption to implement the VMS, which was granted; but in 2008, they were unsuccessful, yet they continued to fish. Masifundise supported the rights holders with their application due to the costs involved. In March 2008, MCM sent a team of inspectors to Struisbaai to investigate if permit holders had installed VMS systems on their boats. Fish were confiscated and fishers were arrested and fined up to R2,500. Soon thereafter, with the support from Masifundise, fishers showed their discontent through a peace march, and subsequently moved to blocking the harbor. The fishers also spoke to the local councillor and to MCM regarding their situation.

The rights holders and fishers in both fishing communities also argue for local economic development to focus on poverty alleviation. For this, they need access to land, Hazard Analysis and Critical Control Points (HACCP) infrastructure and markets. They want to move to ski boats, and to extend their harvesting area. According to one chukkie rights holder, the boat and fishing zone restrictions are, "keeping us behind from benefitting from the economic development. We need and want to go big with the promise to share. We need speed."

The relationship between crew and boat owners is also uncertain as fishers indicated that they are not sure of a job; one day they work and the next day they can lose their place on the boat, and this is entirely within the power of the skipper. They feel exploited and vulnerable as they do not have a right of permit to fish, and all the power is given to the skipper. All fishers have gone through survival training, and according to them this makes them legitimate fishers. The crew and rights holders work according to the principle of a 50/50 split of the earnings.

The 50% that goes to the rights holder includes the maintenance of the boat, fuel and bait. The remainder is split between the crew of eight. Fishers feel that a few people monopolize the value chain, and that they are being exploited as the price per kg ranges from R12 to R30/kg. Rights holders are not transparent, and fishers feel exploited yet desperate for work. They would like a more collective system of allocating rights; modernization of their boats; to be able to fish more species and in other zones (i.e. targeting snoek); they want a more transparent market structure; that fishers get more benefits from the value chain; and state subsidized modernization of boats. Fishers are strongly critical of commercial line-fishers from Cape Town who are allowed to harvest in their area using modern technology (speed boats). This means that within 1 week, they harvest tonnes of fish. If they (traditional line-fishers) were the only fishers allowed to harvest yellowtail and cape salmon during the summer season, they would have more fishing days.

16.4.4 Role of Women

The role of women in fishing communities needs to be understood within social and cultural contexts (Fig. 16.2). Their customary role is mainly to support their husbands/partners in the pre- and post-harvesting activities associated with line fish and beach seine. The women in Struisbaai and Arniston used to gut and wash fish until fishers decided they can get more money by gutting and cleaning it themselves on the boats. They also assisted their spouses in maintaining the fish traps. Today, only one fisher continues to maintain the use of fish traps.

Rights holders, fishers and women in both communities have different perspectives on the role of a small-scale fisheries policy to guide management, rights allocation,

Fig. 16.2 The creation of a new small-scale fisheries policy is an instrument to formalize the informal fisheries, and promote the role of women in the post-harvest activities. Picture taken by Mafaniso Hara

zones and species. They all agree that pronouncements on new MPAs and accessing MPAs should be part of the small-scale policy.

16.4.5 Governance Structures

In both communities, local organizations are constantly being created and recreated to suit the needs of the fishers. There is no one organization that speaks on behalf of the whole community. This also seems to be convenient for some rights holders as they plan which organization to address their specific needs to at the time. This is mainly what the elite rights holders do, while the fishers without rights have no voice. Although there is representation in both communities from the following organizations – Masifundise, Coastal Links, Arniston *Visserunie*, Confederation of South Africa Trade Union (COSATU) and Struisbaai *Vissersforum*, it was difficult to find one community organization representing the needs of all the fishers.

In Arniston, the *vissersunie's* role is mainly to protect the heritage of their land rights and traditional fisher houses from outsider investment. In the past, they used to deal with fisheries issues; but over the last 20 years, the conflicts between different family groups has led to fishers not identifying with this body. In 2002, the Masifundise NGO expanded their focus of advocating land rights to include the rights of traditional fishers who have been excluded from the formal allocation process. The organization promotes the need for fishers to be organized and mobilized to fight for human dignity and livelihood rights.

In 2004, Masifundise launched Coastal Links to strengthen local leaders' knowledge on fisher rights, gender, allocation regimes, management systems and developments in the drafting of the new small-scale fisheries policy through regular workshops in various regions. In Struisbaai, there is a mismatch between the traditional line-fish rights holders and what Coastal Links promotes. Rights holders are clear: "We want to convert traditional line-fish permits to ski-boats." This is contrary to what Coastal Links promotes, which are TURFs and traditional methods of harvesting.

However, in Arniston, Coastal Links has gained many members through the allocation of interim relief rights,[4] and it seems they are making more impact on those who were left outside the formal allocation process. "We would have still been without fishing rights had it not been for the work of Coastal Links," according to a fisher interviewed. Coastal Links represents 70 members of the Arniston community, and this includes young people, women and fishers on the interim relief permits.

Struisbaai *Vissersforum* was established in October 2007 comprising six rights holders. In 2008, the municipality of Agulhas indicated their support to the forum, but rights holders felt they would lose their independence and would be absorbed in municipal structures.

In November 2009, after President Zuma visited the region to listen to the issues facing fishing communities, it was recommended to establish a fishing desk to address problems facing fishing communities. Subsequently, through the local authority, both Struisbaai and Arniston created fisher forums that will have a direct link to the office of the Minister of Agriculture, Forestry and Fisheries. Local authorities have also created livelihood opportunities through government road works, subsidies through the VMS system, and they have established nutritional centers during winter to address poverty alleviation.

The rights holders in both communities use the governance structures in the following ways. Masifundise and Coastal Links in Struisbaai seem useful when they need to organize a protest, i.e. VMS and to get media attention. A Masifundise protest assisted in placing a moratorium on installing VMS in March 2008. They used the local councillor and municipal manager to access funding to purchase the VMS system.

In 2009, the Industrial Development Corporation of South Africa (IDC), through the local government, subsidized all rights holders with installing the VMS system (more details below). Fishers advocate poverty relief during winter in poor households. In 2008, each poor household received R435 in food vouchers.

[4] In May 2007, just before the inquiry in the Equality Court was due to begin, the Minister of Environmental Affairs and Tourism agreed, with the consent of all parties, to find a long-term policy solution for more than 1,000 traditional artisanal fishers. The agreement later became an order of the court, providing for relief mechanisms to allow fishers to subsist, and reserving 120–140 t for poor fishers.

In 2009, nutritional centers were established in both communities to provide one warm plate of food daily during winter. The local authority used the newly formed ANC fishing desk to lobby for the reinstatement of their abalone rights. In January 2010, abalone rights were reinstated in Arniston.

16.4.6 Marine Protected Areas and Fisher Livelihoods

Arniston resides within the Arniston Marine Reserve, while there is a process of declaring Agulhas an MPA, which will affect fishers in Struisbaai. Many fisher households were removed from Skipskop and Ryspunt, 60 km and 50 km, respectively, from Struisbaai (Fig. 16.1). These areas form part of the de Hoop MPA and Arniston MPA declared in 1989. Many fishers claim that fishing in the MPA is their historical right, and that they have not derived any economic benefits or compensation for this move. Furthermore, since the 1970s, the fisheries department has strategically declared their fishing areas as part of marine reserves; and from 1989, they have been allocated only small areas to harvest. In Arniston, fishers state that: "It does not make sense to fish in the same place every day for 20 years; we will only deplete the resource."

Currently, the best fishing areas are within the MPA, and fishers enter the restricted areas regularly. Although it is mandatory to have VMS on the fishing boats, they view it as a policeman, and switch it off when entering restricted areas. Fishers generally feel that MCM is practicing double standards, more keen on protecting the resource than their livelihoods. "How will they replace our fishing grounds?"

In Arniston, fishers' state: "If the MPA is there to protect the marine resources, how can you test weapons in the protected zone? In our area, they are keen to protect the weapons and the fish, but not the fishers and fishing communities." Fishers also question why recreational ski boats are allowed to enter their fishing grounds. "The MPA had a negative impact on the livelihoods to the community. The MPA makes us criminals in our own fishing grounds." In both communities, people questioned why MCM is over-regulating marine resources. As one fisher in Arniston puts it:

> There are too many permits and regulations – during apartheid we had open access to resources in the inshore fishing zone. Then we were more free, there was a sense of community and values of sharing, helping each other.

16.4.7 Poaching

The many new restrictions and permits have given rise to increased poaching activities. The costs of permits, registration, taxes and fuel make the allocation unviable and have also led to poaching. In both communities, there are fishers who poach for subsistence. In recent years, there has also been an increase in youth getting involved in poaching abalone, especially at night. This has also led to an increase in criminal activities in the communities. According to fishers, the poaching of abalone is getting out of control, especially with the younger men who are unemployed. Fishers attribute the rights allocation process, no-take zone in MPAs, the lack of livelihood opportunities and a means of accessing quick and easy money for the younger generation as reasons for the unmanageable state of poaching.

In a focus group interview with fishers who often access the MPA and fish undersize species, there were echoed a number of requirements that need to be in place before they will comply with regulations and participate in the management of the resource. One of the most critical issues is to have open access in the de Hoop marine reserve – a traditional harvesting area which has been a no-take zone since 1989. They feel that scientists regularly conduct research in their fishing zones, but do not include local and traditional knowledge into their research. They would like scientists to engage and consult fishers and communities when conducting research and provide information on the stock assessments.

Currently, scientists are making recommendations to MCM that impact on their livelihoods which they view as "unfair, unjust, unequal." Fishers call for a review in the size restrictions of certain species that could be used as food security, especially during the winter season. When MCM communicates new regulations to communities, it should be in their local language, which is Afrikaans.

When the question was posed if the community will be able to manage the resource, allocate access rights, have access to and harvest sustainably in MPAs and address the poaching, fishers stated that it would be difficult. Poaching has become too big for them to deal with, as organized crime is involved. It has become dangerous because gangs use firearms to access abalone. However, they think that management would have to include the poachers in the allocation of rights in order to make them comply with the regulations.

16.5 Discussion

16.5.1 Small-Scale Fisheries Governance

The informal action space within the small-scale fishery governance system has been occupied mainly by marginalized groups, and fishers who have been constrained by social disadvantages, lack of assets, limited communication skills and low numeracy and literacy skills. Some of these groups have shown their discontent with the governance system by embarking on litigation against the formal rights process.

A national task team representing fishing communities, NGOs, academics and government officials has now developed a new small-scale policy. The Department of Agriculture, Forestry and Fisheries[5] (DAFF) is in the process of following up suggestions from the task team, and consulting with various stakeholders before finalizing it into an official policy document. The task team is promoting collective rights, multi-species allocation and Territorial User Right Fisheries, in a co-management system.

This is contrary to the mainstream ITQ model adopted by DEAT in implementing long-term fishing rights in 2006. However, the government and many rights holders in the inshore zone are not in agreement with the task team on collective rights allocation.[6] It was decided that further and broader consultation with existing rights holders on allocation regimes for the small-scale policy will be necessary before a decision will be taken. In Struisbaai, there is a mismatch between policy and practice; i.e. what the rights holders and fishers would like to see in the small-scale policy.

First, fishers say the allocated species are not found in this area. Second, fishers want to migrate with the fish – move from one zone to another. Since they are traditional line-fishers, they have restricted zones; but commercial line-fishers[7] with speedboats from greater Cape Town and the West Coast regions are allowed to fish tonnes of yellowtail and cape salmon during the season. They operate in the zones allocated to traditional line-fishers. If the small-scale fisheries policy could prevent commercial line-fishers to harvest in traditional line-fishers' zones, then they would have more fishing days. Third, Struisbaai fishers want the west coast rock lobster to be part of the multi-species allocation. Fourth, they want to modernize their vessels to target fish in the 15–18 mile zone. In Arniston, fishers are satisfied to fish in the 3–8 mile zone provided the MPA is open for sustainable harvesting. Finally, rights holders are holding onto the ITQ system, while fishers without rights would like to move to more collective rights allocation. It is clear that fishers are asking for a mix of the two approaches; they prefer a hybrid rights allocation model for the small-scale fishery.

16.5.2 Formal and Informal Institutions

Leach et al. (1999) state that access rights to natural resources are shaped by institutions and, at the same time, institutions are confirmed and reshaped repeatedly by livelihoods. In Struisbaai and Arniston, a number of local organizations are constantly

[5]In April 2010, Marine and Coastal Management split into the two ministries, Department of Environment and Water Affairs and Department of Agriculture, Forestry and Fisheries (DAFF). All the policies, regulations, allocations, management and administration of fisheries will be under the auspices of DAFF.

[6]This was stated in the small-scale fisheries policy workshop with stakeholders on 7 and 8 May 2010.

[7]In Struisbaai, only traditional line-fish permits were allocated with the long-term rights in 2006. However, commercial line-fish permit holders are allowed to migrate from greater Cape Town and the West Coast to target yellowtail and cape salmon.

being created and recreated to suit the needs of the fishers at the time. Although Arniston has had an established *Vissersunie* since 1932, this body currently only deals with the land rights and services to the community. Due to the intergenerational conflicts and family splits, fishers do not recognize this body as truly representative of their needs and interests.

Coastal Links, the community-based organization which links to Masifundise, forms part of the litigation process against the long-term rights allocation process. They have gained more members through the allocation of the interim relief permits in Arniston. The Struisbaai *Vissserforum* (Fisher forum) only represents the rights holders and not the fishers. Since the forum is an initiative of the local municipality, its independence from local authorities is questioned. Coastal Links and Masifundise (NGO) were used to stage a protest march in Struisbaai to place a moratorium on VMS on boats; whereas the African National Congress (ANC) fishing desk[8] and Masifundise were used to lobby for the reinstatement of abalone rights in Arniston. The elite (rights holders and community representatives to NGOs) are constantly scheming on which organization would best serve their needs. In some cases, more than one institution would be used to achieve their goals, i.e. VMS on traditional hand line boats. However, when an organization is established at the local level, the nature and structure of the organization are often set by the elite to serve their needs, and effectively exclude poor fishers with weak agency.

Not having a representative body to deal with fisher struggles, management, development and conservation at the local community level seem to be in the best interests of the rights holders. My findings also indicate that there is a constant struggle between the rights holders (elite) and crew (subordinate), to the extent where the latter feel exploited. The crew would like to move to a more collective system of rights; while it is in the best interests of the rights holders to keep the current ITQ system.

Fishers show their discontent with the system, local elite, regulations and MPAs through poaching. Hence, the action space for small-scale fishers created by the new democracy in South Africa is constantly created and recreated depending on the agency of the groups and individuals. Those with no agency, the vulnerable, marginalized, unskilled and unorganized will resort to what I (with reference to Scott (1985, 1990)) have called "hidden transcripts" and "weapons of the weak" (i.e. poaching). Agrawal and Gibson (1999) also warn us of elite capturing with devolutions of rights and responsibilities, which indirectly reduce access to the poor. What Webster and Engberg-Pedersen (2002) say about lack of political agency of the poor is also true for small-scale fishers in Struisbaai and Arniston:

> While the political agency of the poor in some cases remains only potential – it remains a central element in the process of poverty reduction… The political agency of the poor can be qualified in three ways. First, the poor constitute a very heterogeneous category, only in

[8] The ANC majority party in government formed a group of representatives from local, district and provincial officials to sit on the ANC fishing desk. The goal of this group is to address the issues facing fishing communities along the southern cape.

> extremely rare cases acting in a collective and co-ordinated manner. Secondly, political actions of marginalised groups may have indifferent or detrimental consequences for other poor people, due to the highly structured world of poverty. Finally, the informal, elusive, sometimes questionable nature of the poor's political agency, as land disputes are not only taken to various authorities for settlement, but are also characterised by violence, poisoning and witchcraft accusations (Webster and Engberg-Pedersen 2002, p. 257–258).

Poverty alleviation, inclusive development and environmental protection should be central themes in fisheries policy (Wilson 2003). The fisheries governance process should recognize the economic struggles over allocation of resources and political struggles over management decisions. Moreover, the politics of fisheries governance depends on who participates in working together to achieve common fisheries policy goals. Participation can be achieved through mobilizing one's own stakeholders group to achieve narrow goals, and as a mechanism for holding management systems accountable (Wilson 2003).

The creation of a new small-scale fisheries policy is a necessary step toward addressing poverty in coastal communities in South Africa. It has also created space for fishers to hold management structures accountable for the impacts of long-term rights allocation. However, it is crucial for fisher communities to benefit, especially the marginalized poor; and to create local community institutions that address poverty. Co-management of the marine resources is what government, NGOs and CBOs are advocating in the draft of the small-scale fisheries policy. The question remains whose responsibility it is to create these institutions.

16.5.3 Livelihoods and MPAs

Conservation of marine resources and creation of sustainable livelihoods in coastal communities has been hotly debated in South Africa. However, fisheries managers are simultaneously strongly influenced by fisheries science and the promotion of a strict precautionary approach. In the last few years, fisheries management in South Africa has adopted an ecosystem management approach to fisheries, and declared 20% of the coast as MPAs.

The way the tension between the conservation and social justice imperatives shapes the management of MPAs in South Africa needs to be understood in the context of the vulnerability, structural poverty and livelihood needs of coastal communities in South Africa (see Gustavson et al. 2009). The work of Almudi and Kalikoski (2010, p. 231) in Peixe Lagoon in Brazil contributes to the academic debates about the "incompatibility of no-take protected areas and livelihoods of traditional peoples where they coexist"; and, "there is an urgent need to share control (of marine reserves) with the most vulnerable." The World Summit on Sustainable Development in 2002 promoted the link between conservation, social equity and poverty reduction. Experiences from the Philippines indicate that integrating conservation with promotion of livelihood opportunities adds to the MPA success (Tobey and Torell 2006).

The planning and implementation of MPAs require solid knowledge from ecological as well as socio-economic and cultural disciplines. The use of traditional knowledge is also critical, in addition to science in designing MPAs, which meet the dual objectives of generating local benefits and preserving biodiversity (see Almudi and Kalikoski 2010; Berkes 1999; Tobey and Torell 2006). Gelcich et al. (2009) adds the importance of meaningful local fisher communities' input into MPAs. Sarkar (1998) stresses the importance of traditional societies continuing to exist, as they have developed cultural practices of resource use that maintain their cultural values and local biodiversity.

The use of a mix of modern science and traditional knowledge is imperative in the allocation of fishing rights, agreeing on fishing gear, and regulating fishing activities. This will increase community participation, heighten awareness of benefits from effective management (and co-management) regimes, increase the likelihood that biodiversity conservation and fisheries management will be achieved and thereby increase the efficiency of management decision-making (Sunde and Isaacs 2008).

Research in fisheries in South Africa has been scientifically biased with a concentration on bio-ecological research and very little in terms of social science and human ecology research. With the resurgence of the fortress conservation approaches to protecting biodiversity, South Africa's constitution and democracy created an enabling environment for legislative and strategic frameworks to find consensus between conservation needs and developmental needs of South African citizens.

However, the political will and commitment to implement this must be conveyed to the marine science community; with a fundamental paradigm shift facilitated so that a partnership between local fisher communities and fisheries scientists can be established. The implementation of integrated research projects that harness the indigenous knowledge of the fishers; secure participation; and demonstrate the tangible benefits of MPAs will contribute enormously to gaining support for the objectives of the MPA authorities (Sunde and Isaacs 2008).

MPAs and permit restriction have negatively impacted on the livelihoods of fisher households, especially during winter.[9] Fishers echoed the role of marine scientists in informing MCM management regulations without the fishers in their studies. Since the scientists have impact on the amount and size of the fish that can be harvested, this also impacts on fishers' livelihoods.

Fishers recollect how, in the past, the inability to access rights through formal channels and the quick and easy money to be gained, forced many fishers to resort to poaching, even within the MPAs. With size regulations in place, they are unable to harvest what they traditionally used for their own consumption to subsist. In Arniston, fishers are convinced that the weapons testing sites have had a negative consequence for the habitats and breeding of fish. They question why no research has been done on how this impacts on their livelihoods. Fishers also feel that their local and traditional knowledge of the area is not considered by scientists conducting research.

[9] During summer, fishers mainly harvest migratory species such as yellowtail and cape salmon.

A study by Dennis (2009) found that although fishers in the area understood the importance of protecting marine resources, they are excluded from the conservation and research processes in their fishing zones. Their families have fished in areas for generations and their local knowledge could potentially be beneficial to the research and management processes. Yet, this knowledge is not perceived as relevant by government or others who are involved in the design of the MPAs.

When authorities declared a MPA in traditional access areas of communities, they did not compensate communities for their loss. Neither did they provide communities with viable and sustainable employment alternatives or mechanisms that would reduce vulnerabilities or curb unsustainable harvesting practices. Access to MPAs was restricted, which led to the loss of livelihoods and to low compliance. Many fishers were left with no other alternative than to continue to fish as usual. With very little to no economic opportunities for younger fishers in the area, they resort to harvesting abalone illegally as a means of earning easy money.

In Arniston, fishers are concerned that with abalone poaching, there is now more disposable cash available among the young who resort to abuse of alcohol. With no co-management arrangement assisting MCM; the permit-, species- and zone restrictions in the traditional fishing grounds of small-scale fishing communities; no consultation with fishers with implementing new restrictions; declaring marine protected areas; and imposing restrictions on high-value species such as abalone and west coast rock lobster in their area; all pose tremendous challenges on development of a suitable management framework. It is clear that the current MPA model in South Africa has not been effective, and fishers are increasing their poaching activities.

According to Almudi and Kalikoski (2010), no-take MPAs are incompatible with fisher livelihoods. Fishing communities need long-term rights to resources to be in place before a system of shared decision-making, rule creation, monitoring and enforcement between fishers and authorities would be a possibility. Hence, new legislation for MPAs and small-scale fisheries policy needs a paradigm shift and alignment of social equity, conservation and economic development of affected communities.

16.6 Conclusions

The drafting of the new small-scale fisheries policy in South Africa has created an enabling environment for fisheries that has helped them to formalize their role and struggles in the governance process. The new policy has recognized artisanal fishers as a group, and it seems that the policy will adopt a collective rights system where appropriate, such as TURF zones, community-based management structures and multi-species allocation. However, these gains at the policy level are not necessarily what fishers in Struisbaai and Arniston want most in practice, and this will most probably be the case in many other small-scale communities around the coast.

To address the mismatch between policy and practice, certain conditions need to be in place; locally based management structures that are attuned to the relationship between poverty reduction, environmental sustainability, and the allocation of resources. Also essential is a community structure that promotes social justice and accommodates marginalized poor fishers and commercial rights holders. The small-scale policy should therefore guide the management plans of local communities, and this should be adaptable to the local setting and conditions. Policy should consider a hybrid system of rights allocation.

The ITQ-based approach of a one-size-fits-all model has failed in poor communities over the past 15 years, and has resulted in the elite capturing the rights at the expense of the poor. A cooperative system to manage the processing and marketing of inshore resources should be considered, especially with high-value species that target the export market. This would ensure value chain benefits to the community. The action space for women in the post-harvest activities should be strengthened. DAFF should secure fisheries rights and develop a management policy for MPAs that would benefit local communities. There needs to be clear and strong linkages to the current draft small-scale fisheries policy, especially in the southern cape area where poor communities reside adjacent to MPAs. With the low numeracy and literacy levels of many fishers, there is a need for adult basic education when training fishers or communicating new regulations and policy. To achieve these conditions, more state involvement and support would be necessary, i.e. directed funding from local government for infrastructure to fish harvesting, processing and marketing.

The current ITQ allocation framework has been biologically biased, and allocation decisions were strongly influenced by the general reform processes in South Africa to achieve equity on the race and gender fronts. With regard to poverty alleviation, interim relief measures were, in many instances, allocated ad-hoc to give in to political pressures from fishing communities.

To be more successful in alleviating poverty, promoting local economic development and achieving environmental sustainability, a paradigm shift is clearly needed. There should be trans-disciplinary teams created, comprised of scientists, social scientists and economists, etc. to address these problems and feed their research into management decisions. Moreover, inputs from fishers, based on their local and indigenous knowledge, should form an important part of this process.

Acknowledgments I would like to thank Tracey-Lee Dennis, Stuart du Plessis, and John Granfield who assisted me in the field. Thanks also to the Norwegian Research Council for providing the funding for the PovFish project. In addition, the comments of Ratana Chuenpagdee and Svein Jentoft to my chapter were invaluable, and also the editing of this chapter by Maria-Victoria Gunnarsdottir.

References

Agrawal A (2001) Common property institutions and sustainable governance of resources. World Dev 29:1649–1672

Agrawal A (2002) Common resources and institutional sustainability. In: Ostrom E, Dietz T, Dolšak N, Stern PC, Stonich S, Weber EU (eds) The drama of the commons. National Academy Press, Washington, DC

Agrawal A, Gibson CC (1999) Enchantment and disenchantment: the role of community in natural resource conservation. World Dev 27(4):629–649

Almudi T, Kalikoski DC (2010) Traditional fisherfolk and no-take protected areas: the Peixe Lagoon National Park dilemma. Ocean Coast Manage 53:225–233

Barberton C, Kotze H (1998) Creating action space: the challenge of poverty and democracy in South Africa. Idasa, Cape Town

Berkes F (1999) Sacred ecology: traditional ecological knowledge and resource management. Taylor and Francis, Philadelphia/London

Cape Agulhas Municipality's Draft Integrated Development Report (2009/2010)

Dennis TL (2009) Perceptions of history and policy in the Cape Agulhas area: could history influence policy on small-scale fishing? Unpublished mini-thesis submitted to Institute for Poverty, Land and Agrarian Studies (PLAAS), University of Western Cape

Gelcich S, Godoy N, Castilla JC (2009) Artisanal fishers' perceptions regarding coastal co-management policies and their potential to scale-up marine biodiversity conservation. Ocean Coast Manage 52:424–432

Gustavson K, Koeker Z, Walmsley J, Juma S (2009) A process framework for coastal zone management in Tanzania. Ocean Coast Manage 52:78–88

Hogan JD, Marandola E (2005) Towards an interdisciplinary conceptualisation of vulnerability. Popul Space Place 11:455–471

Isaacs M (2003) Understanding the social processes and politics of implementing a new fisheries policy, the Marine Living Resources Act 18 of 1998, in South Africa. Unpublished PhD thesis, University of Western Cape

Isaacs M (2006) Small-scale fisheries reform: expectations, hopes and dreams of 'a better life for all'. Mar Policy 30:51–59

Isaacs M, Hara M (2007) Transformation in the South African fishing industry and its ability to redistribute fishing rights. American fisheries society symposium, 49, 2007

Isaacs M, Hara M, Raakjær J (2007) Has reforming South African fisheries contributed to wealth redistribution and poverty alleviation? Ocean Coast Manage 50:301–313

Jentoft S (2006) Limits of governability: institutional implications for fisheries and coastal governance. Mar Policy 31:360–370

Leach M, Mearns R, Scoones I (1999) Environmental entitlements: dynamics and institutions in community-based natural resource management. World Dev 27:225–247

Ostrom E (1990) Governing the commons: the evolution of institutions for collective action. Cambridge University Press, Cambridge

Paavola J (2006) Institutions and environmental governance: a reconceptualization. Ecol Econ 63:93–103

Sarkar S (1998) Restoring wilderness or reclaiming forests. Terra Nova Nat Cult 3(3):35–52

Scott JC (1985) Weapons of the weak. Everyday forms of peasant resistance. Yale University Press, New Haven

Scott JC (1990) Domination and the arts of resistance: hidden transcripts. Yale University Press, New Haven

Sunde J, Isaacs M (2008) Marine conservation and coastal communities: who carries the costs? A study of marine protected areas and their impact on traditional small-scale fishing communities in South Africa. SAMUDRA Monograph. International Collective in Support of Fishworkers. www.icsf.net

Tobey T, Torell E (2006) Coastal poverty and MPA management in mainland Tanzania and Zanzibar. Ocean Coast Manage 49:834–854

Webster N, Engberg-Pedersen L (2002) In the name of the poor: contesting political space for poverty reduction. Zed Books Ltd., London

Wilson DC (2003) Conflict and scale: a defense of community approaches in fisheries management. In: Wilson DG, Raakjaer Nielsen J, Degnbol P (eds) The fisheries co-management experience: accomplishments, challenges and prospects. Kluwer, Dordrecht, pp 193–211

Wilson DG, Raakjaer Nielsen J, Degnbol P (eds) (2003) The fisheries co-management experience: accomplishments, challenges and prospects. Kluwer, Dordrecht, pp 193–211

Chapter 17
Building Resilience: Fisheries Cooperatives in Southern Sri Lanka

Oscar Amarasinghe and Maarten Bavinck

Abstract Among the many models proposed to address vulnerability and poverty in fisheries, this chapter takes a social capital approach. It focuses particularly on the role of cooperatives in providing small-scale fishers with linking social capital. The latter allows for the transfer of resources from other societal levels, such as government. The chapter is based on a study carried out in two landing centers in the Hambantota District of southern Sri Lanka. Fishers in this region suffer major problems as a result of weakly developed credit, product and insurance markets, increasing costs of fishing equipment, and deficient educational and training services. Cooperatives have played a positive role in all these fields, improving the resilience of small-scale fishing households significantly. Two qualifications are, however, in order. The first is that not all fishing cooperatives in Sri Lanka function effectively. The research sample, which contrasted a well-functioning with a weakly functioning cooperative, demonstrates the range of results available. Second, cooperatives have been more oriented toward promoting welfare than toward resource conservation, and have contributed to a potentially harmful increase of fishing effort. In order to remain successful over the long term, cooperative leaders will need to start paying attention to resource governance.

O. Amarasinghe (✉)
Agricultural Economics, Faculty of Agriculture, University of Ruhuna, Mapalana, Kamburupitiya, Sri Lanka
e-mail: oamarasinghe@yahoo.com

M. Bavinck
Department of Human Geography, Planning, and International Development Studies, University of Amsterdam, Nieuwe Prinsengracht 130, 1018VZ Amsterdam, The Netherlands
e-mail: j.m.bavinck@uva.nl

S. Jentoft and A. Eide (eds.), *Poverty Mosaics: Realities and Prospects in Small-Scale Fisheries*, DOI 10.1007/978-94-007-1582-0_17,

17.1 Introduction

Fisheries cooperatives have moved out of the limelight. The present dearth of policy interest in the potential of cooperatives for meeting societal challenges contrasts markedly with earlier time periods. The post-independence era witnessed strong confidence in the contribution cooperatives could make toward inclusive development, in particular the field of fisheries (Jentoft 1985, 1986). The publication by the Food and Agriculture Organization (FAO) of the United Nations in 1971 of a manual on fishermen's cooperatives is exemplary of the mood at the time. Admitting that this organizational type presents special challenges, it provides initial and experienced cooperative managers throughout the world with a detailed, stage-wise "how to do."

Although cooperatives have not gone completely out of date since, the fact that this manual never witnessed a reprint or follow up is symptomatic for what amounts to a pervasive shift in attention. This change had two causes. The first reason is the poor performance of state-sponsored cooperatives in developing countries. As Birchall (2004) points out:

> For the first 50 years after the Second World War [cooperatives] were used in a planned, top-down attempt by national governments and international aid agencies to deliver economic growth. This attempt came to an end with the end of the cold war and the era of 'structural adjustment,' leaving an indelible impression in the minds of many policy makers at the UN, World Bank, and other international agencies, which cooperatives had failed.

The second reason is the shift in fisheries away from development to resource governance objectives. With increasing evidence of the contribution of fishing to resource depletion and degradation (World Bank 1997), has come widespread attention for fishery management and control. Although fisher organizations are expected to play an important role in this effort (Kurien 1988), such as in the exercise of comanagement (Amarasinghe 2006), cooperatives are seldom included in the deliberations (cf. Jentoft 1985).

This chapter revisits the role of fishing cooperatives in the context of the first Millennium Development Goal, to halve the proportion of people suffering from poverty by 2015. Departing from the observation that poverty is a common phenomenon in small-scale fisheries (Béné 2003; Thorpe et al. 2007; Daw et al. 2009), and making use of the analytical framework suggested by the sustainable livelihood approach (Allison and Ellis 2001; Allsion and Horemans 2007), it enquires into the position of small-scale fishers in southern Sri Lanka. More specifically, it asks what contribution cooperatives, as providers of social capital, can make to the alleviation of poverty.

The chapter is organized as follows. Section 2 summarily discusses the incidence of poverty in small-scale fisheries, particularly in Sri Lanka. Section 3 considers the role of social capital in alleviating poverty, and explores the role of cooperatives in creating such goods. The next sections present the results of a field investigation on cooperative performance in southern Sri Lanka. These are followed by a set of conclusions.

The chapter draws upon an array of primary and secondary sources of information. The most important source of field material is a study on legal pluralism in fisheries, carried out by the first author in Hambantota District from 2003 to 2005 (Amarasinghe 2006). Additional sources include field work carried out from May to December 2008 under the PovFish Project. The latter fieldwork focused on two small-scale fishing villages called Bata Atha and Rekawa in the Hambantota District, in southern Sri Lanka. The first author was familiar with these locations from previous fieldwork, and therefore aware of the existence of functioning cooperatives. The Bata Atha fisheries cooperative had previously won a number of awards at competitions organized by the Department of Cooperative Development and other institutions; whereas Rekawa fisheries cooperative is considered to be one of the weakly functioning cooperatives in the South (ibid.).

Methodologies employed include a mixture of cross-sectional surveys, focus group discussions, and participant observation. For the 2008 field study, a sample of 80 fishers from each village was randomly selected from the lists of the 2006/2007 Boat Census of the Ministry of Fisheries and Aquatic Resources (MFAR 2007). The samples consisted of male fishers making use of so-called traditional craft and small motorized craft. As half of the membership of the cooperatives consisted of women, however, the information from the samples was complemented by interviews with some women, selected randomly, in order to understand their participation in cooperative activities and the nature of the services they had received. A pretested structured questionnaire was administered to the fishers in the samples to elicit information on various aspects of their involvement in the cooperatives. Officials of cooperatives were subject to informal interviews to obtain information on their organizations' structure, functions, assets, and links to the outside world.

17.2 Vulnerability and Poverty

Vulnerability is a factor that exposes people to various risks and shocks leading to destitution (Dhanani and Islam 2002; Parker and Kozel 2005; Barrientos 2007; Bird and Prowse 2008). A system is "vulnerable" when it is exposed to risks and shocks. The challenge of development not only includes the elimination of persistent poverty but also the removal of vulnerability to sudden and severe destitution (Sen 1999). Although, a standard definition of vulnerability does not exist, a few efforts at defining vulnerability are worthy of note. For some, vulnerability is defined in terms of a set of conditions having impact on a community's susceptibility. For example, the International Strategy for Disaster Reduction (ISDR 2004[1], as cited in Birkmaan 2006, p.12), regards it as: "The conditions determined by physical, social, economic, and environmental factors or processes, which increase the susceptibility of a community to the impact of hazards." Likewise, UNDP (2004, p. 11) suggests that it is a "human condition or process resulting from physical, social, economic and environmental factors, which determine the likelihood and scale of damage from the impact of a given hazard."

[1] International Strategy for Disaster Reduction website http://www.unisdr.org/eng/terminology/terminology-2004-eng.html (accessed May 18, 2011)

For some, vulnerability is a kind of insecurity arising from hazards. It is "lack of security from environmental threats. It results from a combination of processes that shape the degrees of exposure to a hazard, sensitivity to its stress and impacts, and resilience in the face of those effects" (SEI undated).

Overall, the concept of vulnerability has been continuously widened and broadened toward a more comprehensive approach encompassing susceptibility, exposure, coping capacity, and adaptive capacity, as well as different thematic areas, such as physical, social, economic, environmental, and institutional vulnerability (Birkmann 2007, p. 21). All people, ecosystems, and regions confronting environmental or socioeconomic stresses are potentially vulnerable to their impacts, but the level of vulnerability varies widely and is generally higher among poorer people.

Vulnerability cannot be separated from the act of "coping." Coping is a capacity to respond and to recover from shocks – an ability also known as "resilience" (Allison and Ellis 2001). Thus, vulnerability and coping are two sides of the same coin: the more one is vulnerable, the less one has the capacity to cope (WHO/EHA/EHTP 1998). The term coping strategy then emphasizes the ability of households to decide and select appropriate activities in light of their assets and endowments (Bird and Prowse 2008). The latter are of crucial importance: if people's access to various livelihood capitals is limited, they are unable to adopt appropriate strategies to effectively cope with shocks.

The relation between vulnerability and poverty is close, but weakly defined. Traditionally, poverty meant low or inadequate means to meet the basic requirements of life, which was measured in terms of monetary income against a particular standard measure. This has now evolved to include other dimensions of poverty, and today, poverty is considered as a multidimensional concept, encompassing diverse perspectives, such as income, basic needs, and human capabilities (UNDP 1997). Béné et al. (2007) consider poverty as a general lack of economic, political, and institutional development that affects rural areas. The World Bank, in its 2001 *Attacking Poverty* actually adds a "vulnerability" dimension into the analysis of poverty. Aside from those "deprivations that keep them from leading the kind of life that everyone values," poor people "also face extreme vulnerability to ill health, economic dislocation, and natural disasters" (World Bank 2001).

Thus, poverty has entered the equation as one of the dominant social dimensions affecting vulnerability. Poverty is not, however, the only dimension. The poor "are often deprived of informal systems of support and social capital – poverty of social relationships" (Parker and Kozel 2005, p. 14). Some make the point that, livelihoods of poorer people are highly vulnerable in large measure because they have fewer buffers, or because the range and effectiveness of the buffers available to them provide inadequate protection (Chambers and Conway 1992, in Barrientos 2007). As Suryahadi and Sumarto (2001) put it, shocks and crises will generate flows of nonpoor into poverty, as well as ensure the persistence of poverty among those already poor. Vulnerability can thus cause poverty, while poverty creates vulnerability: this is called the Vicious Circle of Vulnerability and Poverty.

The Sustainable Livelihoods Framework (Allison and Ellis 2001) puts at the center five types of livelihood capitals, which people combine to develop various vulnerability-coping strategies. Although there is variation in the way in which livelihood capitals are categorized, it is common (DFID 1999) to distinguish five: natural capital, physical capital, financial capital, human capital, and social capital. In the following pages, we will be focusing especially on the crucial role of social capital in coping with vulnerability.

17.3 The Role of Social Capital

The notion of social capital first appeared in the writings of Hanifan (1916, 1920). Hanifan demonstrated that social capital – accumulated through a person's links with his neighbor, and them with other neighbors – will immediately satisfy the person's social needs and will potentially lead to substantial improvement of living conditions in the community, of which he is part. Since this beginning, there has emerged immense interest in the subject with most noted contributions by Bourdieu (1983), Coleman (1988, 1994), Putnam (1993, 2000), and more recently by Ostrom (2000). Yet, a clear cut definition of social capital is hard to find among the writings of these influential advocates of the concept. All of them stress that "social networks have value"; all accept that individuals by interacting with each other and forming networks can reap higher benefits than by operating alone. These benefits are achieved through trust, and norms (such as reciprocity and values), which shape the behavior of individuals in a community and bring about beneficial outcomes.[2]

Reacting to the extreme individualism suggested by new classical economics, which is considered to be nonrepresentative of the real behavior of economic agents (Di Ciaccio 2005), a selection of economists has promoted the investigation of social capital. Some have noted the ability of individuals to extract private returns from interactions with others (Glaeser et al. 1999; Lin 2001), raise efficiency in social exchange (Durlauf and Fafchamps 2004), and lower the cost of search (Granovetter 1974). Dasgupta (2002) considers social capital as a system of interpersonal networks. These create externalities, which could form an aspect of "human capital" if the externalities are localized, or a component of total factor productivity, if the network externalities are more in the form of public goods.

Scholars generally distinguish three types of social capital, depending on the nature of the social network created (Woolcock 2001). Bonding capital ties people in similar situations (such as family, friends, or neighbors), and bridging capital connects people of equal status at larger physical or social distance. Linking capital finally joins people of dissimilar situations or status. We argue below that cooperatives deliver all types of capital, and are therefore particularly useful for the process of coping with poverty.

[2] This is, in fact, the brighter side of social capital. Durlauf and Fafchamps (2004) has also noted the darker side of it – the possibility of social capital ties leading to immoral or unproductive behavior, and its use to overtake others, generating political tension and inequalities among groups.

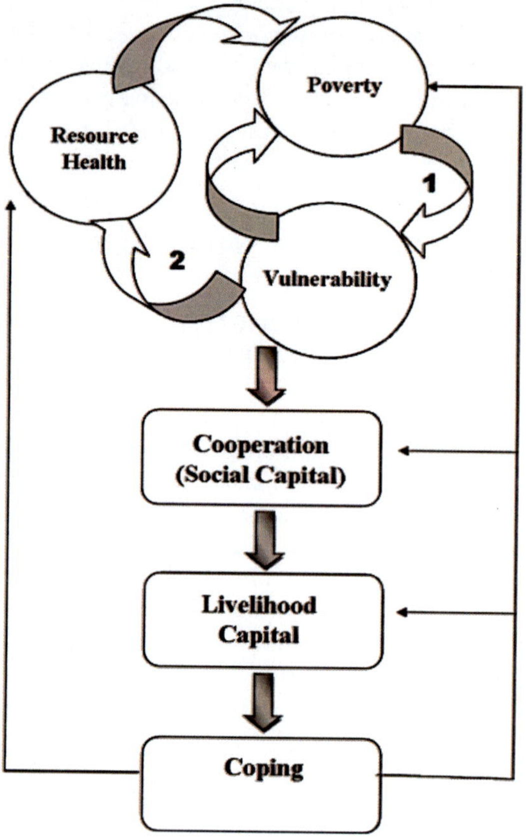

Fig. 17.1 SCACO (Social Capital Approach to Coping). Circles 1 and 2 are, respectively: (*1*) vicious circle of vulnerability and poverty; and (*2*) vicious circle of vulnerability–reduced resource health–poverty

17.3.1 Social Capital Approach to Coping (with Vulnerability)

The Social Capital Approach to Coping with Vulnerability (SCACO), which we are proposing, identifies two vicious circles (Fig. 17.1). Vicious circle (1) suggests that poor people do not have access to adequate livelihood capitals, and thus are unable to adopt appropriate coping strategies to deal with vulnerability. This is the "vicious circle of poverty and vulnerability." From vicious circle (1) follows vicious circle (2): If people are unable to secure livelihood capitals to effectively deal with vulnerability, the only way to survive may be to extract natural resources – such as fish stocks – more intensively. This exerts a negative effect on future returns and may thereby reinforce the state of poverty.

Social capital constitutes a vital input for households in dealing with vulnerable situations. SCACO suggests that social capital may first of all assist households in accessing other livelihood capitals. The result helps them deal more effectively with shocks, by influencing their capacity to cope (or resilience). If resilience improves, then households become more able to escape from vicious circle (1) (reduced vulnerability and poverty). If pressures on natural resources are reduced, then this could also have a positive impact on vicious circle (2) (leading to an improvement in resource health).

17.3.2 Cooperatives as Suppliers of Social Capital

The International Cooperative Alliance defines a cooperative as: "An autonomous association of persons united voluntarily to meet their common economic, social, and cultural needs and aspirations through a jointly-owned and democratically-controlled enterprise" (ICA undated). This definition emphasizes what could be seen as the bonding capital function of cooperative organizations. But, as Birchall (2004, p. 47) points out: "Cooperatives can also be isolated, and may, like other local groups, be high in "bonding social capital," but not be able to find the "bridging social capital" that will link them to others. If they are encouraged to form vertical federations and horizontal networks, and if cooperative laws are passed that recognize their right to do this, they can overcome this problem."

Besides through "vertical federations and horizontal networks," cooperatives may also be linked vertically to governmental agencies. Many of the cooperatives found in developing countries have actually been instigated by government. This is often perceived as a crucial weakness, and even contrary to the essence of the cooperative movement (Jentoft 1986). The Overseas Cooperative Development Council (OCDC 2007, p. 4) thus concludes flatly that: "Government-controlled parastatals are not true cooperatives." We believe that a purist position of this kind is, however, not helpful for the purpose of social science investigation. We therefore take the reality of fishing cooperatives in Sri Lanka as an object of study, and return to their possible deficiencies in the final section.

Fisheries cooperatives in Sri Lanka have been organized by the intervention of the Department of Cooperative Development and the Department of Fisheries and Aquatic Resources, which make them "formal organizations." Many of the fishing cooperatives in Sri Lanka can be characterized as multipurpose, combining functions such as the provision of credit, technology, and insurance; and occasionally, the organization of marketing. The blue revolution initiated in the post-WWII period constituted an important stimulus to the formation of fisheries cooperatives, as the Sri Lankan government strove to channel subsidies and credit to enable asset-poor fishers to adopt new capital-biased technology (Amarasinghe 2005b). In fact, the "cooperative model" was seen by the state as a tool for rural organization aiming at fisheries development. What is important to note is the fact that membership in cooperatives is voluntary, and that individual cooperatives enjoy a great freedom in planning, organizing, and implementing activities aimed at meeting the diverse needs of the community. Today, the activities of fisheries cooperatives are guided by the Cooperative Societies Act No. 5 of 1972, and the Fisheries Cooperative Constitution.

Of a total of 943 fisheries cooperatives currently registered in Sri Lanka, only 439 are considered as active (NARA 2007). Some of the reasons for this state of

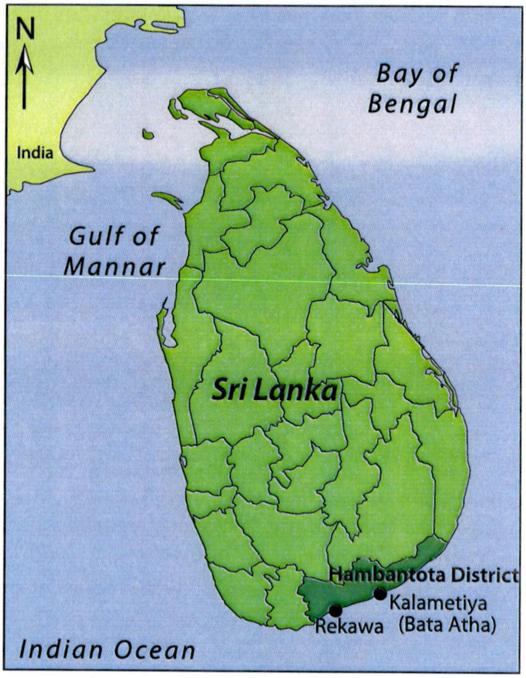

Fig. 17.2 A Map of Sri Lanka, showing the Hambantota District and the two study locations: Rekawa and Kalametiya (Bata Atha)

affairs are poor management, lack of interest among office bearers, inadequate training of personnel in business management, lack of awareness among members of principles of cooperation, and poor loan recovery rates. The total membership of fisheries cooperatives (active and defunct) was 115,700 persons,[3] roughly about one-fifth of the total fishing population. Some of the best functioning fisheries cooperatives, with high rates of savings and lending have been restructured as Fisheries Cooperative Banks (*Idiwara* banks). There were about 120 such banks in operation in 2007.

17.4 The Fisheries Setting

The study area falls in the Hambantota District of Sri Lanka, a coastal belt 151 km in length, located in the southeastern part of the country (Fig. 17.2). Hambantota District has a population of 558,000 people, of whom 89% live in rural areas, and is an area characterized by general poverty. The average per capita monthly income of

[3]Ministry of Fisheries and Aquatic Resources, information obtained from the statistical division.

the people is Rs. 5,789 (51 USD),[4] compared to 10,165 in Colombo, 6,608 in Galle, and 6,463 for Sri Lanka in general. The District has recorded a Poverty Head Count Index of 12.7 (Department of Census and Statistics 2008).

Total annual fish landings of Hambantota District in the year 2008 were 24,520 MT (MFAR 2009), accounting for approximately 8% of the country's total fish landings. Historical data reveal a sizeable drop in fish production in recent years (Amarasinghe 2006). Hambantota has 79 fishing villages, 6,940 active fishers, 5,720 fishing households, and a total fishing population of almost 20,000. The Boat Census of 2006 estimated the number of crafts in the district to be 2,079, of which 87% were coastal crafts.

In the following, we concentrate on small-scale fishers living and operating in two fishing villages, Bata-Atha (South)[5] and Rekawa (East).

17.4.1 The Socioeconomic Context

The two study villages are characterized by high rates of unemployment (20% in Bata Atha, and 30% in Rekawa); much above the national unemployment rate of 5.2% in 2008 (Central Bank 2008). About 11% of the people live in wattle and daub houses; and 28% in Bata Atha and 42% in Rekawa receive Samurdi Benefits (a kind of state assistance scheme for the poor), revealing the general poverty of the population. The fishermen in all three villages use motorized and nonmotorized outrigger canoes (*oruwa*), and small fiberglass boats to exploit inshore fish resources (Fig. 17.3). Each craft carries two to three crew members. The most common gears employed are gill nets (during the monsoon season), longlines, and hand lines (during the non-monsoon period).

As in the case of small-scale fisheries all over the world, fishing incomes vary from day to day and from season to season (McGoodwin 1990). Farming, fish processing, and other livelihood activities (such as the fabrication of coir products, batik, and confectionary products, as well as manual labor) are some of the non-fishing activities available for the fishing population in the research area, which could earn them supplementary incomes.

Our field study reveals that, in coping with poverty, fishers of Bata Atha and Rekawa face three important types of problems. The first is the availability of weakly developed credit, product and insurance markets. With respect to credit, the fishers in the study area have two options: formal lenders (banks) and informal lenders. Formal lenders always demand suitable collateral, against which the loans are lent. In this sense, fishers are put at a serious disadvantage because the assets to which fishers often have perfect property rights – their crafts and gear – are not acceptable

---

[4] 1 USD = 113 Rupees.

[5] The landing center in Bata Atha South is called "Kalametiya" Landing Center, while the fisheries cooperative society is called the "Bata Atha South Fisheries Cooperative Society."

Fig. 17.3 The Kalametiya (Bata Atha) fish landing center: an early morning scene of outrigger canoes and small fiberglass boats landed with their fish catches

to formal lenders because they entail collateral-specific risks. The other option is to turn to informal lenders, such as fish merchants. Most of the problems associated with lending in the formal market do not occur with the same force in informal credit transactions. Yet, as Amarasinghe et al. (2005a) have aptly pointed out, the interest rates incurred for informal loans could be as high as 180%!

Fishers often complain of weak product markets. Fish merchants, who are interested in handling large quantities of fish, often provide credit to craft owners, against the latter's promise to hand over their future fish catches. Craft owners frequently feel that they are often forced to borrow from fish merchants, due to their weak access to other sources of credit and insurance. This form of lending gives rise to credit-product market interlinkages (Platteau et al. 1985; Platteau and Abraham 1987; Amarasinghe 1989) and the phenomenon of "boat-tying."

Finally, due to the high geographical and social distance between fishers and formal insurance agents, there are high informational asymmetries between the two parties. This prevents the emergence of private insurance markets against fishing-related risks and, in the context of Sri Lanka, deposits these risks with the fishers themselves. Fisheries observers have repeatedly noted the high risks of loss or damage faced particularly by small-scale fishers all over the globe (McGoodwin 1990; Amarasinghe et al. 2005b).

The second problem identified by fishers in the research area is the increasing monetary cost of fishing equipment and its operation. Traditionally, fishers in the south depended heavily on coastal fisheries, using canoes and beach seines. In efforts to increase the fish production, however, the government introduced motorized craft, nylon nets, and new fishing techniques, such as long lining and trolling. These new fish catching technologies are characterized by high capital intensity and import dependency. The adoption of motorized fishing technology, unlike in the case of traditional fishing techniques, requires a large capital outlay. Ordinary fishermen frequently do not have savings sufficient enough to meet such heavy capital funds. Not only is modern technology capital intensive, but, in the context of Sri Lanka, it is also highly "import dependent." Moreover, it is characterized by high operating costs, particularly for the purchase of fuel. Recent estimates show that fuel accounts for approximately 60% of the operating costs of small-scale fiberglass boat fishers (Amarasinghe 2005b).

The third problem voiced by fishing households in overcoming poverty is the difficulty of obtaining a school education, which would allow for upward social mobility. Sri Lanka is widely known for its very high literacy level – 93% of the people have received some form of school education and about 38% have passed the secondary school (Central Bank 2005). For Hambantota, the above figures are 93.3% and 39%, respectively, revealing that Hambantota is enjoying educational facilities available to an average Sri Lankan. Average figures, however, do not reveal regional and local disparities. Both Bata Atha and Rekawa are located about 3 km away from the main road, and there is no proper public transport in these areas. Schooling for children has, therefore, always been a problem for parents, who were unable to send children to preschools and national schools.

Hambantota district is also an underprivileged district with respect to the availability of technical institutions and training institutions. Although a fisheries training college is located in Tangalle, the college can be rated as very weak with respect to courses offered and the quality of the instructors. Even most of the seminars and workshops organized by the Department of Cooperative Development and the Department of Fisheries are held in towns – Tangalle, Matara, and Galle – which are miles away from the study area.

17.4.2 Fisheries Cooperatives in the Study Area

There are two fisheries cooperatives in the study area; the Bata Atha South Fisheries Cooperative Society in Bata Atha; and the Rekawa East Fisheries Cooperative Society. Both cooperatives were established in 1989, under a fisheries cooperative reorganization program. The more common services offered by the co-ops include credit facilities, savings schemes, livelihood support, and training. From informal discussions, it became evident that the fisheries cooperative in Bata Atha was quite active in all these spheres of activity, while the Rekawa cooperative seemed weak in almost all respects. Following the methodology of Grootaert (undated), the researcher

Table 17.1 Social capital dimensions of fisheries cooperatives; dimensions are density of memberships, index of heterogeneity, meeting attendance, decision-making index, cash contribution, and work contribution

	Bata Atha	Rekawa
Density of memberships	1.6	0.9
Index of heterogeneity	55.43	42.30
Meeting attendance	3.2	1.6
Decision-making index	85.55	54.04
Cash contribution	212.4	80.6
Work contribution	5.1	1.8

Definitions are:

Density of Memberships: Average number of active memberships per household.
Index of Heterogeneity: Scale (0–100) of internal heterogeneity of the co-op, according to eight criteria (*neighborhood, kin group, occupation, economic status, religion, gender, age, and level of education*).
Meeting Attendance: Average number of times a household member attended a co-op meeting in the last 3 months, normalized for the number of memberships.
Index of Participation in Decision-making: Scale (0–100) of extent of active participation in decision making.
Cash Contribution: Amount of fees (Rupees per month) paid for membership in the co-op.
Work Contribution: Number of days worked per year as membership contribution in the co-op.

Table 17.2 Assessment of the functioning of cooperatives (based on score allocated for five efficiency criteria)

	Score allocated for efficiency criteria (%)[a]	
Assessment criteria	Bata Atha	Rekawa
Transparency	100	70
Diversity of services	100	90
Attention	90	90
Leadership	100	60
Expectations met	100	70
Average	98	76

[a]A score of 100 means 100% efficiency

then made a systematic attempt at measuring the "social capital dimensions" of these organizations. Six variables were taken into account: density of associations, their internal heterogeneity, the frequency of meeting attendance, members' effective participation in decision making, and payment of dues.

It is quite evident from Table 17.1 that Bata Atha co-op has recorded higher values than that of Rekawa for all dimensions of social capital. To find out the reasons why the fisheries cooperative in Bata Atha performed better than the one in Rekawa, an institutional mapping exercise was carried out via focus group discussions. These discussions concentrated on six institutional dimensions: efficiency; transparency, diversity of services provided, attention paid to members' needs, leadership, and ability to meet members' expectations (Table 17.2). In all respects, the Bata Atha cooperative proved to perform better than the Rekawa cooperative.

Table 17.3 Fisher's access to craft, gear, engines, and transport vehicles through fisheries cooperatives

	Fishers reporting procurement of physical capital through assistance provided by co-op (%)			
	Bata Atha		Rekawa	
Access to crafts and gear	NTRB	OFRP	NTRB	OFRP
Crafts	60	65[a]	–	–
Gear	75	70	28	–
Engines	–	–		65
Three-wheel vehicles	25	15	–	–
Motorcycles	–	–	2	

Source: Field studies 2008 (PovFish project)
NTRB Non-motorized traditional boat, *OFRP* Fiberglass boat with outboard engine
[a]craft + engines

The figures in Table 17.2 confirm the earlier findings: Bata Atha cooperative scores significantly higher on all dimensions except one. In the following sections, we will explore how the differential performance of the two cooperatives affects members' ability to access diverse livelihood capitals and deal with vulnerability.

17.5 Dealing with Vulnerability

17.5.1 Access to Physical Assets (Crafts and Gear)

Since we are dealing with open-access fisheries, fishers' access to the sea remains unconstrained. Yet, their access to fish resources depends on whether they can obtain the required physical capital – craft and gear. The field studies in Rekawa and Bata Atha revealed that a significant percentage of fishers obtained their craft and gear from the cooperatives, under various loan schemes. Table 17.3 summarizes the information on this topic. Sixty to seventy-five percent of all craft owners in Bata Atha have obtained their fishing equipment under diverse assistance schemes operated by the Fisheries Cooperative.[6] Although similar assistance was provided by the Rekawa cooperative, the proportion of fishers who benefitted has been small. Cooperatives also supplied members with other types of physical capital, like three-wheel vehicles and motor cycles, thereby revealing their multipurpose character.[7]

[6] Fishing cooperatives were able to borrow from state banks under various loan schemes and disbursed such funds through their own programs, earning an interest income. Co-ops have also been able to access assistance through "linking social-capital" (as explained later in this paper).

[7] Bata Atha cooperative members, in particular, have benefited from these programs. This village is not served by public transport; and the major means of transport is by the "three wheeler" (small scooter), the purchase of most of which have been financed by the Bata Atha co-op. While providing employment to the youth, this assistance scheme has facilitated the establishment of a good transport network within the village.

Table 17.4 Long-term, medium-term, and instant loans from fisheries cooperatives

	Fishers obtaining loans from co-ops (%)			
	Bata Atha		Rekawa	
Type of credit	NTRB	OFRP	NTRB	OFRP
Long- and medium-term credit[a] – crafts[b]	50	45	48	39
Long- and medium-term credit[a] – gear	85	55	64	58
Instant loans[c] (to meet immediate cash needs[d])	48	39	79	42
Long- and medium-term credit[a] (for self-employment activities)	80	65	43	46

Source: Field Studies (PovFish project)
[a]Repayment period varies from 2 to 5 years; annual interest varies from 16% to 20% (maximum amounts: Bata Atha, Rs. 250,000; Rekawa Rs. 40,000)
[b]Including craft repair, engine replacement, hull replacement
[c]Repayment period is 2 months; interest rate is 10% per month
[d]May include gear replacement due to damage

The Bata Atha cooperative also possesses its own physical assets: an ice plant, a vehicle to transport fish, and a fish auction shed on the beach. The Rekawa cooperative lacks an ice plant, but does have two trucks to transport fish. In Bata Atha, the ownership of such assets facilitates the rendering of input and marketing services at comparatively lower prices than those charged by private dealers. The same cannot be said of the Rekawa cooperative, where physical assets remain underutilized.

17.5.2 Access to Financial Capital

We noted in the previous section that the fishers frequently require loans for three purposes: (a) to purchase crafts and gear; (b) to meet operational capital; and (c) for smooth consumption in the event of income shortfalls. Table 17.4 provides information on credit obtained by fishers in Bata Atha and Rekawa to secure their diverse needs.

About a half of the fishers operating traditional craft have taken long- and medium-term loans from the Bata Atha co-op either to purchase the vessels themselves, replace parts, procure engines, or to meet craft repair expenses; while 85% of them have obtained loans to purchase, enlarge, or replace their fishing gear. Approximately half of the owners of motorized craft have obtained loans from this co-op for the same purposes. Similar assistance has been provided to its membership by the Rekawa co-op, but the percentage of fishers served by this cooperative has been lower. It is also evident that provision of credit to the traditional small-scale sector has been higher than to the motorized sector.

Nearly 80% of small-scale fishers in Rekawa obtained instant loans from the co-op, revealing the high risk of catch unpredictability in this subsector, compared to those engaged in motorized fishing. Moreover, 85% of the traditional fishers and 65% of the motorized fishers have obtained loans from the cooperative for other self-employment activities. Field studies thus revealed that credit has been used to

Table 17.5 Fishers obtaining assistance from cooperatives to cope with shortfalls in incomes, illness or death in the family, and natural disasters

	Fishers obtaining assistance from cooperatives[a] (%)	
Vulnerability context	Bata Atha	Rekawa
Shortfalls in incomes	50	25
Illness in family (medical expenses)	37	30
Death of family member (funeral expenses)	87	47
Natural disasters (such as 2004 tsunami)	82	17

[a]Aggregate sample

finance the establishment of small shops, cycle repair workshops, home gardens, fish processing activities, farms, and also the purchase of bicycles, motorcycles, three-wheelers, and the like.

Due to the highly fluctuating and unpredictable nature of fish catches, and the hazardous nature of the marine environment in which craft and gear are put into operation, fishers are likely to confront two types of shocks: idiosyncratic shocks and aggregate shocks – both affecting food entitlements of fishing households. Idiosyncratic shocks are shortfalls in income confronting individuals in a fishing community at random (such as illness and death in the family). Aggregate shocks are caused by changes in weather, fish migratory habits, changes in seawater temperature, and natural disasters (of which tsunamis are an extreme example) – affecting all fishers in a particular area in a similar fashion. Both idiosyncratic shocks and aggregate shocks negatively affect the vulnerability of fishers.

In Table 17.5, we present information on the percentage of fishers obtaining assistance from fisheries cooperatives to cope with four types of shocks: catch shortfalls, sickness, death in the family, and natural disasters. Half of the fishers in Bata Atha, and a quarter in Rekawa have had access to assistance provided by the cooperative to cope with the risk of catch shortfalls (see also instant loans in the section on credit). A death in the family could completely destroy the source of livelihood and funeral expenses could also be excessive. Eighty-seven percent of fishers in Bata Atha, and 47% of fishers in Rekawa have been able to borrow from the co-op to cope with these shocks.

The Tsunami of December 26, 2004, is an aggregate shock that affected several thousands of people inhabiting the coastal zone of Sri Lanka, including Hambantota. The most seriously affected sector was fisheries. Although assistance was channelled through various mechanisms, the most proper thing to do was to channel such aid through fisher community-based organizations, which could be assumed to have "near-perfect knowledge" of the damage caused. More than 80% of the fishers were able to obtain assistance from the Bata Atha co-op to cope with the Tsunami; while only 17% of the fishers in Rekawa could obtain such assistance from their co-op (Table 17.5). Most of the donor agencies active in the region opted to channel assistance to affected families through community organizations that were perceived to be performing well. The Bata Atha co-op had a better reputation than the one in Rekawa.

Table 17.6 Workshops and training camps organized by the Bata Atha cooperative

Type of human capital	Training facility rendered	Institution sponsoring the service
Training	Production of *Maldive* fish using sun driers	Sweden Cooperative Center
	Home gardening	Sweden Cooperative Center
	To design business plans	GEF/RUK project
	Organic farming	Sweden Cooperative Center
	Production of organic fertilizer	Sweden Cooperative Center
	Preparing nursery beds	Sweden Cooperative Center
	Leadership training program	Sweden Cooperative Center and Peoples' Bank
	Awareness programs about cooperative concepts	Cooperative Training College
	Leadership training program for cooperatives	Office bearers and members
	Disaster management programs	Disaster Management Center-Hambantota
	Handicraft courses	Sweden Cooperative Center
Education	Running a preschool	Bata Atha Co-op
	Provision of scholarships for school children	Bata Atha Co-op
	Free seminars organized for school children pursuing their G.C.E. Ordinary Level studies	Bata Atha Co-op
	Provision of free tution classes	Bata Atha Co-op
	Free English classes for school children	Bata Atha Co-op
Health	Eye camps	Hambantota District Hospital
	Clinics for women	Hambantota District Hospital
	Awareness programs – HIV	Hambantota District Hospital
	Awareness programs – Tuberculosis	Hambantota District Hospital
	Child psychological programs	Sweden Cooperative Center
	Awareness program – First Aid	National Institute for Fisheries and Nautical Engineering-NIFNE
	Drug prevention programs	–

17.5.3 *Provision of Human Capital*

Human capital is generally taken to represent education, knowledge, and skills (DFID 1999; Woodhall 2001). If such capital is considered as a resource, which increases labor productivity, then health too could be taken as a component of human capital. A large array of training programs, to develop the skills of members in diverse self-employment activities, has been organized by fisher co-ops in Sri Lanka with assistance from donor organizations (Table 17.6). The advantage to donor organizations is that working with well-functioning cooperatives minimizes transaction costs (such as search and monitoring costs), while ensuring that help reaches the most needy. Moreover, with the help of government hospital staff, health camps are also organized. Training for such activities is provided free of charge by development agencies and NGOs.

The Bata Atha co-op has taken up many of the activities mentioned above. It also operates a student scholarship scheme, whereby talented students are awarded scholarships. Assistance to children has been extended in the form of organizing free private tuition classes (especially in English) and the provision of school books. A preschool is being maintained by this co-op for the children of fisher families. The Rekawa co-op has undertaken none of these activities, which is why it is not represented in Table 17.6.

The Bata Atha co-op has been able to provide the services mentioned in Table 17.6, mainly by building up links with organizations outside the village. Table 17.7 shows the links developed by the Bata Atha co-op with state, non-state institutions and development agencies to provide diverse services to its members. The volume of assistance has generally been large, and the number of links extensive. The success of this co-op can be attributed to its strong leadership, and also to its ability to develop programs to meet the varying livelihood needs of the community.

17.6 Discussion

The preceding section related vulnerability in the fishing villages of southern Sri Lanka to inadequately developed credit, product and insurance markets; and to inadequate access to physical and human capital. In the pages below, we will attempt to describe how, through the accumulation of social capital, cooperatives are able to deliver various goods and services, increasing the resilience capacity of fishers. We will also explore whether cooperatives are able to deal with another societal challenge, namely, the ongoing degradation of marine resources.

17.6.1 Accumulation of Social Capital: The Working of SCACO

In the context of limited access to livelihood capitals, and equipped with a low-resilience capacity, we have seen that fishers in some fishing areas in the Hambantota District have benefitted substantially from the social capital provided by fisheries cooperatives. As not all fishing villages possess viable cooperatives, it is worth investigating, in the locations where cooperatives have been particularly successful, how these forms of social capital have actually been established and maintained.

A first point worth noting is that Sri Lankan fishing villages are, from a social viewpoint, relatively homogeneous. Individual village communities are often made up of one ethnic group (Sinhalese/Tamil), one caste, and one religious category (Buddhist/Hindu/Muslim). As a result, there frequently exists a strong sense of community, which is conducive to collective action. Moreover, shore-based fishing

Table 17.7 Links established by the Bata Atha cooperative with outside agencies, and material and monetary assistance received during the period 1990–2009

Year	Institutional link	Services received
1990	Southern Provincial Council	Rs. 30,000
1996–1996	NORAD Project	• Rs. 30,000 for rotating fund • 9–19 ½ ft. small boats • 10 bicycles • 8 boat engines • Assistance to establish a fuel depot • Two Tool Kits
1991–1992	Ministry of Fisheries	3 multiday boats (Rs. 4,121,736 in value)
2003	GEF/RUK project on biodiversity conservation	Rs. 3 million to be used as a circulating fund to assist those persons withdrawing from environmentally destructive practices
2005–2006 (post-tsunami assistance)	CIC disaster relief fund GOAL	• 7 canoes • Repairing of damaged crafts • 3 beach seines • 25 sets of nets for lagoon fishers
	RED CROSS (Sri Lanka – Netherlands)	Rs. 15 million worth of crafts, engines, and gear
	GTZ	10 boat engines
	PLAN SRI LANKA	Rs. 2 million as circulating fund to grant loans to membership
	FAO	• 1 motor cycle • Fish boxes • 1 safe • Training on use of satellite navigators • 1 cupboard
	ISCOS	Training on first aid
2006–2007	Sweden Cooperative Center	• Modernization of office • Establishment of 20 home gardens • Establishment of two plant nurseries • Provision of assistance to undertake income-generating activities for widows, the disabled, and poor households
2008–2009	UNDP – AGSL	• To network nine fisheries co-ops under leadership of Bata Atha co-op and provide with facilities to undertake fish marketing

activities are concentrated on particular landing sites – a narrow strip of a beach, where all fishers anchor their crafts, land and sell fish, dry and mend nets, and chat over a cup of tea. Therefore, personal anonymity in fishing remains very low. Strong interpersonal relationships contribute to the accumulation of bonding social capital, and pave the way for cooperation, such as through cooperatives.

We noted above that fishing cooperatives in Sri Lanka were established by government in the post-Independence era. In a substantial number of cases – we will return to the exceptions below – these cooperatives have gained local acceptance. Their utility for coastal populations is located in the fact that they have successfully addressed core vulnerabilities in fisher existence. The Bata Atha cooperative is an example of an institution that has generated social capital – both bonding and linking types. The linkages thus created were seen to have contributed in substantial measure to the financial and human capital stocks of member households. As a large percentage of fisher households took part in cooperative activities, the distribution of services was relatively widely spread.

Not all cooperatives have been able to realize this potential, however. Like many other cooperatives in coastal Sri Lanka, the Rekawa cooperative thus had a relatively weak performance in comparison with the one in nearby Bata Atha. Although the same amount of linking capital was available, at least in principle, the services generated were few. This could be attributed to weak leadership, inefficient management, and bad reputation of the Rekawa cooperative.

17.6.2 Can Cooperatives Deal with Resource Health Issues?

We have argued above that fishing cooperatives have contributed to a reduction of fishers' vulnerability. But, do they also address the resource health issues identified under the vicious circle 2: *vulnerability–resource health–poverty?* In trying to answer this question, we draw upon the findings of a previous study carried out by the first author who, using a "fixed price" model, analyzed the resource status of a number of fishing villages – including the present research locations – in Hambantota in the period before the Tsunami of 2004 (Amarasinghe 2006). For the purpose of this analysis, catch and effort data for the inshore fisheries of three landing centers of the Hambantota district were extracted from the files of the Assistant Director of Fisheries (Hambantota). A standard unit for fishing effort was established by setting that of the traditional craft to unity and also by taking into consideration the tonnage (which gives an indication of the capacity of crafts), and number of effective fishing days per year per fishing unit. The results of this study, in respect to the two areas under study, are summarized in Table 17.8.

The Bata Atha fishery revealed an MSY of 1,184 t, and an MEY of 1,166 t, with 18,139 and 15,870 standard units of effort, respectively. But the actual level of effort of 16,812 units revealed that the fishery was economically over-exploited. Although this fishery was capable of generating a rent of Rs. 183 million, the actual rents amounted to only Rs. 3.86 million in 2003. The estimated MSY for the Rekawa fishery was 437 t, with an effort of 9,108 standard units. This fishery yielded its maximum economic yield (maximum rent) at a level of 4,877 units of effort with a catch of 343 t. The maximum rent this fishery was capable of generating was estimated to be Rs. 67 million. The present level of effort of 3,827 units yielding a catch

Table 17.8 Status of coastal resources in terms of maximum sustainable yield, maximum economic yield, and open access equilibrium: Bata Atha and Rekawa landing centers

	MSY		MEY		OAE		Actual 2003		Resource rent (TR – TC)	
Fishery	Effort Units	Catch (tons)	Effort Units	Catch (tons)	Effort Units	Catch (tons)	Effort Units	Catch (tons)	Max (Rs.m)	2003 (Rs.m)
Bata Atha	18,139	1,184	15,870	1,166	31,739	519	16,812	1,178	183	3.86
Rekawa	9,108	437	4,877	343	9,755	435	3,827	290	67	4.51

Source: Amarasinghe 2006

MSY Maximum Sustainable Yield – the maximum catch that the fishery could yield on a sustainable basis; *MEY* Maximum Economic Yield – the levels of catch and effort that generate the maximum resource rents ("profits") from the fishery; *OAE* Open Access Equilibrium – equilibrium in an open access fishery (zero resource rents); *TR* Total Revenue from the fishery (fish catch × price of fish); *TC* Total Cost of fishing effort (cost of input)

of 290 t reveals that the fishery was operating below MEY, and well below the MSY. In general, the fishery remained underexploited or under-fished.[8]

The results of the above study suggest that high rates of resource exploitation (higher levels of effort) occur in fishing villages which have well-functioning cooperatives (Bata Atha). In fact, in these villages, fishers have entered the fishery quite freely and have exploited the resources heavily. Cooperatives have contributed to this situation by providing fishers with the means to access natural resources. Of course, we are speaking about fisheries which are probably not biologically overexploited, and which have not attained open-access equilibrium. Yet, this situation might have changed after the Tsunami, with the growth of the fishing fleet. In a situation of increasing fishing effort, the success of cooperatives in the longer run depends on the will and the ability to introduce required management measures.

17.7 Conclusions

The thrust of this chapter is that social capital constitutes a valuable resource fishers can draw upon in order to deal with vulnerability. We have demonstrated that small-scale fishers in the Hambantota District are confronted with a variety of risks and uncertainties, and that their ability to cope is constrained by limited access to livelihood capitals. New forms of social capital – in this case, the linking capital provided by fisheries cooperatives – can provide an entry point for improving the entire basket.

It should be emphasized that our objective was not to make a detailed assessment of the functioning of cooperatives, which would have brought out both strengths

[8] In Sri Lanka as a whole, it is hard to state with precision that coastal fisheries are operating beyond MSY, because coastal fish landings have not shown a clear-cut decline during the last two decades. This is unlikely to happen under the open-access situation. However, we are speaking of multi-species fisheries, and we must have lost some species (species extinction) in the process.

and weaknesses. In fact, ample evidence is available to show that there is a great diversity among fishing cooperatives in Sri Lanka, and that many are in fact not doing well. Fisheries cooperatives, as organizations, face challenges in their own right, and many have failed due to faulty design, mismanagement, and political interference. The defects that are so clearly present do not, however, contradict the positive influences that have also occurred. Moreover, alternative arrangements are, in the setting of rural Sri Lanka, frequently not available.

Strong interpersonal relationships at the local level have assisted fishers in the two study locations – one more than the other – in dealing with inadequately developed markets and other shocks. Informational problems associated with insurance are minimized by the high degree of personal knowledge that exists among members; and collateral problems in the "credit market" have been resolved through the mechanism of "group guarantees." High unpredictability of fish catches, and risks and damage to fishing equipment causing income shortfalls have been managed through "instant loans." The study points out that various loan schemes performed both a credit and an insurance function, hedging fishers against a host of risks and uncertainties.

It is important to note that in all of these cases, cooperatives constitute channels for outside resources – their success is thus closely associated with access to supra-local agencies: government, NGOs, and development agencies. The linking capital which fishing cooperatives in Sri Lanka have thus created is maintained through efforts of local leadership and through reputation.

Although well-functioning fisheries cooperatives have facilitated entry into fisheries, their concern for long-term sustainability of the resources appears to be weak. More attention on resource governance is required from their leaders, if the cooperatives are to be successful in the long run.

Acknowledgments I wish to acknowledge assistance rendered, during field studies, by Mr. Sumith and Miss Malika of Bata Atha South; and Mr. Jayatilleke and Miss Nashira of Rekawa, all of whom not only helped me to conduct field studies, but also provided me with valuable information on fisheries in the study areas. My thanks are also due to services rendered by Miss Inoka as the data entry operator, and office assistance extended by Mr. A. Somasiri, Mr. Munasinghe, and Mr. Chandana of the Department of Agricultural Economics of the University of Ruhuna. I greatly acknowledge valuable comments made by Professor Svein Jentoft of MaReMa – Centre of Marine Resource Management, University of Tromsø, Norway, from the first draft, until the final product. Thanks are also extended to the Norwegian Research Council for funding the PovFish project.

References

Allison EH, Ellis F (2001) The livelihoods approach and management of small-scale fisheries. Mar Policy 25:377–388

Allsion EH, Horemans B (2007) Putting the principles of the sustainable livelihoods approach into fisheries development policy and practice. Mar Policy 30:757–766

Amarasinghe O (1989) Technological change, transformation of risks, and patronage relations in a fishing community of South Sri Lanka. Dev Change 20(4):684–734

Amarasinghe O (2005b) Technological change in Sri Lanka's fisheries. In: Amarasinghe O (ed.) Modernisation and change in marine small-scale fisheries of Southern Sri Lanka. Navamaga Printers, Colombo, pp 52–102

Amarasinghe O (2006) Cooperation in a context of crisis: public private management of marine fisheries in South Asia. IDPAD working papers series, 5, Indo-Dutch Programme on Alternatives in Development (DPAD), The Netherlands

Amarasinghe O, Wanasinghe WAG, Jayantha SPM, Wishantha Malraj AS, Nishani D, Somasiri A (2005a) Market for fisheries credit: the functioning of the market for fixed capital in fishing communities in Southern Sri Lanka. In: Amarasinghe O (ed.) Modernisation and change in marine small-scale fisheries of Southern Sri Lanka. Navamaga Printers, Colombo, pp 181–252

Amarasinghe O, Wanasinghe WAG, Jayantha SPM (2005b) Risks and uncertainties in fisheries and their management: fisheries insurance. In: Amarasinghe O (ed.) Modernisation and change in marine small-scale fisheries of Southern Sri Lanka. Navamaga Printers, Colombo, pp 103–180

Barrientos A (2007) Does vulnerability create poverty traps? Institute of Development Studies (IDS). University of Sussex, Brighton

Birchall J (2004) Cooperatives and the millennium development goals. ILO, Geneva

Bird K, Prowse M (2008) Vulnerability, poverty and coping in Zimbabwe. World Institute for Development Economics Research, United Nations University, research paper no. 2008/41, 12 p

Birkmaan J (2006) Measuring vulnerability to promote disaster-resistant societies: conceptual frameworks and definitions. In: Birkmann J (ed.) Measuring vulnerability to natural hazards: towards disaster-resistant societies. United Nations University Press, Shibuya-ku

Birkmann J (2007) Risk and vulnerability indicators at different scales: applicability, usefulness and policy implications. Environ Hazards 7:20–31

Bourdieu P (1983) Forms of capital. In: Richards JC (ed.) Handbook of theory and research for the sociology of education. Greenwood Press, New York

Béné C (2003) When fishery rhymes with poverty: a first step beyond the old paradigm on poverty in small-scale fisheries. World Dev 31(6):949–975

Béné C, Macfadyen G, Allison EH (2007) Increasing the contribution of small-scale fisheries to poverty alleviation and food security. Food and Agriculture Organization of the United Nations. FAO fisheries technical paper 481, FAO, Rome

Central Bank (2005) Economic and social statistics of Sri Lanka 2003–2004. Colombo, Sri Lanka

Central Bank (2008) Annual report 2008. Colombo, Sri Lanka

Chambers R, Conway G (1992) Sustainable rural livelihoods: practical concepts for the 21st century. IDS Discussion Paper no. 296, University of Sussex, Institute of Development Studies, Brighton, UK, 6 pp

Coleman JS (1988) Social capital in the creation of human capital. Am J Sociol 94(Supplement): 95–120

Coleman JS (1994) Foundations of social theory. Harvard University Press, Cambridge

Dasgupta P (2002) Social capital and economic performance: analytics, Mimeo. University of Cambridge and Beijer International Institute of Ecological Economics, Stockholm

Daw T, Adger WN, Brown K (2009) Climate change and capture fisheries: potential impacts, adaptation and mitigation. In: Cochrane K, De Young C, Soto D, Tarûb Bahri (eds.), Climate change implications for fisheries and aquaculture – overview of current scientific knowledge. FAO fisheries and aquaculture technical paper 530, FAO, Rome

Department of Census and Statistics (2008) Statistical pocket book-2007. Ministry of Finance and Planning, Colombo

DFID (1999) Sustainable livelihoods guidance sheets. Department for International Development. http://www.nssd.net/pdf/sectiont.pdf. Accessed July 2009

Dhanani S, Islam I (2002) Poverty, vulnerability and social protection in a period of crisis: the case of Indonesia. World Dev 30(7):1211–1231

Di Ciaccio S (2005) Social capital: social relations and economic rationality. Crossroads 5(1):43–77

Durlauf N, Fafchamps M (2004) Social capital. University of Wisconsin, JEL classification codes: E26, 010, 040, L14, Z13

Glaeser EL, Laibson D, Sacerdote B (1999) An economic approach to social capital. Econ J 112(Nov):F437–458
Granovetter MS (1974) Getting a job: a study of contacts and careers, 2nd edn. University of Chicago Press, Chicago
Grootaert C (Undated) Social capital, household welfare and poverty in Indonesia – local level institutions study. Social Development Department, Environmentally and Socially Sustainable Development Network. The World Bank
Hanifan LJ (1916) The rural school community center. Annals of the American Academy of Political and Social Science. In: Smith MK (2007) Social capital, the encyclopaedia of informal education, 67, pp 130–138. http://www.infed.org/biblio/social_capital.htm. Accessed Dec 2008
Hanifan LJ (1920) The community center, Boston: Silver Burdett. In: Smith MK (2007) Social capital, the encyclopaedia of informal education, 67, pp 130–138. http://www.infed.org/biblio/social_capital.htm. Accessed Dec 2008
ICA (International Cooperative Alliance) (undated). http://www.ica.coop/coop/index.html. Accessed 18 Aug 2010
ISDR (2004) International Strategy for Disaster Reduction website http://www.unisdr.org/eng/terminology/terminology-2004-eng.html. Accessed 18 May 2011
Jentoft S (1985) Models of fishery development – the cooperative approach. Mar Policy 9(4):322–331
Jentoft S (1986) Fisheries co-operatives: lessons drawn from international experiences. Can J Dev Stud 7(2):197–209
Kurien J (1988) The role of fishermen's organizations in fisheries management of developing countries (with particular reference to the Indo-Pacific region). In: Studies on the role of fishermen's organizations in fisheries management, Fisheries technical paper 300, FAO
Lin N (2001) Social capital. Cambridge University Press, Cambridge
McGoodwin JR (1990) Crisis in the world's fisheries: people, problems and policies. Stanford University Press, Stanford
MFAR (2007) Census of fishing boat 2006/2007. Ministry of Fisheries and Aquatic Resources, Colombo
MFAR (2009) Annual fisheries statistics. Ministry of Fisheries and Aquatic Resources, Colombo
NARA (2007) Sri Lanka fisheries year book 2007. National Aquatic Resources Research and Development Agency, Colombo 15
OCDC (2007) Cooperatives: pathways to economic, democratic and social development in the global economy. US Overseas Cooperative Development Council, Arlington
Ostrom E (2000) Social capital: a fad or fundamental concept? In: Dasgupta P, Seragilden I (eds.) Social capital: a multifaceted perspective. World Bank, Washington DC, pp 172–214
Parker B, Kozel V (2005) Understanding poverty and vulnerability in India's Uttar Pradesh and Bihar: A Q-Squared Approach. Q-Squared working paper no. 9. Centre for International Studies, University of Toronto
Platteau J-Ph, Abraham A (1987) An inquiry into Quasi-Credit contracts: the role of reciprocal credit and interlinked deals in small-scale fishing communities. J Dev Stud 23(4):461–490
Platteau J-Ph, Murickan J, Delbar E (1985) Technology, credit and indebtedness in marine fishing – a case study of three fishing villages in South Kerala. Hindustan Publishing Corporation, New Delhi
Putnam RD (1993) Making democracy work. Civic traditions in modern Italy. Princeton University Press, Princeton
Putnam RD (2000) Bowling alone: America's declining social capital. J Democracy 6(1):65–78
SEI (undated). Risk and vulnerability programme. Stockholm Environment Institute. http://www.vulnerabilitynet.org/sei-pov/overview.html. Accessed March 2010
Sen A (1999) A plan for Asia's growth. Asia Week, Oct 8, 25. http://www.asiaweek.com/asiaweek/magazine/99/1008/viewpoint.html. Accessed July 2010
Suryahadi A, Sumarto S (2001) The chronic poor, the transient poor, and the vulnerable in Indonesia before and after the crisis. Working paper. SMERU, Jakarta

Thorpe A, Andrew NL, Allison EH (2007) Fisheries and poverty reduction. CAB review: perspectives in agriculture, veterinary science, nutrition and natural Resources, 2 (085)

UNDP (1997) Capacity development and UNDP. Supporting sustainable human development. United Nations Development Programme May (draft paper), New York. http://magnet.undp.org/Docs/cap/BKMORG~1.htm. Accessed Sept 2010

UNDP (2004) Reducing disaster risk: a challenge for development. In: Pelling M, Maskrey A, Ruiz P, Hall L (eds.) A global report. Bureau for Crisis Prevention and Recovery, United Nations Development Programme. . John S. Swift Co., USA

WHO/EHA/EHTP (1998) Emergency health training programme for Africa- Panafrican. Emergency Training Centre, Addis Ababa

Woodhall M (2001) Human capital: educational aspects. International encyclopedia of the social and behavioural sciences. Elsevier, New York, pp 6951–6955

Woolcock M (2001) The place of social capital in understanding social and economic outcomes. Can J Policy Res 6(1). Government of Canada's Policy Research Secretariat, Ottawa

World Bank (1997) Protecting our planet, securing our future. UNEP/U.S. NASA/World Bank

World Bank (2001) World development report-attacking poverty. Oxford University Press, New York

Chapter 18
Moving Out of Poverty: Conditions for Wealth Creation in Small-Scale Fisheries in Mozambique

Ana Menezes, Arne Eide, and Jesper Raakjær

Abstract Over the last few decades, Mozambique has gone through significant political and economic changes moving from a central planning economy to a market economy. The Mozambican government is developing coastal fisheries, enhancing economic productivity and is placing an increasing emphasis on poverty alleviation. Infrastructure is improved and basic common goods have become more accessible in coastal areas. Nevertheless, more than 70% of the population in the coastal areas lives below the poverty line, and effects of recent improvements are still insignificant. Open access to common pool resources in coastal areas has provided the rural population with some degree of food security and shelter during turbulent periods of political changes; the value of the natural resource in terms of being an economic buffer utilized by poor people along the coast, should not be underestimated. New introduced co-management arrangements were targeting the poorest groups, but actually voiced the interests of those who are relatively better off in the coastal communities. New infrastructures related to coastal natural resources, and external groups holding economic interests in the area have resulted in new types of conflicts emerging in the coastal areas. Co-management has focused on solving internal conflicts; in the future, defending local user rights toward other interest groups might become much more important.

A. Menezes (✉)
UNEP, Rua Francisco Barreto N 322, P.O. Box 4595, Maputo, Mozambique
e-mail: anameneze@gmail.com

A. Eide
Norwegian College of Fishery Science, University of Tromsø, 9037 Tromsø, Norway
e-mail: arne.eide@uit.no

J. Raakjær
Innovative Fisheries Management (IFM), Department of Development and Planning, Aalborg University, 9220 Aalborg Ø, Denmark
e-mail: jr@ifm.aau.dk

S. Jentoft and A. Eide (eds.), *Poverty Mosaics: Realities and Prospects in Small-Scale Fisheries*, DOI 10.1007/978-94-007-1582-0_18,

18.1 Introduction

Mozambique is one of the world's poorest countries, with more than half the population living below the poverty line. Since the mid 1990s, the economic performance of the country has, however, greatly improved. Real gross domestic product (GDP) increases at an average rate of 8% per year.[1] Private investment is growing rapidly and the substantial foreign debt has been reduced to a manageable level. The fisheries sector only represents 3% of GDP, but makes up for about 28% of the export value. Most of this is due to the contribution from the shallow water shrimp fishery at Sofala Bay, conducted by semi-industrial and industrial freezing trawlers.

All along the coast, the artisanal fisheries provide food and income for a large number of people. According to the national census for the artisanal marine fisheries, in 2006 there were about 280,000 fishers and 90,000 collectors and divers. The number of people directly dependant on artisanal fisheries is increasing, and some 45,000 people are involved in activities related to fisheries. This includes activities such as fish processing, naval carpentry, mechanics and trading (IDPPE 2003, 2008; Menezes 2008a). Most households involved in artisanal fisheries also depend on other economic activities such as farming, firewood collection, transport and trade. Major constraints to economic development are access to markets and lack of infrastructure. Means of transport are most often lacking in the coastal region, and if available, transportation is first of all done by bicycles.

Over the last 30 years, artisanal fisheries of Mozambique have gone through major changes, primarily caused by societal and economic changes outside the fishing communities. The colonial period of about 500 years ended in a colonial war lasting 10 years (1964–1974), finally resulting in independence and a socialist Mozambican regime gaining power. For different reasons, a civil war broke out and became the main constraint of economic development in Mozambique, lasting for another decade (1981–1992). After a peace agreement in 1992 and a reformed constitution in 1994, economic development in Mozambique has been characterized by increasing foreign investments, market liberalism and significant annual increases in GNP (World Bank 1994).

These rapid changes happened over a relatively short period of time (Table 18.1), making Mozambique an interesting showcase of economic processes following the policy shift from a centrally planned economy to a free market economy. Different stages of economic development and the impact of governmental decisions are also reflected in the fishing communities along the Mozambican coastline. These communities have to develop coping strategies to adapt to social and economic shocks and associated risks, which can be potentially harmful for already impoverished households.

The importance of artisanal fisheries along the coast during periods of war is reflected in the increasing number of people making a living along the coast during that period. People fled to the coast to escape from war actions and to achieve food security.

---

[1] http://data.un.org/CountryProfile; http://www.state.gov/r/pa/ei/bgn/7035.htm.

Table 18.1 Periods of change affecting resource use and fisheries management in Mozambique over the last 50 years. The Gantt chart shows how the four periods are placed in time and intensity. The *darker gray* indicates the most pronounced period of time. A more detailed description is found in Menezes (2008b)

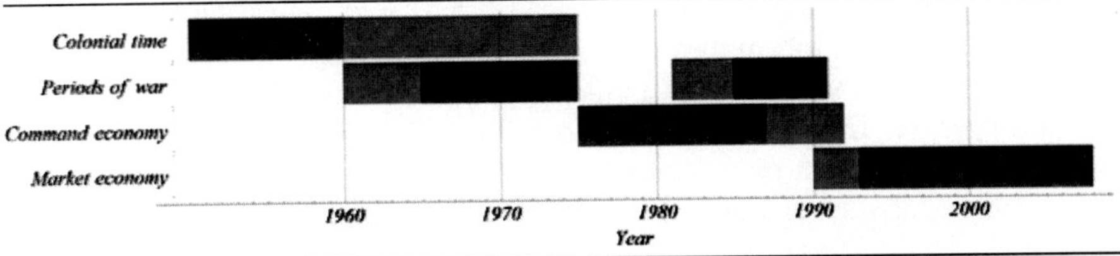

Periods	Utilization of natural resources	Fisheries management
Colonial time	Subsistence economy	No management Local rules
Periods of war	Security valve, increased coastal population	Joint venture companies in industrial fisheries Extension services
Command economy	Policy of small-scale and semi-industrial fleets	Fisheries combines Stock assessment Fisheries regulations
Market economy	New types of interest in the coastal zone	Co-management committees MPA

The statistical figures covering this period are, for understandable reasons, less reliable and it is difficult to quantify the magnitude of the migration. It is also hard to obtain information whether the peace agreement caused people to move back to their places of origin. Normal migration patterns along the coast make it even more difficult to determine the demographic effects of the recent political changes in the country.

In periods when uncertainty increases and the possibility of future planning is called into question, as in periods of war, the properties of fish stock resources become particularly attractive. Farming includes waiting for relatively long periods of time for the crops to reach a suitable harvesting size. Farmers face uncertainties added up over time, and food and income may be lost. In comparison, harvesting fish yields immediate access to food, almost without the need for capital when a fish stock resource and labor are available. In areas far from any market, fishing activities also will be less harmful to the fish stock resource, both because of a poor technological level (almost no capital involved), and because the demand is determined by the local need of food supply.

Coastal communities have always derived their food and income from farming and fishing activities, and it was first after achieving independence in 1975 that artisanal fisheries started receiving governmental support. The governmental policy of developing the artisanal fisheries led to the establishment of fishing cooperatives (combinados pesqueiros) in the early 1980s. This initiative was controlled by the government through the Institute for the Development of Small-Scale Fisheries (abbreviation from the Portuguese name Instituto De Desenvolvimento De Pesca

De Pequena Escala – IDPPE), integrated fishing cooperatives and fisheries extension services. The purpose was to establish a production network for the artisanal sector covering most of the coastal areas where these fisheries were considered to be economically viable. This initiative provided fishers with new fishing technology, cold storage, processing facilities and market access (transportation of fish to rural and urban areas) in the region, where the cooperatives were located.

From 1983, fishing combines had no direct role in fish production, but initiated instead the development of a market network providing production inputs (boats, nets, fishing gears and spare parts). After adhering to the International Monetary Fund's policies and programs, the Mozambican government transformed its centrally planned economy into a free market system, which emerged in 1994. The market liberalization removed state controlled markets, including the infrastructure developed for the purpose of these markets. Market liberalism certainly contributed to increased efficiency and economic growth in many sectors, but it also revealed the lack of basic infrastructure in coastal communities and led to market collapse and even reduced welfare in many communities.

The previous policy of establishing fishing combines had not contributed in developing sustainable market outlets for the artisanal fisheries. The most apparent market failures still impairing the development of artisanal fisheries was lack of public goods (e.g. social policies) and poorly developed common goods such as schools and health care institutions. The situation called for creation of new strategies for the development of coastal communities, finally expressed in the Strategic Plan for the Artisanal Fisheries Sub-Sector (PESPA 2007).

In summary, the new strategic plan reflects a more elaborated understanding of how economic development could be encouraged in the coastal communities, first of all based on the experience gained from some larger development projects in this region. In the next section, we will present the methodological approach of this chapter, including both a general theoretical perspective and a specific focus on the development of small-scale fisheries. The following section presents lessons from the Nampula fisheries project and the last section discusses the findings and draws conclusions on the basis of the Mozambican experiences.

18.2 Methodological Approach

During the last 15 years, the authors of this chapter have been engaged in the management and development of small-scale fisheries in Mozambique both as researchers and consultants. The chapter also draws on Menezes' (2008b) PhD study where she conducted more than 1,000 household interviews in Mongicual, Angoche and Moma districts in 2006 and 2007 (Fig. 18.1). The two other authors (Eide and Raakjær) are coming from Scandinavian countries. Even though they have spent a considerable amount of time in Mozambique, it will be most fair to claim that they have been observing and analyzing the evolution of the development processes of artisanal fisheries from an outsider's perspective. Over the last 15 years, in different

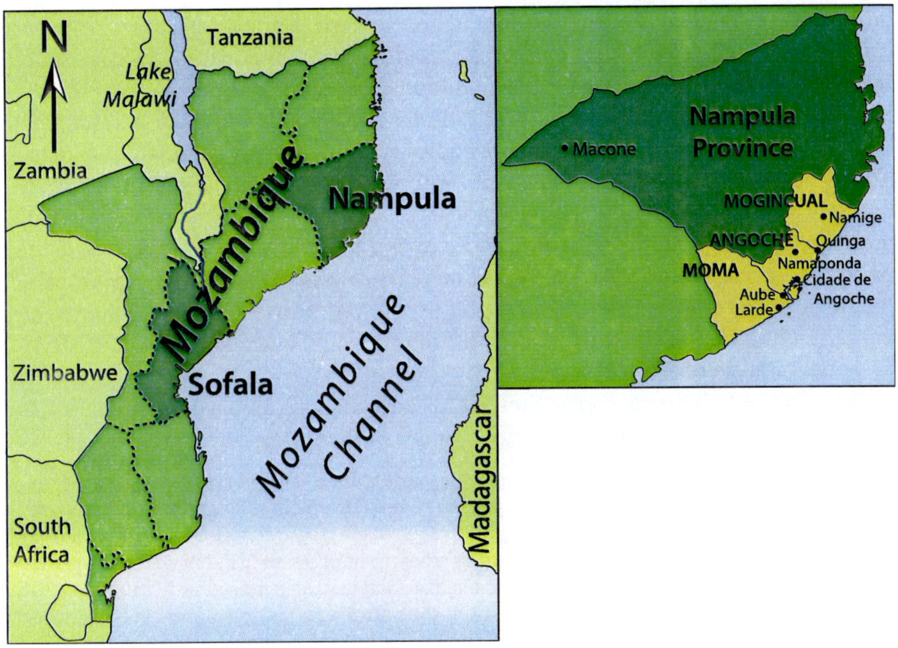

Fig. 18.1 Map of field areas studied in Mozambique – provinces of Nampula and Sofola are highlighted

contexts, the authors have conducted numerous interviews with national and local authorities, major stakeholder groupings, local community co-management committees and also obtained information from official documents. In addition, primarily the Mozambican author has followed the evolution of the development process through reports in the media and by attending public meetings on the subject.

The material, views and ideas presented in this chapter are also the result of discussions with IDPPE staff involved in implementation of the Nampula Artisanal Fisheries Project (1994–2002), and the Sofala Bank Artisanal Fisheries Project (2002–2011). Both are major artisanal integrated fisheries projects, supported through the International Fund for Agricultural Development (IFAD), and implemented by IDPPE.

18.3 Theoretical Perspectives on Development and Wealth Creation

Access to central markets represents a major obstacle in many artisanal fisheries, so also in the Mozambican case. Necessary infrastructure such as roads, electricity and water supply is often absent or in poor condition, and only local markets (if available) can be accessed. Therefore, Mozambican artisanal fisheries, to a large extent, are truly subsistence fisheries without access to commercial markets. A necessary, though not sufficient, condition for market development is access to basic public goods, both for processing and transportation to markets.

Artisanal fisheries are characterized by labor-intensive production systems utilizing relatively simple production technology, and to a high degree rely on excess labor. Few or no other alternatives of employment cause the opportunity cost of labor to approach zero. Even so, when markets are absent, fishing efforts by household members are not likely to increase beyond the point of providing households with only basic food needs. Access to markets also gives access to capital, which could substitute labor in the production of fishing effort. This may enhance efficiency in harvest production, increase fishing pressure, and could eventually lead to overexploitation of the natural resource base.

One of the fish processors expressed his and cooperating businessmen's experience on increased resource exploitation in the following terms[2]:

> [B]efore, I processed my fish by letting it dry on the sand. There was no point for me to improve; my fish would be sold anyway in the community. Now, with good roads I can sell my fish in Nampula, Nacala and even in Lichinga, where they don't have fish. But I cannot sell poor quality fish, because there are many other fish processors in the same route. I asked IDPPE to train us in the techniques to smoke fish, and they also trained us in salt/dry techniques, but the latter is more expensive. I was the only one in the community to implement what we learned during the training; others did not believe we could sell fish at a higher price; some of them still believe in times of shortage people will buy any fish. I did well and now I'm training others, and I provide inputs such as fishing gear and salt, because I need their fish to have more clients in Lichinga and Nampula. Four years ago I lived in a small house made of sticks and clay, today as you can see my house is concrete bricks and I'm working to change the roof. I have my own smoke kiln, and I also rent it to others who want to smoke their fish (Momade, fish processor in Quinga, November, 2006).

The natural resource, providing the subsistence fisher with food security, and being the source of wealth for the fisher with access to markets, in principle is less vulnerable in the non-commercial fishery. A counterworking effect is the ability of wealth creation to stimulate other economic activities outside the fishery, causing fishers to obtain jobs elsewhere. General economic growth may increase demand for fish products, as well as contribute to development of a more efficient fishing fleet, even when the number of fishers decreases. Both reduced costs and increased prices contribute to pushing the market solution toward larger harvest production. This may lead to stock collapses and hence economic crises in the fishing industry. The fortunate situation of having access to a common stock resource, providing coastal communities with food and (by access to markets) income, represents a market failure with increasingly negative effect as the economy develops. Given this, the early stage of economic development when there is no market access or other sources of food, improper regulations aiming to reduce fishing effort (which is the problem of more developed economies) could harm not only the economic development, but also cause increased poverty (Eide 2009).

[2] Names were modified.

Three basic issues related to the livelihood, sustainability and development of rural economies are the ability to:

- Reach the most basic needs for well-being (cover basic needs)
- Maintain well-being (sustain basic needs)
- Develop beyond basic needs (exceed basic needs, wealth creation)

These three abilities also reflect the main focuses at different stages of economic development, employing different sets of coping and adaptive strategies by individuals and communities. Different stages involve different types of constraints and obstacles which may or may not be controlled by governmental decisions and actions.

The ability of satisfying the most basic needs involves coping and adapting capacities of the individual, social constructions and available resources. Obviously, availability of local natural resources is the basic condition of any use of fish stock resources, while other necessary conditions are added while moving from covering the basic needs, toward wealth creation. To obtain the latter, market access and successful fisheries management may be the major constraints for economic development, on the basis of fish stock resource utilization. Given these conditions, wealth is created; hence, capital becomes more available and inexpensive labor scarcer. Consequently, labor is substituted by capital since price on capital is decreasing and price on labor is increasing, following neoclassical growth theory (Solow 1956).

Wealth creation is, however, not only changing the mix of labor and capital inputs in harvest production, but it also affects all other economic activities. Industries not previously developed could emerge as capital becomes available and markets develop on the basis of demand created by increased payoffs to labor. Even though the use of labor in harvest production diminishes, the total production capacity is expected to increase, also within the fishery, as capital becomes more available. The developing labor and capital markets where labor is being more costly and capital more accessible, creates new markets, increased wealth creation and new management challenges.

The new challenges are found in market failures, becoming increasingly apparent as the economy grows. These are management challenges related to preventing overfishing and other negative external effects of increased production capacity. Governmental interaction in terms of participatory decision-making with local groups may contribute in reducing or removing negative effects of market failures; for example, redistributive policies could serve to improve growth processes when capital markets are imperfect (Loury 1981).

However, market failures are not the primary challenges of a coastal economy dominated by subsistence fishing. The main constraint of rural economies dominated by subsistence natural resource use (farming and fishing) is lack of markets demanding the production that is in excess to the subsistence needs of the fishing community. Barter economy, which may develop, can solve part of this problem; barter could create value (e.g. excess goods which are not damaged when stored). But if the value is not converted into capital through a market, it cannot be easily reinvested to enhance production and improve the living conditions within the community.

The main factor hindering market access is missing infrastructure, like roads, electricity and water supply. Essentially, these are public goods at an early stage of economic development, and could only be made available through governmental investments. Providing the communities with the most basic public goods is an essential management responsibility.

This reflects how important it is to consider management challenges within the dynamic perspective of economic growth, under shifting and developing constraints and possibilities. Constraints which relate to market failures could not be omitted without governmental interventions (management actions). But the actions have to target the currently active constraints, not constraints which will be activated in the future. The latter represents just another constraint, effectively reducing the possibilities of wealth creation and economic growth.

18.4 Economic Growth in Subsistence Fisheries

Access to natural resources is a necessary (if not sufficient) condition to cover the basic needs of coastal people. Sen (1985) argues that utilizing the access to resources also depends on an individual's and household's capabilities in terms of human, economic and social capital. Given these conditions, basic needs could be covered, characterizing a typical subsistence fishery. However, without access to markets, excess harvest could not be sold and capital not gained from the fishing activity. Only by access to markets can wealth creation and economic growth take place, developing capital, labor and other markets (Cunningham 1999).

Without roads, fish harvest cannot easily reach markets outside the coastal community, and without access to clean water and electricity, it is difficult to preserve the fish, keeping it at the quality level appreciated/demanded by the market (Fig. 18.2). When governmental institutions replace market dynamics (as in the first year of the new independent Mozambique), information naturally flowing between those operating in the market is reduced. In a market demanding fish, low-quality fish will fetch a reduced price, sending a clear message to the producer.

Following this message, the producer may seek to increase quality (if the cost is supposedly less than the gain collected in the market) and express a demand for ice, improving fish conservation. Given that there is access to water and electricity, ice could be produced through pure market mechanisms to meet the demand (often by home freezers operated by individuals). The ice producer gains income to cover the capital and labor costs, and the fish producer achieves a higher price due to increased fish quality. Given that, at a minimum, the higher price compensates for the cost of buying ice; the ice producer, fish processer and the consumers all benefit from their exchange of market signals.

If the government buys fish and produces ice, price signals from a functional market are lost and so are the potential benefits. On the other hand, if the government did not provide the community with roads, water and power-grid infrastructure allowing a functional market to operate, no fish could be sold and no ice would be produced.

Fig. 18.2 Shrimp catches at the beach outside Pebane, Zambesia in Mozambique

Table 18.2 Properties of three development stages as described in the text

Type of fishery	Use of labor	Use of capital	Infrastructure	Markets	Stock risk
1	High	Low	Poor/No roads, water and electricity supply	No markets, subsistence fisheries	Low
2	Medium	Medium	Roads	Reaching commercial markets by roads	Medium
3	Low	High	Roads, electricity, water	Access to new markets after access to ice and trucks	High

Source: Eide 2009

The assumption is sometimes made that poverty within a fishing community is related to declining stock resources. Since the problem of overfishing is a problem of the highly industrialized fisheries, poverty within the undeveloped fishing communities calls for other explanations. The arguments presented above suggest the problem to be underexploitation rather than the opposite, since there are no market outlets for the fish products.

In Table 18.2, this category of fishery is represented by Type 1, a subsistence fishery with low capital, no markets and hence low stock risk. Capital develops as a consequence of market access (Type 2); a fully developed fishery (Type 3) utilizes a capital-intensive technology and may represent a significant threat to the sustainability of the stock without efficient fisheries management. The fishers of Type 3 fisheries are, however, not typically poor, even though their welfare may also be vulnerable if overfishing causes stock collapses. The general pattern is to find vulnerable fishers and resilient stocks in Type 1 fisheries, versus resilient fishers and vulnerable stocks in Type 3 fisheries. Two or even all three types of fisheries may coexist in a regional context, targeting different species and/or markets. This is also the case in the Nampula Province, where poor, subsistence fisheries of Type 1 coexist with small-scale commercial fisheries (Type 2) and a highly industrialized commercial fishery of Type 3 (the Sofala Bank shrimp fishery).

18.5 Lessons from the Nampula Fisheries Project

Nampula is among the most populated and poorest provinces of Mozambique, and the province having the largest number of artisanal fishermen (Degnbol et al. 2002). The area utilized by artisanal fishers in Nampula is a flat low-lying region along the coast, largely without road access. The fishery takes place within a narrow strip of shallow water, with an average depth of about 20 m. Several fishing methods are used: beach seine, gill net, long line and hand line being most commonly used by artisanal fishers (IDPPE 2003, 2008).

In the mid 1990s, the Mozambican government decided to support a community approach to fisheries development in rural areas, through an integrated development project in Moma and Angoche districts of the Nampula Province. The town of Angoche is the regional centre, where common social infrastructure, services and commerce can be found, including a port and some basic industries (cashew nuts and fisheries). Moma lacks basic infrastructure and services, though there have been some recent improvements. Over the last decade, the number of people having some means of transportation has been doubled, and informal commerce is developed in these two communities (Muchave 2003).

The integrated development project was largely financed through the International Fund for Agriculture Development (IFAD) and included co-management institutions, credit lines, roads and social infrastructure in artisanal fishing communities. Initially, the idea was to focus on fisheries development, but it soon became apparent that the real bottlenecks of this development were linked to general community development issues. It was therefore decided to take an integrated approach to the development of subsistence fisheries. The aim was to tackle fisheries problems from the perspective of the communities. The integrated approach uncovered a range of pressing problems for the communities that needed to be addressed for more effectively being able to deal with constraints and opportunities facing the artisanal fishing sector.

Building infrastructure (roads, water points, schools and clinics) became the major task of the project, in addition to facilitating the organization of co-management bodies in the coastal communities. Activities targeting fishing as well as other activities were supply of input factors to harvest production and fish processing (also including market information), and financial services (facilitating micro credit lines within the communities). Projects such as setting up water points, schools and health care institutions also contributed in gaining confidence in communities, and creating a positive atmosphere, making it easier to put forward new initiatives and to introduce the idea of establishing local co-management committees. The project was also successful in other important policy/legislative initiatives, including a three mile limit for artisanal fishers in Nampula (Degnbol et al. 2002).

18.5.1 The Importance of Infrastructure

Although there have been improvements in public transportation and roads, access to communication by phones, increased supply of education and health care in the coastal areas, there is no evidence that there have been significant positive effects on monetary income or income distribution (Menezes 2008b). Within the communities, it is a common agreement that the development of services and public or common goods is the most important contribution of the local organizations. Both economic and social development goals were given priority over resource conservation (Menezes 2008b).

Throughout this study, the close connection between access to public goods, markets and wealth creation is emphasized repeatedly. As public goods could not be provided by pure market mechanisms, it has to be a governmental responsibility to develop such goods. Sometimes there are, however, shifts in converting previously public or common goods into private goods. Such shifts may have significant market effects, as in the case of phone services and the introduction of mobile phones.

The impact that mobile phones have on development in this and other regions, is already evident. The actual change, and the reason why the effects are dramatic, is that a public good (former telephone systems) – hardly available in rural areas – is converted suddenly into a private good, making new infrastructure available through normal market mechanisms. The situation along the Mozambican coast provides a good illustration of the new development and the impact it may have in the future: The problem of market access does not only concern the fishers but also the fish traders. The traders may be close to the consumer market but far from the fishers. For the remote fisher, the risk of transporting catches over long distances without any secure price information is high; hence trade is often not taking place. Negative factors such as transport on poor roads, and the minimal options for proper fish conservation, are reduced by the exchange of information between trader and fisher. Though the value of such information is substantial, it is not in the range of justifying a large investment in a standard telephone network by the parties involved.

In comparison, the cost of a mobile phone is insignificant, and traders soon discovered the possibilities by distributing mobile phones for free to coastal fishers, in some cases. Emerging trade on the basis of the new means of communication has provided fishers with capital and a new and growing business. This can be seen today on Mozambican beaches; the very small-scale industry of recharging phone batteries by solar energy.

The example above includes two goods that used to be public goods (phone service and electricity) suddenly converted into private goods (though only in small-scale when it comes to the supply of electric power). This development has the capacity of completely changing the set of constraints on economic activities in the area. The final impact of these technical changes is still not fully realized.

18.5.2 Co-management Bodies and the Poor Fishers

Increasing conflicts among artisanal fishers, and between artisanal and industrial fishers, had developed during the period of state-controlled fisheries prior to 1994 (see also Lopes and Gervasio 1997). On the one hand, the artisanal fishers complained about trawlers offending area regulations by trawling in the artisanal zone and destroying their gear. The industrial fleet, on the other hand, claimed that fisheries regulations should also be enforced within the artisanal sector, which they claimed was using destructive fishing methods to a large extent. These viewpoints received sympathy within the Ministry of Fisheries, which had not been able to interact in regulatory terms with the artisanal sector. Coastal fishers found their situation threatened, and the conflicts contributed toward organizing the fishers to demand meetings with the fisheries authorities. Several meetings took place where government representatives met with representatives from local fisheries committees.

The government used this as an opportunity to eliminate the use of mosquito nets by collaborating with local leaders. Mosquito nets are by government and the industrial fleet considered a widespread destructive fishing gear. Mosquito nets are applied directly by fishers or used in the cod-end of beach seines resulting in these gears having no selectivity; nevertheless, it is heavily debated if this gear can be considered destructive in terms of its impact on the fish stocks in question.

The negotiations between the Mozambican government and local fishing committees finally led to the signing of the Larde Declaration (named after a village on the border between the Moma and Angoche districts) in November 1998 (Lopes 1999). This agreement marks the start of Mozambican co-management fisheries policy, asking for collaboration between local organizations on the management of local resources.

The formation of community-based co-management committees were largely linked to community development projects and mobilization processes at the community level. These committees consist of 10–12 members including a president, vice-president, secretary and treasurer. These four representatives are elected by the community, while the other members are volunteers or invited by the board. The board is expected to perform its duties for 3–4 years before new elections take place. In most cases studied, board members are kept for two consecutive terms.

As community-based co-management committees are concerned with the general community development, Fishing Community Councils (FCCs) have been set up having areas of operation specified through an agreement between communities and the government.

The first FCCs were set up in the mid-1990s and by 2006 some 53 had been established. The major management task for the FCCs has been to enforce the mosquito net ban. However, they have also engaged in the licencing of fishing units; have acted as a mediator in conflicts between fishers groups, especially related to the problem of migratory fishers and in some cases have engaged in the formulation and implementation of local rules.

Fisheries co-management in Mozambique operates in a nested system operating on community, district, provincial and national levels. At the national level, co-management is undertaken by the Committee of Fisheries Management (abbreviation CAP from the Portuguese name of the committee: Comissão Administração Pesquera). Here artisanal fishers are represented by representatives from the provincial co-management committee, which comprises representatives from inter-district co-management, which comprises representatives from FCCs.

The importance of the FCCs within the core area of fisheries management is, however, limited in terms of introducing management instruments. Besides being an enforcement body, FCCs were mostly involved in resolving conflicts among artisanal fishers and protecting the rights of artisanal fishers versus industrial fishers. FCCs also became a means through which the government communicated prevailing regulations to fishers along the coast, regarding the role as managers. The FCCs have, since their creation, played an important role representing the previously unspoken viewpoints of artisanal fishers along the coast. FCCs have given communities the possibility to forward their interests, for example, regarding the increasing conflicts between artisanal and industrial fishers.

It is important to mention that the local communities are different; a fishing community is not a homogenous social structure. Therefore, significant changes can be found in recent developments along the coast both between and within communities. Conflicts of interest between different groups, and within and between local communities become more apparent and important. Within the communities, this conflict is between the poorest subsistence fishers, utilizing gears that are perceived to be destructive, and those targeting commercial markets, who are investing in more capital-intensive harvesting technologies, still within the frame of the artisanal sector. Not surprisingly, it is the wealthier group of fishers that dominates the FCCs, and it is seldom that the poor fishers sit on the FCCs.

The wealthier fishers use this power to negotiate directly with the government. Subsequently, it is their voice that is heard by the government at the expense of the subsistence fishers, resulting in larger indifferences within the communities both in relation to economic wealth and political influence. However, from the perspective of the government (as well as many observers), the organizations representing the communities are indeed perceived as representatives of the poor people of the communities. But actually, the poorest fishers are those who are first accused of illegal fishing methods by the FCCs.

Other exposed conflicts can be found between different FCCs, for example, related to the problem of migratory fishers. Together, FCCs have, in general, been able to find local solutions. Nevertheless, some attempts for implementing fisheries management instruments by FCCs can be found, and it is fair to argue that the primary focus of the FCCs can be linked to community development and mobilization processes at the community level. Though, this is primarily for the betterment within the community to protect their interests. No doubt that the FCCs have been giving the communities a voice, which they have previously been lacking, and hereby have been contributing toward creation of a societal consciousness, but this has, only to a minor degree, benefitted the poorest fishers within the communities.

18.5.3 Fisheries Management Challenges

Different government policies have an impact on the fisheries sector, and different interests within the local fishing communities can potentially create problems for the co-management system. The power of each FCC is limited to control and monitor local fishing activities according to central rules.

18.5.3.1 Destructive Gears

The general mesh size regulation (35 mm) is, in principle, applicable in the beach seine fishery. The government has, however, agreed to reduce the mesh size to 12 mm in the beach seine bags; still a large mesh size compared with the commonly used (illegal) mosquito nets. The main argument for banning the use of mosquito nets is the preservation of fish and shrimp juveniles. Even though mosquito nets have been banned for a long time, it was first with the introduction of the FCCs that this regulation could be enforced.

Generally, the FCCs are strong supporters of banning mosquito net usage, even though this gear type is used primarily by poor fishers (representing Type 1 fisheries as referred to in Table 18.2 above) usually operated manually from the beach, exposed to sea for short periods each day, covering only near shore areas. Having limited access to markets, and being primarily subsistence oriented, the risk for overfishing appears rather low. Furthermore, the knowledge about the biological resources within the coastal ecosystem is very poor, also when it comes to possible negative effects from the use of mosquito nets. Thus, scientific knowledge is lacking to support the ban.

According to Normann (2007), many fishers contested the decisions resulting in the ban of mosquito nets in fishing gear. The applied management instrument can be challenged both in terms of relevance and legitimacy. The mosquito net ban is a good example of a management instrument systematically being a disadvantage for the poorest fishers, and definitely representing a management challenge in the coastal zone.

It is crucial to take into account the social heterogeneity within the coastal communities in order to understand how the new co-management bodies could contribute in alleviating poverty and create economic growth. It is striking from the Mozambican examples that nobody is protecting the interests of the poorest fishers. They are excluded from FCCs, and often have conflicting interests with those who are better off in the communities.

Following Normann (2007), poor fishers consider the leaders to only think about themselves, requesting the government to help them. The experiences from Nampula reveal that only the government can defend the interests of the poorest and the lesson is clear: The government cannot rely on local co-management bodies to protect the interests of the poor fishers.

Conflicts are also rising in relation to fishing rights. In the battle for valuable resources, the poorest fishers have a disadvantage as they can do nothing but harvest the resource with the simplest means in order to cover their basic subsistence needs. Their economic importance for the country as a whole is insignificant, and their interests are not defended by any community organization.

In our view, fisheries management is about finding proper means to reach expressed political goals, utilizing fish stock resources. The first and most important challenge therefore is to state which objectives are to be given priority. The Mozambican government has done that, and clearly pointed out poverty alleviation to be goal number one. It is not obvious which management means are best suited to pursue this goal, even after stating what the goal is. Poverty alleviation could, for example, be regarded as a distributional problem or a question of who should be given fishing rights.

The current situation seems to be that management principles based on poor biological knowledge, rather than social context, are given priority; presumably with the basic assumption that this will also benefit poor people along the coast. Some new measures have been included, such as identifying a specific artisanal fishing zone, but the biologically based management principles remain the same as before.

While most fisheries regulations, including the mosquito net ban, are based on biological reasoning, biological knowledge about local resources is weak and the adequateness of some regulations could be questioned. This is a problem not only because the biological basis of the management decisions is poor, but because the regulations may work against the overall goal of poverty alleviation.

On the other hand, there has also been significant investment in providing coastal fishers with access to efficient fishing technologies. Fishers have been encouraged to shift from what government considers destructive gears to technologies improving harvest quality and value. The success of these technologies depends on several conditions being fulfilled simultaneously, including secure and demonstrated markets, available and proven methods of processing and access to material inputs and financial services (Degnbol et al. 2002). Given successful implementation of new technology, the stock resources are more likely to experience a higher fishing pressure than the opposite.

18.5.3.2 Tourism and MPAs

The problems of the poorest fishers are not only toward other fishers, but also due to external interests, such as the expanding tourist industry, increasing environmental concern and the exploitation of other natural resources. In many places, the tourist industry is combined with environmental interests, and new industries targeting the increasing market for environmentally friendly adventure tourism are developing. These are activities in direct competition with artisanal fishers for natural resources – not only for the fish stocks, but also the beach areas as such. NGOs aiming to protect sharks, whales and dolphins have promoted the introduction of marine protected areas (MPAs) in several places along the Mozambican coast. The aim is often to protect resources not even being exploited, only disturbed by fishers operating in the same waters. Some of these areas have developed tourist industries within the context of the MPA, combining significant economic interests and environmentalism. This development is increasingly challenging the fishing rights of poor artisanal fishers, as the poor subsistence fishers usually are the first to be excluded from MPAs.

18.6 Conclusions

MacKenzie is often quoted for his statement: "The fisher is not poor because he is fishing; he is fishing because he is poor" (MacKenzie 1979, footnote 5 on p. 816), referring to observations similar to the situation in subsistence fisheries along the Mozambican coast (MacKenzie 1979). Fishing has been the source of food, not a poverty trap, of the poor fishers. The Mozambican artisanal sector has been playing a crucial role as an economic buffer (security valve) in periods of few or no other sources of food and income. During periods of war, the buffer capacity of the fish resources became evident in Mozambique, offering food and shelter to those fleeing to the coast.

However, to cover and sustain the basic needs is not sufficient for wealth creation. The fishers are still poor. How can poverty end and wealth creation take place? Cunningham (1999, p. 18) expresses it in this way:

> Often there seems to be an assumption that the poor must be helped to become better exploiters of fisheries resources. However, the best way to help the poor may not be to help them to become better fishers, nor to encourage them to become fishers at all, but to use the wealth of the fishery to create alternative employment opportunities for them. In this view, poverty alleviation is considered to be a macroeconomic, and not a sectoral issue.

Wealth creation is obtained by establishing viable conditions for all potential economic activities in the society, by providing the communities with common and public goods: roads, water supply and electricity. Wealth creation will stimulate the demand of other goods through increased salaries. Increased cost of labor and easier access to capital (increased supply) causes labor to be substituted by capital in many production processes, also in fisheries. These are the characteristics of economic growth.

During the last decade, the economic and political environment in Mozambique has changed completely. The conditions in coastal regions have improved in terms of increased access to common and public goods, and the economic growth in the country has created new opportunities outside the rural areas. However, Mozambican industrial fisheries face crises, and they are partly blaming the coastal population for overfishing the resources and blaming the government for allowing this to happen. Environmental groups express their concern and propose to establish marine protected areas in the coastal zone. These areas also catch the interest of a growing tourist industry, wanting to utilize the recreational value of the coastal zone. Finally, economic interests related to the exploitation of other natural resources offshore and inshore along the Mozambican coast are also increasing.

In this situation, the coastal communities face new challenges, very different from previously. The effort of establishing co-management institutions may, in the future, prove to have a greater impact as a means for the communities to defend their interests in the battle of resource rights, rather than being bodies responsible for fisheries management. Given the economic interests involved, the subsistence fisheries along the coast have a difficult case to fight. The fact that there is very poor knowledge about the natural resources they are exploiting represents no advantage for the artisanal fishers. In the perspective of poverty alleviation, the coastal communities have a much stronger case, as the fish stock resource is the single most important potential source of wealth they are able to utilize. It is, however, a danger that the general problem of corruption and private interests of decision-makers at different levels will play a role in developing new conflicts of interest along the Mozambican coast. This may be an increasing problem with the increasing economic interests in the utilization of Mozambican natural resources.

The integration of coastal communities into the national fisheries management scheme also has led to demands from the industrial side that the artisanal fleet should also obey the fisheries regulations.

The consequence for the coastal communities of being heard through the nested system of fisheries co-management has caused a greater pressure from others groups, including the Mozambican government for artisanal fishers to comply with existing rules and regulations. Simultaneously, artisanal fishers have been acknowledged as equal players in the battle of fishing rights and resource exploitation. However, it is important to reemphasize that the created system favors wealthier artisanal fisheries, and not the poor fishers. If the coastal fisheries should be successful in developing the fishing sector and create economic growth, these rights are necessary but not sufficient conditions. Without access to markets, strictly imposed closed season regulation may, for example, have dramatic consequences on subsistence fisheries. The growing political power of the local communities is, however, also increasing demand from other actors to follow the approved fisheries regulations.

In economic terms, the coastal resources are not yet fully utilized, independent of the biological state of these resources. Some essential conditions need to be met in order for the society to take advantage of the resources, and for wealth to be created. Necessary infrastructure is the most basic need before economic development can take place. Without these essentials, fish products will easily be damaged and never reach any market. Given that the resources will remain common pool stock resources, it could still contribute in sustaining coastal communities. But if the population is excluded from utilizing the resource in the future, as some are already today, it may dramatically worsen the poverty situation in rural areas along the Mozambican coast.

Those areas having access to essential infrastructure will very soon become those most exposed to pressure from other interest groups. The irony may be that those areas having the best conditions for wealth creation for the local communities, end up being areas where fish stock resources could not be utilized because it is in conflict with the interests of other groups. This would make distributional issues even more important, as a means of reducing poverty in Mozambican coastal communities.

Acknowledgments The authors would like to thank several Mozambican colleagues from the Ministry of Fisheries and particularly all community leaders who kindly agreed to spend part of their precious time with the authors. Learning the life stories of their families, their communities and their view points allowed this research project to present fisher folks' livelihood stories from a developing country. We would also like to thank the Norwegian Research Council, which awarded the grant to the PovFish project. Particular thanks are extended to Dr. Svein Jentoft and Mr. Kristoffer Kokvold for their unconditional support for this project.

References

Cunningham S (1999) Towards the increased policy relevance of fisheries research. Discussion paper prepared for the fifth fisheries development donor consultation. FAO FIPP/C964, Rome

Degnbol P, Eide A, de Almeida JT, Johnsen V, Raakjær J (2002) A study of the fisheries sector in Mozambique. Report provided for Norad. Norwegian College of Fishery Science, University of Tromsø

Eide A (2009) Economic principles: an economic perspective on fishing. In: Cochrane KL, Garcia SM (eds) A fishery manager's guidebook. FAO/Wiley-Blackwell, Oxford, United Kingdom. ISBN 978-92-5-105963-0

IDPPE (2003) Censo Nacional da Pesca Artesanal, 2002–2003. IDPPE, Mozambique

IDPPE (2008) Censo Nacional da Pesca Artesanal, Aguas maritimas e interiores. Unpublished report. IDPPE, Mozambique

Lopes S (1999) Programa de co-gestao das pescarias artesanais em Mocambique: Sintese dos seminarios distritais e inter-provincais realizados no ambito do PPAN. Mozambique

Lopes S, Gervasio H (1997) As causas dos conflitos entre os pescadores artesanais e frota industrial na zona costeira de Angoche. Um estudo de caso. IDPPE, Mozambique

Loury GC (1981) Intergenerational transfers and the distribution of earnings. Econometrica 49(4):843–867

MacKenzie W (1979) Rational fishery management in a depressed region: the Atlantic ground fishery. J Fish Res Board Can 36:811–826

Menezes A (2008a) Mozambique: fishery country review – big numbers. Report provided for FAO. Washington, DC. Sept 2008

Menezes A (2008b) The governance of natural resources in Mozambique: artisanal fishery. PhD Dissertation, SUNY/ESF

Muchave P (2003) Estudo Sócio-económico de Base na Zona Costeira das Províncias de Nampula, Zambézia e Sofala, preparado no âmbito da definição da situação de referência na área do projecto, versão preliminar, Unidade de Coordenação do Projecto (PPABAS, Projecto de Pesca Artesanal no Banco de Sofala)

Normann AK (2007) Troubled waters, troubled times. Fisheries policy reforms in the transition to democracy in South Africa and Mozambique. Doctoral thesis, University of Tromsø

PESPA (2007) Strategic plan for the Artisanal fisheries sub-sector (PESPA). Ministério das Pescas, Maputo
Sen A (1985) Commodities and capabilities. Elsevier Science, Amsterdam. ISBN 0444877304
Solow RM (1956) A contribution on the theory of economic growth. Q J Econ 70(1):65–94
The World Bank (1994) Adjustment in Africa: reforms, results, and the road ahead. The World Bank, Washington. ISBN 0-8213-2795-X

Chapter 19
The Merits of Consensus: Small-Scale Fisheries as a Livelihood Buffer in Livingston, Guatemala

Hector Andrade and Georges Midré

Mighty is law, but mightier still is need.

Goethe

Abstract Guatemala has among the highest poverty and inequality rates in the Latin American and Caribbean regions. The entire population is affected, but the majority of poor and extreme poor are found among indigenous peoples living in rural areas. It is well documented how privatization and land expropriation have displaced indigenous groups to mountains or scarp terrains sub-optimal for crops, where few income-producing activities are to be found. In these cases, poverty is shaped by marginalization mechanisms related to ethnicity. The Amatique Bay, which is the study site for this chapter, is characterized by semi-open access fisheries available for those who have the knowledge and/or production means to harvest. The lack of resource ownership can be regarded as socially fair, as it does not marginalize groups through allocations of specific access rights. Fisheries may be a social buffer providing work, income, and food, thereby mitigating and/or reducing poverty levels. According to fishers in the area, mechanisms inducing poverty are related to resource degradation due to high fishing intensity, unsustainable fishing practices, and to a lesser extent lack of bargaining power. In this chapter, we explore how "Malthusian overfishing" is the result of both internal and external pressures to the fishery, and how fishers believe it can be overcome by trying to control the prevailing conditions that are leading to it.

H. Andrade (✉)
Norwegian College of Fishery Science, University of Tromsø, Akvaplan-niva, Fram Centre for Climate and the Environment, 9296 Tromsø, Norway
e-mail: hector.andrade@akvaplan.niva.no

G. Midré
Department of Sociology, Political Science and Community Planning
University of Tromsø, 9037 Tromsø, Norway
e-mail: georges.midre@uit.no

S. Jentoft and A. Eide (eds.), *Poverty Mosaics: Realities and Prospects in Small-Scale Fisheries*, DOI 10.1007/978-94-007-1582-0_19,

19.1 Introduction

Poverty rates are high in Guatemala, with malnutrition being among the worst in the world (FAO 2001). The majority of poor and extreme poor live in rural areas, where the population is vulnerable to food scarcity and insecurity (FAO 2001; World Bank 2003). In such areas, where agriculture is the main source of income/subsistence, poverty is directly related to access to arable land. Crop size and soil quality are key factors related to farmers' well-being. Although poverty affects all ethnic groups,[1] historically some of the Criollos and Ladinos have owned the best lands; displacing indigenous people to mountains or scarp terrains sub-optimal for agriculture, and where no other means to make a living are to be found. In this sense, poverty in Guatemala is related to ethnic categories (Midré 2005).

The situation is somewhat different when people have access to fishing grounds. If conditions are adverse on land, people still can go out and fish. In the Guatemalan Caribbean, fisheries are semi-open access; that is, fisheries are only partially regulated as to who can participate (access), and/or how to fish (use). At the same time, law enforcement is seldom applied regarding the rules that do exist. Therefore, new participants can enter the fishery and fish, provided they have the necessary equipment and knowledge. Fishing resources can solve nutritional needs and produce income where markets are developed. However, if the number of entrants to the fishery is too high, overexploitation will follow and economic, social, and biological problems may arise (Charles 2001). This will eventually lead to impoverishment of the fishers.

According to Béné (2003), the causal relationship between poverty and fisheries has been dealt with in the literature focusing on two distinct but related dimensions. The first explains poverty among fishers by the fact that fishing resources although renewable, are limited. To maintain minimum income levels, fishers must keep fishing even at the expense of sustainability, because no other income opportunities exist. Abundance declines over time because new participants keep entering the fishery until resources become economically and/or biologically overexploited. This has been termed "Malthusian overfishing" (Pauly et al. 1989; Pauly 1994). The second causal relationship deals with the distributive dimension of fisheries as a last resource activity; that is, poor people deprived of, or excluded from other economic activities, are attracted to fisheries because of the open access.

---

[1]Defining social and ethnic groups existing in Guatemala is a difficult task, widely covered in specialized literature. For our purposes, *Criollos* are defined as a social class established mainly during colonial times, and comprise the locally born people of pure or mostly European ancestry. *Ladinos* are people of mixed European and Amerindian ancestry. *Garífunas* are an ethnic group with Afro-Caribbean ancestry that arrived to the Coast of Central America in the nineteenth century (González 1995). *Q'eqchis* comprise an indigenous ethnic group whose language is the q'eqchi. We do not have information on when *Hindu* descendants settled in Livingston. There are relatively few and they mix well with other ethnic groups, especially the Ladinos. Still, they consider themselves, and are considered to be, a different ethnic group in Livingston.

This is the case in the town of Livingston, in the Guatemalan province of Izabal. It is populated mainly by four ethnic groups: the Ladino, Garífuna, Q'eqchi, and Hindu descendants; and many of them rely on fishing resources for consumption and income. They compose a multi-ethnic, multi-fleet fishery, operating in the Amatique Bay. The area is a complex ecosystem of coastal lagoons, swamps, marshes, and rivers (Yañez-Arancibia et al. 1994, 1999). While the number of fishers has increased, total landings have declined, eventually threatening fishers' livelihoods and leading many into poverty.

Overfishing has been blamed as the main cause of these developments (Heyman and Graham 2000). Recruitment among relatives and migration from adjacent areas to the town of Livingston have led to an influx of more fishers than can be supported by the available fishing resources. Increased fishing effort resulted in gear conflicts between fleets sharing a common area. To partially solve the problem, fishers decided upon taking collective action and agreed on fishing-zone delimitation. Gill net and shrimp fishers started the process, as explained below, with Garifunas joining at a later stage. A document called "The Gentlemen's Pact" was signed and recognized in the Regulation of the Fishing and Aquaculture General Law, Government ruling 223-2005 (MAGA 2005). This is how an informal bottom-up or root co-management institution started, as local fishers organized themselves with the help of NGOs, to regulate effort in the fishing grounds. Fisher groups organized in associations[2] are now participating in defining the timing of the closed seasons, ahead of the yearly governmental legal declaration. "The Gentlemen's Pact" has been honored to a high degree, resulting in less frequent gear conflicts. However, compliance regarding the closed seasons remains low, in spite of the fact that it has been recognized as beneficial by fishers.

In this chapter, we explore why and how Malthusian overfishing, and lack of compliance with regulations occur in Livingston fisheries, despite the fact that fishers have been involved in the formation and administration of the local resource management regime. As we bring together fishers' opinions, the intrinsic (within fisheries) and extrinsic (outside fisheries) mechanisms that bring about the current overexploitation state are analyzed. Why are people attracted to the fishery? Why do people employ unsustainable fishing practices? The future of the fishery is uncertain, yet capital is invested in new gear, motors, boats, etc. If the set of common rules agreed upon in "The Gentlemen's Pact," and the consultation processes to define closed seasons can effectively address some of the lack of management concerns, why are they not fully honored? Special emphasis is placed on fishers' quests to introduce an effective management regime to avoid overfishing and poverty.

After the Methodology section, the chapter gives a general description of the fisheries in Livingston, emphasizing the conflicts that have occurred between fisher groups and why they have developed. Then the poverty concept in fisheries is explored

[2]It is unknown if all associations are represented periodically in such consultations. For a list, see Hidalgo and Mendez (2007).

by examining how Livingston fisheries are acting as a buffer by providing livelihood to local and migrating people. Subsequently, the importance of fisheries management, and the need for rules enforcement are discussed in light of sustainability.

19.2 Methodology

We recorded 24 semi-structured interviews between September and October 2008 targeting fishers from different ethnic groups where owners, captains, crew, and one of the major seafood intermediaries were addressed. In two of the interviews, a group of three fishers and one intermediary were present. The interviews covered issues regarding fishers' perception on poverty, well-being, fishery resource abundance, and resource management among others. Most informants had more than 10 years of fishing experience. Before starting fieldwork, H. Andrade had previously spent 13 months in the area working on another project.[3] During this time, the researcher participated in routine activities like fishing, processing the catch, selling the fish, etc.; as well as in other non-fishing activities like town festivals, cultural events, etc. Thus, many informants of the present project were acquaintances which facilitated the collection of information. Others were approached randomly at the docks, or targeted because they represented a special group of interests, for example Q'eqchi fishers working in trawlers.

Interviews took place at fish docks, the informants' houses, on boats, etc. Of the fishers who were asked to participate, only one refused the interview. Recordings were also made in a fishers' meeting organized by an NGO whose purpose was to adapt FAO's Code of Conduct for Responsible Fisheries in the Guatemalan Caribbean. During the discussions here, we had the opportunity to record fishers' points of view regarding management policies and needs.

Our study group and area were delimited to fishers operating from the town of Livingston in the Amatique Bay for two reasons: The majority of Guatemalan Caribbean fishers are located here (Heyman and Graham 2000); and the core of the artisanal shrimp trawl fleet which has been heavily involved in gear conflicts, operates from this area. The important Amatique Bay "manjua" or anchovy *Anchoa* sp. fishery was excluded since few fishers are operating from the town of Livingston. It is important to note that we use the term "Livingston fishers" to refer exclusively to fishers operating from the town of Livingston, i.e., where the Municipal Center is located, as Livingston is also the name of a greater region in the Province of Izabal. As historical data regarding fish or fisheries (landings, effort, price fluctuations, etc.) is very scarce and not very reliable, we decided to base much of our analysis on the data from the interviews, personal observations, and informal meetings with institutions related to management.

[3]PhD project "Study of the fish communities and fisheries of the Guatemalan Caribbean: A contribution to the management of the Mesoamerican eco-region." The project was partially funded by the World Wildlife Fund through a Russell E. Train scholarship, and the University of Tromsø.

19.3 Describing the System: Fisheries, Fishers, and Conflicts

The fisheries operating from the town of Livingston can be classified according to employed fishing gear and species targeted (Fig. 19.1). Three groups are well identified among the fishers, especially when gear or area conflicts are discussed:

1. The "*barqueros*" or bottom shrimp trawlers, which use modified yacht-like boats, equipped with inboard truck engines and hand-recovered nets to harvest shrimp. They operate on sandy bottoms with a crew of three.
2. The "*trasmalleros*" or gill net fishers who target both migratory and non-migratory species like snook (*Centropomus* sp.), scombrids (*Scomberomorus* sp.), and catfish (*Bagre marinus*) using small boats with an outboard engine and a crew of two. Some of these boats, especially the ones targeting snook, deploy their fishing gear around noon to be recovered early the next morning. It is a common practice to leave the gear at sea while fishers return to shore. Some fishers operating small boats have also a fishing gear called "*chango*," which is a small trawling net used to catch shrimp near the coast or at low depths. None of the vessels targeting shrimp employ mechanized systems to recover deployed gear. This is a feature fishers use to describe themselves as "artisanal."[4]
3. The Garífuna hook and line fishery employing mainly hand lines and a sole baited hook to catch migratory species like scombrids or non-migratory such as snappers including *Lutjanus synagris*, *L. griseus* among others. Usually, one or two people fish from a single boat in pseudo corals or rocky outcrops locally known as "bajos" (Giudicelli 1971; Hidalgo 2004). In these areas, some species aggregate seasonally at higher densities.

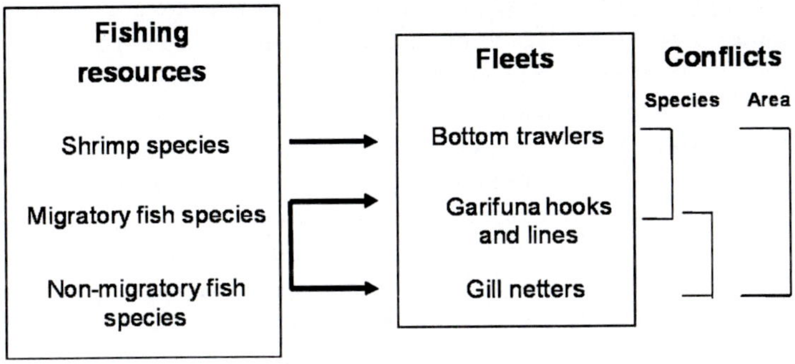

Fig. 19.1 Linkage between fishing resources, fishing groups, and fishermen conflicts in the Amatique Bay. Species conflicts occur when different fleets compete for the same species either targeting it directly, or when a fleet induces juvenile mortality as by-catch. Area conflicts occur when passive and active gears operate concurrently. Migratory species include mainly *Scomberomorus* sp. and *Caranx* sp. Non-migratory include *L. sygrais*, *L. griseus,* and *Bagre marinus*

---

[4]For more detailed vessel information, see ICSED (2000) and Hidalgo and Mendez (2007).

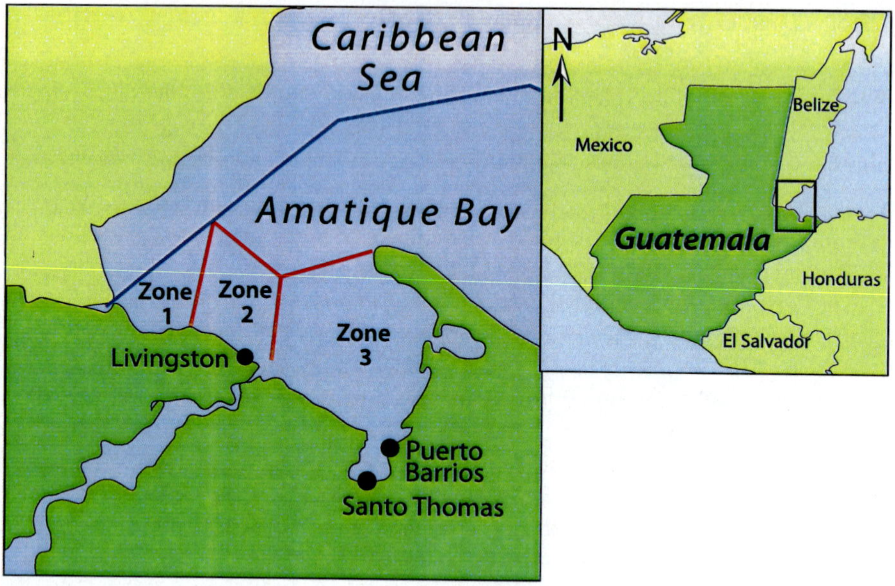

Fig. 19.2 Fishing zones accorded in the "Gentlemen's Pact" and Fisheries Law (Source: Heyman and Graham 2000). Zone 1 can be trawled each year from July 15 to October 31. Zones 2 and 3 are rotated weekly between trawlers and gill netters

Of the aforementioned groups, it is only the Garífuna who are organized on the basis of ethnicity; that is, they are recognized in the Asociación de Pescadores Tradicionales Garífunas (Association of Traditional Garífuna Fishers) as an ethnically differentiated group of fishers. Other fishers' associations are organized regardless of their ethnic background. It is common that boat crews (captain and mariners) are multiethnic. For example, a trawler can have an indigenous (Q'eqchie) captain and Ladino crew. It is very rare that Garífunas take part in multiethnic fisheries, but this mainly has to do with the fact that Garífuna consider their traditional hook and lines fishery among the most sustainable fishing practices; i.e., they do not support what they consider unsustainable fishing like trawling or manjua harvest as discussed below.

Amatique Bay has an aquatic surface of about of 541 km^2 (Yañez-Arancibia et al. 1999). In this area, trawlers and gill netters share a common fishing ground outside the "bajos," where shrimps and fish species are found. The simultaneous operation of both active and passive gears has led to conflicts. Before the "Gentlemen's Pact" was signed, during night bottom trawling, gill nets were accidentally dragged and destroyed. Trawlers did not want to pay back gear losses, and tensions were high between both groups. This agreement divided Amatique Bay into three fishing zones (Fig. 19.2). Zone 1 can be trawled during night time from July 15 to October 31 each year. The other two zones are rotated weekly – when Zone 2 is fished by trawlers, Zone 1 is fished by gill netters and vice versa (Giannini 2006; Hidalgo and Mendez 2007; personal interviews).

Area conflicts still occur between Garífuna fishers and gill netters in the "bajos," where both fleets compete for the same species using different gears (Fig. 19.2). Species conflicts occur in two forms: Fleets are targeting the same species using different gear; or, one fleet is discarding as by-catch juveniles of commercial important species that are targeted as adults by other fleets. Although conflicts occur between fleets, each kind of fishery faces other problems that we could identify during our investigation. Those are described below:

19.3.1 Shrimp Trawlers

Despite strong opposition from trawl fishers, the number of trawlers has increased. The main reason for the complaints relates to the semi-open access regime in which more entrants are allowed into the fishery. The fisheries law establishes a licensing system which, according to trawler owners, only few possess. As one informant mentioned:

> ...people thought that the fishery could sustain everybody and therefore we are so numerous...there are some that abandoned their boats since it is no longer profitable ... About 18 or 19 boats have (fishing) licenses...

The number of trawlers operating in the Amatique Bay is uncertain, but the fishers estimated that there were at least 60. An assessment in the area, conducted by FAO, concluded that shrimp stock size was too small for industrial exploitation (Giudicelli 1971). The license system has not deterred new entrants simply because the rules controlling access are not enforced. Both of the most experienced trawler owners mentioned that they were having problems to pay debts, or were lacking funds to repair their boats.

Trawl fishers also described a general decline in total shrimp landings (the fishery is producing less); or rent dissipation (the fishery produces the same, but is shared between more fishers). In the latter instance, some of the informants believe the yearly total catch among all the boats has not decreased. The problem is that the same quantity is now shared between larger numbers of fishers. Some fishers argue that it is difficult to overfish the shrimp stocks given that the species migrate inside Amatique Bay through marine currents.

The Interamerican Centre for Sustainable Ecosystems Development (ICSED) modeled three different management scenarios to evaluate possible outcomes of restricting access (ICSED 2000):

1. Open Access: The first scenario of open access or no management would still increase the number of trawlers and "changos." Benefits would be dissipated, while total annual catch decreased.
2. MSY: To achieve maximum sustainable yield (MSY), maximizing the number of vessels fishing, trawlers would have to be removed from the fishery, leaving "changos" exclusively.

3. MEY: If the aim was to maximize economic yield (MEY), "changos" would have to abandon the fishery, favoring the more cost-efficient trawlers. Only in the MEY scenario, could control and surveillance costs be covered by the fishery. ICSED's (2000) study together with trawl fishers' perceptions provides insight into the state of shrimp resources. Overexploitation and rent dissipation seem to be the result of lack of regulation.

Some informants mentioned that in the past, to cope with resource degradation, the trawler organization took the initiative of increasing cod-end mesh size to protect small shrimps and improve landings.[5] The price of shrimps is size-dependant; that is, bigger shrimps bring a better price. The results were positive in the sense that landed average size, and shrimp abundance increased after some months, but then gradually more and more fishers returned to the old cod-ends in a race to harvest higher quantities. According to some informants, many trawlers now employ cod-end mesh sizes smaller than those required by law.

Another problem faced by shrimp trawlers is the lack of bargaining power where relatively few markets are developed. At the same time, the fishers are lacking the capacity to store the product and wait for prices to rise. Shrimps are highly perishable and must be sold at the price offered by middlepersons. Seafood prices are subjected to demand laws, and when abundance is high, the price drops. An informant commented:

> The intermediaries have become wealthier at our expense…We are having bad times because we do not have a stable market. This is fundamental. We do not have the economic means to wait for restaurants and hotels to pay us back (referring to credit sales where fishers must leave their product to be reimbursed a posteriori). The higher the shrimp abundance, the lower the price…

Malthusian overfishing and lack of bargaining power are the main problems faced by shrimp trawlers in the Amatique Bay.

19.3.2 Gill Netters

As for trawlers, Malthusian overfishing and lack of bargaining power seem to pose the biggest concerns; also for the gill netters. Total harvest and average fish size landed are decreasing. Since smaller fish are more abundant, fishers invest in new, longer gill nets with smaller mesh size particularly in the snook fishery, to obtain higher harvests. An informant mentioned:

> Before there was more fish: with 200 fathoms (equivalent to 366 m), you could catch 5 to 6 snooks. Now you catch the same with 800–900 fathoms (1450–1650 m) as 10 years ago. The mesh size is decreasing from 6 to 5 inches because it is difficult to catch fish…

[5] The trawler organization does not include all the trawler owners. Hidalgo and Mendez (2007) estimate that only 21 trawl owners representing 30 boats are organized out of 67 boats fishing; i.e., only 45% of the boats are represented in the organization.

For some, catches have decreased to the point where unless in high season, fishing is not profitable. Instead, they work as crew on the shrimp trawlers, despite owning boats and equipment for gill net fisheries. High fuel prices aggravate the situation. Some of these fishers organized into an association; they applied for a loan to acquire boats, engines, and equipment for relatively low monthly payments. Many are not paying back the debt, but it is not clear if this is due to the state of the fishery, the increase in living expenses, or if people are misusing the funds. During a group interview with three fishers and an intermediary, they explained how the first payments were reimbursed; but then the majority of borrowers, including two of these informants, just stopped paying. They expected that the governmental loaning institution would write off the rest of the debt; or use the equipment until it was claimed back as a payment.

Fishers' concerns regarding the health of the snook fishery are well supported by the research literature. Snook is one of the most valuable fish species in the Guatemalan Atlantic (Heyman and Graham 2000). This species is a protandrous hermaphrodite; that is, functional males undergo sexual change becoming functional females (Taylor et al. 2000; Sadovy and Min 2008). Such change is related to age and length and seems to be area-dependant (Taylor et al. 2000). Since gear selective properties determine the size of fish to be retained, by reducing mesh size, fishers are directing fishing mortality toward a higher proportion of males.

Future reproductive output could be compromised in two ways: fewer males are able to develop into spawning females, and fewer spawning males are available to fertilize the output from mature females. Milton et al. (1998) recommend controlling mesh sizes to target fish of intermediate size; that is, to protect proportions of both sexes. Research is needed in such areas for Guatemala. By investing in new fishing gear, fishers are taking high economic risks, while further increasing fishing intensity in a most likely overexploited resource.

19.3.3 Garífuna Hook and Line Fishers

According to Garífuna hook and line fishers, their landings have declined because of unsustainable fishing practices employed by other non-Garífuna fishers. The number of new trawlers has increased by-catch mortality of several juvenile species, including those targeted by the Garífuna. For example, the lane snapper *L. synagris,* is a common trawler by-catch species discarded at sea. Only a small proportion of the biggest fish is landed for crew consumption or put up for sale at very low prices. Garífuna targets the same species at an adult stage, but because of fish size, the price increases about eight-fold. The problem of trawl by-catch mortality induced upon Caribbean commercial fish species has been described elsewhere. The introduction of by-catch reduction devices (BRD), like the fisheye, is recommended (Manjarres et al. 2008). For the Atlantic, the Guatemalan fisheries law does not require the use of any BRD. By-catch conflicts are also common between fleets in the Guatemalan Pacific (Andrade 2003).

Garífunas also claim that the use of gill nets to surround aggregating species in the "bajos" has decimated their landings. When fish aggregations are detected, some non-Garífuna fishers employ this illegal practice; which according to some became banned when the "Gentlemen's Pact" was agreed upon. Vast fish quantities are landed when compared to Garífuna landings. For the Garífuna, this is unsustainable and egoistic. The main species of concern are Scombrids *Scomberomorus* sp. which seem to migrate in aggregations to the reefs during the dry season, the time of the year when Garífuna land more fish. The migratory nature of the species has been described elsewhere (Sutter et al. 1991; Schmidt et al. 1993; Clardy 2008).

Garífunas also indicate that landings are declining due to the low abundance of *manjua* or anchovies, species regarded as the main food item for scombrids. They blame the anchovy fishers for having depleted the anchovy stocks. Many fishers, including non-Garífuna, believe that if there are no anchovies, scombrids migrate to other places instead of the pseudo reefs in the Amatique Bay. Scombrids feed on anchovies, but its relative importance as a food item seems to be area-dependent (see Godcharles and Murphy 1986 for a more detailed review).

For Garífuna, Malthusian overfishing is again a problem but claims are based on what they perceive as unsustainable fishing practices by other fleets, rather than more participants entering the fishery. A Garífuna expressed his view in the general meeting mentioned above, where fishers from the Guatemalan Caribbean were present:

> …I am one of the persons opposing hard, very hard against any fishing practice different from mine. Because if you ask me, what is your opinion regarding the change that must undergo in the [Amatique] Bay, do you know what will I answer? Drop all you fishing gear and buy the cheapest. It will take us all out of the problems we have. You know why? The Garífuna fish with a hook that costs GTQ 1.10, a line that costs GTQ 6-7 (1 USD is about 8 GTQ), and for many years the Garífuna have fished like that, God be praised, without dying of hunger. Some people own many types of fishing gear and still they complain about the fishery, do you understand me? The necessity is one…who of us does not have necessities? We need to cover them, when we think that our families need money and we do not have it…

It is interesting how the argument is built up: By only using what he considers non-destructive fishing gear, conflicts among fleets will be reduced; but everybody would still obtain profits or at least food. Recognition of everybody's "necessity" or right to fish is praised, but without affecting negatively each other's interests. One Garífuna described his ethnic group as "communist" because they shared the catch with those Garífuna who did not have money to purchase fish to eat. Our data are too scarce to further develop the subject and draw any firm conclusions in this regard. Many see Garífuna fishing techniques as the most sustainable, but at the same time, as the least efficient for profit-making.

The Garífuna association also applied for a loan to purchase boats and engines hoping to increase their landings, as traditionally they used paddling and sails. Some informants mentioned they succeeded; but after oil prices increased, many could not pay debts. For some, it was a "mistake" to change their fishing ways.

Garífunas sell their catch directly on the beach, mainly to other Garífuna. If there is still surplus, they sell it in restaurants. If a day's catch is not sold, Garífuna prefer to stop fishing for some time until demand increases. In this way, middlepersons are avoided. In these cases, lack of bargaining power reported from the other categories of fishers does not seem to be a problem.

To summarize, all fishers perceive a decline in landings: either total abundance is lower compared to previous years, or individual harvest has decreased. Many had stories to tell about how abundant the fisheries were before, how easy and quick it was to fish near land, or how common it was to find "rare" species. For example, some mentioned that in the past, it was possible to harvest shrimp with a coconut palm from the shore, or that there was an active shark fishery. Today, very few sharks are landed, if any. According to the fishers, the main reasons for landing decline are a general increase in fishing intensity and unsustainable fishing practices.

Effort has increased in the sense that there are more entrants to the fishery. Many former children of fisher households go into fishing, or people are recruited from adjacent areas near Livingston where agriculture is no longer profitable. Furthermore, effort has increased due to gear modifications; i.e., longer gill nets with smaller mesh size and new and more powerful motorized boats are introduced to compensate for declining harvest levels. These vessels are also capable of transporting fishers to fishing grounds in less time. In addition, effective fishing time has intensified with longer towing times due to overnight trawling.

Unsustainable fishing practices include the high juvenile by-catch mortality produced by both shrimp trawling and the anchovy fisheries using small mesh size, and fishing down the food web (Pauly et al. 1998) in the case of anchovy fisheries. To a lesser extent, contamination is also mentioned. Similar results are presented in Heyman and Graham (2000). Almost a decade after their report, the problems persist. In summary, the main factors leading to reduced income and poverty are: resource degradation through overfishing and unsustainable fishing practices (Malthusian overfishing); lack of bargaining power; and poor management.

When our informants were asked how to improve fisheries to harvest levels of the past, many mentioned the need for a strong management regime. As we have seen, a law including regulations adopted by fishers themselves exists, but is seldom followed. To understand this lack of compliance and boycott of the existing rules, we asked opinions regarding the closed seasons and the "Gentlemen's Pact"; two regulations which are at least partially supported by fishers. The general perception is that the existing regulations are good. However, they are either not respected because people are experiencing poverty (where they cannot cover their basic needs "la necesidad" – i.e., bringing food and money home); or because of the lack of rules enforcement. As living expenses and basic nutritional needs are, to an important extent, covered by income generated through fishing, the informants said that fishers cannot afford to stop, even if this entails breaking the rules. To further explore this issue, we analyzed fishers' perceptions on poverty and its causes. There are intrinsic and extrinsic causes, bringing about the current fishing situation.

19.4 Poverty in Livingston Fisheries

Regarding the closed seasons, when a fisherwoman was asked her opinion on why the low compliance with the law, she responded:

> ...in the world nothing is respected 100%, and if we add that there are no other (income-generating) alternatives, then we cannot ask fishers during the closed season to stop fishing; isn't it true? Because, if during one month you cannot provide for your home, while knowing that there are resources in the sea...Fishers' communities are numerous isn't it true? And that is the cause, I think, that sometimes we do not respect the closed seasons...

Many informants shared the same view arguing that although the existing fishing regulations are needed, they are difficult to comply with due to the current situation where poverty seems to be increasing among fishers. Informants described the fishery situation as "hard," as resources are scarce while facing higher production costs. Fishing is an activity associated with a high degree of uncertainty, where traditional knowledge, availability of different fishing gear types, and "luck" can be decisive in profit-making. As one of the interviewed mentioned: "Going out fishing is like playing lottery. Sometimes you win, sometimes you loose."

With declining landings, the risk of loosing money increases because the probability of covering production costs is reduced. To cope with the situation, people seem to be investing more in human capital (education). All informants were literate and had attended school for at least some years. This is an interesting observation in a country where illiteracy reaches 37.1% of the population above 15 years of age (Porta and Laguna 2007). Many fisher families are sending children to school because they do not believe the fishery will provide for the future. Some fishers stated they are willing to leave the fishery if a secured, fair salary is available on land. Many are skilled laborers: teachers, carpenters, construction workers, etc. They report they are willing to trade the uncertainty of the fisheries for alternative employment with a secure salary. As these jobs are scarce in the area, they remain in the fisheries. Thus, the lack of alternative employment contributes to Malthusian overfishing and eventually to poverty among the population, depending on income from the fisheries.

There are many definitions of poverty. In some cases, poverty is defined by low income. According to Walmsley et al. (2006), income poverty can be analyzed at a national and livelihoods level (income and consumption). In other definitions, a multi-dimensional model is used where people's capabilities, vulnerabilities, participation, and social factors such as education, health, and political empowerment are incorporated (Béné 2003; Walmsley et al. 2006). Following Midré (2005), one might define poverty as limited room for maneuvering. The poor are tied to a social and geographical space that is lacking sufficient resources.

When speaking about poverty, the fishers we interviewed seemed to employ a notion of relative poverty. In other words, they explained what it meant to be poor by making comparisons between local groups and households. At the same time, they used a rather restricted concept of income poverty. We classified responses in two interrelated sub-groups: (1) poverty is linked mainly to food access and (2) poverty is relative when compared to other ethnic groups, places, or economic activities.

19.4.1 *Poverty as Related to Food Access*

For the first case, informants used three circumstantial categories to describe poverty levels: being poor because of a lack of food; being poor, but having food; and not being poor because of having food. The second and third categories can create confusion. These perceptions are dependant on the beholders' eye in the sense that for some, poverty exists even if starvation is avoided; while for others, eating is enough to be considered "not poor."

This is in accordance with the findings of a similar study performed in Guatemala. A lack of basic material needs (food, clothing) and the lack of access or ownership to assets, like land and housing, were ranked first when analyzing people's perceptions on poverty in ten rural villages (World Bank 2003). The strong relationship between poverty and food can be partially explained by the fact that Guatemala has among the highest malnutrition rates in the world (World Bank 2003). In the town of Livingston, fishers mentioned how it is "at least" possible to have something to eat by getting fish at the dock, either as a gift or by fishing with hand lines. The fact that people can avoid starvation makes them better off, compared to other Guatemalan areas.

Through mass media, fishers are informed about what is happening in the rest of the country, where high rates of criminality, land tenure problems, famine, etc. are often reported. Based on this information, the interviewees stated they have a "medium" living standard in Livingston, i.e., there are people that are both better and worse off. When asked about the poor people in Livingston, in particular, many informants indicated they believe Q'eqchi people to be the poorest ethnic group.

19.4.2 *Poverty and Ethnicity*

The relationship between poverty and ethnicity in Guatemala has been widely discussed, especially with regard to land tenure. Through land expropriation and privatization, indigenous peoples have been displaced from the most productive lands since colonial times (Manz 1995; World Bank 2003; Grobakken 2005; Midré 2005). This practice still continues today as a by-product of market-assisted land reform efforts tackling land tenure insecurity (Gauster and Isakson 2007; Ybarra 2008). Land distribution in Guatemala continues to be highly skewed, and among the worst in Latin America (Gauster and Isakson 2007). In Livingston fisheries, however, the relationship between ethnicity, poverty, and access to fish has not been a determinant because no group displacement has occurred in the same regard. While the most productive lands have become enclosed through privatization, fisheries remain relatively open.

19.4.2.1 Q'eqchi Coping Strategies

Ethnic groups are coping differently with poverty and the fisheries situation. Many Q'eqchi living in or near the Livingston district are migrating into the town to work (among other things), as shrimp trawler crew. They are leaving their villages where poverty seems to be higher. Agriculture is no longer profitable and the few jobs available, if any, are poorly paid. Some of the Q'eqchi informants mentioned that becoming fishers has improved their living conditions – first, because it pays around double per day the amount earned on land, and second, the work is neither as physically intensive nor time consuming as agriculture, for example. Many non-Q'eqchi fishers agreed with such a statement. Trawl owners further provide fishers with meals, so basic nutritional needs during the fishing trip are secured. The job is learned easily, and the only occasional requirement, according to some of the captains, is to know how to swim. A Q'eqchi working as crew and/or captain in a trawler explained why and how he became a fisherman:

> Jobs are not found at other places… things are hard… the situation is very hard. We moved to Livingston because the situation back there (Rio Tatin, an adjacent village) was difficult… Here (in Livingston) the situation is a little better, one can defend himself… In Tatin, one has to work very hard to support oneself, to provide for every day… There you find more poverty than here (Livingston)…One has to work much more, everyday, cleaning grounds, planting corn, looking for firewood, planting fruits to survive. Here you can earn at least one night a week and you can make it, but not in the mountain[6] …

The risks that small crop farmers must face when facing bad weather, where harvests can be lost after flooding or drought, may be disastrous for farming households, leading to starvation. In fishing, the costs of not operating during bad weather are much lower. If the boat does not fish for some days due to weather conditions, the yearly income is still relatively safe.

19.4.2.2 Garífuna Coping Strategies

Garífunas, on the other hand, seem to be coping differently. They are moving out of the fisheries, migrating mainly to the United States. Internal and international remittances have become an important component of household incomes in Guatemala (Adams 2004), but the amount of money sent to Garífunas is especially high. For the year 2007, Garífunas received 0.2% of the almost 3.9 billion dollars sent as remittances (UN-Instraw 2007). In 2002, there were 5,040 people listed as Garífunas in Guatemala (INE 2002). By using this number as a demographic reference, we can roughly estimate an income of about USD 1,548 per Garífuna, per year. According to our Garífuna informants, many fishers left years ago because they saw how the fishery situation was deteriorating. When asked if Livingston was as fishery-dependant as before, one Garífuna answered:

> Things have changed. We Garífunas do not live solely from fishing. If we did not have family in the USA that gives a helping hand, I don't know how we would have had it; that is the truth.

[6]This is referring to the fact that the trawler will likely fish at least 1 day per week, thus providing 1 day's salary. In the "mountain," unemployment is relatively long-lasting, with no jobs available for many weeks.

Thus the Q'eqchi and Garífuna seem to have the ability to maneuver differently to improve their well-being and avoid poverty. Both migrate to obtain better lives; but while the former move to the town of Livingston seeking jobs to overcome poverty, the latter are leaving the country. We did not obtain data to explore further why Q'eqchis from the area do not migrate abroad as the Garífunas. However, one possibility could be that they do not have as extended a network already settled in the USA. For Arrivillaga-Cortés (2009), mobility then becomes a coping option to forced displacement. Under present conditions, the opportunity cost defined by Begg et al. (1997) as "the amount lost by not using the resource (labour or capital) in its best alternative use" to stay in the fisheries is different for the two groups. For Garífuna, the opportunity costs are high, since their alternatives in the US are favorable. For Q'eqchis, however, these costs are almost non-existent, due to the extreme poverty and lack of alternative income opportunities where they lived before moving to Livingston.

This analysis is over-simplistic in the sense that not all Garífunas can migrate freely. In fact, many of them enter the US illegally and face the risk of expulsion. There are other costs as well. A Garífuna fisher mentioned how parents that migrated to the US are not recognized by their own children when returning back to Livingston. This was sufficient for him to not consider migration. Still, his income is dependant on remittances as he works as a contractor, building retirement houses for those who want to move back to Livingston in their elder years. The Q'eqchis, on the other hand, when arriving in Livingston face a tight labor market in the trawl fisheries. To a large extent, they lack the opportunity of using family relations and other informal community ties to obtain work.

Summing up, we note that Garífunas and Q'eqchis have different opportunity costs when deciding to participate in the fisheries. The cost to stay in the fisheries is higher for Garífuna, as they seem to have wider room to maneuver in order to improve well-being. Q'eqchis, conversely, try to improve their living conditions by becoming fishers as they migrate to Livingston, due to the lack of alternative employment. This is important when planning management actions trying to stop new entries coming into the fishery or removing high fishing effort, i.e., the number of fishers. As there are no alternative income-generating activities, displaced fishers or possible new entrants are unlikely to stop fishing. Economical development must occur outside the fisheries for management to succeed. Béné et al. (2010) for example, recommended preserving the welfare function of small-scale fisheries providing labor and cash income until macroeconomic conditions allow for management strategies aiming to maximize fisheries rent and redistribution.

19.4.2.3 Ladino and Hindu Descent Fishers

Some fishers of Ladino and Hindu descent are investing in new gear to be able to catch the more abundant smaller resources, or to gain access to a different fishery. The latter group has the advantage that many are granted Belizean fishing licenses because of family ties. For the four ethnic groups remaining in the fisheries, the ability to access distant fishing grounds and/or having the gear necessary to harvest specific species during high seasons has become the key to improve well-being or at least to avoid poverty. In addition, many fishers try to diversify economic activities by combining fisheries with work in agriculture or construction, when possible. If governmental policies are to be successful in fighting against poverty, it is crucial that the importance of the diversity concerning economic adaptations of the population is recognized (Allison and Ellis 2001).

As described before, the bargaining power of the fishers remains at the center of the problem. It works as a marginalization mechanism where fisheries profits are transferred from fishers into the hands of middlepersons. For example, during Easter Holiday, fish prices peak in Guatemala. Middlepersons are able to stock dried salted fish months in advance. During the last months of the year, snook is more abundant. This lowers the price that can be obtained by fishers, who must sell as they lack storage capacity; or because capital is needed to continue fishing. In this situation, the middlepersons greatly increase their profits. Through an international project, a processing plant in the town of Livingston was constructed, the main goal of which was to secure minimum prices for fishers. However after completion, it is not working. According to Hidalgo and Mendez (2007), the explanation is the lack of starting capital among the fishers. Some informants mentioned that the problem is linked to difficulties while negotiating how profit shares should be distributed among fishers' associations. This failure demonstrates how the incipient bottom-up fisheries management institution would greatly benefit from governmental and/or NGO support. Capacity building is needed in areas such as business administration, accounting, and marketing among others, to further empower fishers in their pursuit for a co-management regime.

19.5 The Need for Fisheries Law Enforcement

Our informants explained poor fisheries law compliance with regard to the monitoring, control, and surveillance (MCS) system that should be carried out periodically, to enforce regulations in the Amatique Bay. According to the fishers, MCS activities should be the responsibility of the government and/or environmental NGOs located in Livingston. Such claims are based on previous experiences when governmental authorities decided to implement a closed season and then enforced it, mainly through the Coast Guard. According to governance theory, three forms of governing systems exist: interference, intervention, and interplay (Kooiman 2003; Jentoft 2006). During the recent history of the Livingston fisheries, all of these forms of governing systems are found. They have changed, however, from interference to intervention and now to interplay.

The government made and enforced the rules as a Leviathan; the government acted as a mediator participating when conflicts arose between groups. Now the government seems to prefer to consult fisher groups in some of the decision-making processes; for example, concerning the timing of the closed season. During the interviews with the fishers, however, we noted a general confusion about who specifically should perform MCS activities. Fishers pointed to environmental NGOs, the police, the government fisheries office, and the Coast Guard as responsible for monitoring the area. But, regardless of which of these institutions are responsible, there are not enough effective controls. For example, during the closed season, informants from all the fisher groups reported that they continue to fish, even if they acknowledge that this practice contributes to overfishing and declining incomes. Some fishers want to assume MCS responsibilities themselves, but there are gaps in the law, i.e., a civilian cannot "arrest" another civilian, impose fees, seize a boat, or confiscate fish.

19.5.1 Gentlemen's Pact Versus Closed Seasons

The fact that the fishers to some extent seem to have abandoned the "Gentlemen's Pact" have other explanations. According to Jentoft (2000), in order for a management system to be considered legitimate, it must build on local moral principles and values. Gill netters and trawl owners have reached agreement about the best rational, reasonable, and fair approach to resolve area use conflicts. Both groups recognize each other's rights to fish, but without causing direct harm to each other's interests, i.e., destroying gear. Fishers agree that area delimitation has been successful in significantly reducing confrontations, although they still occasionally occur. Trawler crews are paid either a percentage of the day's catch or a fixed salary per fishing trip. Regardless of payment type, crews are depending on boat landings for income.

If the fishing trip does not cover operational costs excluding salaries, i.e., fuel and provisions for the crew, fishers earnings based on percentage of catch value will not get paid. Moreover, if the trip results in losses, they are accumulated, to be discounted after the next profitable fishing trip. The owner can also decide not to send the boat out if he or she believes that the next fishing trip will be unprofitable. As crews want to secure a salary, they sometimes decide to fish illegally in the zone reserved for the gill netters, hoping for a profitable shrimp harvest. If successful, the owner will continue to send the boat to fish, maybe without knowing that the crew is breaking the "Gentlemen's Pact" rules.

The difference between the relative degree of compliance regarding the "Gentlemen's Pact" (high) and the closed seasons (low) can be explained by the immediate outcomes of infringement of the law, but also how the rules are legitimized among the fishers themselves. Destroying a gill net is affecting a particular person, most likely an acquaintance or even a family member. The person affected by having his gear destroyed suffers considerable economic loss and must buy new expensive equipment to continue fishing. In addition, trawl owners and gill netters have agreed upon the zone delimitation. In this way, regulations are at least partially legitimized. In contrast, by not respecting closed seasons, the cost of rule infringement is not personalized, as in the former case. Fishers seem to be engaged in by-passing a system that is perceived to be less legitimized. For some people fishing with "chango," it is only during the closed seasons, that shrimp harvest is profitable because no trawlers are operating in the area. Therefore, it becomes justified and "fair" to fish.

A similar case occurs with people living in Creek Chino, a village located near Livingston but closer to Rio Dulce, a nearby river with outlet into the Amatique Bay. According to some informants fishing with "chango," shrimps are found in this area only 1 month a year, coinciding with the closed season timing. They believe, "trawl owners are the ones making the law." They do not agree with the no-take timing, as entire families participate in harvesting shrimp and depend on the incomes. This shows that closed seasons are not supported by all the fishers groups.

Apart from the ideas and perceptions of what is moral and fair, Berkes et al. (2001) summarize other factors determining level of compliance: Potential illegal gains; and the severity and certainty of sanctions. This kind of analysis rests on a "rationalist" model (Hauck 2008). If sanctions are higher than gains, compliance increases. When asked how to improve resource abundance, many fishers referred to how the Belizean MCS system works. They believe fish abundance is higher in Belizean waters because among other factors, the Coast Guard patrols regularly. People fishing illegally are sentenced to jail, while boats and fishing gear are confiscated. These severe sanctions on law infringements are lacking in the Amatique Bay because few effective controls are organized by the authorities. Bennett et al. (2001) recognizes that without the support from the Government (policy makers and resource managers), and State institutions (law enforcement, stable markets, and clear political process), sustainable management is not possible regardless of the support from local institutions. In our opinion, the need to apply a management system with an MCS protocol in the Amatique Bay, effectively supported by governmental institutions, is indeed unquestionable.

In the case of the "Gentlemen's Pact," it can be regarded as an important step to co-management. It may be interpreted as part of an empowering process, where user groups participate in regulatory decision-making (Jentoft 2000, 2005). The Pact is an example of how institutions related to management can transform the ways in which people access, use, and derive well-being from natural resources (Leach et al. 1999). Co-management is expected to improve fisheries management in two ways: by incorporating local knowledge into fishery science; and by enhancing law compliance among resource users when the arrangements are seen as legitimate by the actors involved (Jentoft et al. 1998). Due to the fact that there are ongoing consultations between government and fishers regarding local regulations, the experience from the "Gentlemen's Pact" and the formation of bottom-up institutions, a formal co-management regime could be well suited in the Amatique Bay fisheries. There are definitely challenges to overcome. Ethnicity and differing world views, for instance, can play an important role in determining the level of success. Groups may associate attributes and values to nature which are, in principle, antagonistic.

For example, while the anchovy fishery is justified by one fisher group, others regard it as the cause of declining landings for several other species of fish, as explained before. When this occurs, conflicts may emerge. In Livingston, people with the same ethnic background tend to group together, articulating differences that distinguish them from the others (Kahn 2001, 2006; Jacobsen 2007). This could present a challenge, if the aim is to incorporate bottom-up institutions to regulate fisheries. Again, co-management can be effective as an institution bringing together affected groups. For some fishers, a successful management plan must be based both on scientific knowledge as well as traditional knowledge. Many stated, for example, that for some species the closed seasons are set without any biological background; i.e., they are set in times of low species abundance instead of aiming to protect spawning seasons. Scientific research is needed in such basic aspects of fisheries management. NGOs and universities, together with fishers groups could be engaged in such research.

19.6 Conclusions

Processes both intrinsic and extrinsic to the fishery are contributing to poverty among fishers in the town of Livingston. Both the scarcity of available income-generating opportunities, and the labor migration from adjacent areas, are the main factors explaining the increased number of active fishers. Because fisheries remain a semi-open field, we find a tendency toward Malthusian overfishing and resource degradation through overfishing and unsustainable fishing practices. Agreed management regulations are not fully respected because either there is a poor MCS system capable of punishing free riders; or the regulations have not obtained sufficient support and legitimacy by all resource users. A formal co-management regime, including governmental support to assist in local MCS activities while following up the fish processing plant to tackle the lack of bargaining power should be introduced.

Ethnic groups display different coping strategies to overcome poverty. They try to diversify their economic activities, but due to the few opportunities available, many have to migrate abroad. Although ethnically differentiated, Livingston fishers are mainly recruited from the area, the town of Livingston, or adjacent neighborhoods. We have not identified external investors allocating capital to maximize returns. There exists a community bond in the town in the sense that the whole population is partially dependant on the fisheries. The owner of a store and another running a pharmacy affirmed that during the closed season, sales decline and people cannot afford medicines. Thus, what happens in the fishery affects the town of Livingston in general. None of the informants talked about removing existing fishers from the activity as a solution to overcome overexploitation. They know that the lack of jobs is a problem faced by everybody, and fisheries work as a "safety valve," providing food and income for the population. The key issue is then to improve conditions outside the fisheries, while implementing a viable system of fisheries management to overcome both resource degradation and poverty.

Acknowledgments We would like to thank all the people from Livingston who participated in this work by sharing information with us on their daily lives, including what it is like to be a fisher. As we promised in the interviews, no names are revealed, so therefore we cannot extend more personalized gratitudes. With this work, we hope to contribute to the quest to improve Livingston and fishers' livelihoods. We also extend sincere gratitude to the Norwegian Research Council for awarding the grant for the PovFish project, and making this research possible.

References

Adams R H Jr (2004) Remittances and poverty in Guatemala, World Bank Policy Research Working Paper, Series 3418

Allison EH, Ellis F (2001) The livelihoods approach and management of small-scale fisheries. Mar Policy 25:377–388

Andrade H (2003) Age determination in the snapper *Lutjanus guttatus* (Pisces, Lutjanidae) and investigation of fishery management strategies in the Pacific Coast of Guatemala, Unpublished master's thesis, University of Tromsø, Tromsø

Arrivillaga-Cortés A (2009) Informe de Consultoría: Diagnóstico situacional de la población Garífuna emigrante. Comisión Presidencial contra la Discriminación y el Racismo, Guatemala

Begg D, Fischer S, Dornbusch R (1997) Economics. McGraw-Hill, London

Béné C (2003) When fishery rhymes with poverty: a first step beyond the old paradigm on poverty in small-scale fisheries. World Dev 31:949–975

Béné C, Hersoug B, Allison EA (2010) Not by the rent alone: analysing the pro-poor functions of small-scale fisheries in developing countries. Dev Policy Rev 28(3):325–358

Bennett E, Neiland A, Anang E, Bannerman P, Rahman AA, Huq S et al (2001) Towards a better understanding of conflict management in tropical fisheries: evidence from Ghana, Bangladesh and the Caribbean. Mar Policy 25:365–376

Berkes F, Mahon R, McConney P, Pollnac RB, Pomeroy R (2001) Managing small-scale fisheries: alternative directions and methods. International Development Research Centre, Ottawa

Charles AT (2001) Sustainable fishery systems. Blackwell Science, Oxford

Clardy TR (2008) Spatial and temporal variability in the relative contribution of king mackerel (*Scomberomorus cavalla*) stocks to winter mixed fisheries off South Florida. Fish Bull 106:152–160

FAO (2001) The state of food insecurity in the world 2001: food insecurity: when people live with hunger and fear starvation. Food and Agriculture Organization of the United Nations, Rome

Gauster S, Isakson SR (2007) Eliminating market distortions, perpetuating rural inequality: an evaluation of market-assisted land reform in Guatemala. Third World Q 28:1519–1536

Giannini V (2006) Diagnostic of community marine reserve models and methodologies for application in the Mesoamerican Reef Region. Fondo para el sistema arrecifal mesoamericano MARFUND. http://www.marfund.org/diagnosticofcommunitymarinereservemodels.pdf. Accessed 22 May 2009

Giudicelli M (1971) Exploratory and simulated commercial fishing operations in the Western Caribbean Sea, R/V Canopus, May–Nov 1970. CEN/REG/30 Central American Fishery Development Project CCDP-FAO-PNUD. Boletín Técnico 5

Godcharles M, Murphy M (1986) Species profiles: life histories and environmental requirements of coastal fishes and invertebrates (south Florida) – King mackerel and Spanish mackerel. U.S. Fish Wildl Serv Biol Rep 82(11.58). U.S. Army Corps of Engineers, TR EL-82-4. http://www.nwrc.usgs.gov/wdb/pub/species_profiles/82_11-058.pdf. Accessed 22 May 2009

González NL (1995) Los Garifunas. In: Zilbermann de Luján C (ed) Historia General de Guatemala. vol III. Siglo XVIII hasta la independencia. Asociación de Amigos del País, Guatemala, pp 399–406

Grobakken I A (2005) Land reform, poverty and empowerment of women in Guatemala, unpublished master's thesis, University of Oslo, Norway

Hauck M (2008) Rethinking small-scale fisheries compliance. Mar Policy 32:635–642

Heyman W, Graham R (eds) (2000) La Voz de los Pescadores de la Costa Atlántica de Guatemala. FUNDAECO- TIDE, Guatemala

Hidalgo H (2004) Informe final proyecto Arrecifes Garífunas: Caracterización de la pesca ancestral del pueblo Garífuna en Arrecifes de la Bahía de Amatique (Guatemala), Geoposicionamiento y evaluación ecológica de los mismos, propuestas técnicas para su protección y manejo sostenible por parte de la comunidad. English translation: Garifuna Reef project, final report: characterization of ancestral Garifuna fishing in the reefs of the Amatique Bay (Guatemala),

Geopositioning and ecological assessment, technical proposals for their protection and community sustainable management. FUNDAECO-COSTAS, Guatemala. Available under request to fundaeco@quetzal.net
Hidalgo H, Mendez A (2007) Diagnóstico organizacional y de necesidades en las comunidades pesqueras en el sistema arrecifal mesoamericano. Fondo para el sistema arrecifal mesoamericano MARFUND. http://www.marfund.org/diagnosticoorganizacionalydenecesidadesenlascomunidadespesqueras.pdf. Accessed 22 May 2009
ICSED (2000) Desarrollo de herramientas económicas para la preparación de políticas sostenibles en el sector pesquero del Golfo de Honduras Tomo I: Caracterización de las pesquerías de camarón y langosta, aproximación conceptual para la gestión pesquera. PROARCA/CAPAS. http://pdf.usaid.gov/pdf_docs/PNACK091.pdf. Accessed 22 May 2009
INE (2002) Reportes dinámicos censo 2002. http://www.ine.gob.gt/index.php/pxwebcenso2002. Accessed 25 May 2009
Jacobsen C (2007) Garífuna i Guatemala: resignasjon, frykt og avsky i et nominelt demokrati, unpublished master's thesis, University of Oslo, Norway
Jentoft S (2000) Legitimacy and disappointment in fisheries management. Mar Policy 24:141–148
Jentoft S (2005) Fisheries co-management as empowerment. Mar Policy 29:1–7
Jentoft S (2006) Limits of governability: institutional implications for fisheries and coastal governance. Mar Policy 31:360–370
Jentoft S, McCay BJ, Wilson DC (1998) Social theory and fisheries co-management. Mar Policy 22:423–436
Kahn HE (2001) Respecting relationships and Día de Guadalupe: Q'eqchi' Mayan identities in Livingston, Guatemala. J Latin Am Anthropol 6(1):2–29
Kahn HE (2006) Seeing and being seen: the Q'eqchi' Maya in Livingston, Guatemala, and beyond. University of Texas Press, Texas
Kooiman J (2003) Governing as governance. Sage, London
Leach M, Mearns R, Scoones I (1999) Environmental entitlements: dynamics and institutions in community-based natural resource management. World Dev 27:225–247
MAGA (2005) Reglamento de la ley general de pesca y acuicultura 2005. Acuerdo gubernativo 223-2005. MAGA, Guatemala
Manjarres L, Duarte LO, Altamar J, Escobar F, Garcia C, Cuello F (2008) Effects of using bycatch reduction devices on the Colombian Caribbean Sea shrimp fishery. Cienc Marinas 34:223–238
Manz B (1995) Indians, land, and poverty in Guatemala. In: Shields D (ed) The color of hunger: race and hunger in national and international perspective. Rowman & Littlefield, Boston, p 107
Midré G (2005) Opresión, espacio para actuar y conciencia critica. Lideres indígenas y percepción de la pobreza en Guatemala. F&G Editores, Guatemala
Milton DA, Die D, Tenakanai C, Swales S (1998) Selectivity for barramundi (*Lates calcarifer*) in the Fly River, Papua New Guinea: implications for managing gill-net fisheries on protandrous fishes. Mar Freshwater Res 49:499–506
Pauly D (1994) From growth to Malthusian overfishing: stages of fisheries misuse. South Pacific Commission Newsletter of the Special Interest Group on Traditional Marine Resource Management and Knowledge 3:7–14
Pauly D, Silvestre G, Smith IR (1989) On development, fisheries and dynamite, a brief review of tropical fisheries management. Nat Resour Model 3(3):307–329
Pauly D, Christensen V, Dalsgaard J, Froese R, Torres F Jr (1998) Fishing down marine food webs. Science 279:860–863
Porta E, Laguna JR (2007) Present state of education for all: the case of Guatemala. Paper commissioned for the EFA Global Monitoring report 2008, education for all by 2015: will we make it? UNESCO. http://unesdoc.unesco.org/images/0015/001555/155575e.pdf. Accessed 12 May 2009
Sadovy Y, Min L (2008) Functional hermaphroditism in teleosts. Fish Fish 9:1–43
Schmidt DJ, Collins MR, Wyanski DM (1993) Age, growth, maturity, and spawning of Spanish mackerel, (*Scomberomorus-maculatus*) (mitchill), from the Atlantic coast of the southeastern United States. Fish Bull 91:526–533

Sutter FC, Williams RO, Godcharles MF (1991) Movement patterns and stock affinities of king mackerel in the southeastern United States. Fish Bull 89:315–324

Taylor RG, Whittington JA, Grier HJ, Crabtree RE (2000) Age, growth, maturation, and protandric sex reversal in common snook (*Centropomus undecimalis*), from the east and west coasts of South Florida. Fish Bull 98:612–624

UN-Instraw (2007) Cuadernos de trabajo sobre migración: Encuesta sobre remesas 2007; perspectiva de género. http://www.un-instraw.org/jdata/images/files/ Cuaderno%20de%20Trabajo%20 No_%2024.pdf. Accessed 12 May 2009

Walmsley S, Purvis J, Ninnes C (2006) The role of small-scale fisheries management in the poverty reduction strategies in the Western Indian Ocean region. Ocean Coast Manage 49:812–833

World Bank (2003) Poverty in Guatemala. The World Bank, Washington, DC

Yañez-Arancibia A, Zárate-Lomelí D,Terán-Cuevas A (1994) Evaluation of the coastal and marine resources of the Atlantic Coast of Guatemala. EPOMEX. CEP Technical report No. 34. UNEP Caribbean Environment Programme, Kingston, Jamaica, 64 pp. http://www.cep.unep.org/publications-and-resources/technical-reports/tr34en.pdf. Accessed 29 May 2009

Yañez-Arancibia A, Lomeli DZ, Cruz MG, Orantes RG, Fandino VS (1999) The ecosystem framework for planning and management the Atlantic coast of Guatemala. Ocean Coast Manage 42:283–317

Ybarra M (2008) Violent visions of an ownership society: the land administration project in Petén, Guatemala. Land Use Policy 26:44–54

Part V
Imagining

Chapter 20
A Better Future: Prospects for Small-Scale Fishing People

Svein Jentoft, Arne Eide, Maarten Bavinck, Ratana Chuenpagdee, and Jesper Raakjær

> *The call to action and change is compelling. It is to define development as equitable wellbeing for all, to put the bottom poor high on the agenda, to recognize power as a central issue, and to give voice and priority to poor people. It is to enable poor women and men to achieve what they perceive as a better life. These basics underpin efforts to transform the conditions poor people experience, empowering them with freedom to choose and act.*
>
> Source: A call to action: The challenge to change (Chapter 12 in Voices of the Poor: Crying Out for Change)

Abstract Before one can begin to create a better future for small-scale fisheries and those who depend on them, one would first need to *imagine* it. What scenarios are likely and which are preferable to others? One would also need to think about how to get from where small-scale fisheries are now, to where we want them to be. What governance initiatives would be needed? What is more urgent? What should happen first? This final chapter synthesizes the lessons for policy and governance that can be drawn from the case studies that have been presented in this volume. All PovFish participants were invited to submit their own views about what the key messages from their contributions are and what others should learn from them. The chapter builds on their ideas and propositions, and includes excerpts from what they formulated. It also brings back some of the theoretical issues that were discussed in Part 1 - *Positioning*.

S. Jentoft (✉) • A. Eide
Norwegian College of Fishery Science, University of Tromsø, 9037 Tromsø, Norway
e-mail: svein.jentoft@uit.no; arne.eide@uit.no

M. Bavinck
Department of Human Geography, Planning, and International Development Studies, University of Amsterdam, Nieuwe Prinsengracht 130, 1018VZ Amsterdam, The Netherlands
e-mail: j.m.bavinck@uva.nl

S. Jentoft and A. Eide (eds.), *Poverty Mosaics: Realities and Prospects in Small-Scale Fisheries*, DOI 10.1007/978-94-007-1582-0_20,

20.1 Introduction

Small-scale fisheries are too big to ignore. They employ millions of people around the world, and are major providers of food to a growing human population. They play a significant role in alleviating global poverty. Still, small-scale fisheries harbor a lot of poor people whose livelihoods are less than secure. This situation calls for bold governance initiatives. Small-scale fisheries must be elevated on the political agenda. But before we can begin to create a better future for those who depend on them, we need to think about what small-scale fisheries can possibly be and what kind of future they can offer. We also need to be creative about policies and interventions.

This volume offers a wide range of ideas of what can be done. Many come directly from the poor themselves. The case studies present the voices of people who live the small-scale fishing life. In interviews, they talked about the problems they face and the things that are holding them back. They also spoke about what makes them satisfied and happy, and what their aspirations are. They do not believe that change is easy or even very likely. But they have made criticisms of and demands on government, and have suggested what should be done to improve the situation of small-scale fisheries.

Before we can begin to create a better future for small-scale fisheries and those who depend on them, we first need to imagine it. What scenarios are likely and which are preferable to others? We also need to think about how to move small-scale fisheries from where they are now to where we want them to be. What governance initiatives are needed? What is most urgent? What needs to happen first? What are the obstacles and governability limitations? This final chapter synthesizes the lessons for policy and governance that have been drawn from the case studies presented in this volume. All PovFish participants were invited to submit their views about the key messages from their contributions, and what others should learn from them. This chapter builds on their ideas and propositions, and includes excerpts from the responses that they formulated. It also brings us back to some of the theoretical issues that were discussed in Part I.

20.2 Fisheries Development as Freedom

Marloes Kraan states that small-scale fishing can also take place on a large scale (as in Ghana, Chap. 8), and thereby has large-scale implications for – amongst others – fish stocks. In some parts of the world, as illustrated by Bavinck (India, Chap. 9), the number

R. Chuenpagdee
International Coastal Network, Department of Geography, Memorial University of Newfoundland, St. John's, NL A1B 3X9, Canada
e-mail: ratanac@mun.ca

J. Raakjær
Innovative Fisheries Management (IFM), Department of Development and Planning, Aalborg University, 9220 Aalborg Ø, Denmark
e-mail: jr@ifm.aau.dk

of people employed in small-scale fisheries has increased dramatically in recent decades, and the fishing pressure that this sector is now dealing with cannot be overlooked.

Rather than assuming that freedom of the commons causes excessive fishing strain, we argue for critical thinking about what freedom in the fishery commons means and can possibly become in the future of small-scale fisheries. It cannot mean the freedom to ruin the resource, as Hardin (1968), Graham (1939) and others warned against. We suggest, for instance, that we should also consult Amartya Sen (2000, p. 10) who argues that freedom is "not only the primary ends of development, it is also among its principal means." Small-scale fishers may be poor for other reasons than the freedom to overexploit the resource, as Ståle Knudsen holds in his proposition. Rather, in many instances described in this book, it is the freedom of the commons that allows small-scale fisheries to avoid poverty, be it relative or absolute, and it is poverty that drives people to sometimes fish beyond the limits that the resource can sustain.

Notably, the freedom Sen (2000) talks about is not a freedom to overexploit and destroy but a freedom to pursue a better life, built on secure entitlements, proficient capabilities, and social justice that enables people to be resourceful, autonomous, and creative in forming their own institutions. These freedoms are also emphasized by the people interviewed by the PovFish research team members. It is also the freedom that local people need to be more effective stewards of their common resources. This leads Paul Onyango (Tanzania) to conclude that "expansion of freedoms and capabilities of the poor should ideally be the foundation of poverty alleviation."

The small-scale fishers interviewed for this volume all talk about protective security, availability of health services, and education for themselves and their children. They also talk about their need for credit so that they can invest in means of production, and for improved market access that allows them to sell their produce. Many also mention political participation and freedom to organize so that they are collectively capable of managing their own affairs, including their commons and communities. It is true, as Kraan mentions, that small-scale fisheries may have the capacity to overexploit and that growth may therefore need to be kept in check. But such control can be well exercised in the form of "mutual coercion, mutually agreed upon," as Hardin (1968, p. 1247) puts it. This is also the principle of co-management (Jentoft 1989), which several of the PovFish team members recommend for their area.

20.3 Aspirations for a Better Future

What fishing people hope for themselves, their families, and communities is discussed in many of the chapters within this volume. Like anyone else, they want to be more secure, healthy and in a better financial situation. There is nothing worse than not being able to send your children to school or your spouse to a doctor. Small-scale fishing is also hard work, with long hours, and in conditions that are not always safe. In some cases, small-scale fishers are victims of abuse, either from government officials, middlepersons, or others who do not refrain from exploiting their weaker position.

Despite such hardships, small-scale fisheries also come with many rewards; food on the table, a relatively reasonable income, a life together with family and friends in familiar settings. Unlike large-scale fisheries, small-scale fishing allows fishers to be self-employed and to be their own manager. Small-scale fishers appreciate the *positive* freedom that comes with this occupation – the freedom *to* organize their day and pursue their life's meaning. But small-scale fishers interviewed by PovFish team members are also worried about their *negative* freedom, the lack of protective security – the freedom *from* the things that keep them entrapped in poverty. This distinction between negative and positive freedom was made famous by the philosopher Isaiah Berlin (1969).

It is for these reasons that Onyango and Kraan take issue with an often expressed view that small-scale fisheries is an occupation of last resort, and that fishers would prefer to leave if there was an opportunity to do so. Rather they emphasize that small-scale fisheries are, for many, a way of life, and an occupation that comes with a lifestyle that they value and which they were brought up with. In the case of the Anlo-Ewe fishers of Ghana, according to Kraan, fishing "is a thriving self-managed sector" and "has been so for centuries." Both Kraan and Onyango argue that disregarding or disrespecting the deep attachment that fishing people have to their way of life may not only bring harm to them but that it also may reduce the effectiveness of resource management because people will rebel against it. The government's subsequent response may easily worsen the situation and turn non-compliance and enforcement into a vicious cycle.

Boulding (1977, pp. 286–287) argues that "identity is a very powerful source of decisional behavior. People make decisions with regard to their images of the future and the consequences that they anticipate that their decisions have." But they also do so "in accordance with their value structure… which depends very much on the nature of the individual image of identity." Given people's often strong identification with small-scale fisheries, Kraan holds that "policies aiming at offering alternative livelihoods to fishers may prove to be unsuccessful." She thinks that it is better to create supplementary jobs that can be combined with fishing, as people can benefit from increasing their sources of livelihood.

Indeed, this is also how small-scale fishing people survive in many instances, like in the Pearl Lagoon area of Nicaragua, described by Miguel González (Chap. 13). Rather than having all their eggs in one basket, they need to spread them in several. People are able to survive because they rely on multiple sources of food and income. But for this, people need secure access not only to fishing, but also to land and forest; and they need to have access to markets for the produce that they do not consume in their household and local community.

20.4 Well-being as a Priority

The freedom that comes with being a small-scale fisher is something that cannot be taken for granted, as it is often taken away by governments that do not believe in the freedom of the commons. A problem with policy prescription that restricts fishers' ability to cope with their situation is that it easily exacerbates the deprivation of those who are already poor, vulnerable, and without other livelihood alternatives. In poorly developed small-scale fisheries, where the level of technology is low, access to investment capital is difficult, and the simplest infrastructure is inadequate or non-existent, removing freedom of access to the fishery commons is tantamount to confiscating the only entitlement that poor people have to sustain themselves (cf. Béné et al. 2010). In such a case, poor people will not remain passive.

Rather, as is the situation in South Africa, described by Moenieba Isaacs (Chap. 16), they sometimes resort to unlawful practices. Poor small-scale fishers will, as Isaacs argues, employ what James C. Scott (1985) originally coined the "the weapons of the weak"; in this case fishing illegally. Similar situations also occur in other countries discussed in this volume, like Vietnam and Bangladesh. The reason for this is not only that they need to fish in order to feed themselves and their families, but in many instances they fish using traditional methods and gears as they have always done. When it is not clear to them why governments ban certain fishing practices, they are left thinking that they are being treated unfairly. This illustrates that illegal fishing practices have complex motivations and are, as Hauck (2008) submits, often as much about social justice as about criminal justice.

Writing from Bangladesh, Mohammad Mahmudul Islam thinks that providing credit to fishers should be a priority. People need credit to get back on their feet when their assets are wiped out after a cyclone. Without credit, they cannot diversify their livelihood base. Similar to Onyango (Tanzania), and Kim Anh Nguyen and Ola Flaaten (Vietnam), Islam also thinks that policies should be directed toward maintaining (and expanding) the working conditions and physical security of fishers, as they are often at risk when out at sea, for instance because of piracy. He argues that fishing people also need education and healthcare to be productive, as illness in the family, accidents, or loss of gear may easily jeopardize livelihoods. Mafaniso Hara (Malawi) argues that "women and their children are particularly vulnerable when their husband (the breadwinner) dies as this can mean sudden and abrupt loss of income and its source." Another problem in Malawi is that heritage law systematically disfavors women.

For this reason, Islam states that "poverty alleviation strategies and policies for fisheries communities should not only target men, but also aim at uplifting the women and the contributions they make by focusing on female education." Thus, as was also stressed in Chap. 4 by Svein Jentoft and Georges Midré, it is important to recognize that poverty is also gendered. Women's well-being and action space in fisheries are equally essential. Therefore, fisheries development aimed at poverty alleviation must also include policies focused on development *of* and *for* women. Alternative and supplementary job creation must be for women and for the household, and not just for the men who most often draw the fish out of the water. A gendered perspective on fisheries development and poverty alleviation necessarily involves a fisheries chain analysis, as argued by Chuenpagdee and Jentoft in Chap. 3.

20.5 Secure Rights Are Essential

Fisheries development and poverty alleviation must also include rights. Rights to nature can also be seen as innate. For people who are poor, to have access to the resources that they depend on for their food security and livelihood is a matter of survival. People should not be excluded from harvesting the natural resources they need to feed themselves – in this case those of the fishery commons – but they can be helped to do it in a way that is sustainable and leaves room for the next generation. In some instances, they need to be presented with alternatives to current practices so that they can expand their freedom of choice.

This is as much a human rights issue as it is a property rights issue. Human rights in recent years have become more accentuated in fisheries management, particularly with regard to indigenous peoples' fishing rights. In fisheries, human rights and property rights have usually been treated as if they belong to separate spheres; the former as a basic right of people not to be excluded or discriminated against; the latter as a regulatory tool which does exactly that, since property rights always involve exclusion following a demarcation between the "haves" and "have-nots." In resource management, such exclusion tends to be regarded as inevitable collateral damage – unfortunate perhaps, but still justified by the life-boat ethics that Hardin (1977) represents. This conflict of human rights and fisheries rights has been brought into sharper focus since the 2008 Global Conference on Small-Scale Fisheries, organized by the Food and Agriculture Organization of the United Nations (FAO 2008).

González (Nicaragua) emphasizes the need for resource rights for small-scale fishing people that involves more than fisheries. "Land and aquatic rights in small-scale fisheries should be conceptualized in a holistic way so as to better tackle communities' needs in securing a sustainable resource base." Rights-based systems come in different forms, and some are more sensitive to small-scale fisheries, local communities, and human rights than others. Vesting resource rights in communities, as in the case of Nicaragua, is a way of securing livelihoods while at the same time enabling communities to be better stewards of their natural resources and ecosystems. In Nicaragua, the concept of communal rights has both legal and popular backing in domestic and international law pertaining to indigenous peoples. Globally, however, fishing rights conferred to local communities are still an exception (Kurien and Willmann 2009). Instead, the situation is more like that in South Africa where, according to Isaacs, those who are already privileged with individual resource rights are generally opposed to the idea that rights should be vested in legal community entities, as advocated in the new small-scale fisheries policy.

20.6 Dealing with Vulnerability

Even if small-scale fishers are not always the poorest of the poor, they are often vulernerable and therefore at risk of slipping (back) into poverty. To borrow the title of Anirudh Krishna's recent (2010) book, they are in many instances just "one illness away" from poverty. The vulnerability of small-scale fisheries is close to the hearts of several members of the PovFish research team. Based on their Mexican experience (Chap. 10), Silvia Salas and Maiken Bjørkan state:

> While dealing with poverty, many countries have tried to orient their policies toward changing some marginality indicators, without addressing crucial issues related to vulnerability, which can also expose people to poverty. Temporary palliative solutions cannot build resilient systems; it is necessary to recognize this condition and identify the necessary steps to reduce vulnerability and improve population welfare.

Drawing from their Sri Lankan case study (Chap. 17), Oscar Amarasinghe and Maarten Bavinck point out that "the relationship between vulnerability and poverty in fisheries goes both ways, thus forming a vicious circle. To break this circle, the fishers should be provided with diverse livelihood capital to improve their resilience capacity." These authors also focus on the potential stewardship role of cooperatives: "More attention on resource governance is required from their leaders, if the cooperatives are to be successful in the long run." Islam (Bangladesh) is similarly adamant about the functions of cooperative organizations as safeguards against crises. In his opinion, they are an obvious measure that can be established to reduce vulnerability. Chuenpagdee and Juntarashote share this faith: "The examples of the marketing cooperative in Thailand and the good relationship with fishmongers and middlepersons can be replicated elsewhere."

Small-scale fishers often find themselves in fierce competition with other resource users. They are not necessarily poor because they overfish, but because others do so. They are sometimes pushed aside and replaced by other resource users, as when industrial vessels encroach on their fishing grounds, when marine protected areas are introduced, or when tourist developers occupy their beaches. In some instances, they become impoverished because they do not get a fair share of the value of their produce due to their weak bargaining power with middlepersons. Poverty may therefore be alleviated if policies help to spread opportunities and incomes more equitably.

As recognized by Eide et al. in Chap. 2, and by Maarten Bavinck in Chap. 9 about India, the fishing industry is also a place where fortunes are made. Therefore, one cannot consider poverty in fisheries without also considering the wealth that they have created during the industrialization of the twentieth century, and the way it is distributed. Indeed, small-scale fishing people may be poor even if their countries are not, as income generated at the national or industry level, through international trade or foreign assistance, does not always "trickle down" (as Jawaharlal Nehu, the first Prime Minister of India coined it) to local communities (cf. Arndt 1983; Hersoug 2004; Béné et al. 2010). Governance systems that do not grant small-scale fishers the power they need to withstand exploitation, and mechanisms that are not transparent or honest, exacerbate this problem (Robbins 2000; Sumaila and Jacquet 2008).

As PovFish researchers observe in this volume, such vulnerabilities are often due to poor organization and weak institutions that leave small-scale fishing people without bargaining power or insurance mechanisms and, thus, defenseless. Cooperatives may help counteract this predicament, as they did in Norway in the 1930s when a law granted fishers' sales organizations monopoly rights and the authority to fix minimum prices on their produce (see Jentoft and Midré, Chap. 4). Andrade and Midré (Guatemala) argue the same: "Strong fishers' organizations could also be the first step to increase bargaining power." But how such an organization is formed is also important.

Drawing from the experience of the Indo/Norwegian fisheries development project in Kerala that began in the early 1950s, Kurien (1985) points out a major difference between forming such organizations *by* and *for* fishers. Cooperatives have a better prospect of success if they are generated from within the community than when introduced and established from the outside, because they need local support and loyalty. This, however, does not take away the fact that community organizations, including cooperatives, need external support, such as enabling legislation that permits them to enforce management decisions that restrict the freedom of members to overexploit and unites them around a common goal.

20.7 Learning from the Poor

Policymakers and fisheries managers are obliged to understand what small-scale fisheries mean to people, what their lives are all about, what choices they have, and how they would like to improve. This argument is also advanced by Narayan and co-authors (2000). Policymakers should not assume that a fishery not managed by the government is not managed at all. Rather they should clarify and learn from how small-scale fishing people previously or currently are managing their livelihoods and resource use based on how they know their ecological and social systems (Berkes et al. 2001). What precautionary principles and ideas of justice do people abide by? What do people think is necessary in order to be ecologically, economically and socially safe?

The social and ethical values that people ascribe to, form a basis for their fishing and management practices. Ignoring them may easily lead to failure, because people will be unable or unwilling to relate to government initiatives. As also stated by Onyango, fisheries management and poverty alleviation initiatives can only be significant "if they are built upon the meaning and value that fishers attach to the fisheries." Too often, fisheries management and development reforms start without such a deep understanding. In many instances, it is a preconceived idea of the solution that defines the problem, as when policymakers, managers, or NGOs operate as champions of certain management tools derived from international environmental and governance discourse (Degnbol et al. 2006). With globalization, this tendency has increased in recent years. The consequence is that particular concepts and models are imposed on people from the outside, leading to institutional misfit, which in the past has often been the situation with fisheries cooperatives (Jentoft 1986) and is now being repeated with quota systems and marine protected areas (Degnbol et al. 2006).

After generations of experience from working on the water and from dealing with management authorities, small-scale fishers have their own inherited and self-generated ideas of what constitutes sound management, and what the ecological, social, and institutional conditions are and must be. As several PovFish members emphasize in this volume, in many instances small-scale fishers exercise their own customary rules that government can learn from. People have standards and mechanisms for dealing with user-conflicts, irregular fishing practices, and justice. Talking from a Guatemalan perspective, Andrade and Midré believe that customary rules, be they formal or informal, should be incorporated into the governance toolbox. This idea is supported by Bavinck, as these rules "play an important role in limiting the inflow of labor into fisheries, as well as steering the increase of fishing effort. Policymakers therefore need to pay more attention to the possibilities afforded by customary law to govern fisheries for sustainability." This is particularly the case, Bavinck argues, because, "coastal regions are not equally 'covered' by customary law. To the contrary, there are variations in the substance and effectiveness of customary law along the coasts." This calls for sensitivity but also action in order to "reinforce customary law in districts where it is weak and has potential." In other words, there is also a precautionary principle to abide by for the social system.

González, who speaks from his Nicaraguan case study (Chap. 13), shares this view:

> Securing a resource base for communities that depends on small-scale fisheries requires us to understand the complexities and complementarities of local ecological systems. These systems are very often mediated by community norms that strive for sustainability. For the world of poverty alleviation and vulnerability, it is important to nurture conducive policy and institutional environments sensitive to these practices.

He concludes that "poverty alleviation programs in small-scale fisheries should not be disassociated from strategic empowerment. In the long-term, a successful reduction of poverty might be sustained if the concerned community is able to deploy its full capacities to secure its livelihood."

20.8 Involving People

Small-scale fisheries and the poverty and vulnerability that people experience within them need careful consideration and reflection, based on "thick description" (McCay and Jentoft 2010) of the complexity and diversity of local contexts, just as the PovFish authors have provided. For policymakers and public officials to do so, they also need to move away from the image of the small-scale fisher as a person who is only and fundamentally deprived because he/she is short of economic, social, and cognitive assets. Instead, they must begin to recognize fishers' actual and potential resources and capabilities and therefore facilitate opportunities to extend their freedoms.

Poor people certainly have deficits, but they also have adaptive capacity that can be enhanced. They have learned from what they have experienced, and are able to continue to do so. As Chuenpagdee and Juntarashote state, drawing from their Thai study (Chap. 14), fishers are "the true poverty experts, not so much because of their experience with poverty alone, but more because of their deep understanding about their situation and their own potential to overcome poverty." Indeed, there are important lessons to be learned about how to alleviate poverty from how poor people actually cope with it, how they utilize their resources and mobilize the capabilities that they have, and how they, as Ståle Knudsen points out from his Turkish insights (Chap. 11), in some instance are able to break the trap that they are in and move out of poverty.

It is also for this reason that poverty alleviation, however well-intended and warm-hearted, cannot and should not be imposed on people from the top-down. It matters to people how things come about, whether a particular good is handed to them or created by them (Sen 2009). This is also an issue of human dignity and freedom. Therefore it is essential to encourage their involvement and facilitate their empowerment. Indeed, Knudsen argues that lack of capacity to participate can be considered a dimension of poverty, and that the marginalization of small-scale fishers undermines the capabilities which they need to become involved. In a similar vein, coming from the Polish experience (Chap. 7), Boguslaw Marciniak states that sustainable fisheries development "should include from the very beginning the active participation of fishers and other inhabitants of coastal areas in the management of local resources." This is also about freedom that is strengthened by political democracy and social justice, and which led Amartya Sen (2009) to stress the need for not only getting the institutions right, but also the social and political processes that people are engaged in. The common assumption is that poor people are not interested in democracy, that other issues are more urgent, and that they are not yet ready for it. There is nothing in this volume to suggest that this is the case – quite the contrary, as with the support of co-management among the poor fishers in Mozambique, for instance. Involving the poor in participatory democracy does require capabilities and entitlements (such as institutions) but these are, as Krishna (2008) argues, more about education and information than about wealth per se.

20.9 Going Beyond Fisheries

Fisheries poverty is not always and necessarily only or basically a fisheries problem. Small-scale fishers often find themselves at the receiving end of a string of causes and effects that originate outside fisheries. This requires, as Chuenpagdee and Jentoft argue in Chap. 3, a "chain analysis" of poverty and vulnerability, and hence a broad perspective that emphasizes linkages and systemic relationships and interactions of fisheries (cf. Kooiman et al. 2005). Thus, Knudsen (Turkey, Chap. 11) concludes that the analysis should move beyond the vicious circle of poverty and overfishing.

As also described by Marciniak (Poland, Chap. 7) and González (Nicaragua, Chap.13), small-scale fishers are witnessing the degradation of their estuary, and poverty may well be due to unsustainable fishing practices. But in both instances, there is also a severe soil erosion problem related to pesticide run-off from agriculture and up-stream forestry, which are beyond fishing people's control. González argues that in order to sustain their livelihoods, those who fish the lagoon also need to sustain their land and forests and, to do so, they need to secure cooperation from people who make their living in other sectors and the authorities who oversee them. This requires an integrated governance effort where different industries that impact livelihoods are dealt with simultaneously. On a similar note, Chuenpagdee and Juntarashote (Thailand) state:

> Poverty in fisheries cannot be alleviated only through improving the fisheries situation. It requires a broader perspective that includes looking at improving other aspects of people's livelihoods, such as transportation, education, health, and information technology. Improving these other aspects helps de-marginalize small-scale fishers economically and politically, thus making them less vulnerable.

Menezes, Eide, and Raakjær (Mozambique) also call for an "integrated approach where more attention should be given to infrastructure that would boost development of markets and therefore act as the foundation of wealth creation." They further think that fisheries development must include the "provision of public goods assisted to propel community participation, and leadership at central and local levels (co-management and other associations), assisting the creation of a fisheries community identity." They also emphasize the need for awareness of the fact that "policy instruments for one specific sector do not act in isolation, but are part of an intricate matrix of factors affecting the whole social-political system."

The need for supplementary and alternative sources of livelihoods is a recurrent issue among PovFish team researchers. They emphasize the positive impact on the ecosystem and income security for the poor. Nguyen and Flaaten (Vietnam) observe that: "In general it is hardly possible for people to achieve more in open-access fisheries than in alternative income-creating activities." On a similar note, Hara (Malawi) maintains:

> In situations where fishers desperately need to continue deriving livelihoods from fishing even when a fishery might seem to be degraded because of lack of competitive or comparable economic opportunities outside fishing, they will argue against proposals from Departments of Fisheries for reduction in fishing effort. Policy has to be towards broadening competitive economic opportunities outside fishing, which can attract people away from fishing, if reduction in fishing effort is to be acceptable in fishing communities.

Also Andrade and Midré support this view, and suggest that "policies aiming to reduce fishing pressure are likely to fail regardless of the management regime, unless other income-generating activities are developed." In Amatique Bay in Guatemala (Chap. 19), which they base their lessons on, the fishery provides income and food for people who have been displaced from other productive activities. "Land reform could reduce pressure on fisheries because some of those migrating to fishing communities will find livelihoods in traditional agricultural areas." But, as mentioned already, small-scale fisheries as a way of life often sits deep in people's identity and community heritage. If alternatives are provided, people do not necessarily find them culturally advantageous. If possible, therefore, alternatives should be developed within the small-scale fisheries sector, for instance within post-harvest activities, as pointed out by Chuenpagdee and Juntarashote for Thailand.

20.10 Creating Capable Government

To create employment alternatives as a way to alleviate poverty and reduce fishing effort, Nguyen and Flaaten stress the constructive role of government. This relates to common goods in general, including the provision of essential material, social and institutional infrastructure. But as Menezes, Eide, and Raakjær also argue from the Mozambique experience, "poor design and weak implementation make policies sometimes fail, to the dismay of their well-intentioned authors." Partly, this is an issue of "governability" (see Chuenpagdee and Jentoft, Chap. 3; Kooiman et al. 2005); government agencies do not have sufficient capacity and capability to carry out their role. In many developing countries, governments do not have the research capacity to generate the biological and social data they need to play a credible management role. Resource management, the science underpinning it, and the surveillance mechanisms needed to make enforcement effective, are often beyond their financial means. As Hara claims in the case of Malawi, the Department of Fisheries therefore "has no leg to stand on! (as it were)" when they try to enforce management decisions. The department "is always lagging behind the technical developments in the fishery" and is therefore too late in responding to overfishing.

Other members of the PovFish research team also point to the often inadequate capacity of government. Isaacs (South Africa) holds that government staff is not as well trained as it should be. Management agencies also lack the funding they need to build capacity. Salas and Bjørkan (Mexico) therefore believe that "building capacity within the institutions in charge of fisheries and coastal management is a necessary condition to move ahead with the solution of poverty conditions in coastal areas." But for the effective implementation of management and development policies, there must also be capacity at the receiving end of the governance system – within what governance theory terms the "system-to-be-governed" (Kooiman et al. 2005).

Therefore, as Salas and Bjørkan also argue based on their research experience in Yucatan, Mexico (Chap. 10), "…it is essential to increase fishers' adaptive capacity to deal with the increasingly risky conditions that the fishery sector is facing, especially within the small-scales fisheries." This would certainly make governance efforts less demanding, but it would also require initiatives by the government "to help strengthen the capacity of coastal communities using a long-term perspective" (Salas/Bjørkan). Again in governance parlance: to build adaptive capacity within the system-to-be-governed, it also makes sense to build such capacity with the "governing system" (Kooiman 2008). For government to increase the level of knowledge and education in local communities, it must itself first acquire this knowledge. But González argues:

> Policymakers (and NGOs) involved in designing/proposing small-scale fisheries' poverty reduction programs should be attentive to the great diversity in adaptive capacities deployed by fisher communities (resourcefulness) and individual fisher folks. Adaptive capacities to cope with poverty and vulnerability have forced fishing communities to diversify their sources of income and livelihood sustenance. These practices are not always acknowledged by governments, international funding agencies, and NGOs.

20.11 Building Stronger Communities

González's observation calls for more interactive learning. Building governance capacity works better if government and communities learn from each other. This is also in line with what Armitage and co-authors (2007) want to capture with the concept of "adaptive co-management," which they built on the argument that resource management institutions need to be flexible in order to create resilient social and ecological systems. It is therefore important to work with government and local communities in a mutually supportive way. From their case study in Vietnam (Chap. 15), Nguyen and Flaaten conclude: "To attain the goal of poverty and vulnerability alleviation in the area of fisheries, it is necessary to follow a new approach that includes both governmental institutions and the local organizations of fishermen in a cooperative manner. Poor fishermen should be active partners who themselves have to find the causes of their poverty and suggest solutions."

On a similar note, Salas and Bjørkan (Mexico) state that "it is necessary to generate opportunities for people by improving their skills, diversification of livelihoods and assets of coastal communities, as well as by developing contingency programs to overcome the increasing challenges that people in these communities are facing."

PovFish researchers point to the importance of building stronger community organizations as a means of building such capacity. Isaacs (South Africa) stresses the need to "create legal entities that are representative of those fishers who were left outside the formal rights allocation process [as] crucial for the successful implementation of a policy" aimed at poverty alleviation among small-scale fishers. Fishing cooperatives, which have already been mentioned in this concluding chapter, are such legal entities, and they have general application. But to become legal entities, community organizations such as cooperatives need the recognition and support of government.

Building communities is also about building social capital and networks between people based on familiarity, trust, solidarity, and mutual support. Islam (Bangladesh) emphasizes the relevance of networks for sharing information to alleviate poverty from the bottom up. This is also consistent with Tilly's (2007) observation that "most of the world's very poor people, it seems likely, lack favourable categorical memberships and helpful connections."

Amarasinghe and Bavinck (Sri Lanka, Chap. 17) focus on the role that fishing cooperatives play in strengthening social capital for collective action. Cooperatives would benefit from state support and recognition, but may also compensate for the absence of such support as instruments of self-help and community control. Social capital is also among those resources that cooperatives draw from. Thus, cooperatives and social capital can potentially form a virtuous circle, as these two authors illustrate by the SCACO model.

However, given the complexity of cooperatives as institutions that try to balance welfare, business, and resource management functions, one should not be surprised of their often mixed results. They sometimes try to do too much, but without anyone to share responsibilities, in many instances, they are left to do whatever they possibly can. As demonstrated in Sri Lanka, some fishing cooperatives have proven more successful than others and are therefore variably effective as community organizations. This is partly for internal reasons; cooperation is no doubt challenging, their goals are often inconsistent, and cooperative leadership is usually characterized by role conflicts.

Still, for communities to be resilient and sustainable they must, with or without a cooperative, be able to overcome such challenges. They cannot afford much conflict if they are to become effective stewards of the resources that are the foundation of their livelihoods. In the case of Ghana and India, there are traditional institutions in place which, backed by customary law, play important governing roles largely independent of state government. In the absence of government or formal organizations such as cooperatives, local institutions – such as the Panchayat system in Tamil Nadu, India (Bavinck, Chap. 9) and local chiefdoms (Kraan, Chap. 8) – have a long history of addressing collective concerns in their communities. The fact that these institutions have been operative for a long time does not automatically mean that they are not fit for the modern age. Thus, if they exist, it may be wise to help strengthen them.

Replacing customary institutions with some other form of institution, such as cooperatives, is not necessarily a good idea. Neither should government replace them if they are capable of doing the job or if, with support, they have the potential to do so. This is also the reasoning behind "legal pluralism" as put forward by Bavinck (Chap. 9) and Kraan (Chap. 8).

20.12 Governing by Principles

Viewed together, the chapters in this volume provide a diverse portrait of small-scale fisheries, which vary according to circumstances and places. Therefore, governing interventions for small-scale fisheries must always be measured against ecological and social contexts. This implies that governance needs to follow what we refer to as the *dexterity principle*, by which we mean sensitivity to details that differ from one situation to another. It also means taking into account how people are actually living and operating, and what they value. When Kraan (Ghana, Chap. 8) argues that management institutions and development strategies must always be embedded in the reality of "plural normative orders" that exist within small-scale fisheries, she is alluding to this principle. The dexterity principle is also evident in Anirudh Krishna's (2010, p. 5) observation that: "Reducing poverty more effectively in the future will require attending carefully to the minutiae of everyday lives."

Governance designs according to the dexterity principle require broader involvement than that of central government. Decision-making must be brought closer to where the problem is actually experienced and where many of the solutions must be sought. Thus the dexterity principle leads logically to another governance principle, that of *subsidiarity,* which states that management's decision-making authority should be vested with the lowest possible organization. This principle is particularly relevant to poverty alleviation where, according to Mehrotra and Delaminica (2007, p. 212), integrated approaches aimed at creating synergies "between interventions in the spheres of health, education, sanitation, productive health and nutrition with a geographical location" are needed. (In fisheries, we must also include resource management on this list). These authors argue that, "the state is incapable of delivering these services effectively as long as it operates vertically."

Notably, the subsidiarity principle does not necessarily exclude the state from the equation. The state can and should provide macro-economic and institutional policies aimed at alleviating poverty through fisheries development. It also, as has been argued in several chapters, holds a responsibility for providing certain collective goods, which include a constructive climate for investment, business and social entrepreneurship, and growth (Alvord et al. 2004; Mair and Martí 2006). However, the subsidiarity principle also emphasizes the limits of state governance and that the state cannot and, therefore, should not try to do everything. For instance, Nguyen and Flaaten (Vietnam) argue that "the state should concentrate on supporting the poor in small-scale inshore fisheries during a transitional period to improve the conditions of the ecological environment, the inshore fisheries resources, knowledge and infrastructure."

Bavinck and Jentoft (2008) have suggested that the subsidiarity principle should also be applied to fisheries resources distribution and technology; resources that are within the reach of small-scale fisheries should also be reserved for them. Such a principle, which is particularly pertinent from a poverty alleviation perspective, also invokes principles of justice, most importantly the so-called *difference principle* of John Rawls (1971). This principle is applicable to situations when it is fair to treat people differently, as with positive discrimination of those who are most vulnerable to poverty and who have been previously marginalized and excluded from access to natural resources, or from decision-making pertaining to their usage.

Thus, the difference principle could well be applied to fishing rights. Rights-based systems that are supported by fisheries managers and academics all over the world should therefore be subject to the litmus test: Do they benefit those in direst need? (cf. Jentoft 2007). The difference principle for small-scale fisheries leads to a *precautionary principle* that is sensitive to the social and cultural dimensions of people's livelihoods, as people are vulnerable and their communities have tipping points beyond which there is no point of return (Groenfelt 2003).

Governance principles such as those mentioned here are easier to proclaim than to implement and follow in practice. Partly, this is because they involve hard choices by policymakers. Principles, even when drawn from values that are generally shared, often meet resistance from those who have the most to lose. These people often also hold the most power and are therefore best positioned to influence those choices. As Collier (2007, p. 180) argues:

> Reform in these countries has to come from within, and it takes courage. Vested interests can be relied upon to use their power, resources, and ingenuity to oppose change. Although the reformers have truth on their side, truth is just another special interest, and not a particularly powerful one.

Thus, policymakers and administrators are not always free to do what they want to do, and have to do if they follow these ideal principles. But there is hardly any other way to advance small-scale fisheries for the economy as a whole, and for those communities and people for whom this sector is key to their survival. What keeps policymakers from implementing agreed upon meta-principles is a question that begs for further research (cf. Pitcher et al. 2009).

Governments do have means at their disposal to make a better future for small-scale fisheries. And if they do not, they are in a position to acquire them. Not all policies targeting the poor are expensive, but require bold political leadership and resolve (Sachs 2005). Great economic inequities are not only unjust (see Chuenpagdee and Jentoft, Chap. 3), they are a waste of human capital, and are not sustainable. Political expediency is no excuse for inaction. Neither is lack of complete knowledge, as stated by the precautionary principle that countries accepted when they ratified the UN Convention on Biodiversity (www.cbd.int).

So, let us insist on what we will call *the urgency principle*. Given that (a) small-scale fisheries are vulnerable and may easily slip into poverty; (b) that small-scale fisheries are an important contributor to poverty alleviation around the world and can potentially play an even bigger role in feeding the poor; and (c) that small-scale fishing communities can be effective stewards of marine and coastal ecosystems if they are organized and supported for the task, then policies should not be developed only for the long-run. They must also be developed and implemented for the immediate future. The starting point of the chain of actions is "the poor themselves" (Sachs 2005, p. 242). People need to feed their families, bring their children to school, maintain their health, and must prevent their environment from further degradation, and thus cannot wait for policymakers to make up their minds. Poor people have had to learn to be patient, but they deserve initiatives that make a difference in their lives today.

Small-scale fishing communities and cultures are not as resilient as we often tend to believe, or not sufficiently resilient to withstand the new threats that they are now facing, such as climate change and globalization. It does not seem to matter whether they are located in lower or relatively more developed countries, whether they are poor in a relative or in an absolute sense. When people lack the entitlements and individual and collective capabilities they need to protect themselves, their livelihoods and communities are at risk. Without respect for human rights, which must include access rights to the fishery commons, small-scale fisheries are more vulnerable than they have to be. Due to the vagaries of nature and the forces of globalization, small-scale fisheries can never be stable and fully secure, but they have to be resilient and adaptive. They also need the power and the empowerment to be innovative and responsible. Only then can small-scale fisheries become the solution to poverty, rather than the problem.

20.13 Learning by Comparing

Kurien and Willmann (2009, p. 406) argue that it is "neither a legitimate nor a feasible proposition [to make] sweeping generalizations about the characteristics of small-scale fisheries." Globally, they are simply too diverse and complex to do so easily.

This volume confirms their view. The chapters covering 15 countries on 4 continents show a mosaic of situations that cannot be readily equated. What is real for small-scale fishing people on the Chittagong coast of Bangladesh is not identical to what those who live on the coast of Mozambique or the Yucatan coast of Mexico are experiencing. Although small-scale fishing in Nicaragua's Pearl Lagoon and in Guatemala's Amatique Bay occurs in multi-ethnic communities, the conditions are also different. The particular challenges they face are such that no single policy formula would work across the board. But, at a general level, the challenges that small-scale fisheries are confronted with are also remarkably similar.

The experience of being poor is not all that different from place to place. From the point of view of a poor individual, it does not matter so much whether he/she is poor in a relative or an absolute sense. Helping to improve the well-being of small-scale fishing people, while sustaining their natural environment, is a universal challenge. What has been said above about community development, empowerment and gender, the role of the state, and customary law applies everywhere. But it still requires policies and governance mechanisms that are sensitive to, while knowledgeable of, the particular situations on the ground.

Thus, even if sweeping generalizations are futile, there are lessons that can be learned from studying small-scale fisheries at the micro level, as illustrated in this volume, and comparing them with other detailed studies. While different, at a general level these case studies share many of the same characteristics. Regardless of country and place, small-scale fishers draw their livelihood from fresh water, marine, and coastal ecosystems that are susceptible to heavy exploitation. Also, small-scale fishers are exposed to risks from working on the water. They share the fact that they must struggle to survive in circumstances that make them vulnerable to natural and social forces beyond their control. They are often poor and powerless but, as the case studies in this volume show, they are not equally so. We have also seen examples of small-scale fishers who do not consider themselves poor, or who have worked themselves out of poverty into a situation that they feel is sufficient (cf. Chuenpagdee and Juntarashote, Chap. 14).

Small-scale fishing can therefore provide a good life. People in this sector do not have to be poor. They can also be made less vulnerable than they currently are. Many things can be done to increase their well-being, and their many similarities suggest that there is ample room for exchange of ideas about how to move small-scale fishing people out of poverty, and to reduce the chance that they might fall into it. Governments can pay more attention to small-scale fisheries, their economic potential, their environmental impact, and to their human dimensions. Governments should be more conscious of experiences elsewhere, and lessons do not need to come from affluent countries in the north. Comparing small-scale fisheries in the north and south is certainly worthwhile. They are not necessarily all that different. Absolute and relative poverty are often experienced in similar ways, and they often have parallel social and ecological impacts. But the learning process should not be unidirectional, from the north to south, as there are ways of addressing small-scale fisheries challenges in the south that have potential application in the north. Some of the problems that small-scale fisheries in the north are facing may already have found a solution in the south and may provide useful lessons, although we always need to take the social, economic, political, and environmental context into consideration.

This also raises important research questions: When comparing how fisheries-dependent people cope in different social and ecological contexts, what determines how well they are, and how they contribute to poverty alleviation? What makes some communities more resilient than others? What role do institutions and social capital play in building such resilience? Some specific questions pertaining to fisheries cooperatives emerge from the case studies in Sri Lanka (Chap. 17) and Mexico (Chap. 10): What makes cooperatives unequally successful; their internal design or external contexts? Members of the PovFish research team are not short of ideas of what can be done to improve small-scale fisheries within the areas where they have carried out research. But they also have suggestions that are relevant elsewhere. These suggestions should not be taken as policy prescriptions or be accepted uncritically. Still, they should trigger curiosity, promote learning, and spur willingness to adopt new strategies.

References

Alvord SH, Brown LD, Letts CW (2004) Social entrepreneurship and societal transformation. J Appl Behav Sci 40(3):260–282

Armitage D, Berkes F, Doubleday N (2007) Adaptive co-management. Collaboration, learning, and multi-level governance. University of British Columbia Press, Vancouver

Arndt HW (1983) The "trickle-down" myth. Econ Dev Cult Change 32(1):1–10

Bavinck M, Jentoft S (2008) Imagining the future. Samudra 51:41–43

Béné C, Hersoug B, Allison EH (2010) Not by rent alone: analysing the pro-poor functions of small-scale fisheries in developing countries. Dev Policy Rev 28(3):325–358

Berkes F, Mahon R, McConney P, Pollnac R, Pomery R (2001) Managing small-scale fisheries. International Development Research Centre, Ottawa. http://web.idrc.ca/es//ev-9328-201-1-DO_TOPIC.html

Berlin I (1969) Four essays on liberty. Oxford University Press, London/New York

Boulding KE (1977) Commons and community: the idea of a public. In: Hardin G, Baden J (eds) Managing the commons. W. H. Freeman and Company, San Francisco, pp 280–294

Collier P (2007) The bottom billion. Why the poorest countries are failing and what can be done about it. Oxford University Press, Oxford

Degnbol P, Gislasson H, Hanna SS, Jentoft S, Raakjær Nielsen J, Sverdrup-Jensen S, Wilson DC (2006) Painting the floor with a hammer: technical fixes in fisheries management. Mar Policy 30:534–543

FAO (2008) Global conference on small-scale fisheries. Securing sustainable small-scale fisheries: bringing together responsible fisheries and social development. FAO Fisheries and Aquaculture Report, No. 911. Food and Agricultural Organization of the United Nations, Rome

Graham M (1939) The sigmoid curve and the overfishing problem. Rapp P-v Réun Cons perm int Explor Mer 110(2):15–20. Reprinted in Thomasson ED (ed) The study of the sea. ICES 1981. Fishing News Books, Oxford

Groenfelt D (2003) The future of indigenous values: cultural relativism in the face of economic development. Futures 35:917–929

Hardin G (1968) The tragedy of the commons. Science 162:1243–1248

Hardin G (1977) Living on a lifeboat. In: Hardin G, Baden J (eds) Managing the commons. W. H. Freeman and Company, San Francisco, pp 261–279

Hauck M (2008) Rethinking small-scale fisheries compliance. Mar Policy 32:635–642
Hersoug B (2004) Exporting fish, importing institutions – fisheries development in the Third World. In: Hersoug B, Jentoft S, Degnbol P (eds) Fisheries development. The institutional challenge. Eburon Publishers, Delft, pp 21–92
Jentoft S (1986) Fisheries co-operatives; lessons drawn from international experiences. Can J Dev Stud VII(2):97–209
Jentoft S (1989) Fisheries co-management: delegating government responsibility to fishermen's organization. Mar Policy 13(2):137–154
Jentoft S (2007) The litmus test. Samudra 46:30–32
Kooiman J (2008) Exploring the concept of governability. J Comp Policy Anal 10(2):171–190
Kooiman J, Bavinck M, Jentoft S, Pullin R (eds) (2005) Fish for life. Interactive governance for fisheries. Amsterdam University Press, Amsterdam
Krishna A (ed) (2008) Poverty, participation, and democracy. A global perspective. Cambridge University Press, Cambridge
Krishna A (2010) One illness away. Why people become poor and how they escape poverty. Oxford University Press, Oxford
Kurien J (1985) Technical assistance projects and socio-economic change: Norwegian intervention in Kerala's fisheries development. Econ Polit Wkly 20(25/26):A70–A88
Kurien J, Willmann R (2009) Special considerations for small-scale fisheries management in developing countries. In: Cochrane KL, Garcia SM (eds) A fishery manager's guidebook. Wiley-Blackwell, Oxford, pp 404–424
Mair J, Martí I (2006) Social entrepreneurship research: a source of explanation, prediction, and delight. J World Bus 41:36–44
McCay BJ, Jentoft S (2010) Uncommon ground: critical perspectives on common property. In: Rosa EA, Diekmann A, Dietz T, Jaeger C (eds) Human footprints on the global environment. Threats to sustainability. MIT Press, Cambridge, pp 203–230
Mehrotra S, Delamonica E (2007) Eliminating human poverty. Macroeconomic and social policies for equitable growth. ZED Books, London
Narayan D, Chambers R, Shah MK, Petesch P (2000) Voices of the poor: crying out for change. Oxford University Press/The World Bank, New York
Pitcher T, Kalikoski D, Pramod G, Short K (2009) Not honouring the code. Nature 475:658–659
Rawls J (1971) A theory of justice. Oxford University Press, Oxford
Robbins P (2000) The rotten institution: corruption in natural resource management. Polit Geogr 19:423–443
Sachs JD (2005) The end of poverty. Economic possibilities for our time. Penguin Press, New York
Scott JC (1985) Weapons of the weak: everyday forms of peasant resistance. Yale University Press, New Haven
Sen A (2000) Development as freedom. Anchor Books, New York
Sen A (2009) The idea of justice. Allen Lane, London
Sumaila UR, Jacquet J (2008) When bad gets worse: corruption and fisheries. Working paper – Sea around us project. http://www.southchinasea.org/docs/Corruption%20%26%20fisheries.pdf
Tilly C (2007) Poverty and the politics of exclusion. In: Narayan D, Petesch P (eds) Moving out of poverty. Cross-disciplinary perspectives on mobility, vol 1. Palgrave Macmillan/The World Bank, Washington, D.C., pp 45–75

Biographies of the PovFish Team

Oscar Amarasinghe is a professor in agricultural economics at the University of Ruhuna, Matara, Sri Lanka. He has been in the university service for the last 30 years, and possesses nearly 25 years of experience working in fishing communities in Southern Sri Lanka. He has worked in a number of fisheries-related fields, especially in fisheries policy, fisheries management and governance. Dr. Amarasinghe has more than 60 research papers to his credit, and has won a number of international awards. He has served as research partner/consultant for several international and local research and development projects, and has been instrumental in preparing the National Fisheries Policy of Sri Lanka in 2004. Dr. Amarasinghe is a member of the International Collective in Support of Fishworkers (ICSF).

S. Jentoft and A. Eide (eds.), *Poverty Mosaics: Realities and Prospects in Small-Scale Fisheries*, DOI 10.1007/978-94-007-1582-0,

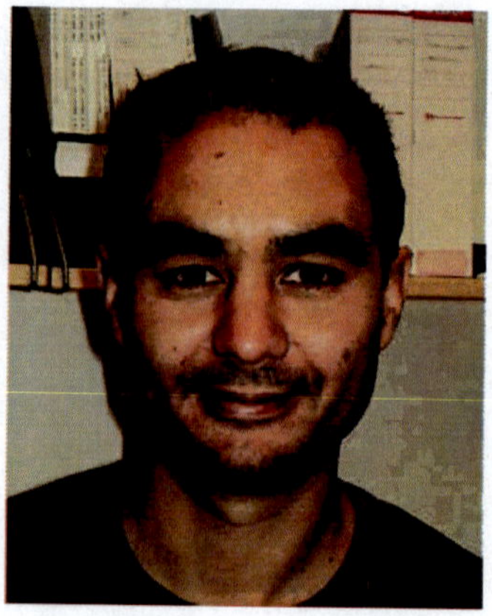

Hector Andrade is a Ph.D. candidate in fisheries management at the Norwegian College of Fishery Science, University of Tromsø, Norway, where he previously obtained a master's degree in international fisheries management. Andrade has worked in natural resource management and natural protected areas in Guatemala, especially within coastal zones and with sea turtle conservation programs. His main research interests are within resource management and biological assessment of small-scale fisheries. Currently, Andrade works as an environmental consultant for Akvaplan-niva AS, Tromsø Norway. Project assignments include impacts of oil and gas offshore production on the benthos.

Maarten Bavinck is an associate professor at the Department of Human Geography, Planning, and International Development Studies at the University of Amsterdam; and director of the social-science Centre for Maritime Research (MARE). Dr. Bavinck's academic career has focused on the dynamics of fisheries in South Asia, about which he has written two books and a large number of articles. The two books are titled: *Small Fry – The Economy of Petty Fishermen in Northern Sri Lanka* (Free University Press, Amsterdam 1984); and *Marine Resource Management – Conflict and Regulation in the Fisheries of the Coromandel Coast* (Sage Publications, New Delhi 2001). Dr. Bavinck conducted his Ph.D. (1998) on the topic of fishing rights and legal pluralism in the fisheries of the Coromandel Coast, India. His most recent research project (2010–2015) is on conflict, environmental sustainability, and bottom-up governance in the fisheries of South Asia and South Africa.

Maiken Bjørkan is a Ph.D. candidate at the Norwegian College of Fishery Science, University of Tromsø, Norway. She holds two master's degrees: one in social anthropology from the University of Bergen; and one in international fisheries management from the Norwegian College of Fishery Science, both in Norway. Her main research interests are within resource management, stakeholder participation, and knowledge. She has published several articles on marine protected areas and bio-prospecting. Her current research is oriented toward the mobilization of fishers' knowledge as expertise for Norwegian fisheries management.

Felipe Bobadilla was a research assistant at CINVESTAV del I.P.N Unidad Mérida, Méxcio, from 2007 to 2010. He holds an M.Sc. in marine ecology. He has worked on several research projects in coastal areas. Bobadilla's research has been focused on fishers' perceptions of vulnerability, and issues associated with community-based fisheries management. His main research interests focus on analysis of socio-ecological systems, resilience, and governance, in order to generate information to improve fisheries policies, and to favor linkages between coastal communities, fisher groups, government, and research institutions to generate such policies.

Miguel Cabrera is a research assistant at CINVESTAV del I.P.N Unidad Mérida, Méxcio. He holds an M.Sc. in marine biology. Cabrera has worked on fisheries assessment and economic valuation in coastal areas for 20 years. He has taught courses in fisheries and economic valuation in Mexico, Chile, and Venezuela. He has extensive experience working with fishers' organizations and private fishing companies in the development of proposals for development projects. Cabrera has participated in several meetings organized by FAO. His research interests are related to population dynamics, bio-economic assessment of small-scale fisheries, and estimation of economic values of coastal resources. He has published several scientific papers and reports based on this research.

Ratana Chuenpagdee is Canada Research Chair in Natural Resource Sustainability and Community Development at Memorial University, Canada. Her research emphasizes interdisciplinary approaches to fisheries and coastal ecosystems management – focusing particularly on small-scale fisheries, marine protected areas, community-based management, food security, and fisheries governance. Dr. Chuenpagdee is also co-director of the Coastal Development Centre in Thailand, which serves as a secretariat for the Coastal Zone Asia-Pacific Association, and is responsible for the organization of its biennial conferences. In addition to conducting research in Thailand and Canada, she has worked in other countries including Cambodia, Malawi, and Mexico.

Arne Eide is an associate professor in resource economics at the Norwegian College of Fishery Science, University of Tromsø, Norway. Together with colleagues, he initiated and has been running the Masters Programme in International Fisheries Management (IFM) at the College since 1998. Dr. Eide has working experience from Norway, Canada, Mozambique, Namibia, South Africa, Cape Verde, Hawaii, Spain, and Vietnam; and has participated in several international projects. Among current and recent projects are the EU-funded BASIS, BALANCE, CEVIS, ATP, and ACCESS. Others are FiMaGloW (financed by the Nordic Council of Ministers), and PovFish (financed by the Norwegian Research Council). Eide's recent research interests and scientific work have been on climate change issues related to Arctic fisheries, economic impacts of global warming, optimal fishing, and harvest control rules. He has also contributed expertise to a number of appraisals, project reviews, and reports for Norad (Norwegian Agency for Development Cooperation), IFAD, FAO, and the World Bank.

Ola Flaaten is a professor of resource economics at the University of Tromsø, Norway. From 1998 to 2001, Dr. Flaaten was Head of the Fisheries Department, for the Organisation for Economic Cooperation and Development (OECD), Paris, and has spent sabbatical years at the University of British Columbia, Canada, and the University of Portsmouth, UK. Currently, he is the advisor to the economic component of the joint project between Nha Trang University and three Norwegian universities funded by the governments of Vietnam and Norway. Dr. Flaaten initiated the Masters Program in Fisheries and Aquaculture Management and Economics (NOMA-FAME) in Nha Trang. He is currently a partner of one EU-project (SPICOSA) and two Norwegian Research Council funded projects. Dr. Flaaten has published two books, book chapters, journal articles, conference proceeding papers, and economic and policy reports, in both English and Norwegian, with some translations to Russian and Vietnamese.

Miguel González holds a Ph.D. in political science (York University, 2008). After graduating with a degree in social anthropology from Mexico in 1995, he worked as Vice Principal at the University of the Autonomous Regions of the Caribbean Coast of Nicaragua (URACCAN) at the Bluefields' campus (1996–2000). He has also worked as an independent consultant for national and international organizations on issues related to the Caribbean Coast of Nicaragua. His research focuses on indigenous social movements, territorial autonomy regimes, and sub-national governance. Dr. González is the co-editor of *Autonomía a Debate: Políticas de Reconocimiento Indígena y el Estado Plurinacional en América Latina. Policies of Recognition of Indigenous Peoples and the Plurinational State in Latin America* (FLACSO 2010); and also co-editor (with Arja Koskinen, Svein Jentoft, and Diala Lopez) of *The Rama People: Struggling for Land and Culture* (URACCAN, Nicaragua & University of Tromsø, Norway, 2006). Currently, he teaches in the International Development Studies Program at York University in Toronto, Canada.

Maria-Victoria Gunnarsdottir is currently a Ph.D. candidate at the Norwegian College of Fishery Science, University of Tromsø. She holds an interdisciplinary M.Sc. in conservation biology and environmental sciences and policy from Johns Hopkins University, USA. Her current research focuses on examining the sociocultural, environmental, and management factors affecting sustainability within the marine fish tourism sector in Northern Norway, applying an interdisciplinary perspective. Prior to returning to academia she worked for the Arctic Council – most recently as Head of the Arctic Council Secretariat, and before that as Executive Secretary for CAFF (Conservation of Arctic Flora and Fauna Working Group). Gunnarsdottir was part of the team that implemented the Circumpolar Biodiversity Monitoring Programme, and the Arctic Biodiversity Assessment. She has been a scientific and technical editor for more than 20 years.

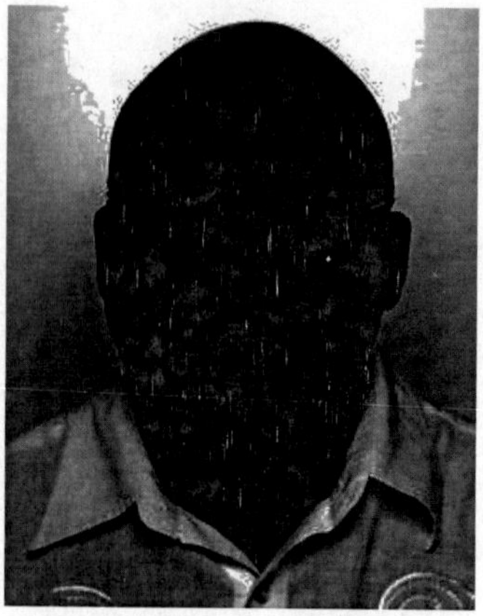

Mafaniso Hara is a social scientist with more than 25 years experience in social research and development work focusing on rural communities in Southern Africa, especially fisheries (both freshwater and marine). Dr. Hara's key qualifications, research and professional experience relate to governance and management of fisheries, natural resources and rural development. He teaches a number of courses at tertiary institutions in South Africa and Malawi in his area of research, knowledge and expertise. Dr. Hara's current research interests include integrated resource management, development of socio-economic indicators for small-scale fisheries, value chain analysis, and poverty in fishing communities. He holds a Ph.D. from UWC (2001), a master's degree from the University of Hull, UK (1990), an honours degree in sociology from Stellenbosch University (2010), a B.Sc. from the University of Malawi (1981), and a diploma in fisheries management from Grimsby College, UK (1982).

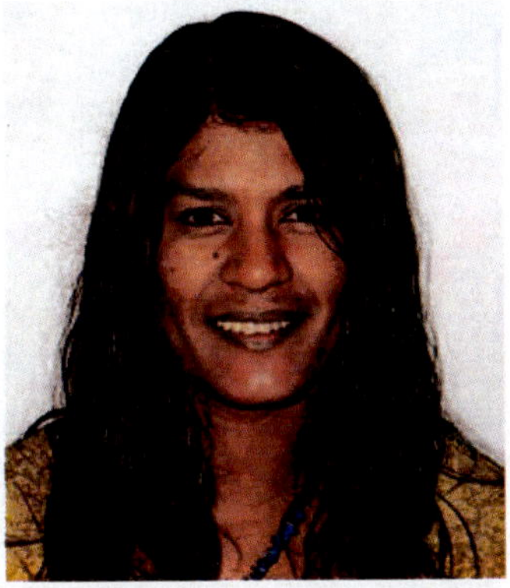

Moenieba Isaacs graduated with her Ph.D. in 2004 from the University of Western Cape in collaboration with the Norwegian College of Fishery Science, University of Tromsø. Dr. Isaacs' research focuses on understanding poverty and small-scale fishers, small-scale fisher rights, fisheries policy reforms in South Africa, poverty alleviation, HIV/AIDS impacts on fisher communities, declaring marine protected areas and the impacts on fishing communities, and co-management in fishing communities in South Africa. Dr. Isaacs also has extensive experience in development work and field research in fishing communities on the west, south, and east coasts of South Africa which cover three coastal provinces.

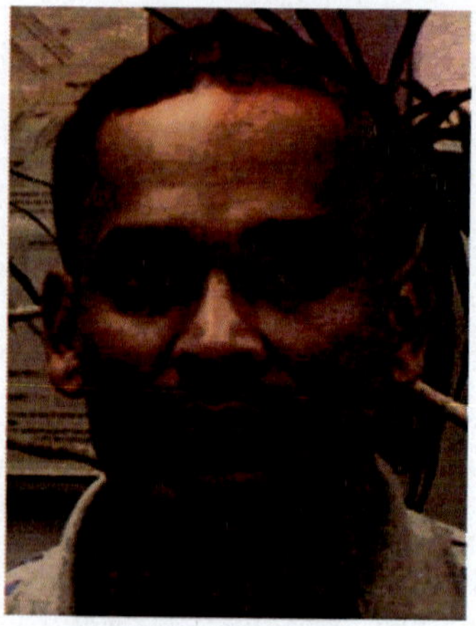

Mohammad Mahmudul Islam received his B.Sc. (Honour's) degree in marine science from the Institute of Marine Sciences and Fisheries, University of Chittagong in Bangladesh. He completed his master's degree from the same institute. In 2008, he obtained another master's degree in international fisheries management from the University of Tromsø in Norway, under a Norad fellowship. Currently (2010) he is a doctoral fellow at the Bremen International School for Marine Sciences (GLOMAR), University of Bremen in Germany. His research interests include poverty in small-scale fisheries in Bangladesh, coping strategies, and resilience capacity of coastal communities in Bangladesh.

Svein Jentoft is a sociologist and professor at the Norwegian College of Fishery Science, University of Tromsø, Norway. He specializes in social and institutional aspects of fisheries and coastal governance and has published extensively within areas such as fisheries co-management, industrial organization, community development, and indigenous peoples. Among Dr. Jentoft's recent books are *The Ulwa People: Identity and Environment in a Multiethnic Context* (co-editor, Managua, Nicaragua: URACCAN 2008); and *In Disciplinary Border Lands: On Interdisciplinarity. Fagbokforlaget* (co-editor, in Norwegian, 2007). Dr. Jentoft was also co-editor of *Indigenous Peoples: Resource Management and Global Rights* (2008), and *Indigenous Peoples: Self-determination, Knowledge and Indigeneity (2008)* (both Eburon Academic Publishing) as well as *Fish for Life. Interactive Governance for Fisheries* (Amsterdam University Press 2005).

Kungwan Juntarashote is an associate professor in fisheries at the Faculty of Fisheries, Kasetsart University, Bangkok, Thailand, and director of the Coastal Development Centre. He has extensive experience in small-scale fisheries, community-based fisheries management, and fisheries development and economics in Thailand and Southeast Asia. Dr. Juntarashote has served in an advisory capacity to the Government of Thailand (Department of Fisheries) and as a technical expert to the Food and Agriculture Organization (Asia-Pacific Region). In addition to his regular teaching duties at Kasetsart University, Dr. Juntarashote regularly gives lectures and training courses at the Southeast Asian Fisheries Development Centre, Asian Institute of Technology, and Walailuk University.

Ståle Knudsen is an associate professor at the Department of Social Anthropology, University of Bergen, Norway. He holds a Ph.D. in social anthropology (2001) and has worked on Turkish Black Sea fisheries for 20 years. This research has covered issues such as knowledge, technology, science, consumption, state policies, and common pool resources. Knudsen has published articles in *Human Ecology*, *Human Organization*, *International Journal of Middle East Studies*, and has written the book *Fishers and Scientists in Modern Turkey: The Management of Natural Resources, Knowledge and Identity on the Eastern Black Sea Coast* (Berghahn 2009). Since 2004, Dr. Knudsen has been involved in inter-disciplinary work relating to the management of European seas, and he is currently leading a working group on the Black Sea as part of an EU-funded project entitled *Knowledge-based Sustainable Management for Europe's Seas*. Dr. Knudsen's main current research interest is the political ecology of biodiversity and introduced species in the Black Sea and beyond.

M. Hakan Koçak is an assistant professor at the Department of Labour Economics and Industrial Relations, Kocaeli University, Turkey. He holds a Ph.D. in labor sociology and labor economics from Marmara University, Istanbul, Turkey. Dr. Koçak has held a position as strategic outreach officer at the workers syndicate Petrol-İş, and continues to support their work as a consultant. Dr. Koçak's publications include the book: *Working Class Formation in the Paşabahçe District: Development and Tendencies in the Glass Workers' Movement* (İletişim 2011, in Turkish). Current research interests focus on the history of labor unions and the development of workers' movements in Turkey, as well as the sociological profile of the Turkish working class. He has worked with Ståle Knudsen on several fisheries-related projects in the Turkish Black Sea region.

Kristoffer Kokvold holds an M.Sc. degree in fishery science from the Norwegian College of Fishery Science, University of Tromsø, Norway. As a researcher, his fields of interest include international fisheries management, and international trade regimes in fisheries. Currently, he works as an international project coordinator for the Faculty of Biosciences, Fisheries and Economics, University of Tromsø. For PovFish, Kokvold served as the international project coordinator, handling all administrative and budgetary issues.

Marloes Kraan studied cultural anthropology at the Free University in Amsterdam, The Netherlands. After university, she became a junior researcher at the Centre for Maritime Research (MARE). Her Ph.D. research (2002–2009) based in Ghana, focused on beach seine fishers and was funded by the University of Amsterdam. During her fieldwork period, Dr. Kraan was affiliated with the Institute of African Studies at the University of Ghana, Accra. In September 2008, she began working at the Dutch Fish Product Board as a policy officer in sustainable fisheries.

Boguslaw Marciniak has a Ph.D. in sociology, and is a researcher at the Sea Fisheries Institute in Gdynia, Poland. For many years, he taught sociology at the University of Gdansk, Poland, and at the Saint Francis Xavier University in Antigonish, Nova Scotia in Canada. He has years of experience in social research in fisheries, especially on the role of cooperation in fisheries as a way toward participatory democracy and self-management. For many years, an area of Dr. Marciniak's research has focused on the social problems of life in northern maritime communities, fisheries organizations, and on the social aspects of the management of natural resources. Dr. Marciniak has conducted research projects in Poland, Norway, Bulgaria, and in Canada. Results of these research projects have been presented in more than 40 articles, as well as at many conferences in Norway, Poland, Finland, Scotland, Denmark, England, Canada, Holland, Tanzania, and Thailand.

Ana Menezes was born and raised in Mozambique. She graduated with an M.Sc. in aquaculture in 1995 from Auburn University, Alabama. In 2006, she received a master's degree in public administration from Maxwell School, Syracuse University, and in 2008 her Ph.D. from SUNY-ESF and Syracuse University. Her areas of doctoral studies were environmental sciences and natural resources policy. Dr. Menezes has a vast education in agricultural engineering, aquaculture, environmental sciences and policy, as well as in public administration. Most of her studies were carried out in the USA. Between 1987 and 2000, Dr. Menezes worked for the Fisheries Research Institute in Mozambique with focus on shrimp culture, and for the Ministry of Environmental Affairs as National Director of Environmental Planning and Research. In 2000, invited by the donor community, she opened a natural resource management and policy analysis unit at the Ministry of Agriculture. She has also worked as a consultant with various national and international institutions. Dr. Menezes currently works for UNEP, and holds the UNEP National Officer position in Mozambique.

Georges Midré is Cand. sociol. from the University of Oslo, and professor of social policy at the University of Tromsø, Norway since 1986. He was Dean of the Faculty of Social Science, University of Tromsø from 1995 to 1999, and was chair of the Department of Sociology at the University of Tromsø from 2006 to 2009. From 1999 to 2008, Professor Midré was academic coordinator of the programme "Maya Competence Building," a cooperation between the Universidad de San Carlos de Guatemala and the University of Tromsø. He has conducted field work in Norway, Russia and Poland, as well as Argentina and Guatemala. Professor Midré is author of books and articles on social history and welfare policies, including research on poverty. Among his latest publications are *Élite ladina, políticas públicas y pobreza indígena* (2002, with Sergio Flores); and *Opresión, espacio para actuar y conciencia crítica. Líderes indígenas y percepción de la pobreza en Guatemala* (2005), both published by the Instituto de Estudios Interétnicos, Universidad de San Carlos de Guatemala. He co-edited the book *Indigenous Peoples. Self-determination. Knowledge. Indigeniety* with Henry Minde, Svein Jentoft and Harald Gaski (2008).

Dr. Kim Anh Thi Nguyen is a senior lecturer at the Nha Trang University (NTU), Vietnam. From 2004 to 2010 she was Dean of the Faculty of Economics, NTU; and worked for a number of development projects on fisheries and aquaculture management, such as PACCOM, and DANIDA. Currently, Dr. Nguyen is leader of the economic component of the joint project between Nha Trang University and three Norwegian Universities, funded by the Governments of Vietnam and Norway. She is academic coordinator of the Masters Program in Fisheries and Aquaculture Management and Economics (NOMA-FAME), Nha Trang. Dr. Nguyen has published journal articles, conference proceeding papers, and policy reports.

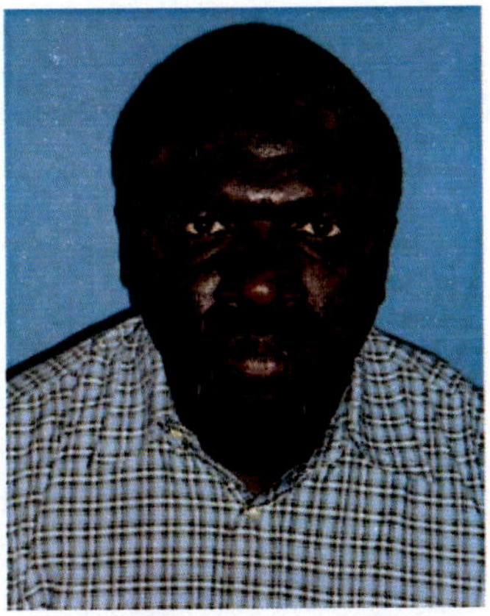

Paul O. Onyango is a research fellow and a Ph.D. candidate at the Norwegian College of Fishery Science, University of Tromsø, Norway. He is originally from the University of Dar-es-Salaam, Tanzania. His research investigates the dilemmas authorities face in confronting poverty in small-scale fisheries. He has 15 years research experience on Lake Victoria and in East Africa. Dr. Onyango's recent publications include: Reconfiguring poverty: the wickedness perspective (*Journal of Tropical Hydrobiology and Fisheries* 2010); Freedom and poverty in the fishery commons (*International Journal of the Commons* 2010, co-author); and Assessing poverty in small-scale fisheries in Lake Victoria, Tanzania (*Fish and Fisheries Journal* 2010).

Jesper Raakjær is a professor and head of Innovative Fisheries Management (IFM), at Aalborg University Research Centre, Department of Development and Planning, Aalborg University, Denmark. He holds a Ph.D. (1992) and D.Sc. (2009). Dr. Raakjær has been involved in numerous international research projects focusing on institutional aspects of fisheries management including ecosystem approach, user/stakeholder participation, legitimacy, and compliance both in developed and developing countries. For 15 years, Dr. Raakjær has been used as a consultant by Danida and Norad. He has extensive practical experience working in the fishing industry, first as a professional fisher, serving on the Board of Directors for pelagic fish processing companies. Dr. Raakjær is presently chairing the Pelagic group within the Danish Seafood Association and is an ExCom member of the EU Pelagic Regional Advisory Committee, representing AIPCE.

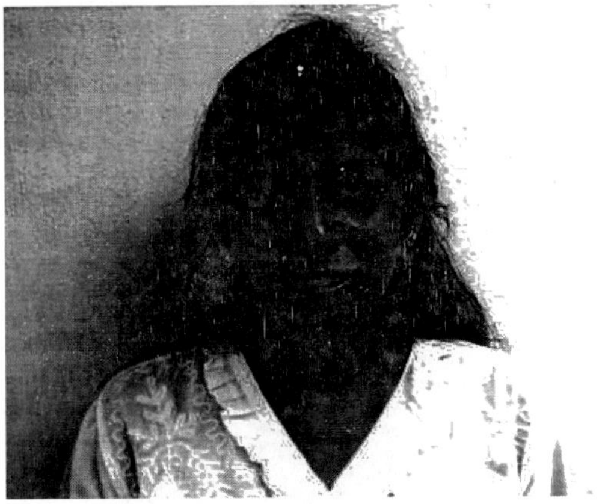

Silvia Salas is a fisheries biologist and economist, with her degree from the University of British Columbia, Vancouver, Canada. She is currently a professor at CINVESTAV del I.P.N Unidad Mérida, Méxcio. Dr. Salas has worked on projects relating to bio-economic analysis and assessment of fisheries in Canada and Mexico. Dr. Salas is engaged in connecting research to management by participating in advisory committees of governmental agencies in Mexico. She is a member of the advisory board for the FAO-WECAF region, and has participated in scientific meetings organized by FAO. Her research involves bio-economic analysis of small-scale fisheries, dynamics of fishing fleets, and evaluation of perceptions concerning the use and management of coastal resources. Dr. Salas has published several scientific papers, books, and chapters of books on these themes.

With special contribution by:

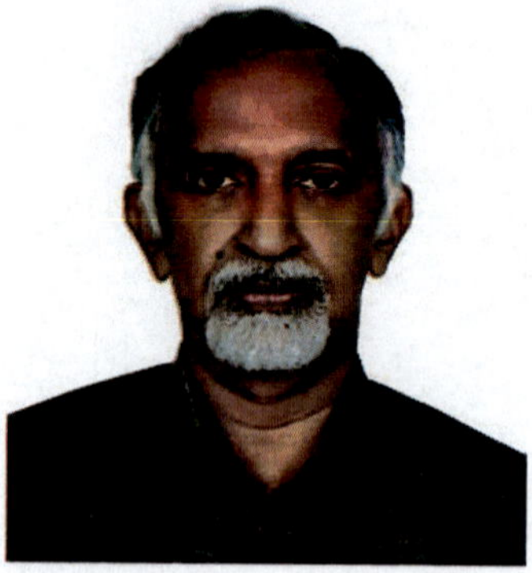

John Kurien is an activist-researcher who has lived and worked with small-scale fishing communities and globally championed their cause – through networking initiatives and research publications – bringing them into the mainstream discourse on fisheries development and management. He was Secretary-General of the first International Conference of Fishworkers and their Supporters held in Rome in 1984, was founder of the International Collective in Support of Fishworkers, and also a professor at the Centre for Development Studies, Trivandrum, India.

Index

O

P

X

Y

Z